Contents

→ CHEMISTRY

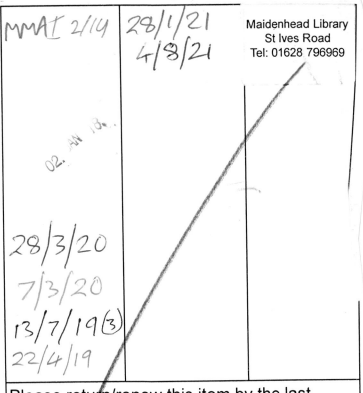
Science Progress 2

Andrea Coates

Michelle Austin

Richard Grimmer

HODDER
EDUCATION
AN HACHETTE UK COMPANY

Although every effort has been made to ensure that website addresses are correct at time of going to press, Hodder Education cannot be held responsible for the content of any website mentioned. It is sometimes possible to find a relocated web page by typing in the address of the home page for a website in the URL window of your browser.

Orders: please contact Bookpoint Ltd, 130 Milton Park, Abingdon, Oxon OX14 4SB. Telephone: (44) 01235 827720. Fax: (44) 01235 400454. Lines are open 9.00–17.00, Monday to Saturday, with a 24-hour message answering service. Visit our website at www.hoddereducation.co.uk

© Andrea Coates, Michelle Austin, Richard Grimmer and Mark Edwards, Sue Hocking and Beverly Rickwood 2014

First published in 2014 by
Hodder Education
An Hachette UK Company,
338 Euston Road
London NW1 3BH

Impression number	5	4	3	2	1
Year	2018	2017	2016	2015	2014

Cover photo © Jeffrey Collingwood – Fotolia

Typeset in 11.5/13 pt ITC Officina Sans by Aptara, Inc.

Printed in Italy

A catalogue record for this title is available from the British Library.

ISBN 978 1 471 801440

→ PHYSICS

Get the most from this book

Welcome to Science Progress Student's Book 2!

This book covers the second half of your KS3 Science course, divided into Biology, Chemistry and Physics sections, each with 6 Topics.

As you work through the year, you will do a combination of the Biology, Chemistry and Physics Topics so you can use the coloured tabs on the right hand side of the pages to easily find the right Topic.

We hope you will enjoy this book as much as our authors have enjoyed creating the content, questions and activities for you!

→ Learn more, show your understanding, and build scientific enquiry skills

This book has been carefully designed to help you build your KS3 Science knowledge, understanding and skills.

Start with an interesting context to see where this topic fits into the real world and the rest of your Science course. See if you can answer the question.

Answer the questions in the photo captions to practice thinking like a scientist. This is called "scientific enquiry".

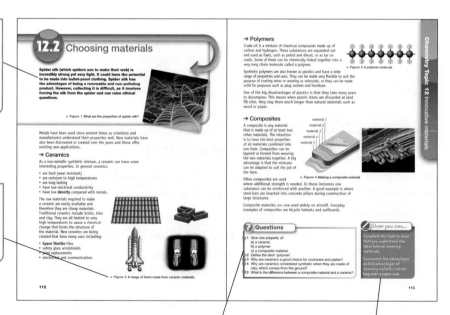

Work through the **Questions** to test your knowledge and understanding. The questions are colour-coded from simple to more advanced so that you can monitor your own progression. Challenge yourself and answer the blue questions.

Complete the **Show you can** task to show that you are confident in your understanding of this topic and that you have learnt the right scientific skills.

→ Build Working Scientifically skills

In this book, Working Scientifically skills are covered in the short questions throughout and explored in more detail in the activity at the end of each Topic.

Work through these activities to build and practice Working Scientifically skills such as Planning and Designing Investigations, Calculating Scientifically and Presenting and Interpreting Data.

Each activity includes a scenario for you to work through with questions to guide you in completing the task. You can complete these on your own or in groups.

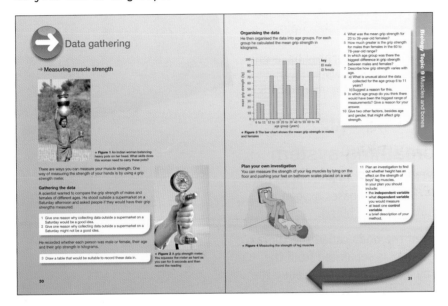

→ Free online extras to help you build your knowledge and understanding

At the back of this book you will find QR codes. You can use a free mobile app to open the online extras. You can also type in the short weblink to find the same content.

- For each set of Questions, you will find Hints to help you answer the questions.
- For all Key Terms, find more detailed explanations and examples in the Extended Glossary
- Once you are confident that you have learnt all the key words in the Topic, complete the Key Word Tests online

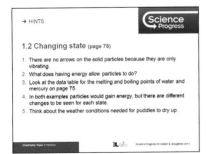

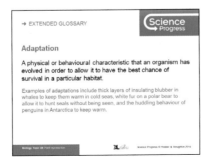

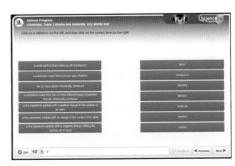

7.1 The breathing system

Many people suffer from asthma, which means that sometimes their airways get narrower so they find it hard to breathe. An image of the lungs of an asthma patient can be seen using gamma scanning. The patient inhales a radioactive gas that can be detected by a gamma camera. The yellow areas in Figure 1 show where the airways are obstructed. The left lung is more badly affected.

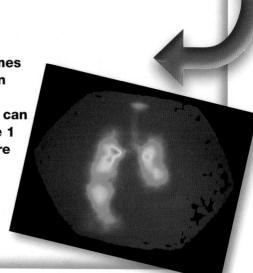

● **Figure 1** A gamma scan showing the lungs of an asthma patient viewed from the back. How do you think it would feel not to be able to breathe properly?

→ The structure of the breathing system

The function of the **breathing system** is to move air in and out of the **lungs**. The function of the lungs is to exchange the gases oxygen and carbon dioxide between the air in the lungs and the blood.

The lungs are situated in the upper part of the body called the **thorax**. They are protected by the ribcage. The thorax is separated from the lower part of the body, the **abdomen**, by a sheet of muscle called the **diaphragm**.

Air enters the body through the nose and mouth. It travels past the back of the throat into the **trachea**. The trachea splits into two tubes called the right and left bronchi. Each **bronchus** carries air into one of the two lungs.

The bronchi divide into smaller and smaller tubes called the **bronchioles**. At the ends of the bronchioles are bunches of tiny air sacs called **alveoli**. It is across the walls of the alveoli that **gas exchange** occurs.

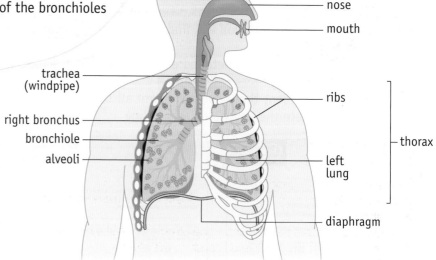

● **Figure 2** The human breathing system

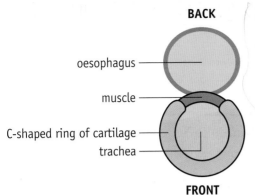

BACK

oesophagus

muscle

C-shaped ring of cartilage

trachea

FRONT

● **Figure 3** C-shaped rings of cartilage hold the trachea open

If you gently rub the front of your throat you can feel hard ridges. These are C-shaped rings of **cartilage** that make sure the trachea always stays open, even when you bend your neck. The gap in the C runs at the back of the trachea. The gap is there so that cartilage does not rub against the oesophagus (the tube that carries food to the stomach).

Specialised cells, called **goblet cells,** line the trachea. They secrete a thick, sticky liquid called **mucus.** Mucus traps dirt and microorganisms.

The cells lining the trachea have many tiny hair-like structures, called **cilia**, on their surface. The cilia are constantly moving to sweep the mucus up and out of the trachea. This helps to keep the lungs clear.

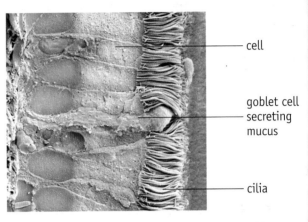

cell

goblet cell secreting mucus

cilia

● **Figure 4** Cells lining the trachea. What would happen if the cilia stopped moving?

→ **Asthma**

People who suffer from **asthma** can find it hard to breathe. An asthma attack is often triggered by house dust mites, animal fur, pollen, tobacco smoke, exercise, cold air or a chest infection. If one of these irritates the lining of the airways in the lungs, the muscles around them contract, which narrows the airway. The lining often becomes swollen and sticky mucus might also be secreted. All of these lead to a narrowing of the airway, making it harder to breathe, causing coughing and wheezing.

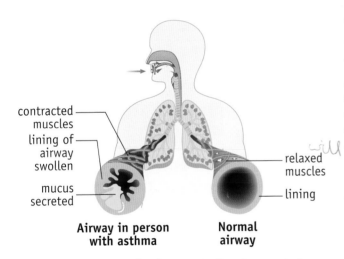

contracted muscles

lining of airway swollen

mucus secreted

relaxed muscles

lining

Airway in person with asthma

Normal airway

● **Figure 5** Changes that happen to the airways during an asthma attack

Show you can...

Complete this task to show that you understand why asthma can make breathing difficult.

Describe all the changes that might occur if the lining of the airways becomes irritated. You may want to include diagrams in your description.

? Questions

1 What is the function of the lungs?
2 Name the sheet of muscle that separates the thorax from the abdomen.
3 Draw a flow diagram to show the structures that air passes through from the nose and mouth to the alveoli.
4 Describe how cells lining the trachea help to keep the lungs clear.

7.2 Gas exchange

During a deep sea dive a diver is put under pressure several times greater than that on the surface of the water. This forces nitrogen gas to dissolve in the blood. If the diver returns to the surface too quickly bubbles of nitrogen gas can form in joints, tissues and the blood. This is very painful and can be fatal.

- **Figure 1** Divers taking a decompression stop. Why is it dangerous to have bubbles of gas in the bloodstream?

Gas exchange is the movement of oxygen from the air into the blood, and the movement of carbon dioxide in the opposite direction. The gases move by a process called **diffusion**. This is the movement of particles from an area where they are in high concentration to an area where they are less **concentrated**. Blood returning to the lungs from the body contains a high concentration of carbon dioxide and a low concentration of oxygen. Air in the alveoli contains a higher concentration of oxygen than the blood and a lower concentration of carbon dioxide.

→ How the alveoli are adapted for gas exchange

Alveoli are tiny air sacs found at the ends of the bronchioles in the lungs. They provide a large surface area for gas exchange. If the lungs were flattened out they would cover the area of a tennis court.

Each alveolus is well ventilated by breathing. This maintains a steep concentration gradient for the diffusion of oxygen into the blood and carbon dioxide out of the blood.

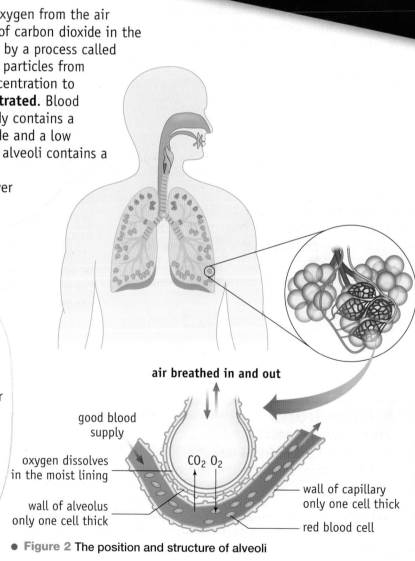

air breathed in and out

good blood supply

oxygen dissolves in the moist lining

CO_2 O_2

wall of alveolus only one cell thick

wall of capillary only one cell thick

red blood cell

- **Figure 2** The position and structure of alveoli

8

The walls of the alveoli and capillaries are only one cell thick, so the gases do not have far to diffuse.

The lining of each alveolus is moist. Oxygen dissolves in this layer, which speeds up diffusion into the blood. Each alveolus has a good blood supply to transport gases and to maintain a steep concentration gradient.

→ Gas exchange surfaces in other organisms

Plants have tiny holes on their leaves called **stomata**. Gases enter and leave the leaf through these stomata.

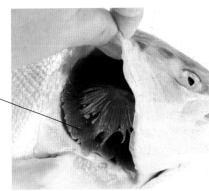

stomata

● **Figure 3** Stomata on a leaf. What factors affect the movement of gases through the stomata?

Fish have to obtain oxygen from the water. Gases are exchanged as water passes over organs called gills. These are made up of many gill filaments, which have a good blood supply and provide a large, thin surface for gas exchange.

gill filaments

● **Figure 4** Fish gills. How do you think fish make water flow over their gills?

Insects have tiny holes along their abdomen called spiracles. Gases move in and out of these holes, which are connected to a network of tubes called tracheae. The tracheae carry oxygen to respiring cells, and carbon dioxide away from the cells.

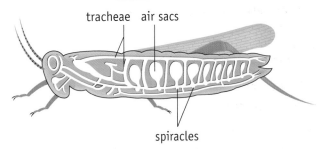

tracheae   air sacs

spiracles

● **Figure 5** The breathing system of an insect

? Questions

1 What is gas exchange?
2 Where in the lungs does gas exchange occur?
3 Describe the process of diffusion.
4 Name and describe the gas exchange surfaces in insects and fish.
5 Explain how a steep concentration gradient for the exchange of oxygen and carbon dioxide is maintained in the lungs.

✎ Show you can...

Complete this task to show that you understand gas exchange.

Make a list of all the ways that alveoli are adapted for efficient gas exchange.

7.3 Breathing

Polio is a disease that can lead to muscle weakness, which means that sufferers can find it hard or impossible to breathe on their own. The iron lung was used in the 1950s to help people breathe, and saved the lives of many people. Today a much smaller and more portable machine is used.

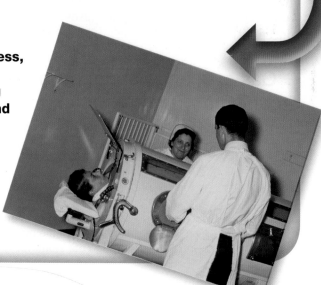

● **Figure 1** An iron lung. Why do you think many polio patients cannot breathe on their own?

→ The mechanism of breathing

Breathing is the movement of muscles to alter the volume of the chest cavity. This results in the movement of air into and out of the lungs. The muscles involved in breathing are the diaphragm and the muscles between the ribs. These muscles contract when you breathe in and relax when you breathe out.

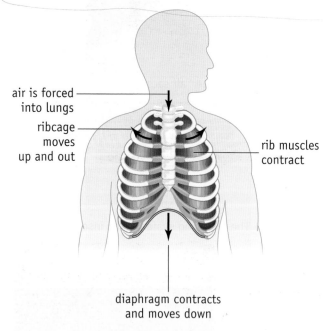

air is forced into lungs

ribcage moves up and out

rib muscles contract

diaphragm contracts and moves down

• the volume in the chest cavity increases
• the pressure in the chest decreases
• so air is forced into the lungs

● **Figure 2** Breathing in

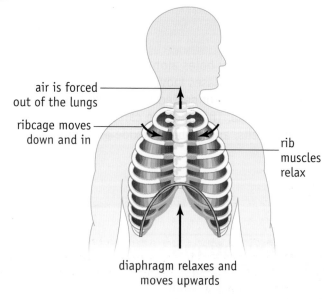

air is forced out of the lungs

ribcage moves down and in

rib muscles relax

diaphragm relaxes and moves upwards

• the volume in the chest cavity decreases
• the pressure in the chest increases
• so air is forced out of the lungs.

● **Figure 3** Breathing out

→ A model of the lungs

You can make a model to show how movements of the diaphragm can move air into and out of the lungs.

The bell jar represents the chest wall, the sheet of rubber stretched across its base represents the diaphragm and the balloons represent the lungs. When the rubber sheet is pulled down the balloons inflate, and when it is pushed up the balloons deflate.

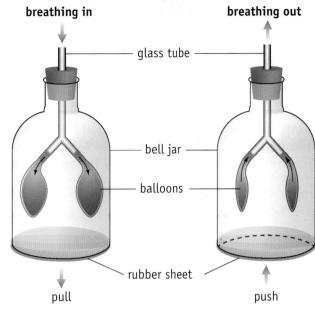

● Figure 4 Model to show the mechanism of breathing

→ Measuring lung volume

You can estimate the volume of your lungs by taking a deep breath, then exhaling through a tube into a container of water. The volume of water that is **displaced** from the container equals the volume of air that you breathed out.

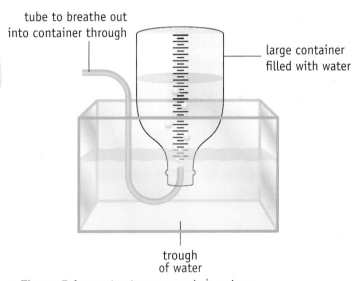

● Figure 5 Apparatus to measure lung volume

? Questions

1 a) Which muscles are involved in breathing?
   b) What happens to these muscles when you breathe in?
2 Look at the model in Figure 4. What do the following parts of the model represent:
   a) balloons
   b) rubber sheet
   c) bell jar
   d) glass tube.
3 Describe how you could measure the volume of your lungs.
4 Suggest factors that could affect the volume of your lungs.

✎ Show you can...

Complete this task to show that you understand how air is moved into the lungs.

Describe the changes that happen in your chest and explain how these changes result in air being inhaled.

The heart and circulatory system

Some people are born with a small hole in the tissue that separates the right from the left side of the heart. Surgeons can carry out surgery to repair this.

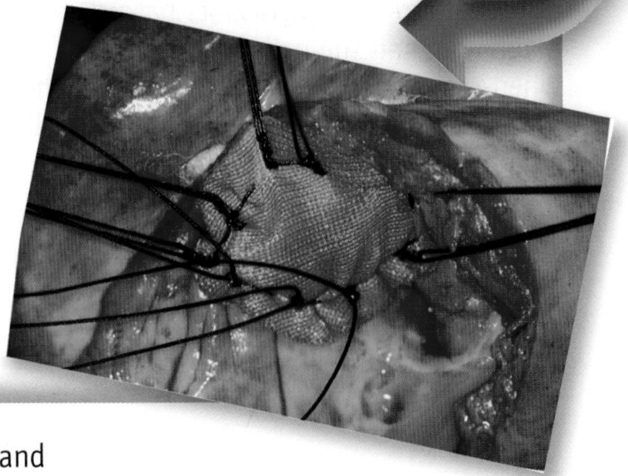

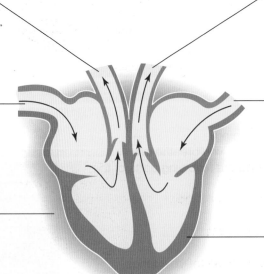

● **Figure 1** Surgery to repair a hole in the heart. Why are people who have a hole in the heart often short of breath?

The **circulatory system** is made up of the heart, blood vessels and blood. Its function is to transport substances around the body. The function of the **heart** is to pump blood around the lungs and body.

→ The heart

The heart is a double pump. The right side of the heart pumps **deoxygenated blood**, which is high in carbon dioxide and low in oxygen, to the lungs. Here carbon dioxide diffuses out of the blood and oxygen is collected from the air in the alveoli.

Blood leaving the lungs is therefore rich in oxygen and low in carbon dioxide. It is returned to the left side of the heart, which pumps this **oxygenated blood** around the body.

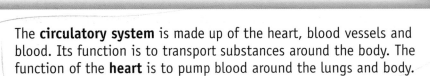

Blood is pumped out of the left side of the heart to go to the body.

Blood is pumped from the right side of the heart to the lungs to pick up oxygen and release carbon dioxide.

Blood returns to the right side of the heart from the body.

Blood returns from the lungs into the left side of the heart, to be pumped around the body.

The lungs are where oxygen in the air passes into the blood, and carbon dioxide diffuses in the opposite direction.

The heart is a muscular pump.

● **Figure 2** The structure of the heart

Blocked arteries

The heart muscle is supplied with blood in the **coronary arteries**. Sometimes arteries get blocked with fatty deposits. If this happens in a coronary artery it can cut off the blood flow to the heart muscle cells. These cells would not get enough food or oxygen so would die. The person might suffer a heart attack. Surgeons can use a piece of vein from an arm or leg and bypass the narrow or blocked artery. This is called a coronary bypass operation.

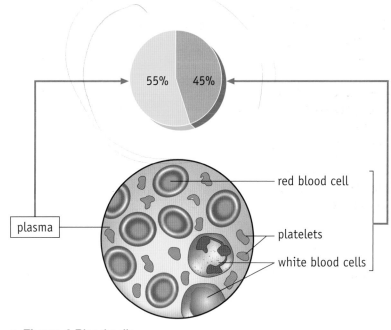

coronary arteries

blocked artery

● **Figure 3** a) The main blood vessels of the heart and b) a blocked coronary artery. What can people do to reduce the risk of having a heart attack?

→ The blood

The main function of the blood is to transport substances around the body. Blood is a made up of a pale yellow fluid called **plasma**. Food, carbon dioxide and other chemicals are dissolved in the plasma. Also in the plasma are **red blood cells**, **white blood cells** and **platelets**.

Red blood cells carry oxygen from the lungs to the cells of the body. White blood cells are important in defending the body against disease. Platelets are fragments of cells that are important in making the blood clot, so a scab forms if you cut yourself.

55%   45%

plasma

red blood cell

platelets

white blood cells

● **Figure 4** Blood cells

? Questions

1 What are the names of the blood vessels that supply food and oxygen to the heart muscle cells.
2 Describe the difference between oxygenated blood and deoxygenated blood.
3 Why do some people need a coronary bypass operation?
4 Suggest what people could do to reduce the chance of having a heart attack.

✎ Show you can...

Complete this task to show that you know what blood is made up of.

Draw a labelled diagram showing the four main parts of the blood, and describe the function of each part.

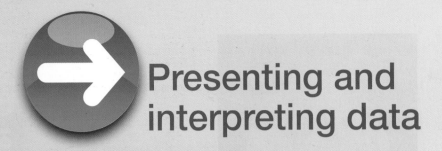

Presenting and interpreting data

→ **Respiratory and circulatory diseases**

● **Figure 1** Most people in the UK die of circulatory diseases

The table shows the causes of deaths in the UK in 2010.

Cause of death	Number of people
Circulatory diseases	158 000
Respiratory diseases	67 000
Cancers	141 000
Digestive diseases	26 000
Other causes	

1 The total number of people who died in the UK in 2010 was 493 000. How many people died from other causes?
2 Calculate the percentage of people who died from each cause of death. Round each answer to a whole number. The percentage who died from circulatory diseases has been done for you:

% people who died from circulatory diseases $= \dfrac{158\,000}{493\,000} \times 100 = 32\%$

3 Plot your answers to Question 2 in a bar chart.
4 What conclusions can you reach from these data?

Exercise and heart rate

Exercise helps to keep the heart healthy. Some students decided to investigate how exercise affects heart rate. They used a sensor to measure their pulse rate at rest and every 2 minutes whilst they ran on a treadmill and after exercise, until the heart rate returned to the resting value.

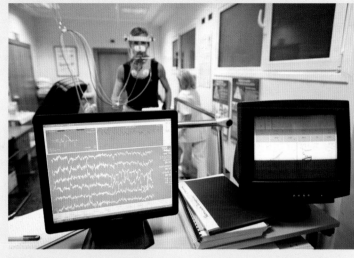

● **Figure 2** Investigating the effect of exercise on heart rate

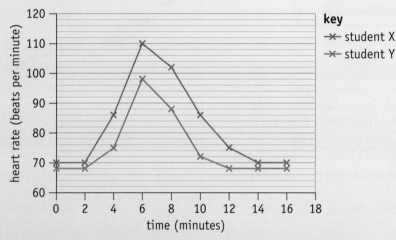

● **Figure 3** The graph shows the results for two of the students

5 For how many minutes were the students running?
6 Which student, X or Y, was the fittest? Give two reasons for your answer.
7 Other factors can also affect heart rate. Describe how you could find out if body mass affects resting heart rate. Your answer should include:
   • what you would measure
   • how you would make it a fair test
   • how many people you would include
   • how you would display your results.

8.1 Aerobic respiration

Animals that live in deserts hardly ever drink water. They can release water from the food they eat or from stored fat. Camels' humps contain fat. This is respired to release energy and water is produced as a waste product. For each kilogram of fat respired, a litre of water is produced.

● **Figure 1** A Bactrian (two-humped) camel. How long do you think camels can survive without drinking any water?

→ What is respiration?

Respiration is the release of energy from food. It happens inside cells. All cells need energy to function, and for growth and repair. Energy is also needed for all the chemical processes necessary for staying alive, such as building larger molecules from smaller ones or enabling muscles to contract. In mammals and birds energy is also used to keep warm.

→ Aerobic respiration

Aerobic respiration needs oxygen. The reaction takes place in lots of little steps, but the overall process is represented by the following equation:

$$\text{glucose} + \text{oxygen} \rightarrow \text{carbon dioxide} + \text{water} \; (+ \text{energy})$$
$$C_6H_{12}O_6 + 6O_2 \rightarrow 6CO_2 + 6H_2O \; (+ \text{energy})$$

Glucose comes from the food we eat. Glucose molecules contain stored chemical energy. Oxygen comes from the air we breathe in. We breathe out air containing more carbon dioxide and water vapour than the air we breathe in. The table shows the differences between **inhaled air** and **exhaled air**.

	Inhaled air (%)	Exhaled air (%)
Nitrogen	79	79
Oxygen	20	16
Carbon dioxide	0.04	4
Water vapour	variable	more than in inhaled air

16

The differences in the air we breathe in and the air we breathe out are due to respiration. You can carry out experiments to show these differences.

Limewater changes colour with carbon dioxide. It is clear and colourless, but if carbon dioxide is bubbled through it, it turns cloudy.

If you gently breathe in and out through the mouthpiece, inhaled air bubbles through the limewater in tube A and exhaled air bubbles through the limewater in tube B. The limewater in tube B quickly turns cloudy, showing that there is more carbon dioxide in exhaled air.

If you take the temperature of the air and then breathe out onto a thermometer bulb several times, you will notice the temperature increases. This shows that exhaled air is warmer than inhaled air. Some of the energy released in respiration is heat energy.

If you breathe out onto a mirror you will see condensation. To show that the liquid that forms on the mirror is water you can use blue **cobalt chloride paper**. It turns pink on contact with water.

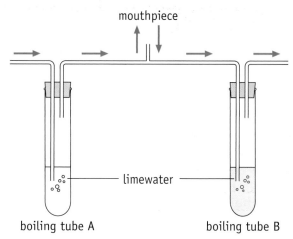

● **Figure 2** Apparatus to compare the amount of carbon dioxide in inhaled and exhaled air

● **Figure 3** Apparatus to compare the temperature of inhaled and exhaled air

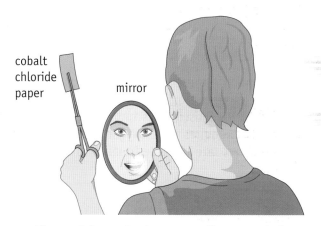

● **Figure 4** Apparatus to compare the amount of water vapour in inhaled and exhaled air

? Questions

1 Why do cells have to respire?
2 Write the word equation for aerobic respiration.
3 Explain why there is less oxygen and more carbon dioxide in exhaled air compared with inhaled air.
4 Describe the test for carbon dioxide and give the colour change for a positive result.
5 Draw a bar chart to compare the percentages of nitrogen, oxygen and carbon dioxide in inhaled and exhaled air.

Show you can...

Complete this task to show that you understand aerobic respiration.

Explain what the process is, write down the word equation and make a list of some uses of energy in the body.

8.2 Anaerobic respiration and exercise

Some athletes train at high altitude to increase the number of red blood cells in their blood. This makes their blood able to carry more oxygen, which is needed for aerobic respiration. However, athletes may use illegal methods to increase their red blood cell count, such as using hormones to stimulate red blood cell production, or by having a transfusion of their own blood just before competing.

● **Figure 1** Testing for the illegal use of hormones in professional sport. Why is the use of hormones in sports illegal? Do you think it should be made legal?

→ Benefits of exercise

Regular exercise is good for the body. It strengthens the heart muscle and improves blood circulation. Exercise also keeps the immune system healthy, the bones, **joints** and muscles strong, and helps prevent obesity. When you exercise you make chemicals in the brain called endorphins that make you feel happy, so exercise is also good for mental health.

If you do gentle exercise like jogging, your muscle cells need more energy. More glucose has to be supplied to the muscle cells to be respired more quickly. This requires more oxygen. Your breathing rate and heart rate increase to meet this demand.

● **Figure 2** A group of people doing yoga exercises. How does gentle exercise such as yoga help you to stay healthy?

If you do more vigorous exercise even more energy is required, so respiration has to happen even faster. Sometimes your breathing rate and heart rate cannot increase enough to supply enough oxygen. If this happens, the muscle cells respire aerobically as fast as they can, but also obtain some extra energy by respiring anaerobically at the same time.

● **Figure 3** This athlete will be respiring both aerobically and anaerobically to release enough energy to keep running fast. How do you think training will affect how long an athlete can respire anaerobically?

18

➜ Anaerobic respiration

Anaerobic respiration is respiration without oxygen. It does not release as much energy as aerobic respiration and produces a toxic chemical called **lactic acid**. A build-up of lactic acid in the muscle cells causes cramp.

The word equation for anaerobic respiration is:

> glucose → lactic acid (+ a little energy)

After exercise, your heart rate stays high and you breathe heavily for a period of time. This is to supply oxygen to the muscles to break down lactic acid. The amount of oxygen needed to break down the lactic acid is called the **oxygen debt**.

> lactic acid + oxygen → carbon dioxide + water (+ a little energy)

Once all the lactic acid has been broken down, all the energy originally stored in the glucose will have been released.

The time from finishing exercise until the heart and breathing rates return to normal is called the **recovery time**. The fitter you are, the shorter your recovery time. Fitter people also have lower resting heart rates.

To measure fitness, measure:

- normal resting pulse
- the pulse at the end of the exercise
- how long it takes for the pulse to get back to normal.

It is also useful to see if your fitness is improving by finding if your recovery times are getting shorter as you train.

Other fitness measures that sports scientists use are strength, suppleness, stamina and speed.

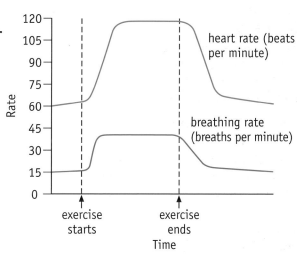

● **Figure 4** Graph showing the changes in heart rate and breathing rate during and after exercise

❓ Questions

1 Write the word equation for anaerobic respiration.
2 What causes cramp?
3 Regular exercise is good for the body. Make a list of reasons why you should exercise regularly.
4 Describe the changes that happen to your breathing rate when you exercise vigorously, and explain why these happen.

Show you can...

Complete this task to show that you understand anaerobic respiration.

Explain what the process is and when it happens, and write down the word equation.

8.3 Anaerobic respiration in micro-organisms

Microbiology is the study of micro-organisms such as bacteria, fungi and viruses. We often associate them with diseases, but we can also use many micro-organisms to produce useful chemicals.

● Figure 1 Some useful products from micro-organisms. What else are micro-organisms used for?

→ Anaerobic respiration in yeast

Yeasts are **unicellular** organisms that respire aerobically when there is plenty of oxygen around, but they can respire anaerobically when there is no oxygen. Some types of yeast are used to make bread, wine and beer because they produce carbon dioxide and **ethanol**, a type of alcohol, when they respire anaerobically. Anaerobic respiration in yeast is also called **fermentation**.

The word equation for anaerobic respiration in yeast is:

> glucose → carbon dioxide + ethanol (+ a little energy)

Bread

Bread is made using flour, water, yeast and a little sugar. The sugar is respired by the yeast and the carbon dioxide produced gets trapped as pockets of gas in the dough and makes it rise. When the bread is baked the trapped carbon dioxide expands, so the bread rises even more. The ethanol produced evaporates due to the high temperature in the oven.

● Figure 2 Making bread. Why doesn't bread contain alcohol?

Beer

Beer is made using barley grains, water, yeast and hops. The barley grains are soaked in water and kept warm. This makes them germinate, and the stored starch in the grains is turned into sugar. The sugary solution is separated from the grains, and yeast is used to ferment it and produce alcohol. Hops are then added to the liquid to give it a bitter flavour. Beer has a slight fizz because some of the carbon dioxide produced during fermentation stays dissolved in the liquid.

● Figure 3 A fermentation vessel of frothy beer. Why do you think a froth forms on top of the beer during fermentation?

Wine

Wine is made from grape juice and yeast. There is plenty of water and sugar in the grapes. Yeast on the skins of the grapes ferments the sugar to produce alcohol and carbon dioxide. Usually the carbon dioxide is allowed to escape, so most wine is flat, not fizzy. However, during champagne production, the carbon dioxide produced during fermentation is trapped in the bottle to make a fizzy wine. That is why champagne corks are fastened on with wire, otherwise the cork could be blown off by the pressure of carbon dioxide in the bottle.

● **Figure 4** Measuring the maturity of grapes before harvesting. Why do winemakers avoid using unripe grapes?

→ Biogas generators

Bacteria can also respire anaerobically. **Biogas** is a mixture of gases produced when bacteria break down plant material or animal wastes. The mixture of gases varies depending on the type of bacteria and the type of waste used, but the main gas produced is **methane**. This can be burnt as a fuel or used to generate electricity.

Biogas generators are used around the world in areas where there is no electricity. They are cheap to build and easy to use. As well as producing fuel, the generators use up plant and animal wastes and the material left in the generator after fermentation can be used as a fertiliser.

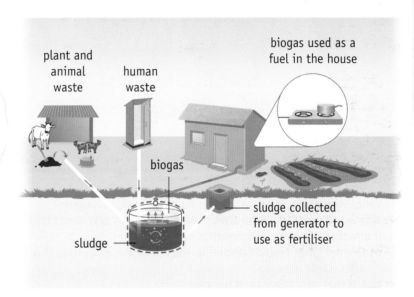

● **Figure 5** A small-scale biogas generator. Do you think we should use more biogas generators in the UK? Why?

? Questions

1 Write the word equation for anaerobic respiration in yeast.
2 Why doesn't bread taste of alcohol?
3 What is the main gas produced in a biogas generator?
4 Make a list of the benefits of using biogas generators.

Show you can...

Complete this task to show that you understand anaerobic respiration in micro-organisms.

Describe how bread is made, and why yeast is used.

Calculating scientifically

→ Do gerbils respire?

An experiment was set up to investigate respiration using a gerbil. Air was drawn through the apparatus as shown by the arrows on the diagram. The gerbil sat and ate some food.

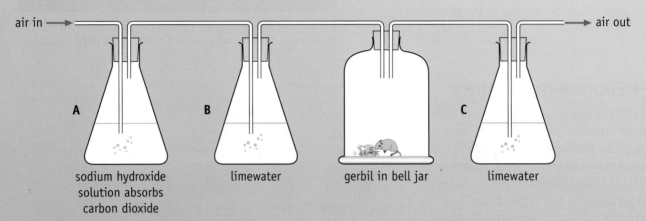

● **Figure 1** A gerbil in a bell jar

1 What was the purpose of the sodium hydroxide solution?
2 Name the gas that limewater is used to detect, and describe how it affects the colour of limewater.
3 **a)** Predict what would happen to the limewater in bottles B and C after a few minutes.
  b) Explain why this would happen.

How much oxygen do we use each day?

Jack wanted to find out what volume of air he breathed in a day. He used the apparatus shown in Figure 2.

He took a normal breath whilst relaxing, and then breathed out through the tube to displace water from the large container. The volume of water displaced gave him the volume of air in one breath. He did this five times.

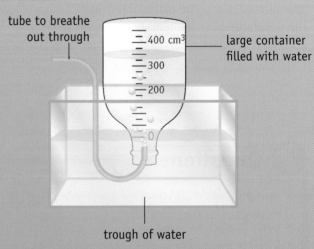

● **Figure 2** Measuring the volume of air in one breath

Jack's results are shown in the table.

Breath number	Volume of air in one breath (cm³)
1	366
2	384
3	335
4	373
5	352

4 Calculate the mean volume of air in one breath.
5 He counted that he took 14 breaths every minute.
   a) What volume of air (in cm³) would he breathe in 1 minute?
   b) What volume of air (in cm³) would he breathe in 1 hour?
   c) What volume of air (in cm³) would he breathe in 1 day?
   d) What volume of air in litres would he breathe in 1 day? Round your answer to a whole number of litres.
6 a) Twenty per cent of the air breathed in is oxygen. How many litres of oxygen would he breathe in during one day? Round your answer to a whole number.
   b) Sixteen per cent of the air breathed out is oxygen. How many litres of oxygen would he breathe out in one day? Round your answer to a whole number.
   c) How many litres of oxygen would he have absorbed in 1 day?
   d) What process would this oxygen be used for in the body?
   e) The figure you have calculated is probably less than the actual amount of oxygen he would absorb in one day. Suggest why.

Energy release in respiration

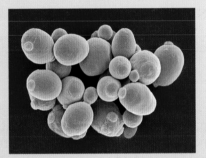

● **Figure 3** a) An athlete and b) yeast cells. What is the difference between the process that powers the athlete and the one that powers yeast cells?

The table shows how much energy in kilojoules (kJ) is released in aerobic and anaerobic respiration.

Type of respiration	Energy released per gram of glucose (kJ)
aerobic respiration	16.1
anaerobic respiration in muscle cells	1.17
anaerobic respiration in yeast	0.83

7 How many joules (J) are in 1 kilojoule (kJ)?
8 How much energy would be released from 3 g of glucose during anaerobic respiration in yeast cells?
9 a) How many more kilojoules of energy are released per gram of glucose in aerobic respiration compared to anaerobic respiration in humans?
   b) Explain why more energy is released in aerobic respiration than in anaerobic respiration in humans.
10 a) How many times more energy is released per gram of glucose in anaerobic respiration in muscle cells compared to anaerobic respiration in yeast.
   b) What does this suggest about the energy content of lactic acid compared to the energy content of ethanol?

9.1 The skeletal system

Between the bones of the spine are discs of cartilage that act as shock absorbers. Sometimes these get damaged and the outer layer of the disc splits or weakens. The contents get pushed out and can put pressure on the spinal cord. This can cause a lot of pain.

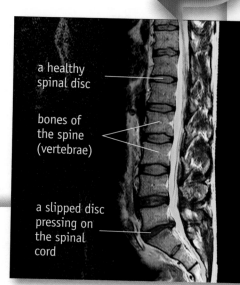

a healthy spinal disc

bones of the spine (vertebrae)

a slipped disc pressing on the spinal cord

● **Figure 1** An MRI scan showing a slipped spinal disc. What do you think causes a slipped disc?

The main parts of the **skeletal system** are bones and muscles.

→ The skeleton

An adult human skeleton is made up of 206 bones. The longest bone in your body is the femur (thigh bone), which is about a quarter of your height. The smallest bones in your body are in your ears.

Bone is hard because it contains calcium compounds. The sternum is not a bone, but is made of a substance called **cartilage**. This is softer than bone because it contains fewer calcium compounds. Cartilage is found in several parts of the body: it covers the ends of the limb bones to reduce friction and act as a shock absorber. It also makes up the ears and nose.

Functions of the skeleton

- to support the body
- for movement, using muscles and joints
- to make blood cells
- to protect body organs:
  - the cranium (skull) protects the brain
  - the ribcage protects the lungs, heart and main blood vessels
  - the vertebral column (backbone) protects the spinal cord
  - the pelvic girdle protects reproductive organs in females.

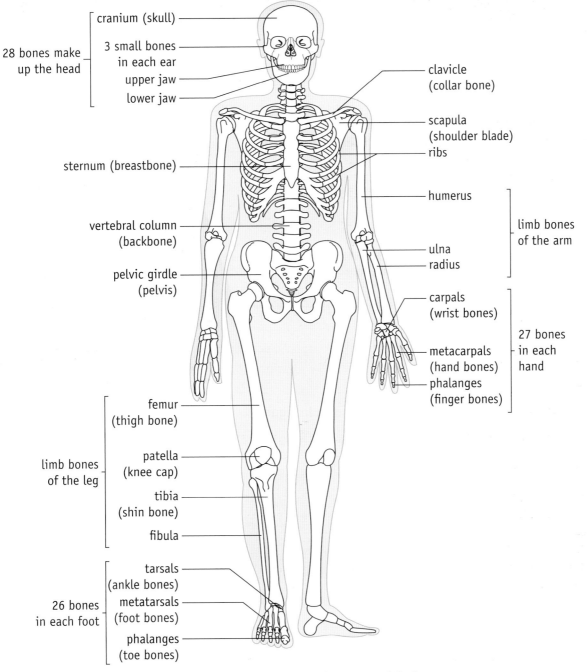

28 bones make up the head
- cranium (skull)
- 3 small bones in each ear
- upper jaw
- lower jaw

sternum (breastbone)

vertebral column (backbone)

pelvic girdle (pelvis)

clavicle (collar bone)

scapula (shoulder blade)

ribs

humerus

limb bones of the arm
- ulna
- radius

carpals (wrist bones)

27 bones in each hand
- metacarpals (hand bones)
- phalanges (finger bones)

limb bones of the leg
- femur (thigh bone)
- patella (knee cap)
- tibia (shin bone)
- fibula

26 bones in each foot
- tarsals (ankle bones)
- metatarsals (foot bones)
- phalanges (toe bones)

● **Figure 2** The human skeleton. You do not need to learn the names of the bones.

? Questions

1 Name the main structures that make up the skeletal system.
2 What does the cranium protect?
3 Describe the function of cartilage.
4 Suggest what you should eat to keep your bones strong.
5 A woman's pelvic girdle is wider than a man's. Suggest a reason for this.

Show you can...

Complete this task to show that you understand the skeletal system.

Make a list of all the functions of the skeleton.

9.2 Muscles and movement

Many people go to the gym to tone their muscles and to stay fit. Other people go to extremes to build large, strong muscles.

● **Figure 1** A body builder. What else besides weight lifting helps to build large muscles like these?

Movement of the body is brought about by muscles.

→ Types of muscle

There are three types of muscle in the body, as shown in Figure 2.

Skeletal muscle	Cardiac muscle	Smooth muscle
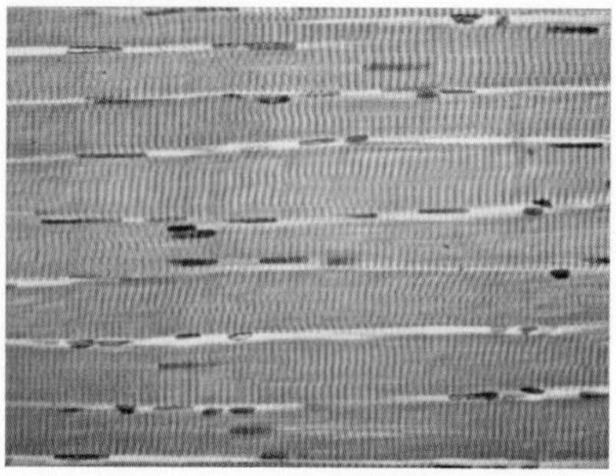	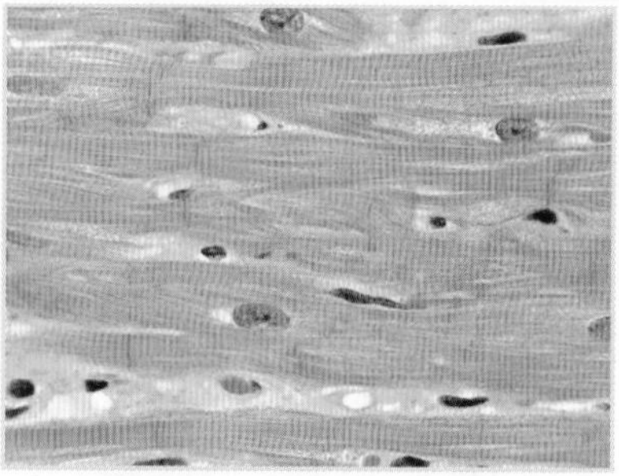	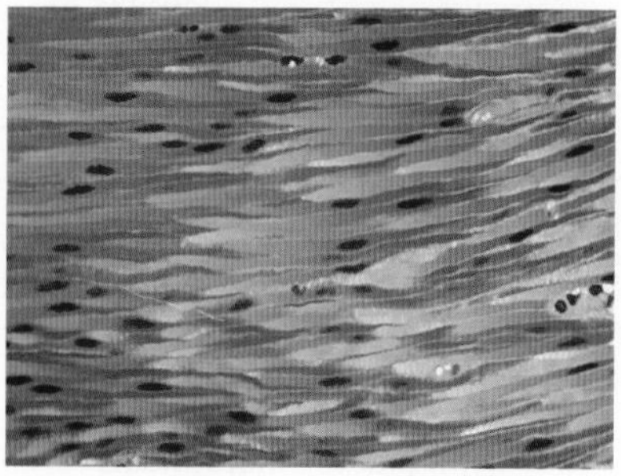
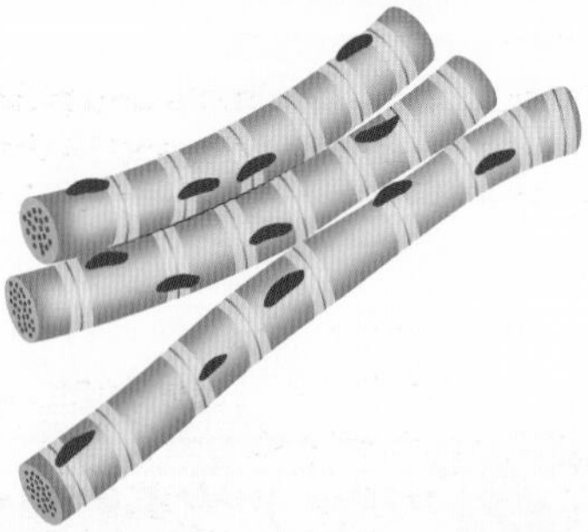	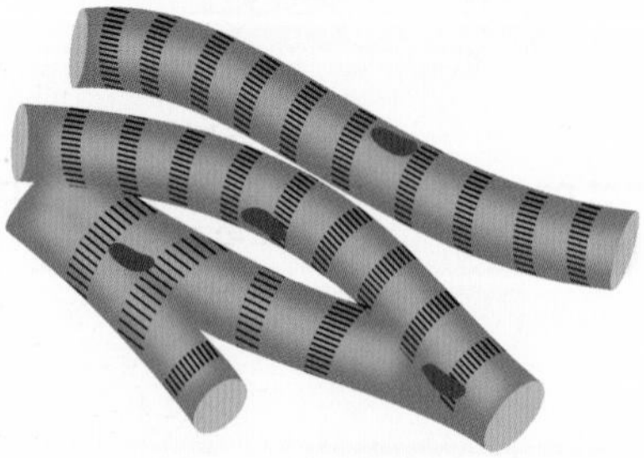	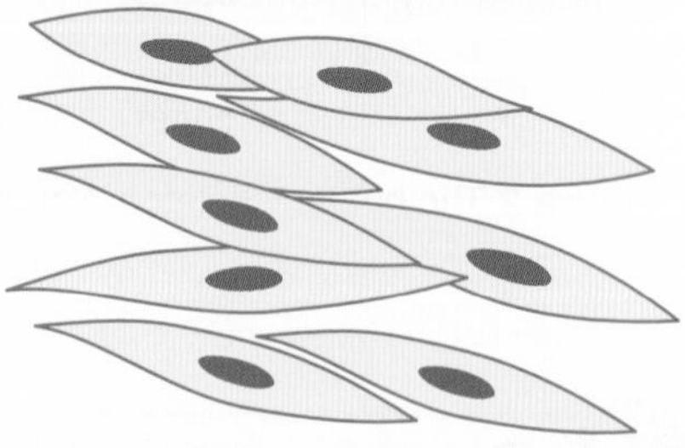
Skeletal muscle causes movement at joints. It is made of long fibres of many cells that have fused together. We can consciously control skeletal muscles. They can contract quickly, but they soon get tired.	Cardiac muscle has branching fibres. This is the muscle that makes the heart keep beating. It never gets tired, otherwise the heart would stop working.	Smooth muscle contacts and relaxes slowly and does not get tired. It lines the intestines, blood vessels and airways in the lungs to move substances along.

● **Figure 2** The three types of muscle

All muscles cause movement by contracting and relaxing. When a muscle cell contracts it gets shorter, when it relaxes it gets longer.

Skeletal muscles are attached to bones by tough cords called **tendons**. When a muscle contracts it pulls on the tendon, which pulls the bone to move it.

muscle cells when relaxed   muscle cells when contracted

● **Figure 3** Muscle cells can change shape by contracting and relaxing

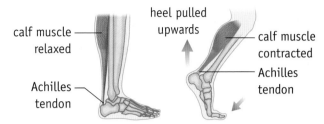

● **Figure 4** The Achilles tendon attaches the calf muscle to the heel. When the calf muscle contracts it pulls on the heel and lifts it upwards

→ Antagonistic muscle action

Muscles cannot push, they can only pull. They exert a pulling **force** when they contract and this causes movement at a joint. Muscles usually work in pairs. As one muscle contracts and pulls, the other relaxes. This is called **antagonistic muscle action**.

Figure 5 shows how the **biceps** and **triceps** muscles in the upper arm work together to bend and straighten the arm.

Muscles all around your body work antagonistically. When you bend forwards the muscles at the front of your body contract to pull you forwards and downwards, whilst the muscles in your back relax. To straighten up, the muscles in your back contract to pull you upwards whilst the muscles at the front of your body relax.

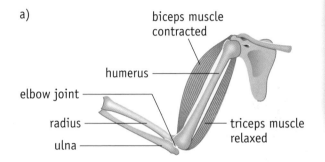

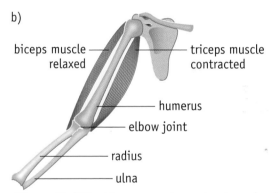

● **Figure 5** a) The biceps muscle contracts and pulls the radius upwards, which bends the arm, and b) the triceps muscle contracts and pulls on the ulna, which straightens the arm

✎ Show you can...

Complete this task to show that you understand antagonistic muscle action.

Sketch diagrams to show how the muscles at the front and back of the thigh work together to bend and straighten the leg. On each diagram, label which muscle contracts and which relaxes.

❓ Questions

1 Name the three different types of muscle.
2 Draw diagrams to show what happens to the length of a muscle cell when it contracts and relaxes.
3 Describe a tendon and its function.
4 Explain what is meant by 'antagonistic muscle action'.

Osteoarthritis is a painful condition caused by overuse of a joint. The cartilage becomes damaged and the bones of the joint rub against each other. Treatment is surgical replacement of the joint.

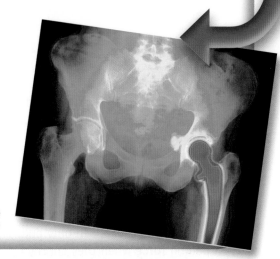

● **Figure 1** What materials do you think are used to make an artificial hip joint?

→ Types of joint

A **joint** is where bones meet. There are several different types of joint in the human body.

Have you ever noticed that the top of a baby's head is soft? This is because the bones of the skull have not grown to meet each other. When they do, the bones fuse together to form a **fixed joint** that cannot move.

Spine and neck joints

The bones in your spine are connected together by discs of cartilage. There is only slight movement between the vertebrae. The head can be moved from side to side because there is a movable **pivot joint** at the top of the spine in the neck.

● **Figure 2** Fixed joints in the skull. Why do you think the bones of a baby's skull do not meet?

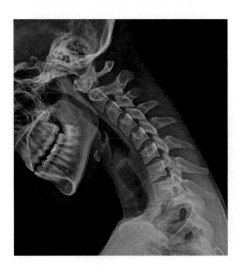

Freely movable joints

Freely movable joints are called **synovial joints** because they contain **synovial fluid**. This acts as a lubricant to make movement easier. Examples include the shoulder and the elbow joints.

Shoulder and hip joints

The shoulder and hip joints are called **ball and socket joints**. The rounded head of one bone fits into a socket in the other bone. The bones are held together by **ligaments**. The ball and the socket are covered with cartilage and synovial fluid fills the joint. These both help to reduce friction between the bones.

● **Figure 3** Slightly movable joints between the vertebrae and pivot joint in the neck

A ball and socket joint can move in all directions.

The elbow and knee joints are **hinge joints**. Hinge joints allow movement in just one direction.

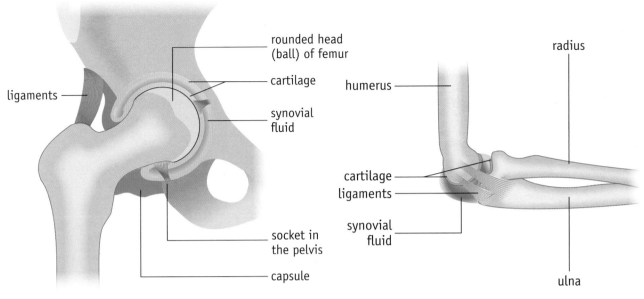

● **Figure 4** The hip joint – a ball and socket joint

● **Figure 5** The elbow joint, a hinge joint

→ Prosthetic limbs

If someone is born without a limb or loses an arm or a leg due to disease or injury, it is possible to replace it with an artificial limb called a **prosthesis**. A prosthesis is actually a replacement for any body part, such as eyes, fingers or ears. People who have prosthetic limbs, such as athletes or climbers, sometimes outperform those with natural limbs.

It's not just athletes who have prosthetics. Many people need an artificial replacement for a missing body part. Most prosthetic limbs are made from a plastic and carbon-fibre material that is lightweight and strong.

Every prosthesis is specially designed for the person who will use it. Sometimes the main aim is cosmetic. Prosthetics can be made to look very realistic and be coloured to match the person's skin tone. For others controllability is more important. Prosthetic hands can be made with individually powered fingers that allow the hand to hold different objects in different ways.

● **Figure 6** A prosthetic hand with individually powered fingers. What sensors do you think are needed for this hand to pick up a delicate glass?

? Questions

1 What is a joint?
2 Describe a fixed joint.
3 What do ligaments do?
4 What is a prosthesis? Give some examples.
5 Suggest why some athletes with prosthetic limbs outcompete those who are able bodied.

Show you can...

Complete this task to show that you understand how the different joints in the body work.

Name five different types of joint and give an example of where each is found in the body.

→ Measuring muscle strength

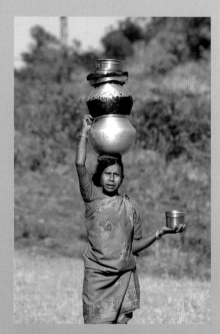

● **Figure 1** An Indian woman balancing heavy pots on her head. What skills does this woman need to carry these pots?

There are ways you can measure your muscle strength. One way of measuring the strength of your hands is by using a grip strength meter.

Gathering the data

A scientist wanted to compare the grip strength of males and females of different ages. He stood outside a supermarket on a Saturday afternoon and asked people if they would have their grip strengths measured.

1 Give one reason why collecting data outside a supermarket on a Saturday would be a good idea.
2 Give one reason why collecting data outside a supermarket on a Saturday might not be a good idea.

He recorded whether each person was male or female, their age and their grip strength in kilograms.

3 Draw a table that would be suitable to record these data in.

● **Figure 2** A grip strength meter. You squeeze the meter as hard as you can for 5 seconds and then record the reading

Organising the data

He then organised the data into age groups. For each group he calculated the mean grip strength in kilograms.

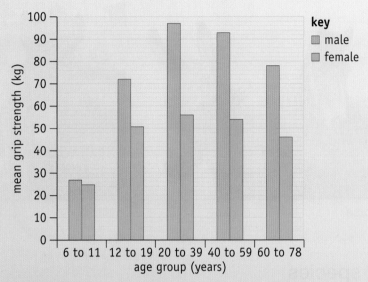

● **Figure 3** The bar chart shows the mean grip strength in males and females

4 What was the mean grip strength for 20 to 39-year-old females?
5 How much greater is the grip strength for males than females in the 60 to 78-year-old range?
6 In which age group was there the biggest difference in grip strength between males and females?
7 Describe how grip strength varies with age.
8 a) What is unusual about the data collected for the age group 6 to 11 years?
   b) Suggest a reason for this.
9 In which age group do you think there would have been the biggest range of measurements? Give a reason for your answer.
10 Give two other factors, besides age and gender, that might affect grip strength.

Plan your own investigation

You can measure the strength of your leg muscles by lying on the floor and pushing your feet on bathroom scales placed on a wall.

● **Figure 4** Measuring the strength of leg muscles

11 Plan an investigation to find out whether height has an effect on the strength of boys' leg muscles.
   In your plan you should include:
   ● the **independent variable**
   ● what **dependent variable** you would measure
   ● at least one **control variable**
   ● a brief description of your method.

10.1 Variation

Horses and donkeys have many similar features, and so it would be easy to think that they are variations of a single species, but horses and donkeys are different species. Although they can breed together the offspring are not fertile.

horse   donkey   mule

● **Figure 1** Female horse + male donkey = sterile mule. What do you think is produced from a male horse and female donkey?

→ Variation between different species

The wide range of different living organisms that exist, from daisies to giraffes, is called **biodiversity**. In Book 1 you learnt that all organisms can be classified into smaller and smaller groups based on their observable features. You may want to look back at this to remind yourself of all the group names.

● **Figure 2** The variety of life. Which groups would you put these organisms into?

If you look at the features of the plant, animal, **fungi**, bacteria and protoctists in Figure 2 you will notice a wide range of variation. Much of the variation is due to **inherited** factors (also called genetic factors), but there is also **environmental variation** due to environmental factors, such as the amount of food available, which would affect how large an animal could grow.

→ Variation between organisms of the same species

It is because there is variation between organisms that we can classify them into groups. Organisms that are very closely related are put into a small group called a **species**. Only organisms of the same species can successfully breed with each other and produce fertile offspring.

However, even members of the same species vary from one to another. Domesticated dogs are all members of the same species, *Canis familiaris*. There is wide variation between the different breeds but they can all interbreed to produce fertile offspring.

Species are named in a particular way, referred to as the scientific name or the latin name, so that the names can be recognised all over the world. Each species name has two parts. The first part is called the genus and the second is called the species.

Stoats and weasels have many similarities, so they are both in the genus *Mustela*. However, they cannot interbreed so must be different species. The main differences are that a stoat is slightly bigger than a weasel, and has a black tip to its tail. The weasel's scientific name is *Mustela nivalis* and the stoat's is *Mustela ermine*.

● **Figure 3** Different breeds of dog. Why do scientists group them as one species?

● **Figure 4** a) A weasel and b) a stoat. How can you tell the difference between them?

? Questions

1 What are the two causes of variation in organisms?
2 How do we know that horses and donkeys are different species?
3 Explain why scientific names are more useful than common names, and state what the scientific name of an organism is made up of.

✎ Show you can...

Complete this task to show that you understand variation.

Look at the different breeds of dog in Figure 3 and identify examples of inherited variation, environmental variation and variation that is caused by both inherited and environmental factors. Record your observations in a table like this, putting a tick or a cross in the inherited and environmental variation columns:

Example of variation in dogs	Inherited variation	Environmental variation

Biology Topic 10 Inheritance and evolution

33

10.2 Genes, chromosomes and DNA

The fruit fly, *Drosophila melanogaster*, has been used in genetic research for over a hundred years, partly because it is genetically very similar to humans. The geneticist Thomas H Morgan (1866–1945) discovered chromosomes and worked out how gender (male or female) is inherited using these insects. Today they are used to study genetic disorders including Alzheimer's, Parkinson's and Huntington's as well as other human diseases.

● Figure 1 A fruit fly. Why is this such a useful species for research into human diseases?

Genes are the units of inheritance. They carry instructions (genetic information) that control the characteristics of an organism. Genes make up the **chromosomes** in the **nuclei** of cells. Chromosomes only become visible under a microscope when the cell is about to divide.

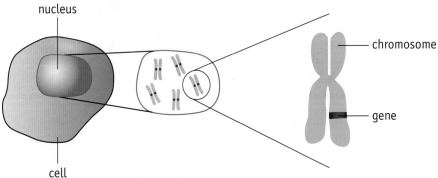

nucleus

cell

chromosome

gene

● Figure 2 Simplified drawing of chromosomes and a gene

Each chromosome is a long molecule of **DNA**. Specific areas of DNA along each chromosome are called genes. Each gene contains information that codes for a specific protein. Some of these proteins are enzymes. Enzymes regulate all chemical reactions that take place in cells.

Different species have different numbers of chromosomes in their nuclei. Hedgehogs have 88 chromosomes, broccoli has 18 and humans have 46 in each of their cells. The chromosomes can be arranged in pairs.

● Figure 3 Chromosomes in the nucleus of a bluebell cell. How many chromosomes are there?

Genetic information is passed from one generation to the next. One chromosome of each pair comes from the mother and one from the father. Knowing this helps us to understand how characteristics are inherited from both parents during sexual reproduction.

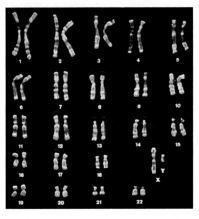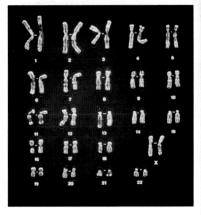

● **Figure 4** a) The chromosomes of a man and b) the chromosomes of a woman. What is the difference between the two sets of chromosomes?

➜ The history of genetic developments

Scientists have not always known about genes.

- Cells were discovered when microscopes were used in the 1600s, and by the early 1800s scientists said all living things were made of cells.
- In the 1860s Gregor Mendel did experiments on inheritance in peas. He suggested that there were things in cells that acted as units of inheritance by being passed from one generation to the next.
- By 1910 the term 'gene' had been introduced and fruit fly chromosomes were being mapped to find the location of genes.
- By the 1940s scientists knew that DNA, found in cell nuclei, was the chemical of inheritance.
- In the 1950s James Watson and Francis Crick, with Maurice Wilkins and Rosalind Franklin, worked out that the structure of DNA was a double helix, like the one shown in Figure 5. In the 1970s the first gene was sequenced and in the 1990s the Human Genome Project began. This project set out to identify all the human genes and work out their sequence on the chromosomes. This was completed in 2003. The human genome is the complete list of human genes. Humans have about 20 600 genes in each cell.

● **Figure 5** The DNA molecule – a double helix

? Questions

1 Name the unit of inheritance and the chemical that chromosomes are made of.
2 How many chromosomes are in a human body cell?
3 Why do we each have some characteristics similar to our mother's and others that are similar to our father's?
4 Which scientists worked out the structure of DNA?
5 Look at the DNA molecule in Figure 5. Describe the shape of a double helix.

Show you can...

Complete this task to show that you understand genes, chromosomes and DNA.

Describe in detail what a gene is, what it does and where they are found inside a cell. You can include a drawing if you wish.

35

Tobacco plants produce a poisonous chemical called nicotinic acid. This kills insects that eat tobacco leaves.

● **Figure 1** Tobacco plants. How do you think tobacco plants evolved to produce the poisonous chemical?

Charles Darwin is often referred to as the father of **evolution.** He lived during the 1800s and made detailed observations of many plants and animals, which led him to put forward his theory of evolution by **natural selection**. He eventually published his ideas in 1859 in a book called *On the Origin of Species*.

→ Evolution by natural selection

- **Competition** between organisms leads to the development of **adaptations**.
- In any **population** there is wide range of variation between organisms of the same species, due to differences in their genes.
- Some variations may be advantageous, making some organisms more suited to their environment. If there is competition between the organisms, those with the advantageous characteristics will be more suited to survive. This is sometimes called **survival of the fittest**.
- The organisms that survive will breed and pass on their genes for the useful characteristics to their offspring.
- Over a period of time only organisms with the useful characteristics will survive.

● **Figure 2** Charles Darwin (1809–1882). This picture was taken in 1874 by his son. How old was Darwin that year?

We can use this theory to explain how tigers evolved to have stripes. Tigers live and hunt in grassland areas of Asia.

- Variation: ancestors of the tiger with striped (rather than plain) coats would be less likely to be seen in the long grass.

- Natural selection: these striped animals would therefore be more likely to catch prey. If they had more food than other unmarked animals they would survive and breed to pass on their genes to their offspring, which would also have striped markings.

- Evolution: over a long period of time only striped animals would exist.

● **Figure 3** A bengal tiger. Which other animals have a survival advantage because of their camouflage?

Evolution usually takes place very slowly, but if a new form of a gene suddenly appears in a species there may be a rapid change in the species if the environment changes.

One example of this is when head lice developed resistance to a particular insecticide. Head lice live on people's hair and feed on their blood. An insecticide was used that successfully killed most of the head lice, but some survived. The survivors then bred and passed on the gene for resistance to the next generation. The head lice had rapidly evolved from head lice that could be killed by a particular insecticide to insecticide-resistant head lice.

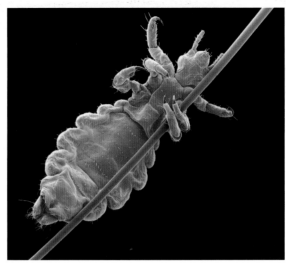

● **Figure 4** A head louse clinging to a human hair. What other organisms live in or on human bodies?

→ Artificial selection

Humans can interfere with nature by selecting animals or plants with characteristics that are useful to them and then breeding from them. This is called **selective breeding** by **artificial selection**. Selective breeding increases the chance of certain variations of genes passing from parent to offspring.

People have been using selective breeding since the start of farming about 10 000 years ago. Farmers selected plants that grew well, gave the highest yield and tasted nice.

Farmers also selected animals that gave lots of milk or meat and were easy to manage. Some animals were selected to do work, like pulling a plough or carrying people.

● **Figure 5** Different varieties of tomato. How were all these different varieties produced?

? Questions

1 Who is well known for his theory of evolution?
2 What causes the wide range of variation in eye colour?
3 How do stripes help tigers to survive?
4 Give two examples of a characteristic that has been created by selective breeding in plants and two examples in animals.
5 What is the difference between artificial selection and natural selection?

Show you can...

Complete this task to show that you understand selective breeding.

Describe what a plant breeder would have done to produce yellow tomatoes.

10.4 Extinction

Dinosaurs roamed the Earth for over 100 million years, but then they became extinct about 65 million years ago.

● **Figure 1** A plant-eating dinosaur called an *Omeisaurus*

→ Extinction

An organism is **extinct** when all the members of the species have died out. It usually occurs when there is a change in the environment. If the species is not as well adapted to survive the change as other species, it will not be able to compete successfully, so will die out. Many species have become extinct, not just dinosaurs.

The dodo was a large flightless bird that lived on the island of Mauritius. It became extinct within a hundred years of Dutch sailors landing on the island.

No one knows exactly what caused their extinction. One suggestion was that the sailors ate the birds. This is unlikely, as reports from the time suggest that they did not taste very nice. It is more likely that the sailors brought other animals to the island that attacked the birds or ate their eggs. As the birds could not fly away they were easy prey for dogs. Dodos laid their eggs in nests on the ground, so these could be eaten by cats and rats.

● **Figure 2** A dodo. What other species do you know that have become extinct?

→ Changes that can lead to extinction

- climate change, e.g. an ice age or **global warming**
- a catastrophic event, e.g. a meteorite hitting Earth
- a new predator
- a new competitor
- disease
- humans, e.g. hunting, deforestation, use of chemicals

● **Figure 3** The black rhino, bluefin tuna and Galápagos penguins are all endangered species. What can be done to prevent these animals becoming extinct?

→ Gene banks

Gene banks are stores of tissue or cell samples from endangered species. They preserve the hereditary material of organisms before they become extinct. The samples are stored at very low temperatures, much colder than a normal freezer. This preserves the material so that it could be defrosted in the future and used to produce more organisms.

Plant materials are fairly easy to store for use in the future. Seeds or cuttings survive well for many years. They can be thawed out to produce plants every few years, and then seeds from these are collected and frozen to maintain a healthy store of material.

● **Figure 4** A store at the Millennium Seed Bank in Sussex. How are plants chosen for the seed bank?

Animal tissues are more difficult to use. Sperm, eggs, blood and other tissues can be frozen, but using them in the future to reproduce the organism requires more research. However, the most important thing is that the DNA of endangered species is being preserved.

? Questions

1 What does 'extinct' mean?
2 Give an example of an animal that is now extinct.
3 Explain how the dodo might have become extinct.
4 Explain what gene banks are and why they are useful.

Show you can...

Complete this task to show that you understand extinction.

Make a list of the causes of extinction and suggest what can be done to prevent more species from becoming extinct.

→ Presenting and interpreting data

→ Human evolution

The first human ancestors to walk on two legs appeared on Earth over 4 million years ago in eastern Africa.

> 1 Suggest why walking on two legs would be an evolutionary advantage.

● **Figure 1** Human evolution: the human body evolved to allow us to walk upright

The timeline shows some human ancestors and when they lived. Modern humans are called *Homo sapiens*. The diagram also shows when stone tools were first used.

Use information from Figure 2 to answer the following questions.

> 2 Which species survived for the longest period of time?
> 3 Which species was the first to use stone tools?
> 4 For how long did *Australopithicus africanus* survive on Earth?
> 5 a) How long ago did *Homo habilis* appear on Earth?
> b) When did *Homo habilis* die out?
> c) For how long did *Homo habilis* survive on Earth?
> d) What word is used to describe an organism that has died out?
> 6 Which species were alive at the same time as *Homo neanderthalensis*?

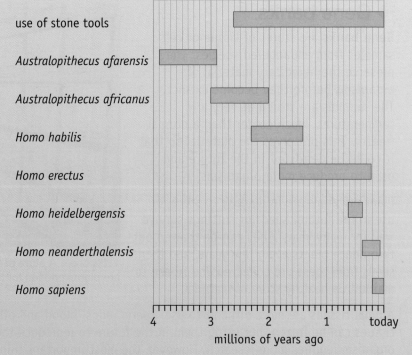

● **Figure 2** A timeline showing some human ancestor species

→ Did birds evolve from dinosaurs?

Many scientists believe that birds evolved from dinosaurs.
Compsognathus was a turkey-sized dinosaur that lived about 150
million years ago. The earliest known bird was the *Archaeopteryx*.
This bird lived around the same time as *Compsognathus* and was
about the size of a raven.

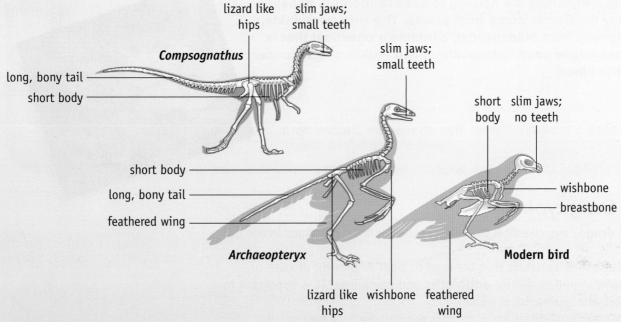

lizard like hips

slim jaws; small teeth

Compsognathus

long, bony tail

short body

slim jaws; small teeth

short body

long, bony tail

feathered wing

Archaeopteryx

lizard like hips   wishbone   feathered wing

short body   slim jaws; no teeth

wishbone

breastbone

Modern bird

● **Figure 3** The changes seen in the evolution from dinosaur to bird

7 a) Use information in Figure 3 to describe the similarities between
   Compsognathus and modern birds.
   b) Describe the changes that have happened between
   Compsognathus and modern birds.
8 Modern birds have a large wishbone. Suggest why this is a useful
   adaptation for birds.

Dinosaurs were a very successful group of reptiles. There were
many different types of dinosaur: plant eaters and carnivores,
some that lived on land or water, and some that could fly. They
died out about 65 million years ago.

9 How do we know that
   dinosaurs once lived on
   Earth?
10 Make a list of possible
   reasons why all dinosaurs
   died out.

● **Figure 4** Some different types of dinosaur

11.1 Medicines and health

Drug companies are helping to save rainforests because many medicines come from plants. The rosy periwinkle, originally from Madagascar, contains a chemical that is successfully used to treat childhood leukaemia, a cancer of the blood.

● **Figure 1** The rosy periwinkle. How did scientists discover that it could be used to treat cancer?

→ Drug types

- A **drug** is any chemical that affects the chemical reactions inside our cells.
- **Medicines** are drugs that are used to treat or prevent disease or pain; examples include **antibiotics** and painkillers. It is important to take the right amount (dose).
- **Recreational drugs** are drugs that people take because they like the effects they have. Caffeine, tobacco and alcohol are all legal recreational drugs. Cannabis and ecstasy are illegal recreational drugs.

Drugs are also grouped by the way that they affect the body:

Drug type	What they do	Examples	
stimulants	• make you feel more alert • stop you feeling tired • can make you feel more competitive work by speeding up the transfer of impulses (messages) from one nerve cell to the next	legal: • caffeine • nicotine	illegal: • cocaine • ecstasy • speed (amphetamines)
depressants	• help you to relax • help you to stay calm • can make you sleepy work by slowing down the transfer of impulses from one nerve cell to another	legal: • alcohol • tranquillisers • anaesthetics	illegal: • cannabis • heroin
painkillers	• reduce or prevent pain work by blocking pain receptors in the brain	legal: • aspirin • paracetamol • ibuprofen • morphine	illegal: • heroin
hallucinogens	• make you hallucinate (imagine things that are not really there) • can lead to confusion work by upsetting the transfer of impulses between nerve cells		illegal: • cannabis • solvents • LSD

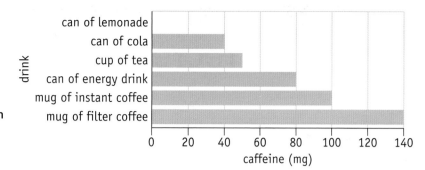

● **Figure 2** The amount of caffeine in different drinks. What effect do you think too much strong coffee might have on the body?

→ Drug testing

The development of a new drug takes many years because drug companies and governments have to be certain that they work and are safe.

The idea for a new drug may come from the traditional use of a plant to treat a certain condition. Scientists analyse the plant to find the active ingredient. In other cases chemists create a new compound based on an existing drug to improve it.

Once scientists think they have produced a suitable compound they have to test it:

Stage of testing	Tested on	Reason for testing
laboratory testing	phase 1: animal and human cells	to see if the drug is safe
	phase 2: live animals	to see if the drug works and to see if there are any side effects
clinical trials	phase 1: a small group of healthy volunteers	to see if the drug is safe in humans and to look for side effects
	phase 2: a small group of patients with the disease	to see if the drug works and to look for side effects
	phase 3: a large group of people with the disease	to work out the best dose to use and to look for side effects

If the new drug passes all these tests it may be given a licence by the European Medicines Agency (EMA). Only then can doctors in Europe prescribe the drug to patients.

? Questions

1 State what a drug is.
2 a) Explain why people use recreational drugs.
  b) Give two examples of legal recreational drugs and two examples of illegal recreational drugs.
3 Explain why drug companies are involved in saving rainforests.
4 Suggest why drug companies test the safety of medicines.

✎ Show you can...

Complete this task to show that you know about the different types of drug.

Create a table with the headings shown and then complete it to classify the following drugs: alcohol, aspirin, caffeine, cannabis, heroin.

Name of drug	Legal or illegal	Medicine or recreational drug	Drug type

11.2 Legal recreational drugs

About 100 000 people in the UK die each year due to smoking. That is about one-fifth of all deaths. Smoking-related deaths are mainly due to cancers, blockages in the lungs and heart disease.

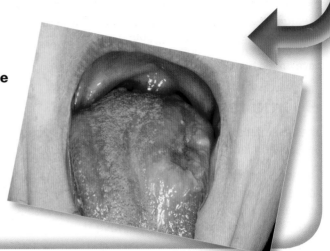

● **Figure 1** Mouth cancer caused by smoking. Why do you think the Government wants people to stop smoking?

→ Smoking kills

It has long been known that smoking causes lung cancer and other illnesses, so why do so many people still smoke? It is even more surprising that young people start to smoke.

Many adults want to stop smoking, but find it difficult because **nicotine** is a very addictive drug. If they do not get a regular fix of nicotine they suffer **withdrawal symptoms** and crave a cigarette. Besides being addictive, nicotine also affects the circulatory system. It is a stimulant so it increases the heart rate and makes the blood vessels narrow, and less blood reaches the cells. This can lead to high blood pressure and heart disease.

● **Figure 2** Tobacco plants produce the toxic chemical nicotine to deter insects and grazing animals from eating it. Why did people start smoking the leaves?

Cigarette smoke contains other harmful chemicals, such as **tar** and **carbon monoxide**. Tar is a sticky, poisonous substance that contains cancer-causing chemicals. It collects in the airways in the lungs and stains teeth, skin and hair a brown colour. Carbon monoxide is a poisonous gas that reduces the amount of oxygen that is carried in the blood. Smokers are often short of breath due to the effects of these chemicals.

A pregnant woman who smokes is putting her baby at risk. The baby does not get enough oxygen due to the carbon monoxide in the mother's bloodstream, so does not grow as healthily as a non-smoker's baby. This means it could be born too early or die.

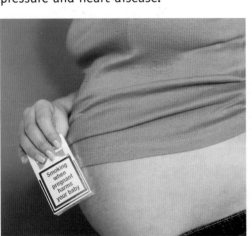

● **Figure 3** Only a very small proportion of women smokers carry on smoking after getting pregnant. Why do you think most manage to give up?

→ Alcohol

Alcohol is a depressant drug. Small amounts make you feel relaxed, but your reactions are much slower. That is why it is illegal to drink and drive. Larger amounts of alcohol have a dramatic effect on people's behaviour. Some people become argumentative, violent or may become unconscious. They also suffer a hangover the next day. Long-term alcohol abuse affects relationships and job security.

Alcohol is also an addictive drug. It is poisonous and excessive amounts cause liver, kidney and brain damage. Many alcoholics die from alcohol poisoning or liver disease.

● **Figure 4** Being drunk can put someone in an embarrassing or dangerous situation. In what ways are these drunk girls at greater risk of harm than if they were sober?

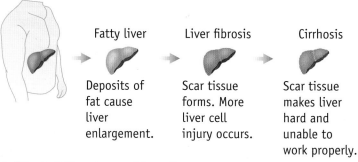

Fatty liver	Liver fibrosis	Cirrhosis
Deposits of fat cause liver enlargement.	Scar tissue forms. More liver cell injury occurs.	Scar tissue makes liver hard and unable to work properly.

● **Figure 5** The stages of liver disease

There are government guidelines as to how much alcohol is safe for men and women to drink in a week. The Department of Health recommends that men should not regularly drink more than three to four units of alcohol a day and women should not regularly drink more than two to three units a day. It is also recommended that everyone should have at least two alcohol-free days a week.

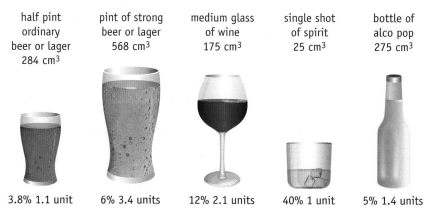

half pint ordinary beer or lager 284 cm³	pint of strong beer or lager 568 cm³	medium glass of wine 175 cm³	single shot of spirit 25 cm³	bottle of alco pop 275 cm³
3.8% 1.1 unit	6% 3.4 units	12% 2.1 units	40% 1 unit	5% 1.4 units

● **Figure 6** The number of units of alcohol in some drinks. How much ordinary beer could a man drink in an evening and still be within the Department of Health guidelines?

❓ Questions

1 Approximately how many people die each year due to smoking-related illnesses?
2 Give three illnesses related to smoking cigarettes.
3 Describe the effects of nicotine on the body.
4 Explain why pregnant women should not smoke.
5 A girl went out to a party and drank three half pints of ordinary lager and three single vodka and cokes. How many units of alcohol did she drink in total?
6 Why do you think the Government is trying to educate people about the risks of smoking and alcohol abuse?

✎ Show you can...

Complete this task to show that you understand the effects of smoking on the body.

Name the chemicals in cigarette smoke and describe how they affect the body.

Cannabis has been used as a painkiller in Asia for about 6000 years, and the Romans used to fry and eat cannabis seeds. Today, some multiple sclerosis (MS) sufferers use cannabis to ease their severe muscle tremors and pain. These people want cannabis to be legalised.

● Figure 1 A person with MS preparing a cannabis cigarette. Why do you think cannabis is an illegal drug?

Illegal drugs are classified into three groups, Class A, Class B or Class C, depending on how harmful they are to users and society. It is illegal to have, produce, give away or sell any of these drugs. Class A drugs are thought to cause the most serious harm, and so carry the greatest maximum penalties for breaking the law.

→ Cannabis

Cannabis has many other names including grass, weed, skunk, pot, marijuana and hash. It is an addictive Class B drug. It is considered so harmful because it contains many different chemicals. Some of these cause cancer and others cause mental health problems.

Cannabis alters your mood, coordination, ability to think and your self-perception. Some people who use cannabis hallucinate or develop an irrational fear. They may become anxious or depressed and have problems with remembering things.

● Figure 2 An illegal cannabis farm. What is the penalty for supplying cannabis to other people?

→ Cocaine

Cocaine is also known as coke, freebase and crack. It is a Class A drug that is very addictive.

Cocaine is produced from the leaves of the coca plant. It is a stimulant, so makes people feel temporarily confident and alert. It is also an anaesthetic and painkiller. People who use cocaine usually sniff it or smoke it. When the feeling of happiness and self-confidence starts to wear off, people can become depressed and tired for several days afterwards.

Coca-Cola was originally made using coca leaves, but today only cocaine-free coca leaf extract is used.

● Figure 3 Sniffing cocaine can make the cartilage in your nose break down. Why would someone continue to sniff cocaine when it starts to deform their face?

→ Heroin

The Class A drug heroin is the most addictive recreational drug used in the UK. It is produced from the opium poppy.

Opium is the sticky liquid that is collected from an opium poppy seed pod. It contains the chemical morphine, which is a legal drug used to treat severe pain. The use of morphine is carefully monitored to prevent patients becoming addicted to it.

Heroin is produced from morphine and is twice as potent. Both of these drugs are depressants, but they make people feel happy and relaxed. They also cause sleepiness. Many heroin addicts inject the drug into veins. This can damage the blood vessels and cause blood clots and infections. If people inject a recreational drug, they also risk developing serious infections, like hepatitis or HIV, from using infected needles. An overdose of heroin can cause breathing to stop, leading to death.

● **Figure 4** A woman injecting heroin into his arm. How do you think scenes like this affect children?

Risks of buying illegal drugs

Any drug bought from a drug dealer has additional risks. You do not know exactly what you are getting, or how strong it is going to be. Some people have died after taking a drug for the first time.

Dealers mix drugs with other ingredients to increase the weight and therefore increase their profits. Sometimes they use sugar, starch or talc, but glass, pesticides and other poisonous substances have also been used.

Dealers only care about making money, so they push people to use more and more. This leads to addiction and a drug dependency that costs a lot of money. Many drug addicts resort to criminal activities to buy more drugs.

? Questions

1 Name the plant that cocaine comes from.
2 Which class of drug, Class A, B or C, is the most harmful?
3 Describe the risks of using cannabis.
4 Describe what heroin is produced from.
5 Describe the risks associated with injecting drugs.
6 Explain why some MS sufferers want cannabis to be made legal.

Show you can...

Complete this task to show that you understand the effects of some illegal drugs.

Make brief notes to describe the effects of cannabis, cocaine and heroin on the body.

➡ Calculating scientifically

➜ How many units?

The Department of Health recommends women should not regularly drink more than two to three units a day and that they should have at least two alcohol-free days a week.

> 1 Based on this recommendation, calculate the maximum units of alcohol a woman should drink in a week.

The number of units of alcohol in a drink is calculated using the following equation:

$$\text{number of units} = \frac{\text{volume of drink in cm}^3 \times \% \text{ concentration of alcohol}}{1000}$$

| 13.5 % alcohol | 4.1% alcohol | 40% alcohol | 4.8% alcohol |

● **Figure 1** Alcoholic drinks showing the percentage concentration of alcohol

> 2 Calculate the number of units in each of the following drinks:
>   a) a 175 cm³ glass of wine
>   b) a 330 cm³ bottle of alcopop
>   c) a single measure (25 cm³) of gin
> 3 a) If a woman drank two gin and tonics and two 175 cm³ glasses of wine in an evening, how many units would she have drunk altogether?
>   b) How many days a week could she drink this amount of alcohol and still be within the recommended limit?
>   c) Suggest why it can be difficult to know how many units of alcohol you drink in a week.

→ The link between smoking and lung cancer

In the 1950s Professor Richard Doll and Dr Austin Bradford Hill thought that smoking might be a cause of lung cancer. They carried out a 5-year study of the smoking habits, health and death rates of around 40 000 doctors. Their data showed a link between smoking and lung cancer.

4 At the time many people did not accept that there was a link between smoking and lung cancer. Suggest possible reasons for this.

Since the 1950s many more studies have been carried out that have produced similar results. The results of one study are shown in the table.

5 How many deaths due to lung cancer in non-smokers would you estimate for a population of 1 000 000 people?
6 Plot the data in a bar chart.
7 What conclusion can you reach from the data?
8 Lung cancer is not the only disease linked to smoking cigarettes. Name one other disease that is linked to smoking.

Number of cigarettes smoked per day	Death rate per 100 000 people due to lung cancer
0	14
1–14	105
15–24	208
25 or more	355

→ Risks associated with recreational drugs

The graph shows the risk of addiction and risk of harm to the body for some recreational drugs.

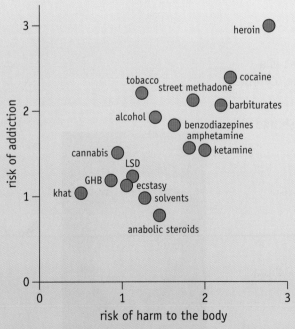

● **Figure 2** The risks of some recreational drugs

9 Which drug shown in the graph is the most addictive?
10 Which drug shown in the graph is the least addictive?
11 Which drug in the graph has the lowest risk of causing harm to the body?
12 Describe the trend shown in the graph.
13 Tobacco and alcohol are both legal recreational drugs, but their use costs the National Health Service far more than for illegal recreational drugs. Use information from the graph and your own knowledge to suggest why.

12.1 Micro-organisms

Louis Pasteur (1822–1895) was the first person to demonstrate that there are micro-organisms in the air that cause foods to go off. Before then no one understood why food went off.

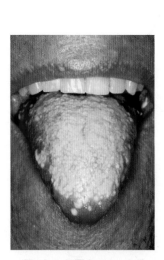

● **Figure 1** Louis Pasteur carrying out experiments in his laboratory in Paris. How do you think Pasteur's discovery affected people's behaviour?

Micro-organisms (also called microbes), are small organisms that can only be seen with a microscope. They include **bacteria**, **viruses** and some **fungi**.

Some micro-organisms cause disease, but other bacteria and fungi are very useful. For example, the fungus yeast is used to make bread and alcoholic drinks, and specific types of bacteria are used to make yoghurt.

Micro-organisms that cause disease are called **pathogens**.

→ Fungi

Fungi can be **unicellular** or **multicellular** organisms. Yeast is a unicellular fungus. Each cell has a cell wall, cell membrane, cytoplasm and a **nucleus**.

Some fungi can infect us and cause diseases such as thrush and athlete's foot.

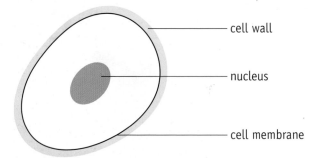

cell wall

nucleus

cell membrane

● **Figure 2** The structure of a yeast cell

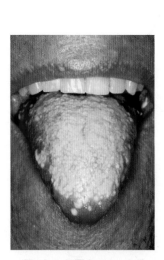

● **Figure 3** This person has a thrush infection on their tongue. How do you think thrush is treated?

50

→ Bacteria

Bacteria are small unicellular organisms. They have a cell wall, a cell membrane and cytoplasm, but they do not have a proper nucleus. They just have a strand of DNA in the cytoplasm.

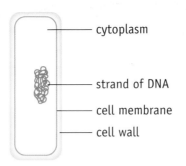

● **Figure 4** The structure of a bacterium

Staphylococcus bacteria live on our skin where they can cause spots or boils, but they do not normally harm us. If we eat food containing *Staphylococcus* bacteria they can cause food poisoning. However, if they get into a deep wound, for example during surgery, they can make us very ill by causing blood poisoning.

Some **strains** of *Staphylococcus* are very difficult to treat with drugs, so this means it is important for surgeons to operate under sterile conditions, where no microbes are present.

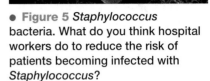

● **Figure 5** *Staphylococcus* bacteria. What do you think hospital workers do to reduce the risk of patients becoming infected with *Staphylococcus*?

→ Viruses

Viruses are very simple organisms, even smaller than bacteria. They have a strand of DNA surrounded by a protein coat.

All viruses cause disease. Viruses can only reproduce if they enter a living cell and take it over. During this process the cell they have entered is killed.

Human diseases caused by viruses include colds, flu, measles, cervical cancer and HIV/AIDS. These are all serious illnesses that can kill. There is no cure for the disease caused by HIV. The virus is transmitted in blood, **semen** and vaginal fluids.

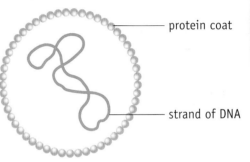

● **Figure 6** The structure of a virus

● **Figure 7** Micrograph showing a blood cell, stained green, infected with HIV, stained red. How could the spread of HIV be reduced?

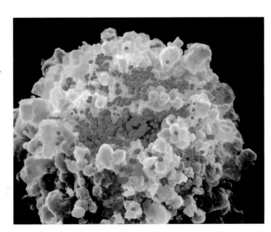

? Questions

1 What word describes all organisms that cause disease?
2 Which organism is the smallest: a bacterium, a fungus or a virus?
3 Which type of organism causes flu?
4 Suggest how you could reduce the chance of being infected by HIV.

Show you can...

Complete this task to show that you know the different types of organism that can cause disease.

Name the three types of organism that can cause disease and give an example of a disease caused by each type.

12.2 Defence against disease

If micro-organisms enter the body they can cause a disease. For this reason it is very important that surgeons and hospital staff wash their hands properly before entering an operating theatre. All the surfaces and equipment have to be sterilised, and gowns, gloves and face masks are worn. All these precautions reduce the risk of transferring infectious organisms into the person having an operation.

● **Figure 1** Surgeons washing their hands before entering the operating theatre. What precautions are taken around hospitals to reduce the spread of pathogens?

→ Protection from disease

Micro-organisms can enter our bodies through our lungs, through cuts in our skin, in the food we eat and during sexual intercourse. To reduce the risk of becoming infected, or infecting other people, we need to follow some basic rules:

- cover your mouth when you cough
- use a handkerchief when you have a cold
- clean cuts and cover them with a plaster
- keep your body clean
- wash your hands before preparing or eating food
- keep food wrapped up and in the fridge
- use a condom during sexual intercourse.

The body has various ways of reducing the chances of micro-organisms entering the body. Figure 2 shows some of these mechanisms.

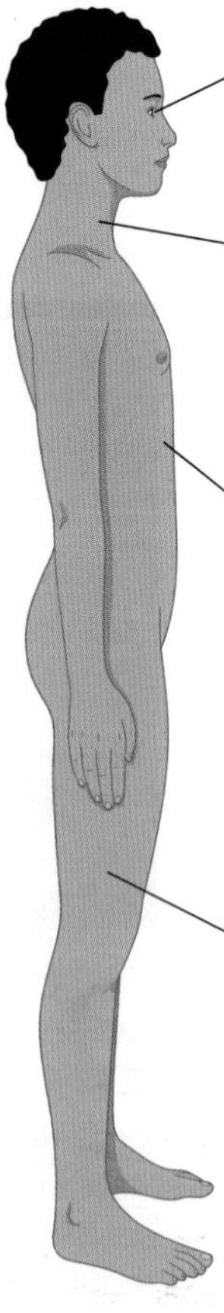

→ Defending the body

If micro-organisms do enter the body the white blood cells work in various ways to kill them.

Tears help wash away microbes and contain a chemical that kills some pathogens.

Cells lining the windpipe produce mucus that traps microbes. Ciliated cells have tiny hair-like structures that move the mucus up and away from the lungs.

The stomach produces hydrochloric acid that kills pathogens that enter our body in the food we eat.

The skin is a barrier that stops pathogens entering the body. If it gets cut a scab forms to block the wound.

lymphocyte that produces antibodies to kill microbes

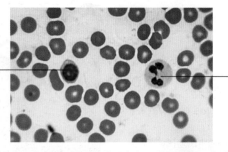

phagocyte that engulfs and digests microbes

● **Figure 3** Human blood cells. The pale pink cells with a dark nucleus are white blood cells. What are all the other cells in this micrograph?

● **Figure 2** Mechanisms to reduce the chances of pathogens entering the body

Phagocytes are white blood cells that engulf and digest microbes. This process is called **phagocytosis**. You can recognise phagocytes because they have a lobed nucleus.

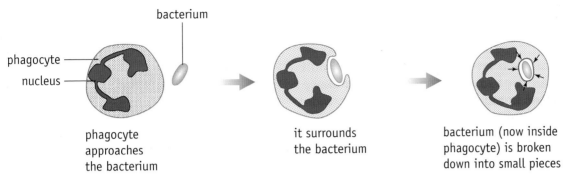

phagocyte — nucleus —

phagocyte approaches the bacterium

it surrounds the bacterium

bacterium (now inside phagocyte) is broken down into small pieces

● **Figure 4** The process of phagocytosis

Lymphocytes make chemicals called **antibodies**. Antibodies kill microbes, or make them clump together so that the phagocytes engulf many microbes at the same time. Some of this type of white blood cell, called memory cells, stay in the blood after the infection. If the same type of micro-organism infects the body again, the memory cells are ready to act quickly, and you do not get any symptoms. You have **immunity** to that infection.

Lymphocytes also produce **antitoxins**. These are chemicals that neutralise the poisons produced by microbes that make us feel ill.

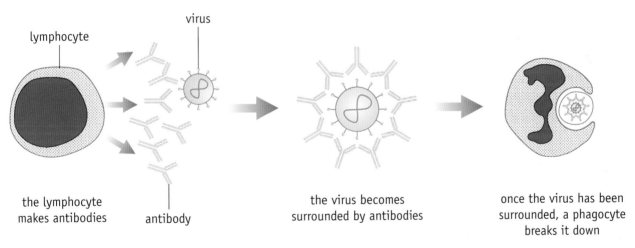

the lymphocyte makes antibodies

antibody

the virus becomes surrounded by antibodies

once the virus has been surrounded, a phagocyte breaks it down

● **Figure 5** One way that antibodies work to kill microbes

Questions

1 What type of blood cell protects the body against disease?
2 List the ways microbes enter our body.
3 List four things you can do to reduce the risk of becoming infected by a micro-organism.
4 Suggest how the precautions taken in operating theatres reduce the chances of a patient contracting an infection.

Show you can...

Complete this task to show that you understand the function of phagocytes.

Draw labelled diagrams to show how phagocytes protect the body against disease.

MRSA is a *strain* of bacterium that is very difficult to treat and has caused many deaths, particularly in hospitals. This is because it is resistant to most antibiotics, meaning that it cannot be killed by them.

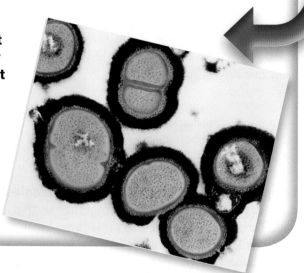

● **Figure 1** MRSA bacteria. What research do you think drug companies are doing related to MRSA bacteria?

→ Vaccinations

Once your body has been infected by a particular microbe you usually do not get the disease a second time. This is because you are **immune** to it.

Vaccination can also prevent you from getting a disease. You are given a **vaccine** that contains a dead or weakened form of the pathogen, and this stimulates the white blood cells to produce the correct antibody against that pathogen. If you are exposed to the pathogen again, the memory cells multiply rapidly and produce the antibodies in large amounts to kill the pathogen quickly. Vaccines are usually injected into the body, but some can be taken by mouth.

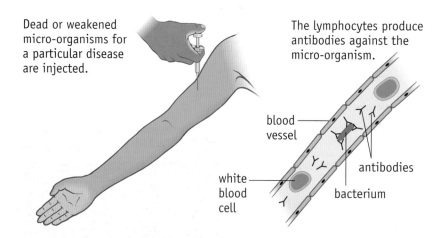

Dead or weakened micro-organisms for a particular disease are injected.

The lymphocytes produce antibodies against the micro-organism.

blood vessel

white blood cell

antibodies

bacterium

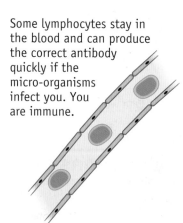

Some lymphocytes stay in the blood and can produce the correct antibody quickly if the micro-organisms infect you. You are immune.

● **Figure 2** How a vaccine works

The body reacts to the vaccine by producing antibodies. Each antibody is only effective against one microbe, so we need to have a different vaccine for each microbe.

MMR vaccine

Children in the UK are immunised against diseases such as measles, mumps and rubella (combined in the **MMR vaccine**) and other diseases such as polio and tetanus.

Some parents chose not to have their children vaccinated with the MMR vaccine after one doctor suggested it could cause autism. This claim has since been disproved, but because of the scare fewer children were vaccinated. There has since been an increase in the number of cases of mumps and measles, and some people have died.

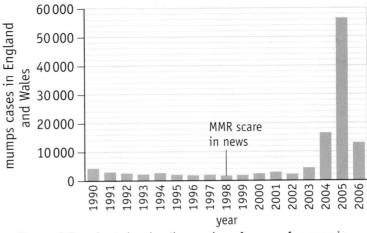

● **Figure 3** Bar chart showing the number of cases of mumps in England and Wales

→ Antibiotics

Antibiotics are chemicals produced by fungi that either kill bacteria or stop them growing. Viruses are not affected by antibiotics, so antibiotics cannot be used to treat a viral infection.

Sir Alexander Fleming first discovered antibiotics but it was by pure chance. He had been growing bacteria on **agar plates**, and when he came back to work after a holiday in 1929 he noticed that one of his plates was contaminated with a fungus called *Penicillium*. The bacteria around the fungus had been killed, so Fleming thought the fungus might have produced something that killed the bacteria.

Unfortunately Fleming did not realise the importance of this discovery. It was not until the 1940s that the drug **penicillin** was developed as a medicine. Since then it has been widely used to cure bacterial infections, and many other antibiotics have been developed.

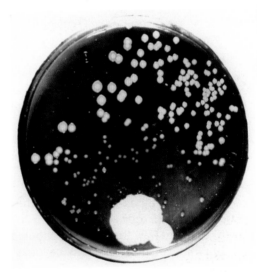

● **Figure 4** Fleming's photograph of his agar plate contaminated with *Penicillium*

Some strains of bacteria, like MRSA, are resistant to antibiotics, so drug companies are carrying out research to develop new ones.

? Questions

1 Give three examples of diseases that children are immunised against in the UK.
2 **a)** Look at Figure 3. How many cases of mumps were there in England and Wales in 1998?
  b) How many cases were there in 2005?
3 Suggest why the number of cases of mumps increased dramatically in 2005.
4 Explain why taking antibiotics does not cure flu.

Show you can...

Complete this task to show that you know about antibiotics.

Describe what an antibiotic does, and how antibiotics were first discovered.

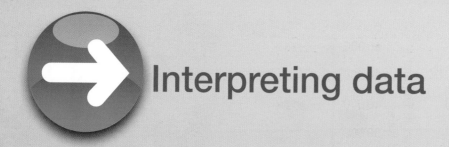

Interpreting data

→ Modern medicines and disease

Antibiotics

Students were testing different antibiotics to see which one was the best to use against *E. coli* bacteria.

The students used an agar plate with *E. coli* growing on it and placed five paper discs soaked in different antibiotics, A, B, C, D and E on top of the agar. They put a sixth disc soaked in water on the plate as well.

They left the plate for several days and then observed the effects of the antibiotics. Their results are shown in Figure 1. A clear area around the disc shows that the bacteria were killed.

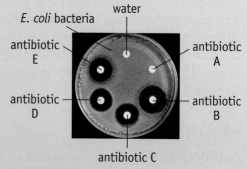

● **Figure 1** The effects of different antibiotics and water on *E. coli* bacteria

1 Explain why a paper disc soaked in water was put on the agar plate.
2 Which antibiotic, A, B, C, D or E, was the most effective at killing *E. coli*? Give a reason for your answer.
3 Which antibiotic, A, B, C, D or E, is *E. coli* resistant to? Give a reason for your answer.
4 The students then wanted to find the best dose of this antibiotic to use against *E. coli*.
   a) Suggest how the students should set up their investigation.
   b) What would the students be looking for in their results?

Smallpox

Smallpox was a terrible disease that killed most of its victims. A type of vaccination against it was common in Turkey, and was introduced to England by Lady Mary Wortley Montague in the early 18th century. Then in 1796 Dr Edward Jenner noticed that milkmaids who had had cowpox (a mild illness) did not get smallpox. He put some cowpox pus into a scratch on the arm of a boy called James Phipps, who then caught cowpox. A few weeks later Jenner injected James with smallpox pus, but James did not get smallpox. The cowpox had made him immune to it.

● **Figure 2** Edward Jenner vaccinating James Phipps

5 Explain why Dr Jenner would not be allowed to carry out this experiment today.
6 What conclusion can you reach about the cowpox virus and the smallpox virus? Give a reason for your answer.

The table shows the number of deaths due to smallpox in people up to the age of 30 in London during 1844.

Age in years	Number of deaths
0	2 235
1	1 524
2	1 197
3	869
4	628
5	1 122
10	226
15	226
20	240
25	148
30	98

7 Plot the data as a line graph.
8 In which age group were there the most deaths from smallpox? Suggest a reason for this.
9 Describe the pattern shown in the graph. Can you suggest a scientific reason for this?

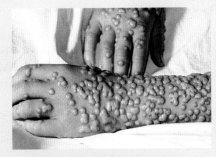

● **Figure 3** Smallpox infection

MMR vaccine

The MMR vaccine gives immunity against measles, mumps and rubella (German measles). Measles can be a serious disease that can lead to death.

The 1998 scare that the MMR vaccine caused autism was later proved to be untrue, but at that time many parents decided not to have their children vaccinated.

10 a) Describe the change in the percentage of children vaccinated with the MMR vaccine between 1996 and 2007.
   b) In what year was the lowest percentage of children vaccinated with the MMR vaccine?
11 Up until 2003 the numbers of cases of measles in the UK population was extremely low. Suggest a reason for this.
12 Since 2005 the number of cases of measles has increased rapidly. Suggest a reason for this.
13 In 2012 there were 2016 cases of measles reported. The Government introduced an MMR catch-up programme, to vaccinate all children up to the age of 18 who had not been fully vaccinated. Explain why this was important.

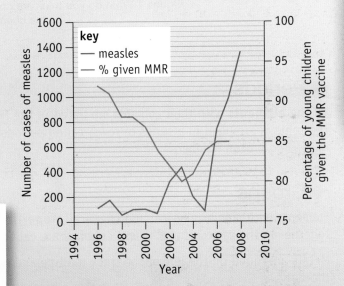

● **Figure 4** Graph showing the number of cases of measles and the percentage of young children vaccinated with the MMR vaccine between 1996 and 2008

7.1 A guided tour

All professional's have tools that they use to carry out their work. Chemists are no different. Alongside Bunsen burners and various pieces of glassware one of the most important tools is the Periodic Table.

● **Figure 1** What do you think are considered to be the chemist's tools?

→ Periods

All the **elements** with their symbols are listed in the Periodic Table. The table is arranged in a series of rows, called periods, in order of increasing proton number.

→ Groups

Elements with similar physical and chemical **properties** are grouped together in the same column. These columns are called groups.

Metals are found on the left of the thick red line and non-metals on the right.

Many scientists helped to find the patterns to organise the Periodic Table. The most famous is Mendeleev.

Patterns in properties of elements

The scientists started by organising the elements in order of their mass. Then they looked for patterns in their properties, placing elements with similar properties into columns. Mendeleev thought there might be some undiscovered elements. He decided to leave 'gaps' to predict the properties of these elements. When these elements were discovered, their properties matched his predictions.

Similar elements are close to each other in the Periodic Table. For example gold (Au), silver (Ag) and platinum (Pt) are all unreactive and shiny, so are used for jewellery and decoration.

Germanium (Ge) and silicon (Si) are next to each other vertically. Though germanium is a metal and silicon is a non-metal, they have similar properties. Both are used as semiconductors in electronic circuits.

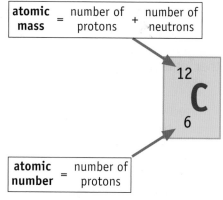

● **Figure 2** Using the data from the Periodic Table

Hydrogen is a gas. It is the lightest element.
In chemical reactions, hydrogen can behave like a metal, even though it is not one.

The first two columns (groups) are the reactive metals. They quickly form a dull coating (a metal oxide) when the metal reacts with oxygen in the air.
The most reactive metals are at the bottom of each column with reactivity getting less as you move up.
These metals are not hard like typical metals.

All the elements to the right of the thick red line are non-metals.

The inert gases or noble gases are all gases and are very unreactive. It is very difficult to get any of these gases to react with anything.

These are the other metals. They all have metallic properties (such as hardness, good electrical conduction, high melting point) but are not chemically reactive like the reactive metals.
Any element with a name that ends in 'ium' is a metal.

1 2   **3 4 5 6 7 8**

1 **H** hydrogen 1	

| 7 **Li** lithium 3 | 9 **Be** beryllium 4 |
| 23 **Na** sodium 11 | 24 **Mg** magnesium 12 |

| 39 **K** potassium 19 | 40 **Ca** calcium 20 | 45 **Sc** scandium 21 | 48 **Ti** titanium 22 | 51 **V** vanadium 23 | 52 **Cr** chromium 24 | 55 **Mn** manganese 25 | 56 **Fe** iron 26 | 59 **Co** cobalt 27 | 59 **Ni** nickel 28 | 63.5 **Cu** copper 29 | 65 **Zn** zinc 30 | 70 **Ga** gallium 31 | 73 **Ge** germanium 32 | 75 **As** arsenic 33 | 79 **Se** selenium 34 | 80 **Br** bromine 35 | 84 **Kr** krypton 36 |

4 **He** helium 2

11 **B** boron 5, 12 **C** carbon 6, 14 **N** nitrogen 7, 16 **O** oxygen 8, 19 **F** fluorine 9, 20 **Ne** neon 10

27 **Al** aluminium 13, 28 **Si** silicon 14, 31 **P** phosphorous 15, 32 **S** sulfur 16, 35.5 **Cl** chlorine 17, 40 **Ar** argon 18

| 85 **Rb** rubidium 37 | 88 **Sr** strontium 38 | 89 **Y** yttrium 39 | 91 **Zr** zirconium 40 | 93 **Nb** niobium 41 | 96 **Mo** molybdenum 42 | 98 **Tc** technetium 43 | 101 **Ru** ruthenium 44 | 103 **Rh** rhodium 45 | 106 **Pd** palladium 46 | 108 **Ag** silver 47 | 112 **Cd** cadmium 48 | 115 **In** indium 49 | 119 **Sn** tin 50 | 122 **Sb** antimony 51 | 128 **Te** tellurium 52 | 127 **I** iodine 53 | 131 **Xe** xenon 54 |

| 133 **Cs** caesium 55 | 137 **Ba** barium 56 | 139 **La** lanthanum 57 | 178 **Hf** hafnium 72 | 181 **Ta** tantalum 73 | 184 **W** tungsten 74 | 186 **Re** rhenium 75 | 190 **Os** osmium 76 | 192 **Ir** iridium 77 | 195 **Pt** platinum 78 | 197 **Au** gold 79 | 201 **Hg** mercury 80 | 204 **Tl** thallium 81 | 207 **Pb** lead 82 | 209 **Bi** bismuth 83 | 209 **Po** polonium 84 | 210 **At** astatine 85 | 222 **Rn** radon 86 |

| 223 **Fr** francium 87 | 226 **Ra** radium 88 | 227 **Ac** actinium 89 |

■ reactive metals   ■ other metals   ■ non-metals   ■ halogens   ■ inert gases

The halogens are very reactive non-metals. The most reactive is fluorine.
Reactivity decreases as you move down the column.
All the halogens have names that end in 'ine'.

● **Figure 3** The modern Periodic Table

? Questions

1 Which element is in period 2 and group 6?
2 What does the red line drawn on the Periodic table show?
3 List all the information you can about bromine.
4 Using the table on the right, compare and contrast magnesium and rubidium.
5 Compare and contrast francium and fluorine.

	Magnesium	Rubidium
group		
period		
mass number		
atomic number		
metal/non-metal		
reaction with oxygen		
other observations		

✎ Show you can...

Complete this task to show that you understand how the Periodic Table is arranged.

Copy and complete the table.

Element	Symbol	Period	Group	Metal/ non-metal	Other information
gallium					
	S				
barium					
chlorine					
	Kr				

7.2 Developing the Periodic Table

Scientists group things together to help study them. Common features can be tested and ideas about why they match can be formed. The discovery of the chemical elements was just the same. Several chemists worked to study newly discovered elements and identify the best way to arrange them.

● **Figure 1** Which groups could this mixture be put into?

Many chemical elements had already been discovered by the early 1800s. Some were discovered in ancient times, such as gold and copper; others more recently, such as lithium (in 1817) and silicon (in 1824).

→ Döbereiner

A German scientist named Johan Döbereiner started trying to put elements into smaller sub-groups. In 1829, he observed through careful experimentation that some elements had similar properties to each other. These similarities fell into groups of three elements, which he called triads.

One of his groups included chlorine, bromine and iodine, as they behaved in similar ways during chemical reactions. He also calculated that the atomic weight of bromine was approximately the average of chlorine and iodine.

His triads worked well for many of the elements. However, once new elements were discovered such as caesium (1860), fluorine (1866), and germanium (1886), they did not fit into his groups and so his ideas were not accepted.

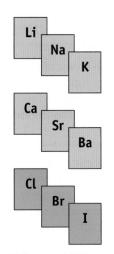

● **Figure 2** Some of Döbereiner's triads

→ Newlands

An English chemist, John Newlands, started looking again at the grouping of the chemical elements. In 1864, he lined the elements up in order of atomic mass from lowest to highest and spotted an interesting repeating pattern. Every eighth element had similar chemical and physical properties.

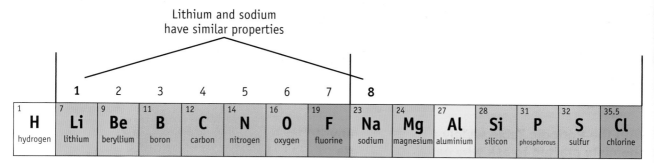

Lithium and sodium have similar properties

	1	2	3	4	5	6	7	8						
1	7	9	11	12	14	16	19	23	24	27	28	31	32	35.5
H	**Li**	**Be**	**B**	**C**	**N**	**O**	**F**	**Na**	**Mg**	**Al**	**Si**	**P**	**S**	**Cl**
hydrogen	lithium	beryllium	boron	carbon	nitrogen	oxygen	fluorine	sodium	magnesium	aluminium	silicon	phosphorous	sulfur	chlorine

● **Figure 3** Newlands' octaves

There were also some problems with Newlands' ideas. Firstly hydrogen did not fit in his patterns at the start of his list and after the element calcium his repeating pattern did not always fit. He had to fit two elements in some spaces and other groups showed the repeat in properties before the eighth element.

→ Mendeleev

In 1869 a Russian chemist named Dmitri Mendeleev took another look. He also ranked the elements according to their atomic mass, and looked for repeating patterns. However, Mendeleev made a key decision – where there were gaps he was so confident in his ideas that not only did he leave a space and predict that an element was yet to be discovered, but he even predicted the physical and chemical properties of the missing elements.

The discovery of germanium, in 1886, was shown to be very close in value to the predication made by Mendeleev. This gave his ideas the validity needed for the scientific community to accept them. It is his model of the Periodic Table you see here in this book and on the walls of chemistry classrooms and laboratories around the world.

The last group of elements to be discovered were the noble gases. This is because they were so unreactive that they were not included in any of the chemical compounds being studied. Argon was discovered first, in 1894, followed rapidly by helium (1895), and then by neon and xenon (1898). The discovery of this new group fitted easily in Mendeleev's model and so gave his idea even more support.

? Questions

1 Which group is missing from all three chemists' tables?
2 Why did Döbereiner call his groups triads and Newlands call his octaves?
3 Both Döbereiner and Newlands based their groups on atomic mass. Describe how each used this piece of data differently.
4 Why was putting two elements in one space such a problem for Newlands' idea?
5 Explain how each of the three chemists mentioned here managed to include new elements that were discovered into their ideas.

Show you can...

Complete this task to show that you understand the key stages in the development of the Periodic Table.

Create a timeline, showing discovery of key elements as mentioned here and the times that the three key scientists were working on their ideas. Add details about why each scientist's ideas were either ignored or accepted.

7.3 Spotting patterns

With the Periodic Table widely accepted, many chemists started to ask why certain elements behaved in similar ways. Once the structure of the *atom* was understood the pattern in the properties of elements began to make more sense.

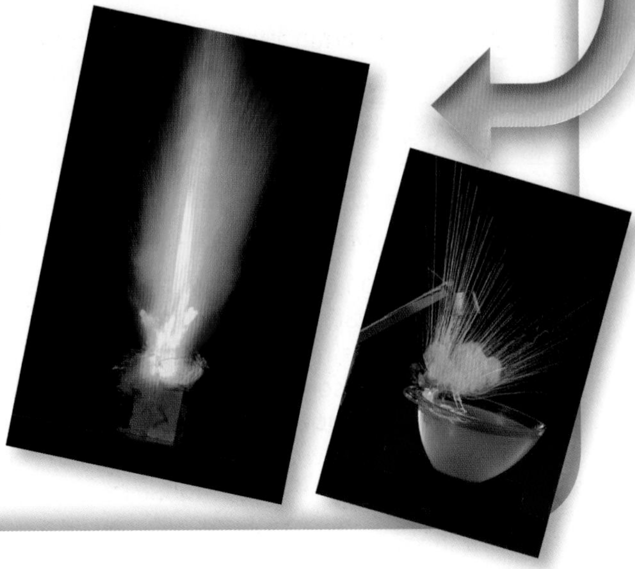

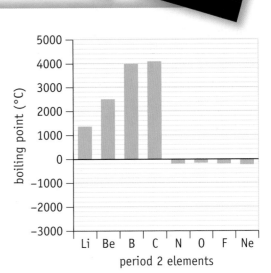

● **Figure 1** What can you see here that indicates these elements are chemically similar?

The groups in the Periodic Table were identified because of their observable physical properties, including atomic mass, melting and boiling point, and appearance. By comparing neighbouring elements, trends and patterns in the properties could be identified. Patterns can be spotted across periods and down groups.

A key factor was realising that Mendeleev had actually arranged the elements in increasing atomic number. Generally this led to an increase in mass. However, some of the elements that did not fit in the patterns of the early Periodic Tables just needed swapping round. The periodic increase in atomic mass was due to total number of protons and neutrons in the nucleus of the **atoms**.

● **Figure 2** The trend in boiling point for the period 2 elements

→ Electron arrangement

To explain the patterns in the chemical properties we need to look at the **subatomic** particle, the electron. You will remember that electrons are arranged in shells. For a chemical reaction to take place, the electrons on the metal atoms are removed and it is the electrons in the outer shell that are affected.

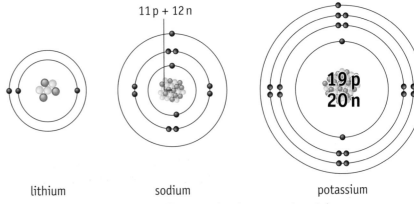

lithium   sodium   potassium

● **Figure 3** Atomic structure diagrams for the group 1 metals

The electron shells can only hold a fixed number of electrons each. The first can hold only two whilst the second and third can hold up to eight. With this information we can begin to visualise the atoms for the first of Mendeleev's groups.

When the electron arrangements for the group 1 metals are compared a clear pattern is seen, as each member of the group has one electron in their outer shell. It is these outer shell electrons that take part in chemical reactions, so it gives the reason why this group behave in such similar ways, such as their reactions with water.

lithium	+	water	$\rightarrow$	lithium hydroxide	+	hydrogen
sodium	+	water	$\rightarrow$	sodium hydroxide	+	hydrogen
potassium	+	water	$\rightarrow$	potassium hydroxide	+	hydrogen

→ Trends in reactivity

There is also the trend in how reactive the metals are, with francium being the most reactive and lithium the least. Therefore, the reactivity decreases down the group. This can be explained by how strongly the outer electrons are held in place.

Lithium's outer electron is held more tightly and so is less reactive. The group 1 metals are more reactive as they only have to lose one electron. Group 2, however, need to lose two electrons in every reaction and so react more slowly than group 1.

We can see this most easily when we observe the chemical reactions of these metals with cold water.

Element	Observation during reactions with water
potassium	metal moves around surface of water rapidly, bumping into the sides of the container with force; you can hear a fairly loud hissing noise as there is lots of effervescence and visible gas around the metal; the metal catches fire to give a purple flame
lithium	metal moves slowly across surface of water, gently bumping the side of the container; no sound but some effervescence and a small amount of gas is visible
calcium	metal does not move and no sound is heard; slight effervescence but no visible gas
magnesium	no visible change

It is the similarities in these reactions that explain why they are in the same group and defined as reactive metals. It is the differences in reactivity that give patterns within the groups.

 Show you can...

Complete this task to show that you understand the sorts of patterns shown in the Periodic Table.

Describe and explain the patterns in reactions with water for some group 1 and group 2 metals. You could include chemical equations and atomic structure diagrams.

? Questions

1 Name the least reactive group 1 metal.
2 How many outer shell electrons are there in:
   a) group 1
   b) group 2?
3 Draw an atomic diagram for magnesium.
4 Write a word equation for the reaction of calcium with water.
5 Describe the pattern shown in Figure 2 for the boiling points of period 2.

Once patterns in physical and chemical properties were identified chemists could use these to predict behaviours of new substances. When predictions were shown to match newly discovered elements, both Mendeleev's arrangement and the atomic model were more widely recognised.

● **Figure 1** What state of matter do you think fluorine will be at room temperature?

gas   liquid   solid

The greater the atomic number the greater the mass – this is the very hypothesis that supported Mendeleev's development of the Periodic Table. From this pattern he could suggest the mass of missing elements.

→ Physical behaviour

Plotting physical data on a graph allows you to see trends and make predictions. The physical properties can also be linked to observations.

→ Chemical behaviour

We can also predict chemical behaviours based on observed trends and patterns in the Periodic Table. By looking at the atomic structure we can explain the patterns and predict the behaviour of other members of the group.

Group 7 elements have seven outer shell electrons. When non-metals react with metals they have to gain an electron. This is easiest for fluorine, which makes it the most reactive halogen and therefore the most reactive non-metal on the Periodic Table. This trend is the reverse for the group 1 and 2 metals, where reactivity increases down the group, whereas the halogen reactivity decreases down the group.

Figure 2 graph showing atomic number vs atomic mass.

● **Figure 2** A graph to show how atomic number affects atomic mass – this data will allow a chemist to predict the state of the element at room temperature

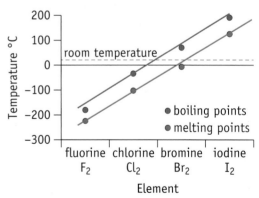

● **Figure 3** A graph to show the melting and boiling points of the halogens

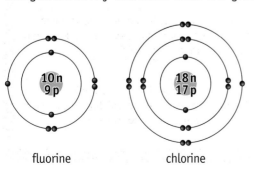

fluorine   chlorine

● **Figure 4** The atomic structures of the group 7 non-metals

Knowing the pattern of how group 7 compounds react with metals, we can predict the formula of products for other metal and halogen reactions. Remember that when a non-metal reacts, the product has a change of name and the ending is changed to –ide. These products are almost always solids as the bond between metals and non-metals is so strong.

sodium + chlorine → sodium chloride

The balanced symbol equations will give us even more information:

To investigate reactivity it is important to only change one of the substances each time. For example, if you wanted to observe the trend in the reactivities of the halogens you would react each with the same metal.

$$2Na + Cl_2 \rightarrow 2NaCl$$
$$Mg + Cl_2 \rightarrow MgCl_2$$
$$2Al + 3Cl_2 \rightarrow 2AlCl_3$$

Halogen	Metal	Product	Observation
fluorine	iron	iron fluoride	cold fluorine gas is passed through iron wool – iron catches fire instantly
chlorine	iron	iron chloride	iron wool is heated and dropped into a gas jar of chlorine – red-brown iron chloride is produced
bromine	iron	iron bromide	bromine liquid is put into the bottom of a boiling tube with some iron wool pushed in above it; the bromine is heated to turn it into a gas, then the Bunsen heat is moved to the iron wool and a bright glow can be seen – red-brown iron bromide is produced

In the table above iron has been used as the second reactant. This is because if you combine two very reactive elements the reaction may become unsafe and could cause an explosion.

? Questions

1. Which halogen would you predict to be more reactive: iodine or chlorine?
2. What would you call LiBr?
3. Predict the melting point and boiling point of the group 7 element astatine.
4. What would you expect the observations to be when iodine reacts with iron?
5. Write a balanced symbol equation, including state symbols, for the reaction between aluminium and bromine.

Show you can...

Complete this task to show that you understand about predicting patterns based on the Periodic Table.

Copy and complete the following table.

Reactant 1	Reactant 2	Observation
chlorine	sodium	A piece of hot sodium is lowered into a jar of chlorine; the sodium burns with a bright yellow flame. Once cooled a white solid is visible on the sides of the jar; this is the salt sodium chloride.
bromine	sodium	
magnesium	hydrochloric acid	The piece of magnesium metal is dropped into the acid. Rapid effervescence is seen from the hydrogen gas being produced. The gas is released fairly quickly and so the metal moves around slowly under the surface of the acid.
calcium	hydrochloric acid	
sodium	oxygen in air	When sodium is freshly cut you see a very shiny surface. This surface will immediately begin reacting with the oxygen in the air to form a dull layer of sodium oxide (Na_2O). After approximately 1h, the surface will no longer be shiny.
potassium	oxygen in air	

→ Building scientific awareness

→ Mendeleev plays patience

You are going to follow in Mendeleev's footsteps and model how he developed his ideas about the Periodic Table. You will need a partner or two and some scrap paper.

It was accepted, in the mid-1800s, that elements could be grouped together based on similar physical and chemical properties.

> 1 Give an example of a physical property.

Dmitri Mendeleev, a Russian scientist, wanted to investigate this idea. He reviewed all the data available for the elements and began making cards.

Scientific ideas develop by looking at known data. You are going to look at a sample of the known elements at the time of Mendeleev. On your pieces of scrap paper, write out a set of cards for the elements in Table 1.
Look up their names and atomic number on the Periodic Table and include the data shown here, which was known at the time of Mendeleev.

Mendeleev put these elements in order of their atomic mass. Repeat this step now with your cards.

Discovery date	Element	Symbol	Atomic number	Atomic mass	Known compounds
Ancient		Sb		122	SbH_3, $SbCl_3$
Ancient		Sn		119	SnO_2, $SnCl_4$
1250		As		75	AsH_3, $AsCl_3$
1774		Cl		35.5	HCl, NaCl, $CaCl_2$
1782		Te		128	H_2Te, $TeCl_2$
1807		Na		23	Na_2O, NaCl,
1807		K		39	K_2O, KCl
1808		Sr		88	SrCl, SrO
1817		Se		79	H_2Se, $SeCl_2$
1824		Si		28	$SiCl_4$, SiO_2
1826		Br		80	HBr, NaBr, $CaBr_2$
1827		Al		27	$AlCl_3$, Al_2O_3
1861		Rb		85	Rb_2O, RbCl
1863		In		115	$InCl_3$, In_2O_3

● **Table 1**

> 2 When scientific ideas develop, several scientists will do the same things. Which other scientist put the elements in order of increasing atomic mass to look for patterns?
> 3 Which of these elements is first on your list? Which is the last?

Now, just as Mendeleev did, you need to focus on the information on the elements. When scientific ideas develop, the scientists need to look at all the data they have and start to identify any patterns and trends. Through changing raw data into trends, useful predictions can be made. Can you spot any similarities between any of the cards? Arrange the cards into groups where similar elements line up in vertical columns. Do not worry about how many you have in each group.

Rank each group in order of increasing mass and line them up next to each other.

4 Explain why there are some gaps.

As Mendeleev built up his table he was not worried about gaps. When scientific ideas are developed, scientists must accept they may not have all the information yet. Gaps give an idea about where to go looking for more data to support their ideas.

Some additional elements were also known to Mendeleev, see Table 2. Create cards for these and put them into the gaps you have left. Work out the missing information from the Periodic Table.

Discovery date	Element	Symbol	Atomic number	Atomic mass	Known compounds
ancient				32	H_2S, SCl_2
1669				31	PH_3, PCl_3
1775				24	$MgCl_2$, MgO
1808				40	$CaCl_2$, CaO
1811				127	HI, NaI, CaI_2

● Table 2

Two more elements were discovered, see Table 3. These were called gallium and germanium. Complete cards for these two elements and add them into your table.

Discovery date	Element	Symbol	Atomic number	Atomic mass	Known compounds
1875	Gallium	Ga		70	
1886	Germanium	Ge			GeO_2, $GeCl_4$

● Table 3

5 What did you predict as the atomic mass for germanium?
6 What did you predict were the known compounds for gallium?

Making predictions is a key part of scientific discoveries. Finding data to support the prediction is what makes ideas believable. Having a model that can work for new discoveries was vital to Mendeleev's success.

7 Suppose a new element X is discovered. It forms a compound with chlorine, and the formula of this compound is XCl_4. What group or family do you think this element would belong to?
8 How do we explain today the fact that tellurium comes before iodine in the Periodic Table, even though tellurium has a higher atomic mass than iodine?

8.1 The reactivity series

By observing what happen when metals react we start to recognise patterns. Magnesium is always more reactive than copper. By observing metals in several different reactions a list or series of reactivity can be determined.

● **Figure 1** Which of these metals is more reactive?

→ The reactivity series of metals

Very reactive substances combine with others easily to form compounds, and can be difficult to extract from compounds. By comparing how easily they react, scientists have produced an order of reactivity for metals. This is called the **reactivity series**.

- Silver and gold are not very reactive so they are good metals to make jewellery from. They do not react easily with sweat on your skin.
- Magnesium is reactive. It burns easily in air with a bright white light. This makes it good to use in distress flares and fireworks.
- Very unreactive metals, like gold, can be dug out of the Earth as **pure** metals (elements). They do not form compounds easily.
- More reactive metals like calcium and aluminium are only found naturally as compounds in minerals.

most reactive

Potassium
Sodium
Calcium
Magnesium
Aluminium
Zinc
Iron
Lead
Copper
Silver
Gold

least reactive

● **Figure 2** The reactivity series of metals

→ Metals reacting with water

Observing what happens when metals react with water helps place them in the reactivity series.

- Potassium reacts violently and bursts into flames as the hydrogen produced catches fire. You have to use a safety screen to view this reaction even when only a tiny piece of potassium is used.
- Magnesium produces bubbles of hydrogen very slowly, but much more quickly if it is heated in steam.
- Copper does not react noticeably with water or steam.

When metals react with water, a metal hydroxide and hydrogen are formed. For example:

potassium + water → potassium hydroxide + hydrogen
$2K(s)$ + $2H_2O(l)$ → $2KOH(aq)$ + $H_2(g)$

● **Figure 3** Potassium reacting with water

→ Metals reacting with oxygen

Observing what happens when metals react with oxygen in the air helps place them in the reactivity series.

- Sodium and potassium are shiny when freshly cut but immediately tarnish on exposure to air.
- Magnesium burns in air with a bright white flame – to protect your eyes it is sensible to look away after initially observing the flame. A dark glass filter may also be used.
- When copper is heated in air a dull grey coating of copper oxide forms on the surface.

When metals react with oxygen a metal oxide is formed. For example:

magnesium + oxygen → magnesium oxide

copper + oxygen → copper oxide

→ Displacement reactions

A reactive metal can displace a less reactive metal from a solution of a **salt** of that metal. So, if you add zinc to copper sulfate solution:

zinc + copper sulfate → zinc sulfate + copper

Zn + $CuSO_4$ → $ZnSO_4$ + Cu

This type of reaction is called a **displacement reaction**. Zinc is more reactive than copper so it can displace copper from a solution of a copper salt.

A metal cannot displace another metal that is more reactive. For example:

copper + zinc sulfate → copper + zinc sulfate

no reaction happens!

Nothing happens because the copper is not reactive enough to displace the zinc from a solution of zinc sulfate.

You should be able to make predictions about whether a displacement reaction will occur, based on each metal's position in the reactivity series.

Even though aluminium is quite reactive, it is often used in metal window frames. When aluminium is in contact with the air, a coating of white aluminium oxide quickly forms. This oxide creates a protective layer over the aluminium and prevents any further reactions from taking place, so the window frame does not corrode away.

● **Figure 4** Zinc and copper sulfate reacting

 Show you can...

Complete this task to show that you understand about the reactivity series of metals.

Plan the following investigation. You have been asked to find out whether an unknown metal is silver or magnesium. You have some iron sulfate solution available. How could you use the iron sulfate to identify the unknown metal?

? Questions

1 Put these metals in order of reactivity: iron, sodium, silver, potassium.
2 Two metals react with acid to produce hydrogen gas. What difference would you observe in the reactions because of their different reactivities?
3 Why do you think we clean pieces of metal using sandpaper before we use them for reactions in the laboratory?
4 Why are the group 1 metals stored under oil?
5 Explain the products you would expect if a piece of zinc and a piece of magnesium were placed in copper sulfate at the same time.

8.2 Extracting with carbon

Rocks that contain metals are called metal ores. They are a very valuable resources, so finding ways to extract or remove the metals from the rock to get a pure sample is very important. As the metals are mostly found as compounds, a chemical reaction is needed to separate the metal from the other substances.

● Figure 1 Where do metals come from?

Some very unreactive metals, such as silver, gold and platinum are found as pure samples. They do not tend to combine to make compounds and so are easier to collect.

Other metals are found combined. For example:

- haematite is an **ore** of iron; it is mainly iron oxide
- malachite is an ore of copper; it mainly contains copper carbonate
- bauxite is an ore of aluminium; it is mainly made of aluminium oxide
- galena is an ore of lead; it is mainly lead sulfide.

→ Extracting metals from their ores

There are two main stages in obtaining a pure sample of the metal: mining the metal ore and **extracting** the metal from the ore.

For all ore samples mining is very similar. The rocks that contain the metal ores must be obtained from underground. Large machinery is used to dig the rock out in large areas called quarries. Sometimes drills have to dig deep underground to get to particular valuable ore sites.

Once taken from the rocks the ore samples need to be physically separated from other waste rock. This will give a concentrated sample of ore. An ore must contain enough of the metal compound to make it worthwhile extracting in terms of cost.

Extracting the metal from the ore requires a chemical reaction and the principles of the reactivity series are used.

You will notice the addition of a non-metal – carbon. This is because it would not be sensible to use one metal to extract another, as all metals are valuable resources. Carbon is readily available as coal or charcoal and is also very cheap.

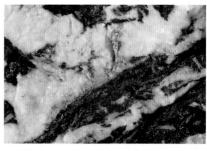

● Figure 2 A sample of gold in a mine

most reactive

Aluminium

(Carbon)

Zinc

Iron

Lead

Copper

least reactive

● Figure 3 A part of the reactivity series of metals

→ Extracting metals with carbon

You will remember that a more reactive substance can displace a less reactive substance from its compounds. That is exactly what happens when carbon is reacted with the compounds formed by the metals below it in the reactivity series.

For example, the ore malachite contains the compound copper carbonate. When it is heated it decomposes to make copper oxide. This can then be reacted with carbon to extract the copper.

copper oxide + carbon → carbon dioxide + copper

$$2CuO + C \rightarrow CO_2 + 2Cu$$

This reaction does not happen at room temperature. It requires very strong heating to enable the reaction to start. In addition it must happen away from a source of oxygen, otherwise the hot metal will simply react with the oxygen in the air and reform the metal oxide.

In the laboratory, this can be investigated by mixing the copper compound with charcoal and covering it with a layer of coal to keep oxygen out, or by using a crucible lid.

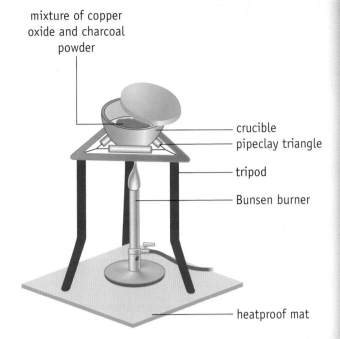

mixture of copper oxide and charcoal powder

crucible
pipeclay triangle
tripod
Bunsen burner
heatproof mat

● **Figure 4** Extracting copper in the laboratory

● **Figure 5** A steel worker taking a sample form a blast furnace, which extracts iron from its ore

On an industrial scale this would happen inside a blast furnace – reaching around 1500 °C. Coke, a pure sample of carbon, is used to extract metals such as copper or iron.

When metals are extracted, it is a type of **decomposition** reaction. When the ore contains the metal oxide, we use a **reduction** reaction to extract the metal. The compound containing the metal is reduced or broken down to release the metal and loses oxygen. In these examples, carbon is acting as the **reducing agent**.

You will see more about reduction reactions in Topic 10.

? Questions

1. Why do we not use carbon to extract gold?
2. Why do we not use carbon to extract aluminium?
3. Write a word equation for the extraction of tin from tin oxide using carbon.
4. Why is a chemical reaction needed to extract a metal from its ore?
5. Complete the balancing of this equation, which shows the reaction between iron and carbon:

   $$2Fe_2O_3 + C \rightarrow Fe + CO_2$$

✎ Show you can...

Complete this task to show that you understand how metal ores can be extracted using carbon.

Explain the steps involved in extracting the metal from the ore haematite. Include a word equation.

For more reactive metals, extracting with carbon will not work so an alternative was needed. Aluminium was first extracted in the early 1900s through the use of electricity. The process, called electrolysis, required the aluminium ore to be molten (as a liquid) – a temperature of around 2000 °C.

● **Figure 1** How many uses of aluminium can you name?

→ Extracting metal from molten ore

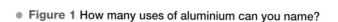

A compound formed between a metal and a non-metal has the potential to conduct electricity if it can be dissolved in water or is molten. Once this happens, it can be called an **electrolyte** due to its conducting properties.

A metal ore is a compound formed between a metal and a non-metal. When an electrical current is passed through a molten sample of such a compound it can be split up and so release the metal.

In order to pass a current though a sample **electrodes** are needed. These are often made of graphite, a special type of carbon that can conduct electricity. Carbon is used because it is less reactive than the metal being produced so will not affect the process.

Once attached to an electrical supply, the electrodes will become charged. If they are then placed in an electrolyte the circuit will be complete and the compound in the electrolyte will become reduced by the electrical current.

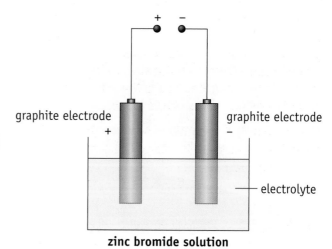

graphite electrode

graphite electrode

electrolyte

zinc bromide solution

● **Figure 2** A complete electrical circuit with electrode and electrolyte

→ Industrial-scale extraction

On an industrial level there are more challenges to consider. Melting the metal ore requires high temperatures and is quite dangerous. The whole system is therefore encased in steel so none of the molten electrolyte escapes.

To maximise production, several electrodes are used to pass more current through the sample. To give maximum surface area the vessel is lined with the negative electrode. As aluminium is separated out from the compound it falls to the bottom of the vessel and can then be removed.

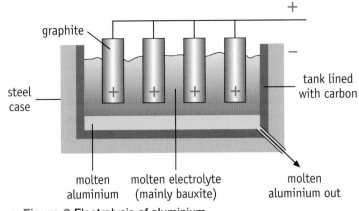

● **Figure 3** Electrolysis of aluminium

Remember, bauxite consists mostly of aluminium oxide. The oxygen atoms from the oxide form molecules of oxygen gas and effervescence can be seen at the positive electrodes.

Disadvantages of industrial-scale electrolysis

All of the most reactive metals can be extracted in this way but there are disadvantages. Electrolysis takes huge amounts of energy from electricity, making the process very expensive. Generating electricity by burning fossil fuels produces greenhouse gases such as CO_2, which contribute to the **greenhouse effect** and global warming. Luckily the heating effect of the electricity keeps the electrolyte molten so no other heat sources are needed.

At very high temperatures oxygen is generated at the positive electrodes. Over time the carbon electrodes react forming carbon dioxide. This can be released as a waste gas but this reaction also damages the electrode; replacing it adds to the cost.

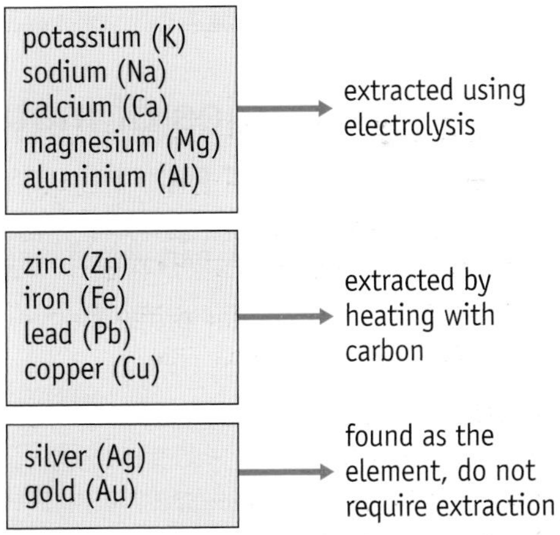

Despite all the challenges and the costs involved it is still worthwhile extracting metals, which shows just how valuable they are.

potassium (K) sodium (Na) calcium (Ca) magnesium (Mg) aluminium (Al)	→ extracted using electrolysis
zinc (Zn) iron (Fe) lead (Pb) copper (Cu)	→ extracted by heating with carbon
silver (Ag) gold (Au)	→ found as the element, do not require extraction

● **Figure 4** A summary of extracting metals from their ores

? Questions

1 How is pure aluminium removed from the electrolysis tank?
2 Which compound does bauxite mostly contain?
3 Which electrode (positive or negative) do each of the following form at:
   a) the metal   **b)** the non-metal?
4 Why is there no flame to heat the bauxite to keep it molten during electrolysis?
5 List the challenges with extracting metal using electrolysis and suggest ways in which these are overcome.

✎ Show you can...

Complete this task to show that you understand how metals can be extracted using electricity.

Sketch out a diagram similar to Figure 3 but showing what you think would happen during the electrolysis of potassium from its ore sylvite, which is mostly potassium chloride.

8.4 Using metals

Metals are used for many different purposes and are one of our most valuable resources. But what influences the wide range of properties? How do scientists and engineers choose the best metal for the job?

● **Figure 1** Which properties of iron make it suitable to be used to build a bridge and a horseshoe?

→ The properties of metals

We choose materials for particular uses because of their properties. There would be no point making a frying pan out of something with a low melting point – the pan would melt when you put it on the hob!

Most metals have similar properties:

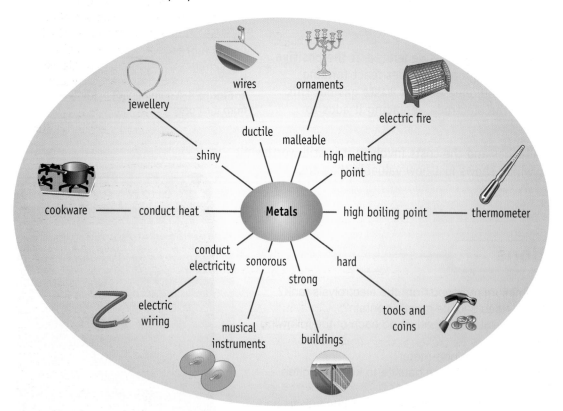

● **Figure 2** The properties and some uses of metals

Many of the properties of metals are special ones that most non-metals do not have. In general only metals:

- are easy to shape (**malleable**), such as when hit with a hammer (non-metals often break or shatter when hammered)
- can be pulled into thin strands like wires (**ductile**)
- are sonorous (make a ringing sound when you hit them)
- conduct electricity and heat well.

The only magnetic materials are the metals iron (and steel), cobalt and nickel.

All these properties make metals useful.

→ Metallic bonding

To explain the properties of metals we look to see how they bond together. Although we represent metals with a single symbol, such as Cu for copper, we know that it will be made up of many copper atoms.

When metal atoms bond together they are held tightly by electrons. These electrons come from the outer shells of the metal atoms and hold the atoms together like a glue.

All of the properties of metals can be explained based on this model.

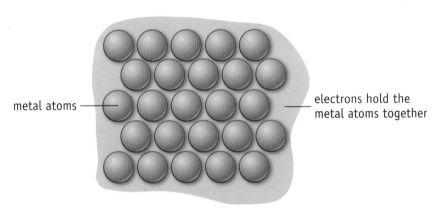

metal atoms — electrons hold the metal atoms together

● **Figure 3 How metals bond together**

For example:

- metals are hard and strong because the atoms are held together so strongly
- they conduct heat well because the atoms are held closely together
- they conduct electricity because the electrons in the glue are free to move and carry the electrical current
- they are malleable and ductile because you can push against the layers of the atoms and the electron glue will move with them, holding the atoms together in the new arrangement.

You will remember that pure samples of metals do not always give the best set of properties. Metals can be combined to make alloys. This means that two or more different types of atoms are held together by the electrons. Review Topic 4 in Book 1, to see how alloying metals affects their properties.

? Questions

1 Give two properties that make metals useful for making jewellery.
2 Give two reasons why metal, rather than plastic, is a good choice for a saucepan.
3 Explain, based on the bonding, why metals have such high melting and boiling points.
4 Remembering that when particles are heated they begin to move around, explain why metals are good conductors of heat.
5 The electrons holding the metal atoms together come from the outer shell electrons. Explain, using your knowledge of atomic structure, why the group 1 metals are much softer than other metals and hence why they can be cut with a knife.

Show you can...

Complete this task to show that you understand the different reasons for using metals.

Explain each of these choices:

- using copper to make wires
- using iron to make bridges
- using gold and silver to make jewellery.

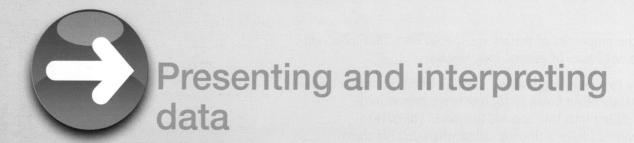

Presenting and interpreting data

→ Extracting copper

Mark and Michelle were asked to do some research on the uses of copper. They had to create a report explaining their findings. You will be following them on their research and checking they are presenting and interpreting the data they have found correctly.

First they went to find some data. The report needed to have three pieces of data presented in graphs that the students had drawn themselves. Before they started they decided to quiz each other on the type of graph you should use for different independent variables (i.e. the ones that you decide to change).

Michelle knew that if the independent variable was categoric, in that it was labelled as a word and not a value, that would mean a bar graph was needed. Mark knew that if the independent variable was continuous and could be any number, or if it was discrete and could only be whole or half numbers, a line graph would be needed with a line of best fit.

Independent variable	Type of data	Type of graph
country		
mass		
year		
type of copper ore		
cost		

> 1 Copy and complete this table to show which type of graph you would draw for each type of data.

Mark was the first to find some data. He also found a graph to go with it and thought they could just copy and paste it.

Country	Mass of copper used in 2009 (millions of tonnes)
China	4.9
Japan	1.2
Africa	0.7
Russia	0.9
India	0.5

Michelle wanted to make sure the graph would get full marks so she checked the teacher's guidance, see Figure 2.

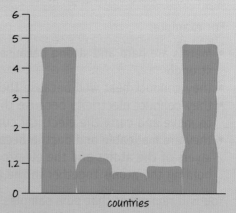

● **Figure 1** Bar graph for Mark's data

> **Bar graphs** – both scales must go up in even spaces (they can be different but each one must be scaled correctly). The bars should be drawn with a ruler and should have a gap between them. The graph must have a title.
>
> **Line graphs** – both axes must be scaled correctly, points should be plotted carefully using a sharp pencil, a line of best fit should be drawn, axis must be labelled with units and the graph should have a title'.

● **Figure 2** Marking guidance for graphs

2 Which piece of information did Michelle not record for the bar graph instructions?
3 Look closely at the bar graph Mark found. Suggest improvements to make it fit the marking criteria.
4 Draw your own bar graph, making sure you include all the improvements.

Now they had one graph drawn, they needed to find some data that would show off their line graph drawing skills. They decided to draw a graph each and thought it would be really good to look at how the demand for copper had changed over the years compared with the cost of copper.

Michelle went to find out about the demand for copper over the last 8 years. She got some data and plotted her graph but struggled to find a line of best fit.

5 Which line of best fit would you suggest and why?
6 Describe the trend in copper demand over these 8 years.

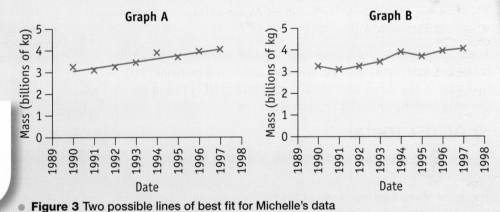

● **Figure 3** Two possible lines of best fit for Michelle's data

Mark found the following data on the cost of copper.

Year	Cost of copper (€ per kg)
1999	1.5
2000	2.0
2001	1.9
2002	1.7
2003	1.6
2004	2.5
2005	2.9
2006	5.1

7 Plot Mark's data on a suitable line graph and draw a line of best fit.
8 Describe the trend in Mark's data.
9 Give as many reasons as you can why Mark's and Michelle's data cannot be compared?
10 If you wanted to say that as the year went by copper prices increased, what other data would you want to find?

Very dilute acids in the rain can cause damage to plant life and statues. Very concentrated acids are used in car batteries. They would corrode our skin, tissues and bones if we were to come into contact with them unprotected. Acids are a part of everyday life so we must understand their key reactions.

● **Figure 1** How long do you think it would take for **acid rain** to damage a statue like this?

There are patterns in the ways **acids** react with different chemicals. The reactions on this page are all chemical reactions; new substances are made.

Reactions using acids can also produce dangerous products, such as the toxic gas sulfur dioxide. You must follow safety rules carefully whenever you work with acids.

Acids react with many substances, but one common product is formed, a **salt**. Salts always have the same type of name; the first part is the name of a metal and the second is based on the acid that has reacted to form it.

corrosive   moderate hazard

● **Figure 2** Acids can be 'corrosive', or 'a moderate hazard' depending on their concentration

Acid	Formula	Salt
hydrochloric acid	HCl	chloride salts
nitric acid	HNO_3	nitrate salts
sulfuric acid	H_2SO_4	sulfate salts

→ Acid + metal

An acid always reacts with a metal to produce a salt and hydrogen gas.

There are three key reactions of acids:

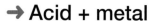

acid + metal
acid + metal carbonate
acid + alkali (metal hydroxide)

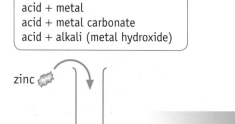

zinc

hydrochloric acid

zinc chloride forms

bubbles of hydrogen gas

● **Figure 3** The reaction of zinc metal with hydrochloric acid

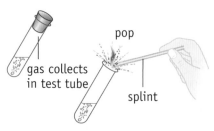

pop

gas collects in test tube

splint

● **Figure 4** To confirm that the bubbles of gas formed are hydrogen you could perform the squeaky pop test

The word and balanced symbol equations for this reaction are:

zinc + hydrochloric acid → zinc chloride + hydrogen
$Zn(s) + 2HCl(aq) \rightarrow ZnCl_2(aq) + H_2(g)$

→ Acid + carbonate

An acid always reacts with a carbonate to produce a salt, water and carbon dioxide gas.

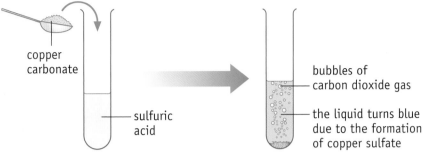

● **Figure 5** The reaction of copper carbonate with sulfuric acid

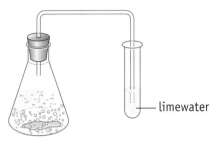

● **Figure 6** To confirm that the bubbles of gas formed are carbon dioxide test the gas in limewater and observe the limewater becoming cloudy

The word and balanced symbol equations for this reaction are as follows:

copper carbonate + sulfuric acid → copper sulfate + water + carbon dioxide
$$CuCO_3(s) \quad + \quad H_2SO_4(aq) \rightarrow \quad CuSO_4(aq) \quad + H_2O(l) + \quad CO_2(g)$$

→ Acid + alkali

An acid always reacts with an **alkali** such as a metal hydroxide, to produce a salt and water.

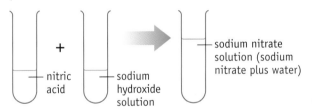

● **Figure 7** The reaction of nitric acid and sodium hydroxide

The word and balanced symbol equations for this reaction are as follows:

nitric acid + sodium hydroxide → sodium nitrate + water
$$HNO_3(s) \quad + \quad NaOH(aq) \quad \rightarrow \quad NaNO_3(aq) \quad + H_2O(l)$$

This is a **neutralisation** reaction. A solution of the product has a pH of 7. You can predict the name of the salt made in a neutralisation reaction if you know the names of the acid and alkali reactants:

nitric acid + sodium hydroxide → sodium nitrate + water
sulfuric acid + calcium hydroxide → calcium sulfate + water

 Show you can...

Complete this task to show that you understand the reaction of acids.

Magnesium metal was added to some sulfuric acid.

1 What would you expect to see in this reaction?
2 What are the names of the products?
3 How could you collect the gas produced for testing?
4 How could you measure how much hydrogen was produced?

? Questions

1 Which substances react with acids to make hydrogen gas?
2 A gas produced from an acid reaction turned limewater cloudy. What gas was made?
3 What products would you expect if potassium hydroxide reacts with nitric acid?
4 Which two chemicals reacted to make these three products: calcium sulfate, water and carbon dioxide?
5 Write a balanced symbol equation, including state symbols, for the reaction of calcium with hydrochloric acid. (The formula of calcium chloride is $CaCl_2$.)

Compounds and acidity

When fuels burn the chemicals in them react with oxygen in the air. The oxides formed rise into the atmosphere where they react and dissolve in the rainwater. Many of these oxides are formed with non-metal elements. When this happens the resulting solution is acidic.

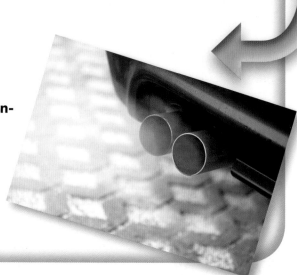

● **Figure 1** What is the link between car exhaust fumes and acid rain?

→ Reactions with oxygen

Hydrochloric, nitric and sulfuric acids are not the only chemical compounds that dissolve in water and form solutions with a pH less than 7. A group of compounds called oxides are formed when substances react with oxygen. When we consider their acidity an interesting pattern appears.

Metals

Metals can react with oxygen in the air readily when heated. For example:

magnesium + oxygen → magnesium oxide
$$2Mg(s) + O_2(g) \rightarrow 2MgO(s)$$

Non-metals

Non-metals can also react with oxygen in the air when heated, for example.

sulfur + oxygen → sulfur dioxide
$$S(s) + O_2(g) \rightarrow SO_2(g)$$

carbon + oxygen → carbon dioxide
$$C(s) + O_2(g) \rightarrow CO_2(g)$$

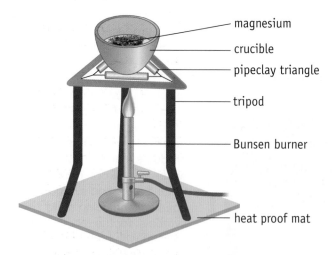

● **Figure 2** Creating magnesium oxide by heating magnesium metal in air

- magnesium
- crucible
- pipeclay triangle
- tripod
- Bunsen burner
- heat proof mat

● **Figure 3** Coal (carbon) burning in oxygen to produce carbon dioxide

→ Acidity of oxides

In terms of acidity, oxides formed from metal elements have pH values greater than 7. Some dissolve in water forming alkaline solutions. Others do not dissolve (they are insoluble) but will react with acids.

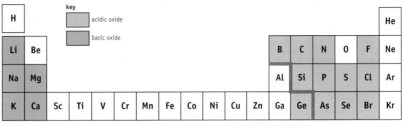

● **Figure 4** Trend in the acidity of oxides linked to the position of the elements on the Periodic Table

Oxides formed from non-metal elements will react with water to form acidic solutions with pH less than 7. It is these acidic gases that react and dissolve in rainwater to form acid rain.

Another way to determine acidity is to see if the substance reacts with acids or with alkalis. Some oxides however, can do both.

> aluminium oxide + hydrochloric acid → aluminium chloride + hydrogen
> aluminium oxide + sodium hydroxide → sodium aluminate + water

Along with aluminium oxide, beryllium, gallium and tin also react with both acid and alkali. If you look again at Figure 4, these are those elements that are very close to the metal/non-metal divide line.

→ Acid + metal oxide

When acids react with metal oxides a salt and water are formed. This is another type of neutralisation reaction and could be considered alongside the other three typical reactions of acids studied so far.

> magnesium oxide + sulfuric acid → magnesium sulfate + water
> $MgO(s)$ + $H_2SO_4(aq)$ → $MgSO_4(aq)$ + $H_2O(l)$

As some metal oxides, such as magnesium oxide, are insoluble they are called **bases**. A base is a substance that has a pH above 7 but will not dissolve in water. In practical terms, these reactions will often be heated to speed up what is otherwise a very slow reaction.

The advantage of using a base is that you can easily see when the acid has all reacted as some of the solid will remain unreacted at the bottom of the flask. This would allow you to ensure the salt produced is not contaminated by unreacted acid.

Show you can...

Complete this task to show that you understand the acidity of metal and non-metal oxides.

Copy and complete this summary table.

Element	Formula of oxide	Equation for the formation of the oxide	Acid, alkali or both?
carbon			
	MgO		
		$4Al(s) + 3O_2(g) \rightarrow 2Al_2O_3(s)$	
	SO_2		

? Questions

1 What would you call the substance with the formula CaO?
2 Predict the pH of sodium oxide and explain your choice.
3 Predict whether bismuth oxide will be acidic or alkali.
4 Is sulfur dioxide more likely to react with sodium hydroxide or hydrochloric acid?
5 When nitrogen gas reacts with oxygen in the air several different nitrogen oxides can form. Write a balanced symbol equation for the formation of nitrogen monoxide (NO) and predict the pH of this gaseous oxide.

Atoms and molecules in neutralisation

You will have met neutralisation reactions many times in your studies of chemistry. Have you ever wondered what is really happening? You have looked at the atoms and molecules involved in general chemical reactions. You now need to understand the changes during a neutralisation reaction.

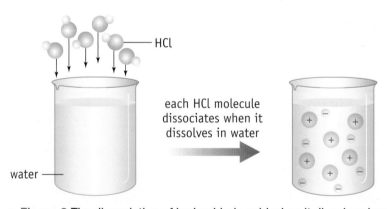

● **Figure 1** What is happening in this reaction between vinegar and sodium bicarbonate?

Look again at our common laboratory acids, shown in the table.

You should notice that all of these acids have one particular atom in common, the hydrogen atom. When these substances dissolve in water to make an aqueous solution, the hydrogen atoms become separate from the other atoms. It is these free hydrogen particles that make the solution acidic. This can be linked to the scale of pH, as the more hydrogen ions there are in a solution the lower the pH will be.

Acid	Formula
hydrochloric	HCl
nitric	HNO_3
sulfuric	H_2SO_4

→ Hydrogen ions

These hydrogen particles are in fact hydrogen ions. An **ion** is a charged particle. The hydrogen ion has a single positive charge. This is shown by adding a + in the top right-hand corner of the formula, as in H^+.

As one hydrogen ion combines with one chloride ion to make a neutral compound, the chloride particle must also be an ion with a single negative charge, Cl^-. So when hydrochloric acid dissolves in water we see **dissociation** of the acid.

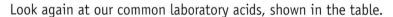
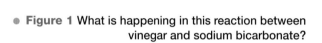

HCl

each HCl molecule dissociates when it dissolves in water

water

● **Figure 2** The dissociation of hydrochloric acid when it dissolves in water

$$HCl \rightarrow H^+ + Cl^-$$

The hydrogen ion is also known by another name – a proton. This is because of the atomic structure of the hydrogen atom. To form the ion it must lose its single outer electron. When this happens all that remains is the single proton in the nucleus.

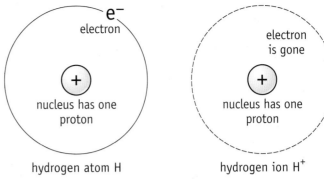

● **Figure 3** The formation of the hydrogen ion, also known as a proton

→ Alkaline substances in water

In a similar way, we can look at the behaviour of alkaline substances when they dissolve in water. Metal hydroxides, such as sodium hydroxide, NaOH, will also dissociate.

$$NaOH \rightarrow Na^+ + OH^-$$

This means that when an acid and a metal hydroxide are combined the solution will contain four ions.
As these ions rearrange, the products form.

$$H^+ + OH^- \rightarrow H_2O$$
$$Na^+ + Cl^- \rightarrow NaCl$$

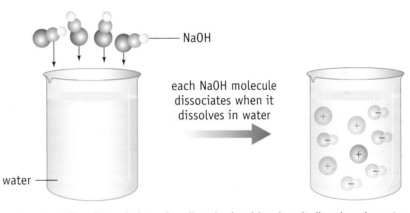

● **Figure 4** The dissociation of sodium hydroxide when it dissolves in water

The first of these will happen in any neutralisation reaction, which is why the general equation includes water every time. The second equation shows how the salt forms from the metal ion and the negative ion in the acid. This is the basis for the rules on naming salt.

It is from these reactions that we have a more formal definition of an acid as a **proton donor** and an alkali as a **proton acceptor**.

? Questions

1 Why is a hydrogen ion also called a proton?
2 What is the charge on the potassium ion in potassium hydroxide, KOH?
3 Why is an acid called a proton donor?
4 Which ions would you expect to be produced if solutions of hydrochloric acid and potassium hydroxide were combined?
5 Write a dissociation equation for sulfuric acid. You will need to consider the charge on the sulfate ion and how to best represent the hydrogen ions that form.

✎ Show you can...

Complete this task to show that you understand the atoms and molecules involved in neutralisation reactions.

Create the following particle diagrams for the reaction between nitric acid and potassium hydroxide:

1 A particle diagram showing the ions formed when nitric acid dissociates ($HNO_3 \rightarrow H^+ + NO_3^-$).
2 A particle diagram showing the ions formed when potassium hydroxide dissociates.
3 A particle diagram of the products that are formed.
4 A balanced symbol equation with state symbols for the reaction.

Salts are used for many different things: rock salt is thrown on the roads for gritting in the winter; we eat salt in our foods and it is used in industry as a starting material for other chemicals. For all these purposes we need to look at how pure the salt needs to be. The steps needed to achieve a pure sample of salt are essential knowledge for any chemist.

● **Figure 1** What evidence is there from this image of salt crystals that they are a pure sample?

In the laboratory we can make samples of salts in several different ways. Any of our acid reactions met in this topic would be suitable. But if we are making a pure sample of salt crystals, we need to look at possible sources of impurities.

There is a laboratory technique that will allow you to make a pure sample of copper sulfate crystals. All salts made from group 1 and group 2 metals are colourless. Copper is in the middle block of metals on the Periodic Table and these can form coloured salt solutions and crystals. Copper sulfate is a blue compound.

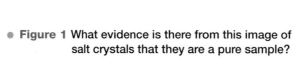

copper oxide + sulfuric acid → copper sulfate + water
$$CuO(s) + H_2SO_4(aq) \rightarrow CuSO_4(aq) + H_2O(l)$$

→ Step 1 – reacting an insoluble base with hot acid

As we want to make copper sulfate, the acid must be sulfuric acid. In order to react with an insoluble copper compound the base copper oxide is used. To speed up the reaction the mixture is heated and copper oxide is added to the hot acid until no more will react. This will be visible because unreacted copper oxide powder will be seen at the bottom of the beaker. This will indicate that all of the acid particles have reacted.

Additional care must be taken by wearing safety goggles instead of glasses.

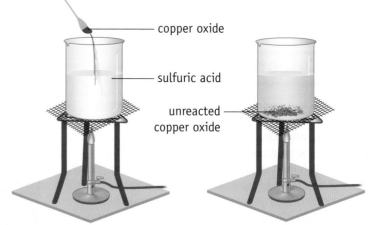

copper oxide

sulfuric acid

unreacted copper oxide

● **Figure 2** Reacting copper oxide with hot sulfuric acid

→ Step 2 – filtering unreacted oxide

To remove the unreacted copper oxide, let the solution cool and use a filter funnel and paper to separate the black powder from the copper sulfate solution.

unreacted copper oxide

copper sulfate solution

● **Figure 3** Filtration of unreacted copper oxide

→ Step 3 – evaporation of salt solution

To separate the salt product from the water the solution must be evaporated. As the solution is heated, the water particles will become gaseous and leave the dish. As this happens the copper and the sulfate ions will combine together making pure copper sulfate crystals. You must stop heating as soon as you see crystals forming.

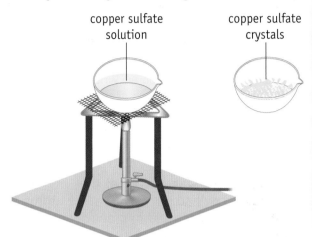

copper sulfate solution

copper sulfate crystals

● **Figure 4** Evaporation of copper sulfate solution leaving pure copper sulfate crystals

The most common impurities will be unreacted substances. This can pose a problem, because in a neutralisation reaction the substances can be hazardous and we would not want our salt sample contaminated with either the acid or the alkali. It is important, therefore, that we can work out the exact amounts needed to react so that the end sample is pure.

There are some advanced ways to do this. Future studies of chemistry will show you how you can calculate the exact amounts needed and ways to very accurately determine when the reaction mixture has reached neutral. You will then know that all acid and alkali particles have reacted.

? Questions

1 Why does the acid have to be hot in step 1?
2 Why does some black copper oxide fall to the bottom of the beaker in step 1?
3 How does using an insoluble base, such as copper oxide, help make a pure salt sample?
4 Why would the colour of the resulting solution start as colourless, become very pale blue and then become a deeper blue as oxide is added?
5 Explain some of the changes you would expect to see if this method were followed using the insoluble magnesium carbonate.

✎ Show you can...

Complete this task to show that you understand about making pure samples of salts.

Explain how you could use **Universal indicator** paper to help achieve a pure sample from reacting an alkali and an acid. For example:

sodium hydroxide + hydrochloric acid → sodium chloride + water

Think about how the indicator could show when the reaction is neutral and hence that all the acid and alkali have reacted.

Interpreting observations

→ Monitoring reactions

A group of chemistry students, David, Aaron and Petra, had recently been studying reactions of acids. In this unit they had seen lots of different reactions and one element kept being mentioned. That element was copper, so when their teacher asked them to plan an independent investigation, they decided to come up with the title:

Investigating copper compounds

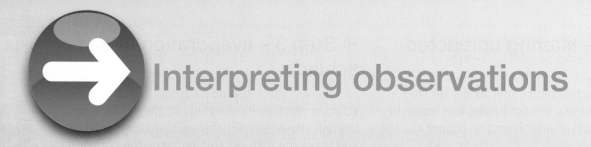

● **Figure 1** Some reactions between copper and varying concentrations of acid

This would allow them to use all the knowledge they had gained about the atoms and molecules involved in reactions and to practice carrying out chemical reactions.

The first thing they wanted to do was to check that copper compounds gave the expected observations when reacted with acids. They put in the request below to the chemistry technicians:

- copper
- copper oxide
- copper carbonate
- copper hydroxide
- **dilute** hydrochloric acid.

1 What safety precautions do they need to take with this set of chemicals?
2 If they want to test the gases produced in these reactions what else will they need?

The first thing they noticed was that the copper metal had been provided in very fine shavings, labelled as copper turnings. They began their reactions by adding the copper turnings to the acid. After 5 minutes they became very disappointed – nothing had happened at all – unlike the image in Figure 1 and on their teacher's presentation where the tube on the right had shown a reaction. They became convinced that the technicians had not given them acid at all. Aaron decided to check by asking for some Universal indicator.

3 Where is copper in the reactivity series?
4 What could explain the reactions in the photo in Figure 1?
5 The solution was hydrochloric acid – what would you expect to see when Aaron added the indicator?

Using the Universal indicator proved that they had been given acid, and told them that copper metal will not react with this acid. They then moved on to their other experiments. There were two powder samples, one black and one green, and two labels that had fallen off, one for copper oxide and the other for copper carbonate. The students saw this as a challenge to see which powder was which.

They added an equal amount of each powder to the hydrochloric acid. The green powder plus acid began to **effervesce**. The black powder did not seem to react at all and just settled to the bottom of the boiling tube. As they had the Bunsen burner available they decided to warm both reactions. Petra told the boys that as they would be warming the acid they needed safety goggles on, instead of just the safety glasses.

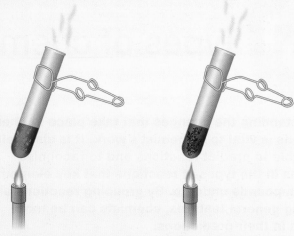

● **Figure 2** Heating the mystery copper compounds with acid

6 Why did Petra suggest they change eye protection?
7 Why would warming the mixture help the reaction?
8 What gas test would you suggest for the effervescence from the green powder plus acid reaction?
9 Identify
   a) the green powder, and
   b) the black powder.

Once heated, both reactions started and the students saw a colour change. David remarked 'I thought copper sulfate was blue – these reactions are going green'.

10 Why would the product of these reactions not be copper sulfate?
11 The solutions of the salt were green – what is the name of this salt?
12 The chloride ion is Cl^- and the copper ion is Cu^{2+}. What is the formula of the salt?

When they came to the reaction of copper hydroxide and acid they saw something very interesting. Copper hydroxide had been made in a boiling tube for them.

13 How would you describe the copper hydroxide?
14 What would you expect to see happen when the hydrochloric acid is added to this beaker?
15 Write a balanced symbol equation with state symbols for this reaction.

● **Figure 3** Copper hydroxide $Cu(OH)_2$

Types of chemical reaction

Understanding the changes that take place in chemical reactions is vital to a chemist's work. It is also helpful to be able to predict reactions and to recognise patterns in the types of reactions that key elements and compounds undergo. By grouping reactions and spotting general features, chemists can be more precise in their predictions.

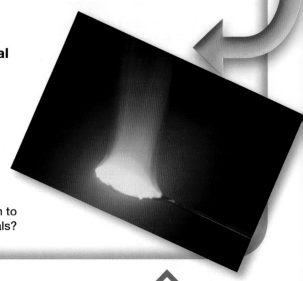

● **Figure 1** What do you observe about the burning of sodium to support its position in the reactivity series of metals?

→ Combustion

Burning – or **combustion** – is a chemical reaction. When oxygen reacts with a substance, energy is transferred, mainly as heat. This is called combustion.

A Bunsen burner uses methane as a fuel. The air hole controls the amount of air (oxygen) that mixes with the methane. When the air hole is open, more oxygen reaches the fuel and the flame is at its hottest (the blue flame).

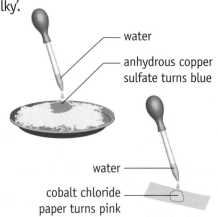

● **Figure 2** This symbol tells us that the material burns easily in air (flammable).

Carbon dioxide and water vapour are produced when a Bunsen burner is used, and these are the typical products when a fuel burns. Notice here that water has the state symbol (g) because the temperature reached when methane burns is above 100 °C.

methane + oxygen → carbon dioxide + water
$$CH_4(g) + 2O_2(g) \rightarrow CO_2(g) + 2H_2O(g)$$

Figure 3 shows the way to collect the products of combustion. As the gases form they are pulled through the apparatus by the suction pump. Due to the ice bath around the middle tube the water vapour produced will condense inside this tube. Limewater is a clear, colourless liquid. When carbon dioxide is bubbled through the limewater, it turns 'milky'.

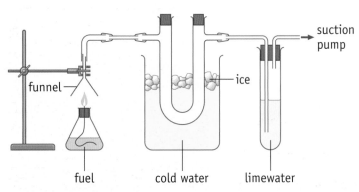

● **Figure 3** Collecting and testing the products of fuel combustion – water and carbon dioxide

water

anhydrous copper sulfate turns blue

water

cobalt chloride paper turns pink

● **Figure 4** You can use one of these tests for water

→ Thermal decomposition

A second type of reaction, called **thermal decomposition**, also involves heat but in this instance the heat causes the chemical substance to break down (decompose).

copper carbonate → copper oxide + carbon dioxide
$$CuCO_3(s) \rightarrow CuO(s) + CO_2(g)$$

Again the product of this reaction could be tested using limewater. You will notice that there is only one chemical written on the left. Heat is not a chemical so does not get included in the equation.

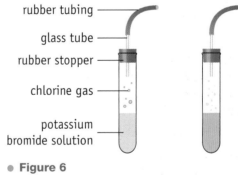

● **Figure 5** The thermal decomposition of copper carbonate

→ Oxidation and reduction reactions

Oxidation occurs when a substance reacts with oxygen. Combustion is one example of an oxidation reaction but there are others. Metals react with oxygen when heated to form metal oxides, as in the example of zinc on the right.

zinc + oxygen → zinc oxide
$$2Zn(s) + O_2(g) \rightarrow 2ZnO(s)$$

The opposite of oxidation is a reduction reaction, where substances lose oxygen.

These two types of reaction occur at the same time, for example, when we extract metals from their ores. In the example on the right, iron oxide has lost oxygen and therefore has been reduced whilst carbon has gained oxygen and therefore has been oxidised.

iron oxide + carbon → iron + carbon dioxide
$$2Fe_2O_3 + 3C \rightarrow 4Fe + 3CO_2$$

→ Displacement reactions

In a **displacement** reaction a more reactive element takes the place of a less reactive element in a compound. This can be used to test the reactivity series of the metals.

It can also show the periodic trends in reactivity of the elements. Shown here is the reaction between chlorine and potassium bromide. Chlorine is more reactive then bromine and so displaces it from the salt. The orange colour seen in the second test tube in Figure 6 is a solution of elemental bromine, Br_2. The chlorine has combined with the potassium making potassium chloride.

rubber tubing
glass tube
rubber stopper
chlorine gas
potassium bromide solution

● **Figure 6**

potassium bromide + chlorine → bromine + potassium chloride

? Questions

1 What type of reaction is this?
   $$2Cu(s) + O_2(g) \rightarrow 2CuO(s)$$
2 Describe the results of two chemical tests for water.
3 Thermal decomposition is a type of reaction in which only one chemical is written on the left. Explain why.
4 Explain what you would expect to see if copper were placed into magnesium sulfate.
5 Displacement reactions can also be oxidation and reduction. Use examples of reactions to show this.

✎ Show you can...

Complete this task to show that you understand the different types of reaction.

Create your own summary for each type, including an example reaction as a word and balanced symbol equation.

10.2 Energy transfer in reactions

When chemical reactions take place one way to show this has happened is to identify a change in temperature. Some reactions release heat energy whilst others take it in. Some chemical reactions have such dramatic energy changes that explosions occur.

● **Figure 1** The thermite reaction can reach a temperature of 1500 °C. What purpose do you think this reaction could be used for?

→ Exothermic and endothermic reactions

Some chemical reactions produce heat. This energy is transferred from the chemicals to the surroundings. This is an **exothermic** reaction. The temperature in the reaction mixture usually rises.

Examples of exothermic reactions include:

- combustion
- explosions
- fireworks exploding; the temperature of most fireworks is at least 1500 °C. Even sparklers have a temperature of about 2000 °C.

Some chemical reactions need energy for the reaction to take place. Energy – usually heat – is taken in from the surroundings. This is an **endothermic** reaction. The temperature in the reaction mixture usually falls. Examples of endothermic reactions include getting metals from their oxides.

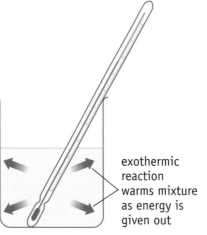

exothermic reaction warms mixture as energy is given out

● **Figure 2** Measuring an exothermic reaction

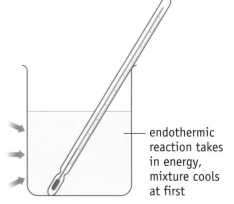

endothermic reaction takes in energy, mixture cools at first

● **Figure 3** Measuring an endothermic reaction

To monitor a reaction where energy changes are taking place a thermometer can be used to measure the temperature of the surroundings. There is no way to measure the energy changes within the chemical reaction and the changes in the surroundings are a reliable indication of energy transfer. Glassware used in the laboratory is an excellent **conductor** of heat – otherwise we could not use it to heat the contents. As the reaction releases heat energy some of this will be lost beyond the beaker.

● **Figure 4** Steps should be taken to insulate the beaker to reduce energy loss

When an energy transfer takes place, this is evidence that a chemical reaction has happened even if the reactants themselves do not appear to change. The energy transfer in or out is not always heat. Sound or light energy can be transferred in chemical reactions.

- Photosynthesis is a type of endothermic reaction that needs light energy, not heat, to be transferred to the reacting chemicals.
- Glow sticks produce light when the chemicals inside react together.

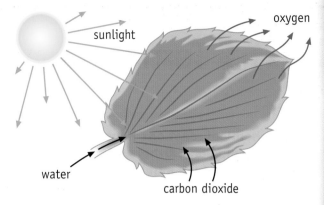

● **Figure 5** Photosynthesis – a reaction that needs energy

➔ Bond breaking and making

Energy is needed to break bonds between atoms in the reactants and to make new bonds between atoms in the products.

For example, when potassium reacts with water:

- the bonds between the hydrogen and oxygen atoms in the water molecule have to be broken
- new bonds have to be formed between the potassium, oxygen and hydrogen atoms.

Not all bonds are the same – they vary in how much energy is required to break or make them. As a chemical reaction involves rearranging atoms the type and number of bonds broken will not be the same as those made.

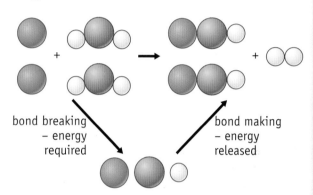

● **Figure 6** Bond breaking and making during a chemical reaction

Show you can...

Complete this task to show that you understand some of the energy changes that accompany chemical reactions.

Decide whether these reactions are endothermic or exothermic and give reasons for your decision.

1 Methane from the Bunsen burner is ignited to heat some water.
2 When sodium metal is put into a gas jar of chlorine a bright light is seen and a drop of water placed on the flask afterwards immediately vaporises.
3 A teacher combines two chemicals – the beaker is wet on the outside. As the reaction progresses the outside of the beaker frosts up – it is also stuck to the desk due to the formation of ice.
4 Two chemicals at room temperature are mixed in a fume cupboard – the teacher steps back. A few seconds later the chemicals begin to fizz, a gas is seen, and very soon after the mixture catches fire with a bright purple flame.

❓ Questions

1 What is the clue in the terms 'endothermic' and 'exothermic' that indicates that these are linked to heat energy?
2 What is the thermometer measuring in Figures 2 and 3?
3 Why is an energy change evidence that a chemical reaction has taken place?
4 Potassium reacts with water in an exothermic reaction. What does this suggest about the energy needed to break the bonds compared with energy released when they are made?
5 Why are some reactions exothermic and some reactions endothermic, if they both involve breaking bonds and then making bonds?

10.3 Temperature and catalysts

Some chemical reactions happen almost instantly, such as explosions. Others, like rusting or acid rain corrosion, take much longer. Several factors can affect how fast or slow a reaction occurs and the particle theory of matter can give us the explanation for these factors.

● **Figure 1** How fast do you think the reaction in a firework is?

→ Reaction rates

Reactions can take varying amounts of time from start to finish, and chemists need a way to monitor these. They need to show the amounts of chemical used or produced in a set time period.

It is easiest to monitor reactions involving a gaseous product. This can be done in two ways. See Figure 2.

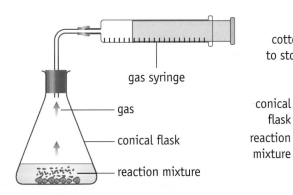

a) the amount of gas collected in a gas syringe in a fixed period of time is recorded

b) the gas is allowed to escape from the flask and the loss of mass is recorded

● **Figure 2** Two ways of monitoring reaction rate

To explain **reaction rates** we revisit the particle theory. Particles are moving all the time and bump into each other frequently. However, they do not always react. For particles to react they must **collide** with enough energy to break the existing bonds in the atoms or molecules. Once bonds are broken atoms can rearrange to make the products.

There are many factors that affect the behaviour of particles and all of these will have an overall impact on the rate of a reaction.

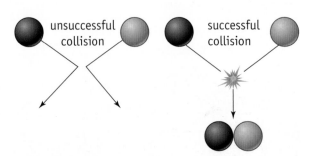

● **Figure 3** Reactions happen when collisions are successful – that is, when the particles have enough energy to react

→ Temperature and rate

As particles gain kinetic energy they begin to move more quickly.

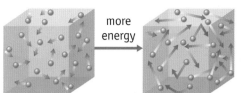

more energy

Collisions occur where the arrows cross. Particles with greater speeds are more likely to collide.

● **Figure 4** As particles gain energy they move with greater average speed

Particles have more energy at higher temperatures so when they collide they will be more likely to have enough energy to react. This combined effect means that a small change in temperature can have a big effect on the rate of reaction. In some reactions the rate can be doubled by increasing the temperature by just 10 °C.

A fast reaction will cause lots of gas to be produced quickly. This will give a steeper curve when plotted on a line graph.

When the temperature is reduced the volume of gas measured become less, giving a less steep graph and indicating a slower rate.

The rate is not the same throughout a reaction. The initial rate will be its fastest and the best to use for a reliable comparison.

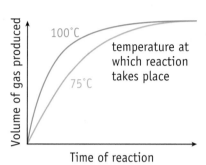

100 °C

temperature at which reaction takes place

75 °C

Volume of gas produced

Time of reaction

● **Figure 5** A graph to show effect of temperature on volume of gas produced in a chemical reaction

→ Catalysts and rate

Catalysts are substances that are added to a reaction to change the rate. They lower the amount of energy required for the reaction to occur. This means that more of the collisions will be successful and increases the overall rate. As the catalyst does not take part in the reaction directly it does not get used up. This is important as many catalysts are extremely expensive.

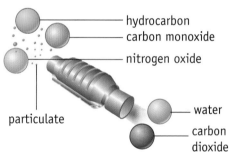

hydrocarbon
carbon monoxide
nitrogen oxide

particulate

water

carbon dioxide

● **Figure 6** Catalytic converters speed up the reactions that reduce the harmful emissions from car engines

Show you can...

Complete this task to show that you understand the effect of temperature on rate of reaction.

Plot the data below on a line graph, describe the pattern shown and explain why the rates vary.

	Volume CO$_2$ produced	
Time (s)	20 °C	30 °C
0	0	0
30	6	8
60	12	18
90	17	22
120	21	24
180	24	25
240	25	25

? Questions

1 Which essential piece of equipment is needed for all rate investigations?
2 Why would it be difficult to monitor the effect of temperature on an explosive reaction?
3 Why is the graph in Figure 5 showing a curved line of best fit and not a straight line?
4 Why does temperature have such a big effect on rate of reaction?
5 Think of the example of an acid and carbonate reaction. Write a word equation and explain why the methods in Figure 2 would be appropriate. Describe how you would vary the temperature safely.

10.4 Concentration and surface area

The chemical reaction when gunpowder is ignited has caused devastation since its discovery by the Chinese over 1000 years ago. By using finely ground chemicals the reacting particles can come into very close contact and set off a very rapid exothermic reaction. The heat generated accelerates the reaction even further and increases the force of the explosion.

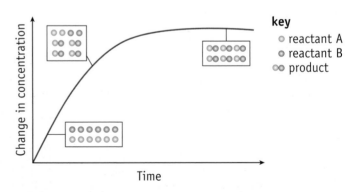

● **Figure 1** A fireball from a gunpowder explosion. Why do you think gunpowder is so explosive?

→ Concentration

Many reactions occur with substances in solution. From our understanding of reactions of acids we know that all solutions are not the same. When a solution is more concentrated, there are more reacting particles in the same volume. This will lead to a greater chance of collisions and therefore a greater rate of reaction.

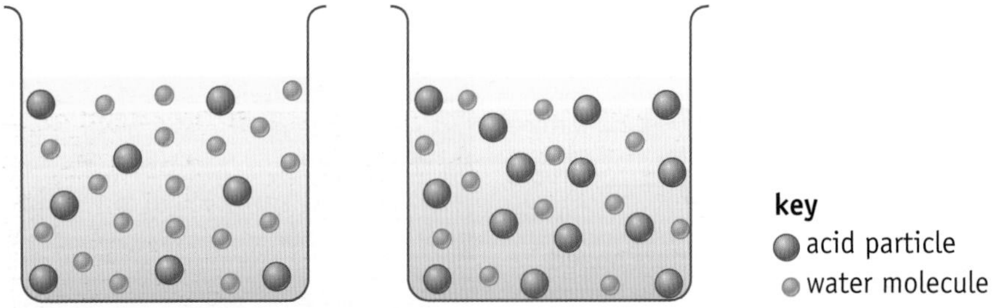

dilute   concentrated

key
● acid particle
● water molecule

● **Figure 2** The difference between concentrated and dilute solutions

By understanding concentration we can explain the change in rate during a reaction. At the start of the reaction the particles are in the highest concentration. As the product begins to form, the concentration of the reacting particles decreases, which gradually slows the rate down until all particles have reacted and the reaction stops. As the product forms it **dilutes** the reacting particles, reducing collisions and slowing down the rate.

Change in concentration

Time

key
● reactant A
● reactant B
●● product

● **Figure 3** The change in concentration of product throughout a chemical reaction

→ Surface area

When a large piece of solid material is cut into smaller pieces the surface area increases. This means that a powdered sample will have a much greater surface area than a single large lump.

Remember that chemical reactions happen when particles collide, so by increasing the surface area the chance of collision has greatly increased.

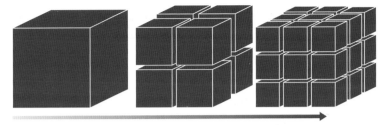

increasing surface area

● **Figure 4** Increasing surface area

As the chance of collision is increased the rate of reaction increases. An important control variable when investigating surface area is that the total mass of material must stay the same, as well as the temperature of each reaction and the concentrations of any solutions used.

a single piece of metal

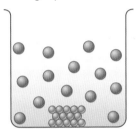

smaller pieces of metal

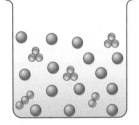

key
● acid particle
● magnesium atom

● **Figure 5** The effect of increasing surface area on particle collisions

? Questions

1 Using these ideas about rate of reaction, explain why concentrated acids are more dangerous.
2 Which sample has the greatest surface area: 5 g of magnesium powder or 6 g of magnesium ribbon?
3 It can sometimes be observed that doubling the concentration can double the rate of reaction. Explain why this could happen.
4 Sketch a graph of time against volume of gas produced for the reaction between calcium carbonate and acid. Include a line for large pieces and small pieces and label the lines.
5 Why does increasing the temperature have a greater effect than concentration on the rate of a chemical reaction?

✎ Show you can...

Complete this task to show that you understand how factors such as concentration and surface area affect the rate of a chemical reaction.

Put the following reactions in order of increasing rate.

1 The reaction between dilute acid and small pieces of calcium carbonate at room temperature.
2 The reaction between dilute acid and large pieces of calcium carbonate at room temperature.
3 The reaction between dilute acid and powdered calcium carbonate at 80 °C.
4 The reaction between concentrated acid and small pieces of calcium carbonate at room temperature.

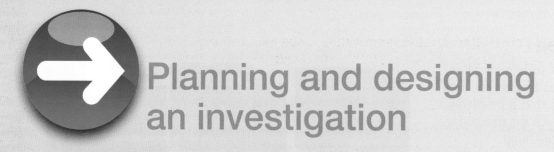

Planning and designing an investigation

→ Investigating rates of reaction

A class teacher gave out the following hypothesis to her class:

'An increase in 10 °C will double the rate of a chemical reaction.'

Kathryn and Theresa knew that the first step in planning an investigating was to consider all their knowledge and identify the variable that might affect the outcome. They then needed to select their independent and dependent variables.

1 What factors affect the rate of a chemical reaction?
2 What does independent variable mean and what would they select to investigate this hypothesis?
3 What does dependent variable mean and what would they select to investigate this hypothesis?

The students knew that to investigate the rate of a reaction you have to first select a reaction and that would help decide which apparatus to use. Rate cannot be measured directly – you have to be able to measure the change in either reactants being used up or the products being formed.

The students decided on the reaction between magnesium metal and hydrochloric acid. Kathryn wanted to measure the mass lost from the reaction every 30 s but Theresa wanted to collect the gas and measure the volume made every 30 s.

4 Write a word equation for the reaction.
5 Write a balanced symbol equation including state symbols for the reaction.
6 Where is the gas that is being made on the Periodic table. What is its relative mass?
7 Why might Theresa's idea give more measurable results?

● **Figure 1** Hydrochloric acid and magnesium ribbon

The next step was to consider the range of readings. Theresa thought that looking at cold acid, warm acid and hot acid would test the hypothesis. Kathryn was not sure: 'With those choices we would have to draw a bar chart – I do not think it would test the hypothesis at all.'

8 Why would Theresa's suggestion have to be drawn on a bar graph?
9 a) Why would testing only three different values not be very reliable?
  b) What number would be better?
10 Why would these choices not allow the hypothesis to be tested?
11 What range of values would you suggest?

Their teacher wanted to check their chosen temperature range for safety reasons. They had chosen 20 °C, 40 °C , 60 °C, 80 °C and 100 °C and she had to ask them to look again.

12 Why was this range presenting a safety risk?

Once the range was decided, they started preparing their metal. They had decided to control the mass of the metal to ensure reliable data. When they went to collect the metal they had a choice between long pieces, an assortment of different sized small pieces or powdered. Kathryn said 'Let's use the small pieces. Powder may be too fast and the large piece may take too long.'

13 Why did Kathryn suggest this?

Back at the bench they realised the small pieces were all different lengths. Theresa knew this would make the test unreliable but could not think of any way around it.

14 How could the students use small pieces and make the test reliable?

Part way through their investigation their bottle of acid had run out. They were sent to the technician to ask for some more. On arrival they were asked 'what concentration do you need?'

15 Why do you think the technician asked them this?

Here is the students' plan:

1 Measure out 25 cm^3 of 1 M hydrochloric acid into a concial flask

2 Cool in ice or warm the acid on the Bunsen burner to the correct temperature.

3 Place five pieces of 2 cm long magnesium into the acid and put the gas syringe in.

4 Measure the gas produced every 30 s for 4 min.

5 Repeat each temperature three times.

16 Write an apparatus list.
17 Draw a blank results table for the investigation.

11.1 The Earth and its atmosphere

Our planet is surrounded by layers of gases called the atmosphere. We use these gases to sustain life and to combust fuels. As you move away from the Earth's crust the atmosphere changes and has some very important roles to play.

● **Figure 1** How did we know about the structure of the Earth before we could take pictures from space?

The Earth is built up in several layers from the inner **core** to the upper **atmosphere**. The uniqueness of our planet's environment compared to all other planets in our **solar system** is the reason why life was able to evolve here.

→ The crust

As the outermost layer of the mantle cools and solidifies, the **crust** forms. Two distinct types of rock form the ocean floors and the land.

The land masses, or the lithosphere, are sections of rock formed in plates. These are slowly moved around the outer edge by the flowing currents in the mantle. The crust is made up of several different types of metals and rock containing metal ores. This is the source of metals for human use.

The oceans, or the hydrosphere, which cover just over 71% of the Earth's surface, give rise to many dissolved compounds and gases.

→ The core

The core is made of a solid inner and a molten outer layer. Temperatures here reach up to 6000 °C. The composition of iron and nickel in the core gives rise to the Earth's magnetic fields.

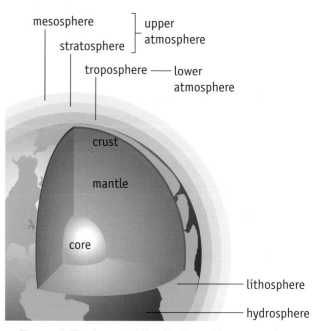

● **Figure 2** The layers of the Earth and its atmosphere

→ The mantle

The **mantle** is made up of semi-molten rock called **magma**. **Convection currents** formed from the rise and fall of the molten rocks produce flowing movements that are considered to be responsible for the movement of the solid layer just above (the crust).

→ The lower atmosphere

Also known as the troposphere, this is the first layer of the atmosphere. It contains the air that we breathe. It is the amount of free oxygen in this part of the atmosphere that allows life on Earth.

It is also the source of all weather. As the Sun warms the Earth's surface the gases here are warmed. They rise and cooler air is pushed down creating wind currents. As these gases are moved around pressure is also affected. Depending on the season this gives rise to a range of cloud formations and other weathers such as rain, sleet and snow.

The troposphere is also responsible for maintaining the temperature of the planet. Heating radiation from the Sun can pass through the atmosphere, but is blocked from reflecting back, warming the planet. This was an essential part of providing conditions for life but human activity has upset the natural balance causing global temperatures to increase. This will be studied further in Topic 11.4 Human activity (page 104).

→ The upper atmosphere

Made of two layers, the stratosphere and the mesosphere, the upper atmosphere protects life on Earth.

Just above the lower atmosphere, in the stratosphere, we see an increase in the presence of a substance called **ozone**. This is formed when high energy ultraviolet radiation from the Sun provides enough energy for oxygen molecules to be split apart. These single oxygen atoms are extremely reactive and can then combine with another oxygen molecule to make ozone (O_3).

Ozone is important because it absorbs the ultraviolet radiation from the Sun, which if it reaches us can cause skin cancer. You will have heard of the hole in the ozone layer – the reason this causes concern is because more ultraviolet radiation will get through the atmosphere and reach the surface of the Earth.

The gases in the mesosphere layer protect us from meteors. As they collide with the particles in this layer the meteors burn up and so never reach the Earth's surface.

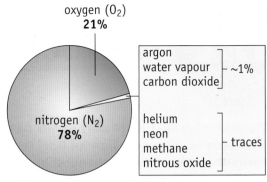

● **Figure 3** The gases in the lower atmosphere

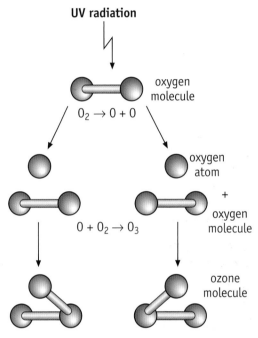

● **Figure 4** Making ozone in the upper atmosphere

? Questions

1 How many layers make up the Earth and its atmosphere? List them all.
2 Why is the atmosphere so important for life?
3 Explain how the oceans used to be in the atmosphere.
4 Explain why the hole in the ozone layer is a cause for concern.
5 There are two layers where convection currents are occurring. Explain what a convection current is and discuss the effects of these currents in the two locations.

✏ Show you can...

Complete this task to show that you understand the Earth and its atmosphere.

Write a series of descriptions that can become 'Where am I?' questions for each part of the structure of the Earth and its atmosphere. Include clues that link to the chemistry of each part and try them out on your partner.

Rocks make up a large part of our natural environment. Studying rocks tells us about how the Earth was formed and how it has changed over millions of years. They are also versatile building materials, from kitchen work surfaces to huge statues, so understanding their properties and the processes that mould them is essential.

● **Figure 1** What evidence is there that the small pieces of rock were once part of the main rock face?

There are three types of rock:

- igneous rock
- sedimentary rock
- metamorphic rock.

→ Igneous rocks

Molten rock beneath the Earth's surface is called magma. Magma forms in areas where the Earth's crust is so hot that the rock is given enough energy for it to partially melt. When magma rises upwards through a crack or volcano to reach the surface, it is called lava. When magma or lava cools it forms igneous rocks.

The igneous rock formed when magma or lava cools is **crystalline**. The rock texture (size of the crystals) depends on how quickly the lava cooled down:

- slow cooling → large crystals
- fast cooling → small crystals.

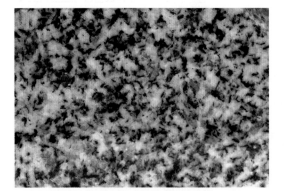
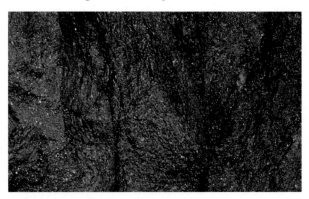

● **Figure 2** The different crystal sizes in igneous rocks (left = granite, right = basalt)

Every time a volcano erupts, another layer of igneous rock is formed on top of earlier layers. These layers could be formed hundreds or thousands of years apart – depending on how frequently the volcano erupts.

→ Sedimentary rocks

Weathering is when rocks at the Earth's surface are broken down due to physical, chemical or biological processes. This can slowly change the appearance of the landscape, or of buildings made from rock or brick. **Erosion** is the removal and transportation of small pieces of rock. Transportation means moving the pieces of rock from their original site to somewhere else.

Sedimentation happens when small pieces of rock are deposited in their final position. Later, a second layer is deposited on top of the first. Even later a third layer is deposited ... and so on. The build-up of thick layers takes place slowly over millions of years.

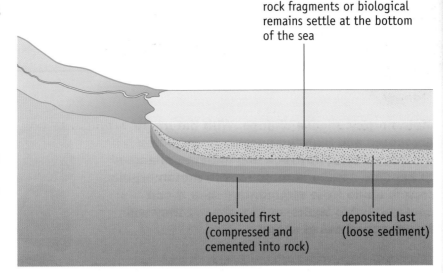

rock fragments or biological remains settle at the bottom of the sea

deposited first (compressed and cemented into rock)

deposited last (loose sediment)

● **Figure 3** Formation of sedimentary rock

As each new layer of sand or mud is formed, it presses down on the layers below it, compressing them. The particles become cemented together. The sediment is now a sedimentary rock.

As the rocks form in layers the remains of living things may become trapped and turn into fossils. As the rock builds up over millions of years the types of fossil present can give an indication of the types of living things alive at different times in Earth's history.

? Questions

1 Which type of rock may contain fossils?
2 Which type of rock forms when molten magma cools at the Earth's surface?
3 Describe how an igneous rock could form a sedimentary rock.
4 Why are fossils found only in one type of rock?
5 Explain why rock that cools deep underground would have large crystals.

● **Figure 4** A fossilised fern formed in the layers of sedimentary rock

✎ Show you can...

Complete this task to show that you understand the process of sedimentation and cementation.

Draw a column of rock with four layers and add some fossils. Select and label two of the layers as 'oldest rock' and 'newest rock'.

Use the diagram to explain how a column of sedimentary rock could give clues about the living things alive at different times in the Earth's history.

The rock cycle 2

Metamorphic means change. With the great pressures and high temperatures under the Earth's surface, any rock can undergo changes to become a metamorphic rock. If metamorphic rock is made from sedimentary rock it can contain visible layers and fossils but they will be squashed out of shape.

● **Figure 1** Which of these metamorphic rocks may have been formed from sedimentary rocks?

→ Metamorphic rocks

Metamorphism means change. It happens when rocks are heated to very high temperatures and/or when they are under very high pressures. Both igneous rocks and sedimentary rocks can be changed into metamorphic rocks under the right conditions.

→ The rock cycle

The rock cycle explains how rocks are produced and change over time. Some parts of the cycle can happen very fast – such as when a volcano erupts. Other rocks can take millions of years to form and reform.

Melting, metamorphism and extreme heat and pressure do not happen all the time, only at certain times and in certain places. The rock cycle is not a continuous cycle.

● **Figure 2** The sedimentary rock limestone (left) can form the metamorphic rock marble (right)

- Weathering and erosion of rocks produces sediments that form sedimentary rocks.
- Some of these are gradually eroded to form new sedimentary rocks.
- Rising magma causes igneous rocks to form.
- These are eroded over time to form new sedimentary rocks.
- Sedimentary or igneous rocks can be changed into metamorphic rocks by very high heat or pressure.
- These are eroded over time to form new sedimentary rocks.

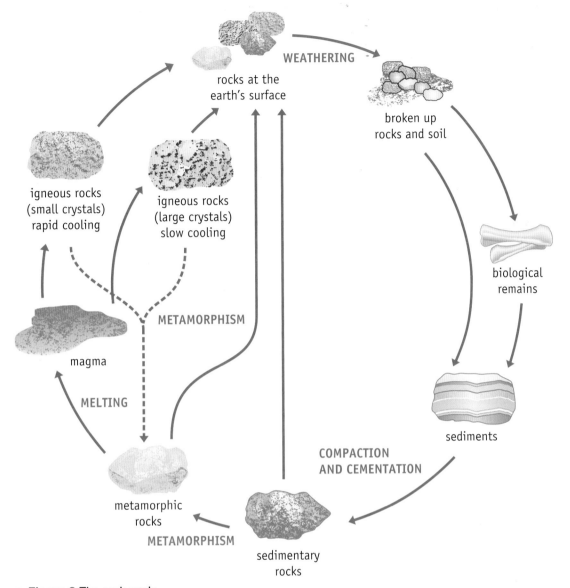

WEATHERING

rocks at the
earth's surface

broken up
rocks and soil

igneous rocks
(small crystals)
rapid cooling

igneous rocks
(large crystals)
slow cooling

biological
remains

METAMORPHISM

magma

MELTING

sediments

metamorphic
rocks

COMPACTION
AND CEMENTATION

METAMORPHISM

sedimentary
rocks

● **Figure 3** The rock cycle

Questions

1 What conditions are needed to form marble?
2 Which two processes in the rock cycle happen under pressure?
3 How can metamorphic rock contain fossils?
4 Explain how sedimentary rocks may be made up of small pieces of metamorphic rocks, igneous rocks and other sedimentary rocks.

Show you can...

Complete this task to show that you understand the rock cycle.

Create a simplified version of Figure 3 based on the following information.

• The igneous rock granite forms the metamorphic rock gneiss under heat and pressure.
• The sedimentary rock limestone forms the metamorphic rock gneiss under heat and pressure.
• Any rock can be turned into sediment through weathering and erosion and over millions of years can become cemented into sedimentary rock.
• If rocks melt and form magma, when volcanoes erupt they form igneous rocks.

Global warming and climate change are recognised by scientists as having a devastating effect on our planet. Rising sea levels and changes to severe weather patterns are just some of the effects. It is also accepted that human activities are one of the main cause of these changes.

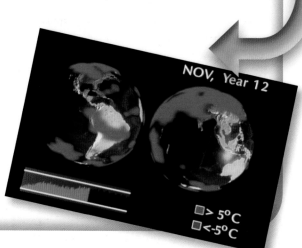

● **Figure 1** Why would the Earth be getting hotter?

→ The greenhouse effect

The temperature of the planet is one of the main reasons it can sustain life. The gases in the atmosphere are responsible for ensuring our planet remains at a steady temperature. The gases that are particularly effective at trapping the Sun's heat include carbon dioxide, methane and water vapour. This is known as the **greenhouse effect**.

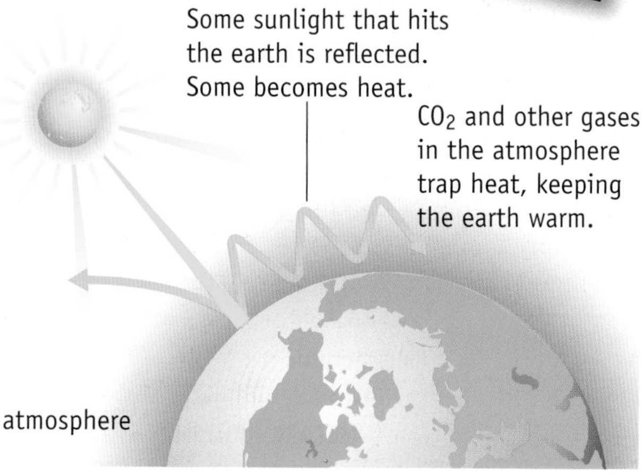

Some sunlight that hits the earth is reflected. Some becomes heat.

CO_2 and other gases in the atmosphere trap heat, keeping the earth warm.

atmosphere

● **Figure 2** The atmosphere traps heat from the Sun, warming the planet's surface

→ Global warming

The Earth's surface temperature has been changing over the last 500 years, with some areas increasing and others decreasing. This can be difficult to identify if average temperature is the only measure. There are many useful ways to look at the effects of temperature changes.

Maintaining the surface temperature is a matter of balance; the gases in the atmosphere need to remain constant and for most of the last 10 000 years have done so. One particular gas causing concern is carbon dioxide (CO_2). It once made up approximately 96.5% of our atmosphere. Its decline and replacement with oxygen was what enabled life to evolve.

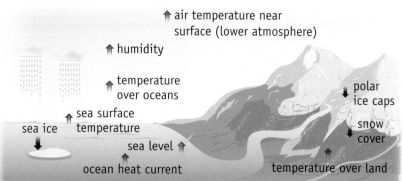

⬆ air temperature near surface (lower atmosphere)

⬆ humidity

⬆ temperature over oceans

⬆ sea surface temperature

sea ice

⬇ (sea ice)

⬆ sea level

ocean heat current

⬇ polar ice caps

⬇ snow cover

⬆ temperature over land

● **Figure 3** All 10 of these signs of global warming have been identified

→ Reasons for the increase in CO₂

The reasons for the sudden increase in CO_2 can be linked to human activity. Fossil fuels, which have trapped carbon for millions of years, have been burnt at a rapid rate, either to provide heat directly or to produce electricity. This has been made even worse with the destruction of vast areas of rainforests. Thousands of huge trees, which would have been able to remove carbon dioxide from the atmosphere through photosynthesis, have been felled.

→ The carbon cycle

Since the evolution of green plants and animals there has been a cycle of carbon through the environment. Plants would use carbon dioxide for photosynthesis and it would be released through respiration from both plants and animals. The carbon in living materials would be released back to the atmosphere when they died and decomposed.

The amount of carbon dioxide taken from the atmosphere was globally equal to the amount released, leading to the same amount being present over many thousands of years. As humans have developed technology and found more and more efficient ways to burn fuels and use electricity they have upset the balance. Large amounts of carbon dioxide are released into the atmosphere through the combustion of fossil fuels. This problem is caused by human activity so it rests with humans to reflect on their impact and make changes to reduce it.

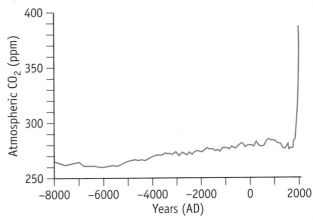

● **Figure 4** Atmospheric carbon dioxide over the last 10 000 years

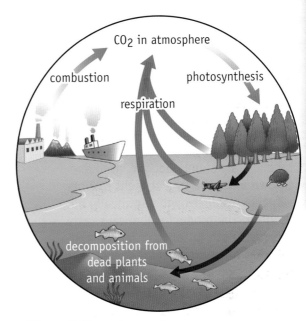

● **Figure 5** The carbon cycle

? Questions

1 How is carbon dioxide removed from the atmosphere by green plants?
2 List the human activities linked to increased carbon dioxide levels in the atmosphere.
3 Using evidence from our neighbouring planets, explain how the atmosphere influences the Earth's temperature.
4 How many parts per million was the increase in carbon dioxide levels from 1000 AD to 2000 AD?
5 Why would removing an area of forest to plant other fast-growing crops, which will be harvested and sold each year, increase carbon dioxide levels in the atmosphere?

✎ Show you can...

Complete this task to show that you understand the human impact on global temperatures.

Draw a cartoon strip to summarise the greenhouse effect and how it can lead to global warming:

1 Carbon cycles between plants, animals and the atmosphere
2 Humans burn fossil fuels, run cars and generate electricity
3 Increased amounts of CO_2 in the atmosphere
4 Increased global temperatures

11.5 Recycling

The result of human activity on the environment is clear. To take the responsibility seriously, people take steps to reduce waste, reuse wherever possible or recycle materials, with the aim of reducing our impact on the environment. Does it make a difference?

● **Figure 1** How do you sort rubbish for recycling at home? What about at school?

→ Life cycle of a product

All products go through various steps and processes during their existence. If products are simply bought, used and thrown away this is a straight line and valuable resources will continually be used up. The product will always carry the cost, both in terms of money and in terms of environmental damage, of extraction and transportation of raw materials. **Recycling** changes this process from a straight line to a circle.

When companies make plans for a new product they must consider the impact on the environment. They could choose to reduce pollutants instead of making the most profit. Recycling products can reduce the cost of manufacture which makes it an attractive option.

→ Carbon footprint

Any stages in a product's life that release carbon dioxide into the atmosphere affect the environment. The total of these effects is often calculated as the carbon footprint of a product. It is also important not to use up all the sources of raw materials, otherwise there will be none left for future generations.

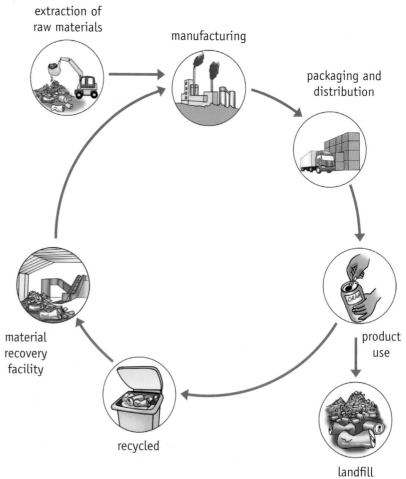

extraction of raw materials

manufacturing

packaging and distribution

product use

landfill

recycled

material recovery facility

● **Figure 2** The stages in the life of a drinks can

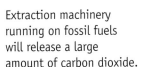

Extraction machinery running on fossil fuels will release a large amount of carbon dioxide.

Manufacturing machinery may run on electricity, or involve the direct burning of fossil fuels if, for example, metal has to be melted down.

Transportation of raw materials and finished goods also adds to the carbon footprint.

● **Figure 3** Extraction of raw materials starts the product's carbon footprint

There is a direct link between electricity production and carbon dioxide emission due to the output from power stations. Aluminium production is an excellent example. Remember that due to the reactivity of aluminium, it has to be extracted from its ore using electricity which requires a lot of energy. Recycling an aluminium can uses only 5% of the energy required to manufacture one from raw materials.

→ Recycling

Recycling also reduces the need to extract raw materials. Remember the chemical law that states that matter cannot be destroyed or created, only changed from one form to another. If a product ends its

● **Figure 4** New from old – recycling reduces the need to extract more raw materials

life on a landfill site its atoms are no longer useful. If it can be reused or recycled the atoms can remain useful for longer and less new material will need to be extracted.

However, recycling also has an environmental impact. Used products still need to be transported, sorted, cleaned and re-made; so the recycling process is not without impact. For each product we need to compare the processes of new production and recycling to reduce the impact of carbon dioxide emissions.

? Questions

1 What does the term 'recycle' mean?
2 Which greenhouse gas is often released during human activities?
3 Why do processes using electricity add to a product's carbon footprint?
4 Describe any steps in the life cycle of an aluminium can in which transportation is required: for example, the transport of aluminium sheets to the factory making the cans.
5 Even though recycling has its own carbon footprint, why is it considered to be beneficial in reducing the human impact on the environment?

✎ Show you can...

Complete this task to show that you understand the impact a product can have on the environment.

Create a promotional poster to encourage people to change their habits. Choose one of these themes:

· separating waste at home for recycling
· buying local goods
· buying recycled products.

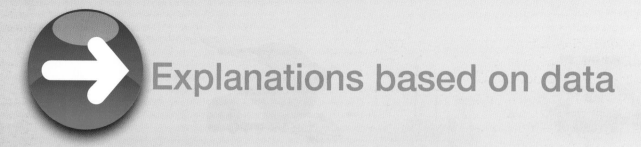

→ Are humans responsible for global warming?

Aaron and Alice have been asked to prepare a report for the class. The title of their report is in the form of a question: 'Are humans responsible for global warming?'. They have been studying global warming so have a good understanding of the scientific ideas behind it. They have also been looking at the human impact and how activities such as recycling could reduce this impact.

They set about planning for their report. They started with an internet search to see what opinions were available on climate change. They collected the quotes shown in Figure 1.

1 Put these quotes into a table with the columns 'for' human impact and 'against' human impact.

When claims such as these are made scientists must look for evidence to support them. This is why data is collected.

2 What sort of data would you need to support the claim that 'The recent warming is within normal ranges over the last 3000 years'.
3 Sketch the graph you would expect if this claim was true.

Aaron found the graph in Figure 2 on the internet.

4 Which opinion is shown to be untrue based on the data in the graph alone?

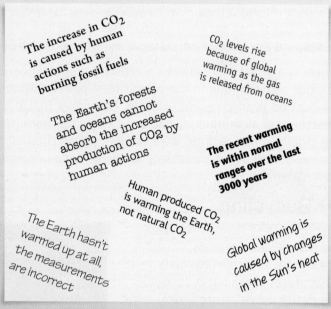

The increase in CO₂ is caused by human actions such as burning fossil fuels

CO₂ levels rise because of global warming as the gas is released from oceans

The Earth's forests and oceans cannot absorb the increased production of CO₂ by human actions

The recent warming is within normal ranges over the last 3000 years

Human produced CO₂ is warming the Earth, not natural CO₂

The Earth hasn't warmed up at all, the measurements are incorrect

Global warming is caused by changes in the Sun's heat

● **Figure 1** Opinions on climate change

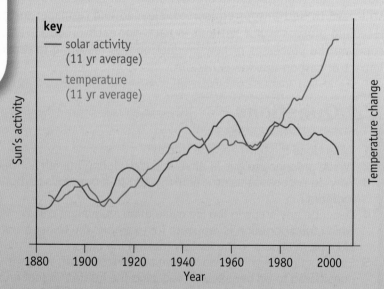

● **Figure 2** A graph to show the Sun's activity and Earth's temperature changes

Alice wanted to find some information that would support the opinion that human produced CO_2 is warming the Earth. She uncovered the fact that at the start of the industrial revolution in 1750, coal was discovered. People realised that you could release a lot of energy from a small mass of coal through burning the carbon-based coal in air. Alice thought that if she could find some data that showed carbon dioxide levels before and after this date she could decide if this opinion was correct or not.

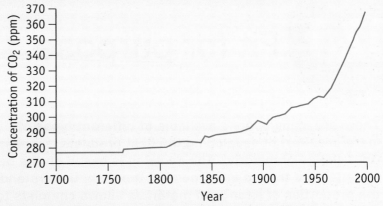

● **Figure 3** Carbon dioxide levels before and after 1750

5 Explain why the industrial revolution would increase the amount of carbon dioxide. Include a word and balanced symbol equation.
6 Explain why this additional carbon dioxide cannot be absorbed by the oceans or by plants through photosynthesis.
7 Explain why the increase in carbon dioxide is linked to increasing global temperatures.
8 Even if the opinion that 'The Earth has not warmed up at all, the measurements are incorrect' were true, what other evidence could be collected to support the view that global warming is occurring?

Their teacher came to check on their progress. She was pleased to see the data they had collected but was concerned that it was all coming on the side of the 'for' arguments. She asked them to balance the views out by finding some data that could support the arguments against.

Aaron found this graph.

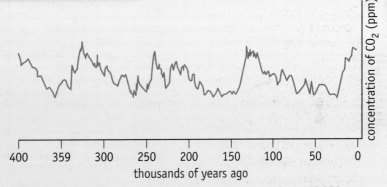

● **Figure 4** Changes in carbon dioxide levels over last 400 000 years

9 Why is it important to give a balanced view when reporting on this type of issue?
10 What does the data in Figure 4 show?
11 Based on the data presented here and throughout this topic create a summary table for Aaron and Alice set out as below. Remember, evidence is based on data and the explanation is the scientific reasons for the pattern in the data.

Claim	Evidence	Explanation

There are many options available of different types of materials. Part of developing the best products is making the right choice of material. The decision is based on the best materials for the job or purpose of the item. Understanding the properties of a range of materials allows chemists to offer advice.

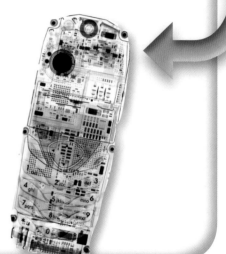

● **Figure 1** What sort of materials is your mobile phone made from?

→ Classifying materials

A material is a substance out of which things are made.

We can use the properties of materials to classify them into different groups. For example:

- solids, liquids and gases
- elements, compounds or mixtures
- metals and non-metals.

Natural or synthetic

We can also classify materials depending on whether they are natural or **synthetic** (i.e. manufactured).

Name	Solid, liquid or gas	Element, compound or mixture	Metal or non-metal	Natural or synthetic
iron	solid	element	metal	natural
crude oil	liquid	mixture	non-metal	natural
helium	gas	element	non-metal	natural
steel	solid	mixture (alloy)	metal	synthetic
wood	solid	mixture (composite)	non-metal	natural
glass	solid	compound	non-metal	synthetic
quartz	solid	compound	non-metal	natural
china (plates and bowls)	solid	mixture (ceramic)	non-metal	synthetic

→ Properties of materials

Properties that can make materials useful include:

- elasticity
- **absorbency**
- flammability
- reactivity with water and other chemicals.

We choose materials for particular uses because of their properties. There would be no point making a frying pan out of something with a low melting point – the pan would melt when you put it on the hob!

Databases of materials can hold a large amount of information, and can be searched to find materials with particular properties.

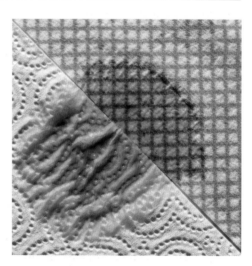

● **Figure 2** Which of these materials is more absorbent?

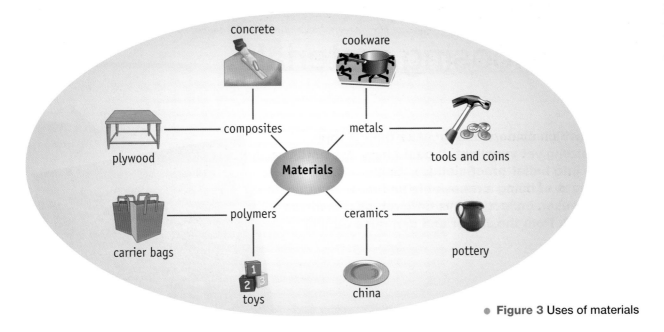

● **Figure 3** Uses of materials

→ Choosing materials fit for purpose

When decisions are made about the best material to use there are some key questions.

1 Will the material last?
2 Will the materials be easy to process and use?
3 Will the materials look right for the item?

Manufacturers also need to balance these ideas against cost, but by understanding how to investigate the properties of different materials they can make a choice based on data.

The advantage of using synthetic materials is that the mixtures used to make them can often be adapted for the specific purpose they are intended for. For example, steel is an alloy – made from iron, carbon and a variety of other elements. The exact amount of each substance in the alloy will determine its properties and therefore its suitability for a particular job. Steels can be grouped according to the percentage of carbon in the alloy.

Type of steel	Percentage of carbon	Properties	Uses
low carbon (mild steel)	0.07–0.25	easily cold worked	car bodies
high carbon (carbon tool steel)	0.85–1.2	strong and wear resistant	cutting tools, railway lines
cast iron	2.5–3.8	easy to cast but brittle	pistons and cylinders

? Questions

1 Why is metal a good choice for cookware?
2 What percentage of carbon is there in high carbon steel?
3 For what purpose would flammability be a useful property?
4 Name an advantage of a synthetic material.
5 For each of the following examples discuss the answers to the three key questions that are asked when deciding on a material to use:
   a) using metal to make a hammer
   b) using wood to make a desk.

✎ Show you can...

Complete this task to show that you understand about types of materials.

Regroup the examples of materials given in the table on page 110 into three new tables:

• solids, liquids and gases
• elements, compounds and mixtures
• metals and non-metals.

12.2 Choosing materials

Spider silk (which spiders use to make their web) is incredibly strong yet very light. It could have the potential to be made into bullet-proof clothing. Spider silk has the advantages of being a renewable and non-polluting product. However, collecting it is difficult, as it involves forcing the silk from the spider and can raise ethical questions.

● **Figure 1** What are the properties of spider silk?

Metals have been used since ancient times so scientists and manufacturers understand their properties well. New materials have also been discovered or created over the years and these offer exciting new applications.

→ Ceramics

As a non-metallic synthetic mixture, a ceramic can have some interesting properties. In general ceramics:

- are hard (wear resistant)
- are resistant to high temperatures
- are long lasting
- have low electrical conductivity
- have low **density** compared with metals.

The raw materials required to make a ceramic are easily available and therefore they are cheap materials. Traditional ceramics include bricks, tiles and clay. They are all heated to very high temperatures to cause a chemical change that forms the structure of the material. New ceramics are being created that have many uses including:

- **Space Shuttle** tiles
- safety glass windshields
- joint replacements
- electronics and communication.

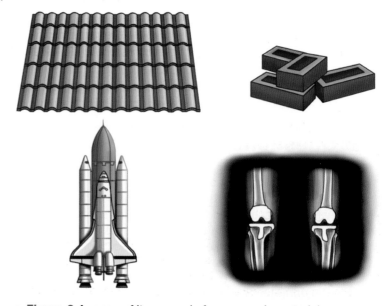

● **Figure 2** A range of items made from ceramic materials

→ Polymers

Crude oil is a mixture of chemical compounds made up of carbon and hydrogen. These substances are separated out and used as fuels, such as petrol and diesel, or as tar on roads. Some of them can be chemically linked together into a very long chain molecule called a polymer.

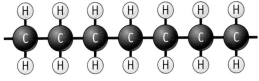

● **Figure 3** A polymer molecule

Synthetic polymers are also known as plastics and have a wide range of properties and uses. They can be made very flexible to suit the purpose of coating wires or wearing as raincoats, or they can be made solid for purposes such as plug sockets and furniture.

One of the big disadvantages of plastics is that they take many years to decompose. This means when plastic items are discarded at land fill sites, they stay there much longer than natural materials such as wood or paper.

→ Composites

A composite is any material that is made up of at least two other materials. The intention is to have the best properties of all materials combined into one item. Composites can be layered or formed from weaving the two materials together. A big advantage is that the mixtures can be adapted to suit the job of the item.

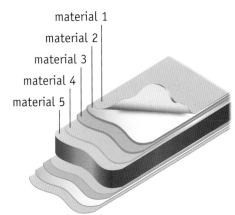

material 1
material 2
material 3
material 4
material 5

● **Figure 4** Making a composite material

Often composites are used where additional strength is needed. In these instances one substance can be reinforced with another. A good example is where steel bars are inserted into concrete pillars during construction of large structures.

Composite materials are now used widely on aircraft. Everyday examples of composites are bicycle helmets and surfboards.

? Questions

1 Give one property of:
   a) a ceramic
   b) a polymer
   c) a composite material.
2 Define the term 'polymer'.
3 Why are ceramics a good choice for cookware and plates?
4 Why are ceramics considered synthetic when they are made of clay, which comes from the ground?
5 What is the difference between a composite material and a ceramic?

✎ Show you can...

Complete this task to show that you understand the ideas behind choosing materials.

Summarise the advantages and disadvantages of choosing a plastic carrier bag over a paper one.

As new materials are designed with properties that have not even been imagined, scientists become very excited about new possibilities. What if we could deliver a medicine directly to a body cell or change the colour of a substance by passing an electrical current through it? Two groups of materials took technology to a new level.

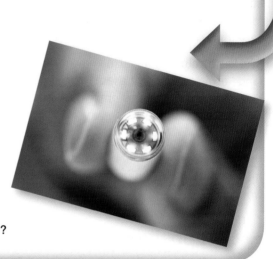

● **Figure 1** What could you see with a camera this small?

→ Smart materials

Smart materials are substances that can change their properties based on the surrounding environment.

Some will react to temperature – such as smart stitches that shorten at body temperature. The stitches can be made loosely and once at body temperature will shorten and close the wound. They will also dissolve after a set period of time, meaning you do not have to go and have them removed.

A special group called the shape memory alloys are smart materials that when bent out of shape can be heated to a high temperature to return them to the original shape.

Some can be affected by force – they may not conduct electricity when untouched but if squeezed they will conduct. This is useful in alarm sensors and speed controllers.

Others are changed by light – changing colour under certain lighting conditions or producing light when a small electrical current is passed through them.

1. glasses made from memory alloy

2. bend out of shape

3. place in hot water

4. returns to programmed shape

● **Figure 2** Shape memory alloys have many applications, including glasses frames and dental braces

● **Figure 3** A £10 note under ultraviolet light

→ Nanotechnology

The camera in Figure 1 is swallowed and can take real time video on its journey through the human body; tiny particles can create water-repellent clothing that will never get wet; and scientists are now able to create tiny balls of carbon atoms that could contain any chemical for delivery to a body cell.

Nanotechnology uses nanoparticles, which are tiny. They are measured in nanometres (nm) and there are 1 million nanometres in a millimetre. There are many advantages to be able to create particles of this size, but also some concerns.

Nanoparticles have special properties:

- They will not necessarily have the same properties as those of the substance when it is present in greater quantities, for example their colour.
- They have a very large surface area compared to their volume so act as catalysts for many reactions.
- They are so small that they do not reflect light so they cannot be seen.
- They can be added to fibres or used in composites.

These properties give rise to a variety of uses:

- cosmetics – UV protective nanoparticles can be added to face creams and sunblock without adding any colour
- medicines – nanoparticles will be absorbed more quickly and are used in some cancer treatments and vitamin supplements
- food packaging – nano-sized silver particles have an antibacterial agent and are added to food packaging to keep the food fresher for longer
- clothing – nanoparticles can repel water and dirt, making clothes stain and water resistant.

● **Figure 4** Waterproof fabric – nanoparticles in the cloth repel the water, causing it to bead and run off the material

Some people are concerned about using nanoparticles. The main reasons are that if they make good catalysts for many reactions, what effect could they have on our bodies? As they have different properties from the original substance in greater amounts, we cannot be sure of our predictions on their properties.

? Questions

1 How many nanometres are there in a millimetre?
2 List three changes in the environment that could change the properties of smart materials.
3 How do smart materials that are affected by light protect shops from taking fake money?
4 How can smart materials help with dental braces?
5 Explain why nanoparticles are easily absorbed and make good catalysts.

✎ Show you can...

Complete this task to show that you understand some of the uses of smart materials and nanotechnology.

Explain how smart stitches and sunblock containing nanoparticles could improve people's lives.

12.4 Designer materials

The adhesive used in Post-it® notes was created in 1970 when chemists were trying to make a new, strong adhesive. Instead they made a very weak adhesive that had no use. Four years later, one of the scientists involved needed a marker to keep his place in a book. He tried putting a little of the new adhesive on a piece of paper – and the Post-it note idea was formed!

● Figure 1 What are the advantages of a Post-it® sticky note?

→ Developing new materials

Scientists use their knowledge of chemistry to make new materials. They can produce materials with particular properties suitable for a specific job – with a specific technical purpose.

Specialists – such as polymer scientists or chemical engineers – work closely with designers, physicists, engineers or doctors to develop a new material.

New materials are made by:

- chemical reactions – making new compounds
- mixing two or more materials – making composite materials.

Examples of materials made from chemical reactions include:

- plastics (made from crude oil)
- synthetic fibres such as Lycra® and nylon
- medicines (pharmaceuticals)
- pesticides and fertilisers.

→ Materials for space travel

Tiles used on the Space Shuttle had to be strong, smooth, light and able to resist very high temperatures. A new type of ceramic compound was designed and produced.

When space travel was first developed, scientists had to develop materials that could withstand the stresses, strains and high temperatures involved. This led to the development of many new materials. One of these was Teflon® – now used as a non-stick coating for cookware.

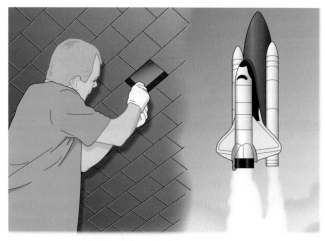

● Figure 2 Shuttle tiles had to be strong, smooth and light

→ Thermochromic and photochromic pigments

Thermochromic pigments change colour depending on their temperature. Photochromic pigments are sensitive to **ultraviolet light** – clothing coloured with these pigments can have interesting effects! Kettles that change colour as the water inside gets hotter are made from compounds containing thermochromic pigments.

● **Figure 3** What other jobs could these colour-changing pigments be used for?

Sometimes the new material does the same job as the material it replaces, but has an advantage such as:

- being less toxic to the environment
- using less energy in manufacture
- using a renewable resource.

→ Materials from crude oil

Many new materials are made from crude oil. The world's supply of crude oil is limited and will run out before long. The more oil we use to make new materials the faster it will run out and the less we will have for other uses.

Instead of using crude oil, renewable biological sources can be used to make bioplastics. Polylactide is made from sugar cane and can be used in packaging and disposable nappies. It is **biodegradable** – polymers made from crude oil are not.

Making any new material is likely to use large amounts of energy. This energy will probably be provided by power stations that burn fossil fuels. This will add to the greenhouse effect.

? Questions

1 What was Teflon® originally designed for?
2 Why would being biodegradable be a useful property?
3 Lycra® is an elastic fibre added to many clothes. What are the advantages adding this to fabrics?
4 Why might scientists want to make new medicines?
5 Explain why each of these features would be useful in a new material:
   a) being less toxic to the environment
   b) using less energy in manufacture
   c) using a renewable resource.

Show you can...

Complete this task to show that you understand the importance of designing new materials.

Outline how photochromic pigments could help protect people from the harmful effects of sunlight. Design the item and explain how it would work.

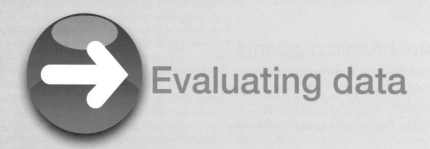

Evaluating data

→ Effective innovation

Innovation is when new solutions are generated to solve new problems. It often involves creative thought and, historically, many new materials come out of accidental discoveries.

Once a new material is discovered, the scientists have to set about showing it will meet the need. You are going to read some information about the discovery of a new material and review some of the scientist's data for yourself. You will be asked to say whether their conclusions match the data and what questions you would want to ask next to investigate the new material fully.

The material Lycra® was discovered in 1958 when scientists were looking for a substance to replace the natural polymer rubber. It is in fact a fibre that can be woven with any other material to add elasticity to an item of clothing.

1 Define 'elasticity'.

To investigate elasticity the company looked at percentage fabric growth. They stretched the fabric for a fixed distance and time and then measured the length after a 1-hour recovery period.

● **Figure 1** Quick-drying and figure-hugging swimwear has made Lycra® a well-known fabric

2 Sketch a series of diagrams to show this test.
3 What would you expect the results to be for a material that has good elasticity?

The results for percentage fabric growth:

Fabric	Percentage fabric growth			
	Try 1	Try 2	Try 3	Average
Lycra®	0	3	1	
polyester	28	35	30	
elastane	6	4	8	

4 Calculate the average percentage fabric growth.
5 Why do you think stiff fabrics like cotton are not included here?
6 Why would polyester not be described as elastic?
7 There are claims that when washed, Lycra® loses its elasticity. How could you adapt the investigation to test this idea?
8 Why would it be important to show it keeps its elasticity after washing?

Another selling point for Lycra® is that it is quick drying. In order to investigate this a moisture sensor can be used, which will measure the amount of liquid in a fabric. These work by measuring the electrical conductivity of the fabric, which directly relates to the moisture content. The meter converts the reading into a percentage moisture content.

These are the results:

| | Moisture content | | | | | | | |
| | Lycra® | | | | Wool | | | |
Time (min)	Try 1	Try 2	Try 3	Average	Try 1	Try 2	Try 3	Average
0	100	100	100		100	100	100	
5	60	59	63		90	85	91	
10	23	32	30		83	80	80	
15	2	8	3		71	62	72	
20	2	2	2		65	62	61	
25	2	1	1		62	62	60	
30	2	2	1		61	61	60	

Copy the table and complete the following analyses:

9 Identify any outlier results that do not fit the pattern by circling them in the table.
10 Calculate the averages.
11 Plot a graph of time on the x-axis and moisture (%) on the y-axis.
12 Describe what the graph shows.
13 Explain whether the data supports the claim that Lycra® is quick drying.
14 What other data would you need to support this claim?

7.1 Domestic electricity and power

Electricity is very useful – it can run everything in our homes from washing machines to games consoles. But whether you are doing the laundry or killing zombies, you are using energy. The energy used per second is the machine's power.

- **Figure 1** How many Christmas lights would double your household's power use?

→ Calculating power

The power of an electrical device (the energy it uses per second) is equal to the **electric current** flowing through it multiplied by the **potential difference** (p.d.) across it:

$$
\begin{array}{ccccc}
P & = & I & \times & \text{p.d.} \\
\text{watts (W)} & & \text{amps (A)} & & \text{volts (V)}
\end{array}
$$

For example, a toaster that runs on a current of 5 A when connected to the 230 V mains supply will use:

$$
\begin{aligned}
P &= 5A \times 230V \\
&= 1150W, \text{ or } 1.15\,kW
\end{aligned}
$$

When you buy a new toaster, its power rating will be printed on the box. In this way you can choose a high-power toaster (e.g. 2.3 kW) that would toast your bread faster.

→ Calculating current

One thing to remember in calculations about mains electricity is that the p.d. is always 230 V.

To calculate the current flowing through an appliance, you need to rearrange the power equation:

$$
I = \frac{P}{\text{p.d.}}
$$

Because the p.d. is the same for all mains appliances, the current is directly proportional to the power of the appliance. A high-power appliance draws more current from the mains than a low-power one. Electricians need to know this when wiring up a new house, because thicker wires are needed to supply larger currents.

- The wires that supply the light bulbs can be thin, because low-power bulbs draw a small current.
- The wire supplying an electric shower would be much thicker – a 9 kW electric shower would draw almost 40 A of current.

→ Calculating energy

The power of a device is the energy it uses per second, but we often need to find out the total energy it has used in the time it has been turned on.

> energy used   =   power   ×   time
> joules (J)   watts (W)   seconds (s)

If the power rating is given in kW (thousands of watts) then the energy is simply calculated in kJ (thousands of joules).

The time must be in seconds for the equation to give the right answer, so if you are given time in minutes, you must first multiply it by 60.

● **Figure 2** Which would use the least energy: a 2 kW kettle that took 3 minutes to boil, or a 1 kW kettle that took 6 minutes to boil?

? Questions

1 Calculate the power of the following appliances. All run on 230 V.
   a) a table lamp that draws a current of 0.04 A
   b) a laptop computer that draws 0.13 A
   c) a fan heater that draws 8 A
2 What current would a cable have to take if it was used to wire up:
   a) a radio rated at 1 W
   b) a television rated at 230 W
   c) a microwave oven rated a 1150 W?
3 What is the power rating of these appliances?
   a) a table lamp that uses 16 kJ in 30 minutes
   b) a hairdryer that uses 960 kJ in 8 minutes
   c) a lawnmower that uses 1200 kJ in 20 minutes

✎ Show you can...

Complete this task to show that you understand how the power rating of a device and the total energy that it uses are related by the time that it is switched on.

State why it is not a good idea:

a) to leave the lights on in an empty room
b) to leave a television on standby for long periods of time
c) to charge your phone overnight.

7.2 Paying for electricity

Electricity is not free. But you do not buy it in cartons at the supermarket, and it is not delivered by the milkman. So how do electricity companies find out how much you have used and charge you for it?

● **Figure 1** When you place a coin in the slot, are you paying for the ride or the electricity that powers it?

→ Pay for energy, not for power

When we pay for electricity, we pay for the energy that we have used, not the power of the device. So although a 2 kW hairdryer has around 10 times the power of a television, the total amount of energy we pay for is around the same as for the television. This is because we only use a hairdryer for a short time each day, while a television may be on for much longer.

We use the same equation for energy as on page 121, except that the units are different:

$$
\begin{array}{ccc}
\text{energy used} & = & \text{power} \times \text{time} \\
\text{kilowatt-hours (kWh)} & & \text{kilowatts (kW)} \quad \text{hours (h)}
\end{array}
$$

Household appliances are usually rated in kilowatts, and are often on for hours. Calculations using watts and seconds would give very unwieldy answers in joules.

For example, a 1000 W electric oven that is on for 1 hour would use:

$$
\begin{aligned}
\text{energy used} &= 1000\,\text{W} \times (1\,\text{h} \times 60\,\text{min} \times 60\,\text{s}) \\
&= 3\,600\,000\,\text{J}
\end{aligned}
$$

To simplify things we multiply the power in kW by the number of hours used. In this way, electricity companies measure energy in **kilowatt-hours** (kWh). They often refer to these simply as 'units' of electricity.

$$
\begin{aligned}
\text{energy used} &= 1\,\text{kW} \times 1\,\text{h} \\
&= 1\,\text{kWh}
\end{aligned}
$$

1 kWh of energy is equivalent to 3 600 000 J.

● **Figure 2** Does the meter record power or energy?

→ Reading the meter

Somewhere in your home is an electricity meter like the one in Figure 2. To find out how many units of electricity you have used, your electricity company needs to take a reading from your meter every 3 months. Usually a person does this, but meters are increasingly being read wirelessly, with no one needing to visit your home at all.

→ Counting the cost

The meter shows how many units of electricity have passed through it since it was installed. The electricity company takes the current reading and subtracts the previous reading. This tells them how many units have been used since the previous reading was taken. They then multiply this by the cost of each unit of electricity and send you a bill for this amount.

Example
Current meter reading: 23 188 units
Previous meter reading: 22 130 units
Energy used: 23 188 − 22 130
    = 1058 units
Cost of electricity
    = number of units used × cost per unit
    = 1058 units × 15 pence per unit
    = 15 870 pence
    = £158.70

? Questions

1 How many kWh would the following use?
  a) a 1 kW lawnmower used for 1 hour
  b) a 1 kW drill used for 0.5 hours
  c) a 2 kW heater used for 4 hours.

2 The table below shows meter readings for the Wells household over 1 year.

Date	Present reading (kWh)	Previous reading (kWh)	Units used (kWh)
March	31 057		1696
June	32 265	31 057	
September	33 319	32 265	
December		33 319	1674

  a) Copy and complete the table with the missing values.
  b) In which part of the year did the household use most electricity? Suggest a reason for this.

3 Brogan came home from work and made a cup of tea, using the kettle for 3 minutes. She washed her hair, then used her hairdryer for 15 minutes. She then used the kettle for another 3 minutes and watched television for 2 hours. The light was on in the living room for 4 hours during the evening. The following table shows the power of the things she used.

Appliance	Power rating (W)	Time used (h)	kWh
kettle	2000		
hairdryer	1500		
television	200		
light bulb	20		

  a) Copy and complete the table with the times each appliance was on and the number of units they each used.
  b) Calculate the total cost of the electricity used that evening (price of electricity = 15 pence per unit).
  c) Do your results suggest why people are slow to adopt energy-saving measures? Explain.

Show you can...

Complete this task to show that you understand when to use the different units for energy in an electrical context.

Give an example of a situation in which you would want to calculate energy in:

a) joules
b) kilowatt-hours.

7.3 Static electricity

The ancient Greeks knew that if you rubbed a bracelet made of amber with fur it would attract small objects, or even produce a spark. You can see sparks jump for yourself by taking off a polyester jumper in a darkened room.

● **Figure 1** Could you charge up a modern bracelet?

→ Charging by friction

When two objects rub against each other, electrons can transfer from one object to the other. If this happens, both objects become charged with **static electricity**.

- Electrons have a negative charge, so the object that gains electrons becomes negatively charged.
- The object that has lost electrons becomes positively charged.

This is called **charging by friction**. The charged objects exert a force on each other, called the **electrostatic force**. Only one in a million atoms in an object may lose an electron this way, but this is enough to exert a very strong force.

→ Attraction and repulsion

Negative charges repel other negative charges but attract positive charges. Positive charges also repel each other but attract negative charges. In summary, like charges repel while unlike charges attract.

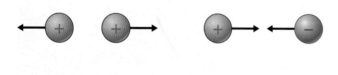

● **Figure 2** Like charges repel, unlike charges attract

● **Figure 3** Why does a balloon that has been rubbed on your head attract your hair?

→ An electric field

Do charged objects need to touch each other to attract or repel? You could try an experiment to find out. Rub a polythene rod with a nylon cloth and then suspend the rod from a string, halfway along its length. When you bring the cloth close to one end of the rod, this end will be attracted to the cloth before they actually touch. This is because the electrostatic force is a **non-contact force**. A charged object creates an **electric field**, which reaches out across space to attract or repel other objects at a distance.

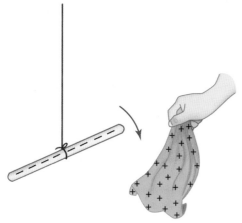

● **Figure 4** The electrostatic force is a non-contact force

● **Figure 5** Why can you charge a plastic ruler by rubbing it, but not a metal one?

Remember when answering questions on static electricity: only the negative charges (the electrons) move. The positive charges are fixed in place and cannot move. The only way we can make something positive is to remove electrons from it.

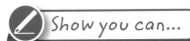

? Questions

1 If you rub a glass rod with a cotton cloth, electrons will jump from the glass to the cotton. Which object becomes negatively charged?
2 In Figure 4 the rod and the cloth have opposite charges. What would happen if they:
   a) were both positive
   b) were both negative?
3 A person jumping on a trampoline often becomes charged. Why does this make their hair stand on end?
4 When the trampolinist in Question 3 climbs down a wooden ladder, their hair will stay standing up until they reach the ground. If the ladder is made of metal, their hair will fall down as soon as they touch the top of the ladder. Explain why.

Show you can...

Complete this task to show that you understand static electricity.

Draw a diagram to show which way electrons move when a woollen cloth is rubbed against a nylon rod. (The cloth becomes negative and the rod becomes positive.) Will the rod now attract or repel the cloth?

Lightning is one of the most visible effects of static electricity, and one of nature's greatest spectacles. A lightning bolt is hotter than the surface of the Sun, and around 50 lightning strikes happen worldwide every second.

● **Figure 1** Does lightning really never strike the same place twice?

→ Charging by induction

If you run a plastic comb through your hair and then hold it near some tiny scraps of paper, they should jump up to the comb. In this experiment the scraps of paper are uncharged – the number of positive charges in the paper equals the number of negative charges. So why do the pieces of paper move?

The positively charged comb pulls some electrons towards the top of each piece of paper, making the tops negative and leaving the bottoms positive. This is called **charging by induction**. The comb then attracts the negatively charged top of the paper, so the paper jumps up to the comb.

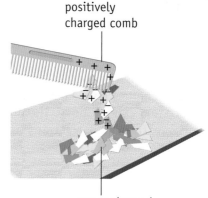

positively charged comb

paper charged by induction

● **Figure 2** The comb attracts the negative top of the paper more than it repels the positive bottom, so the paper jumps up to the comb

→ Using static electricity

Clingfilm is clingy because of static charges.

● When it is unrolled, some electrons leave the roll and stick to the sheet of clingfilm that has been removed.
● This electric charge induces a charge on anything it is wrapped around.
● The attraction between the charge on the film and the induced charge makes the clingfilm cling.

● **Figure 3** Will clingfilm stick to a metal bowl?

→ Lightning

When a storm cloud forms, air currents carry ice particles up to the top of the cloud. On the way up they rub against water droplets that are falling downwards. When these rub together, electrons move from the ice to the water. The ice particles carry positive charge to the cloud top and the water droplets carry negative charge to the cloud base.

The charged base of the cloud induces a charge in the Earth's surface. The potential difference between these two increases until it becomes so large – typically 100 million volts – that a giant spark jumps across the gap. This is lightning.

● **Figure 4** When lightning strikes sand it makes glass tubes called fulgurite. How do you think this happens?

The current in a lightning bolt is also very large – up to 30 000 A. This current heats the air to around 50 000 °C, making it expand so suddenly that a crack of thunder is heard.

→ Current and charge

Electric current is the rate of flow of charge. If the current from a battery changed from 1 A to 2 A it would mean that twice as much charge was flowing each second. You can work out the total charge that has flowed in a circuit by multiplying the current by the time for which it flowed.

> charge = current × time
> coulombs (C)   amps (A)   seconds (s)

The symbol for charge is Q, so we write this in symbols as:

> $Q = I \times t$

? Questions

1 Two charged objects repel each other. Can you tell if they are positively or negatively charged?
2 Will the storm cloud mentioned above induce a positive or a negative charge in the Earth's surface?
3 **a)** If a current of 1.5 A flows through a light bulb for 20 s, what is the total charge that has passed?
  b) If a charge of 90 C takes 30 s to pass along a wire, what current flows through the wire?
4 If you rub a balloon on your hair, you can stick it to the ceiling by static electricity. How does it stick when the ceiling has not been charged?

✎ Show you can...

Complete this task to show that you understand the difference between conductors and insulators.

Explain why a sheet of clingfilm gains a static charge when pulled off the roll, whereas a sheet of aluminium foil does not.

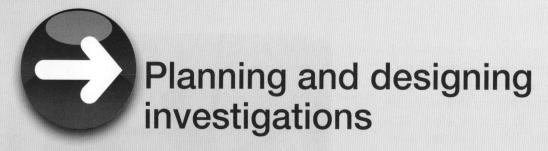

Planning and designing investigations

→ A shocking experiment

● **Figure 1** Are rubber soles conductors or insulators?

Sarah wanted to know why she always got a shock when touching the library door handle. Jamie suggested that she became charged up when her shoes rubbed against the library carpet, and the shock happened when this charge jumped to the metal door handle. Sarah decided to test this by rubbing a selection of materials against other materials.

Materials

The following is a list of materials that Sarah and Jamie found to test.

Cloths:
● nylon
● polyester

Rods:
● glass
● polystyrene (used in plastic cutlery)
● acrylic (used to make fake fingernails and aquariums)
● polypropylene (used to make food containers and carpets)

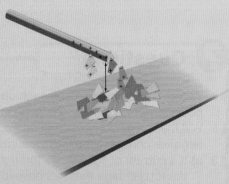

● **Figure 2** What combinations of cloth and rod could you try?

Your task

Sarah and Jamie wanted to find a simple way to test whether the rods were charged. They knew that a charged rod could attract small pieces of paper.

1. Draw a diagram to show how they could use this fact to see how much each rod charged up. (Hint: the less charged the rods were, the closer to the paper they had to be to pick it up.)

2. Draw a table that could be used to record how charged the rods became. Remember that Sarah and Jamie would want to test each different rod with each cloth.

Once they had found which combinations of rods and cloths gave them the greatest charge, Jamie wanted to know which were charged positively and which were charged negatively. Sarah said that the paper would attract to both positive and negatively charged rods, so they needed another method.

Sarah knew that if you rubbed an acrylic rod with a nylon cloth, the rod would always become negatively charged. They decided to use this to test the other rods.

3. Draw a diagram and write a short paragraph to describe how you could use the acrylic rod to tell if the other rods were positively or negatively charged. (Hint: You might like to suspend one of the rods from a stand and bring the other rod near. What would you see?)

Jamie thought that metal objects would still gain or lose electrons when rubbed with these cloths, but neither he nor Sarah managed to see any effect on an aluminium rod that they had.

4. Suggest how you might successfully test a metal rod to see what kind of charge it gains. (Hint: metals are conductors.)

Waves are not only found at the seaside – from light and sound, to the microwaves that carry our phone calls, they are constantly around us. You can even make a Mexican wave in the classroom.

● **Figure 1** The people move up and down, but which way does the wave shape travel?

→ Waves transfer energy

If you stand on a beach and look at the **waves**, you will see them continually coming towards you. But the waves break at your feet and the water comes no further. The waves may have travelled across the ocean but the water they travelled through stays put. If it did not, you, and the land behind you, would very soon be under water.

So what do waves transfer? It is not the **medium** (the substance that the waves travel through – water in this case). Waves transfer energy. This is why many companies are developing wave power machines, like the Pelamis.

● **Figure 2** The Pelamis generates electricity, but where does the energy to do this come from?

→ Transverse waves

In a water wave the medium **oscillates** up and down, while the wave pattern travels along horizontally. Waves like this are called **transverse waves**. You can make a transverse wave by stretching a slinky across the floor and shaking one end from side to side. In any transverse wave the oscillations of the medium are at 90° to the direction of motion of the wave.

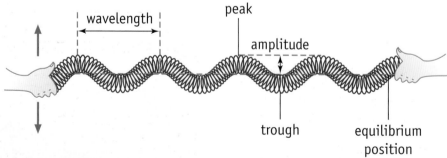

● **Figure 3** The **wavelength** is the distance between two **peaks** or two **troughs**. The **amplitude** is how far a peak moves from the **equilibrium position**

→ Wave reflection

When you shake the slinky you will see the wave reflecting from your friend's hand and coming back toward you. Waves on the sea can be reflected as well – if you stand at the top of a sea wall at high tide, you will see waves coming in to the wall, reflecting off it and travelling back out to sea. The angle between the incoming **wavefront** (or wave crest) and the sea wall is called the **angle of incidence** (i). It is always equal to the angle between the sea wall and the reflected wavefront, or the **angle of reflection** (r).

incoming waves

reflected waves

sea wall

● **Figure 4** The angle of incidence always equals the angle of reflection

→ Rogue waves

When two waves cross they pass through each other. Each continues on its way as if the other wave had not been there. But at the point where they cross, their amplitudes combine. If the peak from one wave meets the trough from the other, they can cancel each other out. But if the peak from one meets the peak from the other, a freak or **rogue wave** many metres high can form, which may capsize a ship.

● **Figure 5** A rogue wave. Could the Bermuda Triangle be explained by rogue waves?

? Questions

1 Imagine that you are on board a ship that is struck by a rogue wave. Tell your story, from the calm before the wave hit to its aftermath.

2 a) Look at the wave below, and find:

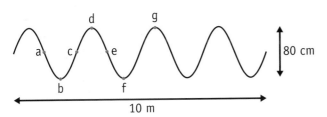

80 cm

10 m

   i) the amplitude of the wave, in cm
   ii) the wavelength of the wave, in m.
 b) Which points are peaks and which are troughs?
 c) Give the name for the horizontal distance from:
   i) b to f
   ii) d to g
   iii) a to e.
 d) Give the name for the vertical distance from:
   i) d to e
   ii) e to f.

3 a) Draw a diagram of a water wave. Label a peak and a trough, the wavelength and the amplitude. Add two more arrows, one to show the direction that the water oscillates, and another to show the direction that the wave transfers energy.
 b) What change would you expect to see in the wave if more energy was being transferred each second?
 c) What happens when a water wave hits a solid object?

✎ Show you can...

Complete this task to show that you understand the properties shared by all waves.

Explain whether a Mexican wave is a real wave or not.

8.2 Sound waves

If you put your fingers on your throat, you can feel your voice box (*larynx*). When you sing a note, you will feel your voice box vibrating.

● **Figure 1** What do you think an artificial voice box would sound like?

→ Making a noise

To produce sound you need to make something vibrate. There are many ways you can do this – musicians can pluck strings on a guitar, blow over a reed in a clarinet or hit the bars on a xylophone.

● **Figure 2** How do the vibrations travel from the instrument to your ears?

→ Spreading out

When an object vibrates it collides with the particles in the medium surrounding it. In this case the particles are air molecules. The particles in the medium then collide with each other, passing the vibrations along. So the vibrations (not the particles themselves) travel outwards from the sound source. This is what we call a **sound wave**.

The particles collide at the same **frequency** that the source collides with them, so the frequency of the sound wave is equal to the frequency with which the source vibrates.

Frequency is measured in oscillations per second, or **Hertz** (Hz). For example, an oscillation that occurs 50 times per second has a frequency of 50 Hz.

Sound needs a medium such as air to travel through. In a vacuum there are no particles to pass the vibrations along. You can show this by pumping the air out of a bell jar that has a ringing alarm clock inside.

● **Figure 3** Light can travel through a vacuum but sound cannot, so you can see the clock ringing but cannot hear it

→ Longitudinal waves

Sound waves are a series of **compressions** (areas where the particles are closer together than usual) and **rarefactions** (where the particles are further apart). You can make a similar wave move along a slinky if you hold one end and shake it towards and away from your partner at the other end. Because the medium oscillates along the direction the wave travels these are called **longitudinal waves**.

The sound wave has the same frequency as the oscillations that produced it. So if the source vibrates faster, it produces a sound with a higher frequency.

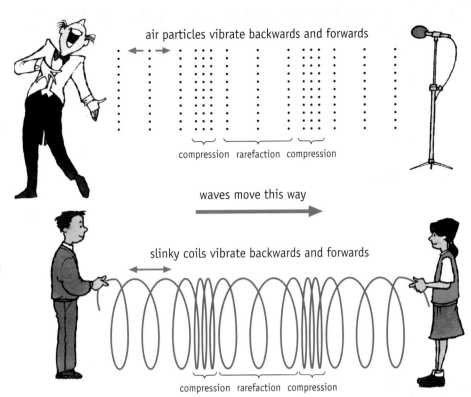

air particles vibrate backwards and forwards

compression   rarefaction   compression

waves move this way

slinky coils vibrate backwards and forwards

compression   rarefaction   compression

● **Figure 4** In longitudinal waves the particles vibrate parallel to the direction of wave motion

→ Detecting sound

You can think of sound as a transfer of energy from the source to the receiver. We hear the sound because vibrating air particles collide with our **eardrums**, making them vibrate. Our ears convert these vibrations into nerve impulses that can be interpreted by the brain. The **diaphragm** of a microphone does a similar thing, turning the vibrations into an electrical signal. Because sound energy spreads out in all directions, the sound becomes quieter the further you are from the source.

? Questions

1 What vibrates to produce sound in a drum, a harp and a cymbal?
2 In 1969 the first astronauts landed on the Moon. They wore space suits because there is no air on the Moon.
   a) Why did they need radios to talk to each other, even though they were very close?
   b) If their radios had broken, what could they have done to hear each other speak?
3 Sound clearly travels through gases, such as air. How could you prove that it also travels through solids and liquids?
4 Sound is a longitudinal wave, whereas water waves are transverse. Explain, using diagrams if necessary:
   a) how longitudinal waves resemble transverse waves
   b) how longitudinal waves differ from transverse waves.

Show you can...

Complete this task to show that you understand what determines the frequency of a sound wave.

Explain why the sound produced by a ruler changes when you 'twang' it over the edge of a desk and shorten it at the same time.

133

Have you ever noticed how a violin, a cello and a double bass vary both in size and in pitch? Larger instruments produce sound waves with longer wavelengths, giving lower notes.

● **Figure 1** Which organ pipes make the highest notes and which make the lowest notes?

→ Musical notes

The **pitch** of a musical note is related to the frequency of the sound wave. The frequency is the number of complete waves that are produced each second. The higher the frequency, the higher the pitch. If the frequency of a sound doubles, our ears hear the same note, but one **octave** higher. When we make or listen to music, our brains are doing complicated maths without us realising it.

(a)

(b)

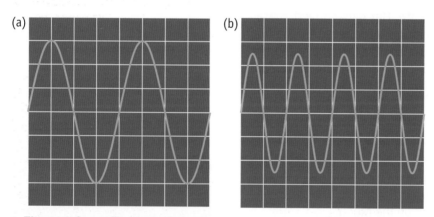

● **Figure 2** An oscilloscope shows how many waves occur in a certain time

On an **oscilloscope** screen, the space between two peaks is not the wavelength, but the time between one wave and the next. So the number of waves on the screen tells you how many waves pass in a set time. The more waves you see on the screen, the higher the frequency.

➜ How high can you hear?

A loudspeaker connected to a signal generator can play sounds over a large range of frequencies. If you try this, you should start to hear a low humming sound at around 20 Hz. As the frequency gets higher, so does the pitch. The sound starts to become harder to hear at 15 000 Hz (15 kHz). Your teacher may not hear it beyond this because sensitivity to high frequencies drops off as you age. At around 20 kHz you should no longer hear the sound. Sound above 20 kHz, the upper limit of human hearing, is called **ultrasound**.

If you have a dog, you may also own a 'silent' dog whistle. This makes a noise above the range of human hearing, at an ultrasonic frequency. Dogs can hear up to 60 kHz, but cats can hear even higher, up to 80 kHz.

Sounds with frequencies below 20 Hz are called **infrasound**. Elephants are thought to communicate with distant herds through infrasound.

➜ Sound and energy

When sound or ultrasound is absorbed it can increase an object's thermal store of energy. Ultrasound is used in this way by physiotherapists to treat damaged tissue deep within the body.

The energy absorbed can also increase an object's kinetic store of energy, so ultrasound can dislodge dirt or dust from delicate objects. The objects to be cleaned are simply placed in a water bath through which ultrasound is passed.

The Mosquito is a loud noise beyond an adult's range of hearing but within the range of younger people. It was meant to prevent teenagers from loitering outside shops. Teenagers had the last laugh, though – installing the Mosquito ring tone on their mobile phones meant that teachers could not hear them ring in lessons!

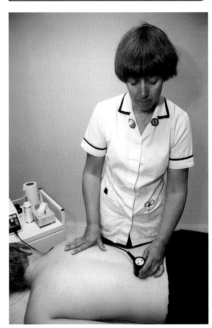

● **Figure 3** A physiotherapist using ultrasound. What other uses of sound can you think of?

? Questions

1 a) What is the range of human hearing?
  b) Name four animals that can hear higher frequencies than humans.
2 Look at Figure 2.
  a) Which screen shows:
     i) the higher amplitude
     ii) the higher frequency?
  b) How does the frequency of Figure 2b compare with Figure 2a?
  c) How does the pitch of the note in Figure 2b compare with Figure 2a?
3 An organist played a note that had a small amplitude and a large wavelength. Copy and complete the following sentence in a way that describes the note best:
  The note was (quiet/loud) and (low/high) pitched.
4 State how many octaves a musical note goes up by if the frequency:
  a) doubles
  b) quadruples
  c) multiplies by eight.

 Show you can...

Complete this task to show that you understand how the Mosquito 'youth deterrent' was intended to affect young people but not adults.

Describe what might have happened when the following customers tried to enter a shop that had the Mosquito:

1 a customer with a baby or a toddler
2 a blind customer with a guide dog.

Like other waves, sound travels at a finite speed. This is different in solids, liquids and gases. The speed of sound in air also changes due to air pressure, humidity and temperature.

● **Figure 1** Should you start your stopwatch when you see the smoke, or when you hear the bang?

→ Speed of sound

In a thunderstorm, we see the flash of lightning before we hear the roll of thunder. The light reaches us almost instantaneously, but the sound takes some time to reach us because it takes time for the compressions and rarefactions to pass through the air.

→ Echoes

An **echo** occurs when a sound wave reflects off a surface and returns to its source. Some animals use **echolocation** for hunting.

A bat emits ultrasonic clicks, which bounce off its prey, such as moths. When it hears the echo, its brain works out how far away the prey is, and builds up a detailed picture of the world around it – even in total darkness! We do not hear bats hunting because the clicks are ultrasonic; bats can hear frequencies up to 90 kHz.

Dolphins do the same thing to catch fish, using frequencies up to 150 kHz.

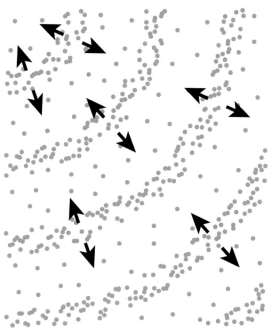

● **Figure 2** A sound wave travels because the particles bump into each other. The further apart they are, the longer this takes

→ SONAR

SONAR stands for SOund Navigation And Ranging, and can be used by ships to find the depth of the sea, or how far a shoal of fish is beneath them, see Figure 3. A **transducer** (a device that converts electrical signals into sound, or vice versa) sends out a pulse of ultrasound and detects any echoes. A computer measures the time to receive the echoes and calculates the distance from the equation:

distance = speed × time
m   m/s   s

Note that the speed of sound in water, which is 1.5 km/s, must be used. This is five times faster than the speed of sound in air. Also, the time measured by the computer is the time for the sound to reach the object and return, so half of this value must be used in the calculation.

→ Ultrasound scans

Ultrasound scanners can show images from within the womb in amazing detail. Ultrasound waves are sent out by a probe, and reflect off everything they encounter inside the mother's body. A computer builds up an image from the reflected waves in a similar way to a SONAR system.

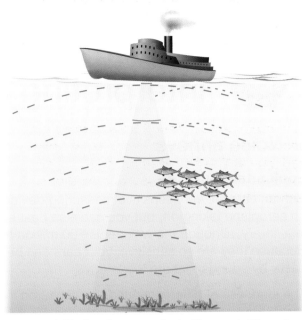

● **Figure 3** Each pulse can reflect off several different objects. The echo that takes the longest time to return is always from the seabed

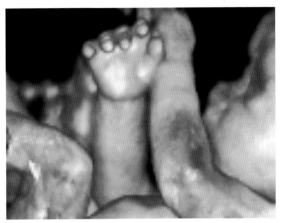

● **Figure 4** How else can you take pictures of the inside of the human body?

? Questions

1 In a thunder storm, the flash of lightning and sound of thunder are made simultaneously. Explain why we always see the lightning before hearing the thunder.

2) The sound of a thunderclap travels at approximately 330 m/s.
   a) If you count 3 s between seeing the lightning and hearing the thunder, how far away is the storm?
   b) How many seconds would you count if the storm was 3 km away?

3 A ship's SONAR system sends a pulse of ultrasound to the seabed; 0.02 s after the pulse leaves the ship, an echo is detected. Then 0.004 s later a second echo is detected. The speed of sound in water is 1500 m/s.
   a) Which echo came from the sea floor, and which came from a shoal of fish?
   b) Calculate the depth to the sea floor.
   c) Calculate the depth to the shoal.
   d) Why do you think the speed of sound in water is so much faster than the speed of sound in air?

✎ Show you can...

Complete this task to show you have understood how SONAR can find the depth of the sea floor using echoes and the equation:

$$speed = \frac{distance}{time}$$

a) Explain how the equation should be rearranged to calculate a distance.
b) Explain why you cannot use 330 m/s for the speed in this equation.
d) Explain why the value you use for time must be half of the value measured by the SONAR machine.

Presenting and interpreting data

→ Noise annoys

Look after your ears

Human ears are very sensitive measuring devices. You can hear an aeroplane taking off, but you can also hear someone whisper in a library, although the whisper is 100 million times quieter! Because they are so sensitive, your ears can be damaged very easily by sounds that are too loud.

The loudness of different sounds is measured in decibels (dB):

● **Figure 1** Why must you wear ear defenders if you work in a noisy environment?

Source of noise	Sound level (dB)	Number of times as loud as threshold of hearing	Comment
threshold of normal hearing	0	1	
breathing	10	10	barely audible
whisper, rustling leaves	20	100	
bird calls	40	10 thousand	
normal conversation	50	100 thousand	
laughter	60	1 million	
hairdryer	70	10 million	
lorry	80	100 million	possible damage to hearing with 8 hours continuous exposure
MP3 player at half volume	90	1 billion	likely damage to hearing with 8 hours continuous exposure
tractor	100	10 billion	serious damage to hearing possible with 8 hours continuous exposure
MP3 player at maximum volume	110	100 billion	average threshold of pain
ambulance siren, thunder	120	1 million million	
shotgun firing	130	10 million million	
jet engine at 30 m	140	100 million million	permanent damage caused by even short-term exposure; loudest permitted exposure with use of maximum ear protection
banger (firework)	150	1 thousand million million	

Measuring loudness

Because the noises we experience can be millions or **billions** of times louder or quieter than each other, a manageable unit is needed for measuring loudness. The table above compares various sounds with the quietest sound that the human ear can hear. A sound that is ten times as loud would be measured at 10 dB, but 100 times as loud would be 20 dB, and 1000 times as loud would be 30 dB.

To ensure that nobody suffers any hearing damage, schools, factories, nightclubs and other organisations employ Health and Safety Consultants to explain the dangers. The consultant would present a selection of information from this table in a form that their clients could understand. One day they might be advising staff in a scientific laboratory, the next they might be in a factory, and another day they might be designing posters to put up at a nightclub. They would always need to put the correct information across, but in ways that would enable each different audience to understand the risks of not wearing appropriate ear protection in a particular environment.

● **Figure 2** What other places can you think of where the noise might be dangerously loud?

The louder the noise, the less time you should endure it:

Sound level (dB)	Maximum daily exposure time (h)
90	8
95	4
100	2
105	1
110	0.5
115	0.25

Your task
Presenting the information

1 Imagine that you are a Health and Safety Consultant. Present the data in these two tables in a selection of the ways listed to the right.

In each case, decide who the audience is (who your graph, poster or leaflet is aimed at) and how much scientific training are you assuming they have.

Remember that you do not have to include all of the data in the table in your presentations – you might wish to pick out the most relevant for your particular audience.

If you are working as a group, you can divide these tasks among yourselves.

Possible ways of presenting the data:

A 'Sound Level (dB) v Source of Noise' could be presented as:
  • a bar chart
  • a pie chart
  • a line graph
  • a bubble chart*.
B 'Number of Times Louder v Source of Noise' could be presented as:
  • a table.
C 'Maximum Daily Exposure Time (h) v Sound Level (dB)' could be presented as:
  • an X-Y plot with a curved best-fit line.
D All of the above could be presented as:
  • a table
  • a piece of creative writing, which could include stories or poetry, warning of the dangers
  • an FAQ page on a website.

* To draw a bubble chart, draw circles with sizes in proportion to the sound levels in the table. Colour code and label each bubble with the name of the sound source.

Comparing the presentations

2 When your group has finished preparing your graphs, charts and creative writing, compare these different ways of presenting the data with each other. Each member of the group should say why their method is a good way of presenting the data. After all group members have spoken, decide on which methods of presenting this data the Health and Safety Advisor should use.
3 Why would a graph of 'Number of times as loud as threshold of hearing' on the vertical axis plotted against 'Source of Noise' on the horizontal axis be of no use?

9.1 Travelling light

When you look at the night sky you see some objects that are sources of light and others that are not. The Sun and other stars are sources, but the planets and the Moon only reflect sunlight.

● **Figure 1** Name two other objects that are sources of light, and two that are only seen by reflected light.

→ Particles or waves?

Since the ancient Greeks, people have argued about whether light is a stream of particles or a wave. It was not settled until 1800, when Thomas Young discovered he could make the amplitudes of two light waves combine to give a brighter light, or cancel out, leaving darkness. This is similar to two waves on the sea combining to form a rogue wave (see Topic 8 page 131). Also, like sound waves travelling through the air, two beams of light can pass straight through each other without either being affected by the other beam. Could this happen if light was a stream of particles?

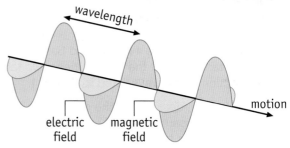

● **Figure 2** A light wave consists of oscillating (continuously varying) electric and magnetic fields, known as an 'electromagnetic wave'

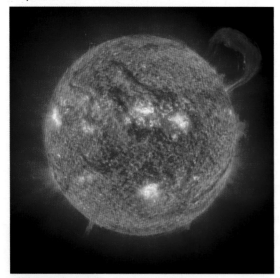

● **Figure 3** What would it be like if you could hear the Sun as well as seeing it?

→ Seeing stars

Light waves are transverse, like water waves (their oscillations are at right angles to the direction the wave travels) but they do not travel by the oscillation of particles. So, unlike water and sound waves, light waves do not need a medium (a substance) to travel through – light can travel through a vacuum, as it does in space.

This means that light can reach us from the Sun, whereas sound does not – we do not hear the enormous explosions that appear on the Sun's surface.

→ Reflection

Light travels in straight lines away from a **source**. It is useful to think of a light ray as a very narrow beam of light. A ray diagram with arrows shows the path of light rays from a source to an object, and what happens after they touch the object.

When a light ray hits the surface of a mirror it bounces off at the same angle. We call this reflection.

When we talked about reflection of water waves in Topic 8 we called the angle between the incoming wavefront and the reflector the angle of incidence. With light we measure angles to the light ray, which is at 90° to the light's wavefronts, so we construct a '**normal**' line, at right angles to the mirror, and measure our angles of incidence (i) and reflection (r) from this. ('Normal' means 'perpendicular' i.e. at right angles to the mirror.) Just as in Figure 4 on page 131, which shows water waves reflecting, the angle of incidence is equal to the angle of reflection.

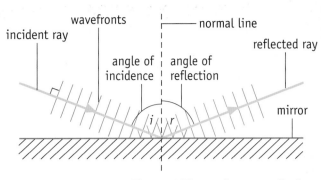

● **Figure 4** The ray is perpendicular to the wavefronts, so the normal line is perpendicular to the mirror

→ A cosmic speed limit

Light also travels much faster than any wave needing a medium. The speed of light is 300 000 km/s, which Einstein proved is a 'cosmic speed limit' that cannot be broken.

Light travels a million times faster than sound. It would take a light beam 0.003 s to travel the length of the UK.

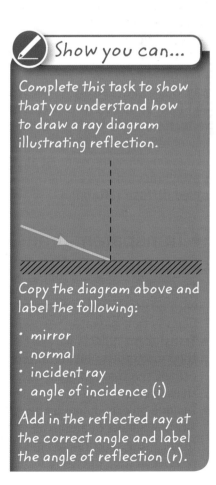

Show you can...

Complete this task to show that you understand how to draw a ray diagram illustrating reflection.

Copy the diagram above and label the following:

- mirror
- normal
- incident ray
- angle of incidence (i)

Add in the reflected ray at the correct angle and label the angle of reflection (r).

? Questions

1 State the angle of reflection in the following situations:
   a) when light strikes a mirror with an angle of incidence of 30°
   b) when light strikes a mirror with an angle of 70° between the incident ray and the mirror surface
   c) when light strikes a mirror with an angle of 40° between the incident ray and the normal line.

2 Light travels at 300 000 km/s. How long does it take light to reach us from these objects:
   a) the Moon, 384 400 km away
   b) the Sun, 149 600 000 km away?

3 a) Light is an electromagnetic wave. Do some research to find out what other electromagnetic waves there are. Then write a short paragraph explaining how they relate to visible light, especially in terms of their frequency, wavelength and speed, and their uses.
   b) Which two types of electromagnetic wave are closest to visible light?

When you clean your teeth in the morning, your reflection holds its toothbrush in the opposite hand to you. Your image is back to front (*laterally inverted*).

● **Figure 1** In what ways is your image in a mirror the same as you?

→ Specular and diffuse reflection

Both shiny surfaces and white surfaces reflect nearly all the light that falls on them, but images only form in shiny objects such as mirrors, and not in white objects such as a sheet of paper.

The reason is that shiny surfaces are very smooth. They reflect the rays in a regular way. Some of these rays enter your eye and form an image. This is called **specular reflection**. Light hitting the paper obeys the law of reflection at every point it hits, but the points on the surface are all at different angles so the reflected rays are not parallel. This is called **diffuse reflection**.

specular reflection   diffuse reflection

● **Figure 2** On a microscopic level a sheet of paper is rough. The reflected rays scatter in lots of different directions

→ Transparent, translucent or opaque?

- **Transparent** objects allow light to pass through them with no scattering. You can see through them easily. Glass is transparent.
- **Translucent** objects let most of the light through but they scatter it in lots of different directions, so you cannot see through them clearly. Tracing paper is translucent.
- **Opaque** objects do not let any light through at all. They reflect or absorb all of the light that lands on them. If they absorb the light, their thermal, chemical or electric store of energy increases. Wood is opaque.

● **Figure 3** Which of these materials are transparent, which are translucent and which are opaque?

→ Images in mirrors

Images in **plane** (flat) mirrors are back to front, as you will have seen. They are also always the same size as the object being reflected, and the same distance behind the mirror as the object is in front of it.

The rays from the object reflect at the same angle that they strike the mirror. If you trace the reflected rays backwards you will find the position of the image.

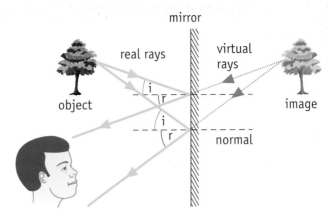

● **Figure 4** The image in a mirror is known as a **virtual image** because the light rays do not actually reach the image

→ Bending light

When a light ray passes from one medium to another it can change direction. We call this **refraction**.

When the ray passes from air to glass (or water) it bends *towards* the normal line. When it passes back into air it bends *away from* the normal line. You can see both of these effects when a ray of light passes through a rectangular glass block.

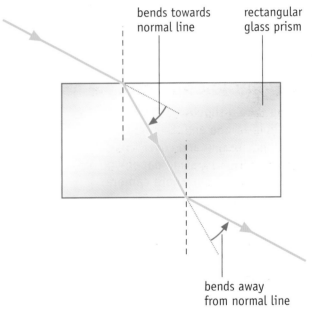

● **Figure 5** A light ray passing through a glass block refracts (bends) as it slows down and speeds up again

? Questions

1 Why is the sign on the front of an ambulance written in reverse?
2 a) A boy stands 50 cm in front of the bathroom mirror to brush his teeth. How far away from him is his image?
  b) He finishes brushing his teeth and walks away from the mirror at 2 m/s. How fast does his reflection move away from him?
3 A motor boat cruised across a calm lake. Its motor churned up the water into foam. Why did the foam look as white as snow, while the lake reflected like a mirror?
4 The bathroom mirror in Question 2 hangs on a solid wall. Use this fact to explain why the boy's reflection is a virtual image and not a real image.

✎ Show you can...

Complete this task to show that you understand the difference between the terms 'transparent', 'translucent' and 'opaque'.

Write a paragraph to explain the three terms, using different examples to those given in this topic.

9.3 Focusing light

A camera obscura is a darkened room with a small hole in one wall. Light from outside enters the room through the hole, which projects an upside-down image onto the wall opposite.

● **Figure 1** How could you make the image in a 'camera obscura' permanent?

→ Pinhole cameras

If you take a shoebox and make a tiny hole in one end, then cut out the opposite end and cover this larger hole with tracing paper, you will have made a **pinhole camera**. Any light that enters the box through the pinhole will end up on the tracing paper, which acts as a screen, as in Figure 2.

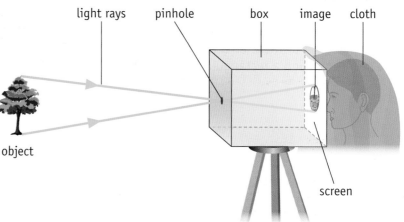

● **Figure 2** The image is formed where light from the object reaches the screen

To use your pinhole camera, look at the screen as you would look at the back of a digital camera. Cover your head and the camera with a cloth, and point the pinhole at a bright object such as a light bulb or the window.

You should see an upside-down image on the tracing paper. This is a **real image**, because the light really does reach it. How else could you describe the image?

You can investigate what happens to the image when you make the hole bigger.

→ The convex lens

When parallel rays of light pass through a **convex lens** they bend so that they come together and cross. The lens is also called a converging lens because the light converges at the **focal point**.

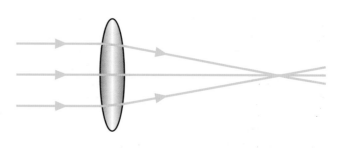

● **Figure 3** A convex, or converging, lens brings light together at the focal point

→ Cameras with lenses

Take your pinhole camera and enlarge the pinhole until it is a few centimetres across. Hold a convex lens over the hole and look at the screen as you did before. How does the image differ from last time?

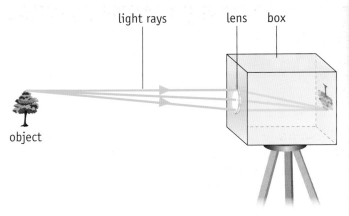

light rays   lens   box

object

● **Figure 4** All rays from the top of the object are focused to the same point on the screen, forming a bright, sharp image

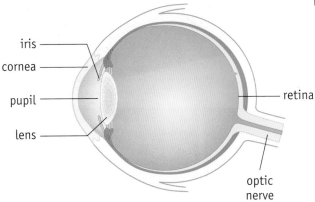

iris
cornea
pupil
lens
retina
optic nerve

● **Figure 5** The human eye has all the features of a lens camera

→ The eye

The human eye resembles a lens camera. In both, a lens focuses light to form an image. In a digital camera, the image is detected by a sensor and causes a change in its electrical store of energy. In an old-style camera, the image is formed on a light-sensitive film, causing a change in its chemical store of energy. In the eye, the image is formed on the **retina**. Retina cells absorb the light, increasing their chemical store of energy and sending signals along the **optic nerve** to the brain.

❓ Questions

1 If you stood inside a camera obscura and walked between the hole and the screen on the opposite wall, what would happen?

2 Complete this sentence about the pinhole camera: A smaller hole gives a .... image; a larger hole gives a .... image.

3 Copy and complete the following sentence:
In both the camera and the eye, the image is .... than the object, it is .... or ...., and in full ....

4 A reflection in a mirror is a virtual image, whereas the image on the screen of a pinhole camera is a real image. State whether the following are real or virtual images and explain why.
   a) The picture on a cinema screen
   b) The image on an interactive whiteboard

5 Do some research to find out the purpose of the following parts of a camera (on the left). Then match each with the part of the human eye (on the right) which does the same job.
   a) lens cap   **i)** iris
   b) aperture   **ii)** pupil
   c) diaphragm   **iii)** eyelid
   d) lens   **iv)** retina
   e) sensor or film   **v)** lens
   f) shutter

✏️ Show you can...

Complete this task to show that you understand how a pinhole camera or a **camera obscura** forms an image.

State and explain what would happen to the image if either camera was made bigger, i.e. the screen was moved further away from the pinhole.

145

9.4 Coloured light

Shine a ray of white light through a triangular prism and you see all the colours of the spectrum (rainbow). The prism splits the light, refracting the higher frequency (shorter wavelength) violet light more than the lower frequency red light.

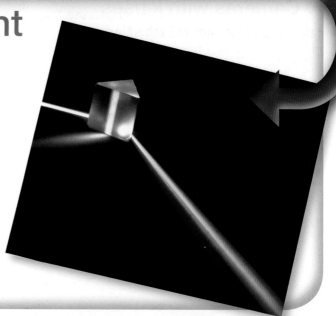

● **Figure 1** What would you see if you recombined these colours?

→ Combining colours

The **dispersion** of white light through a **prism** shows us that white is made up of seven different colours. In fact, our eyes can only really see three pure colours – red, green and blue. That's why we call red, green and blue the **primary colours of light** (different to the primary colours of paint). All of the other colours that we see are just combinations of the primaries at different brightnesses.

If you shine red, green and blue lights on a screen and overlap them, you get white light. Where two primary colours overlap, you get **secondary colours**. Red and green produce yellow, red and blue produce magenta, and blue and green produce cyan.

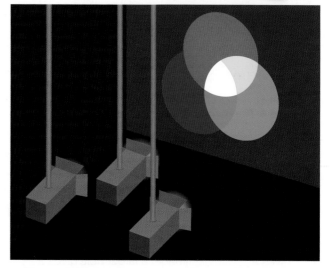

● **Figure 2** Mixing the primary colours of light gives white – not what you get from the primary colours of paint

→ Colour vision

There are three types of specialised cell in your retina called **cones**. Each type detects one of the three primary colours of light. The cones send signals to your brain, which interprets the actual colour of the object you are looking at.

If your retina lacks one type of cone you will be colour blind. The sets of dots in Figure 3 test for colour blindness.

Your cone cells do not function well in dim light, so you see in black and white.

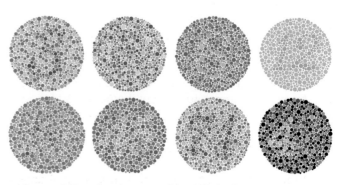

● **Figure 3** People who are colour blind will see these differently from those who are not

→ Reflecting colours

Objects appear the colour they do because they absorb some of the colours of white light but reflect others.

- A red, green or blue surface only reflects its own colour.
- A yellow, magenta or cyan surface reflects its own colour and the two primary colours that make it up.
- A white surface reflects all colours of light.
- A black surface does not reflect any light.

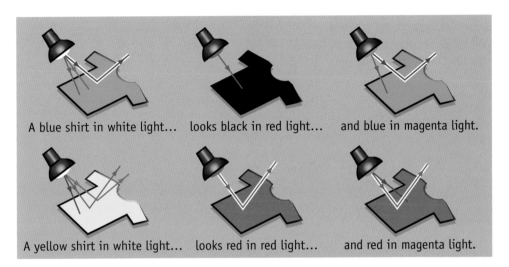

A blue shirt in white light...   looks black in red light...   and blue in magenta light.

A yellow shirt in white light...   looks red in red light...   and red in magenta light.

● **Figure 4** The shirts appear differently under different coloured lights because of the colours they reflect or absorb

A blue shirt under white light absorbs the red and the green but reflects the blue. This enters your eyes so the shirt looks blue. Under red light it would look black, because it would absorb the red light and no other light would be reflected. Can you see why it would appear blue under a magenta light?

? Questions

1 'Richard of York gave battle in vain' is a famous mnemonic (memory aid) for remembering the colours of the **spectrum**: red, orange, yellow, green, blue, indigo and violet. But if you are not that good at History, you might not remember the mnemonic. Invent a new one, which you will find easier to remember.

2 What colour would the following appear:
   a) green grass under yellow light
   b) red tomatoes under blue light
   c) yellow bananas under red light?

3 Why does mixing the primary colours of light give you white light, whereas mixing the different colours of paint gives you dark brown paint?

✏ Show you can...

Complete this task to show that you understand how the three primary colours of light combine.

1 Copy the three overlapping circles from Figure 2.
2 Label the primary colours (1) and the secondary colours (2).
3 Label the colours with their names.

Building scientific awareness

→ Insights into sight

Finding out what people think

You are going to build a model to describe how we see things. This process will take several stages. You should be open to changing your ideas as you progress.

To prepare
If you are doing this task as a class, copy Figure 1 onto a blank page in your book.

If you are doing this task for homework, copy the diagram onto several sheets of paper, or several blank pages in your book.

How do we see the dog?
Think about how you see the dog. Add to your diagram anything you like to illustrate this. You could use lines or arrows, for example. You must not look at anyone else's ideas until you have finished drawing your own.

If you are doing this task for homework, put your own ideas on the first of your diagrams, then ask other members of your family to complete a diagram of their own. Remember not to let anybody see anyone else's ideas.

Comparing ideas
Now compare your ideas with those of your classmates, or of your family. Write descriptions of the different ideas that people have come up with.

● **Figure 1** Copy this diagram and use it to explain how people think we see

Attention to objectivity

Scientists must be objective in working out their theories. The chances are that some of these ideas are a bit like the ancient Greeks' ideas. In the 5th century BC, the philosopher Empedocles stated that we can see objects because light shoots out of our eyes and touches them.

A few simple tests, however, would have shown that this idea was wrong. Answer all of the following questions in your books.

1 If light comes out of your eyes and touches the object you are looking at, would you be able to see in the dark?

● **Figure 2** Why do a cat's eyes shine in the dark?

Empedocles did actually think of this, and he modified his explanation, as any good scientist would. He said that light enters our eyes from the Sun, and then shoots back out to the object we are looking at.

2 If light travels from the Sun to your eyes, and then to the object you are looking at, would you be able to see it if it was placed inside a dark cave?

3 What if you were in the cave? If light from the Sun could not reach your eyes, but you could look out of the cave entrance, would you be able to see out of a dark cave?

● **Figure 3** If the Sun shines on our eyes, then would rays from our eyes enable us to see into the cave?

● **Figure 4** If we need the Sun lighting up our eyes to see, how could we see out of a dark cave?

Holding on to ideas and models that are shown to be false by evidence and experiment is not being objective.

Scientific theories develop

The camera obscura is like a giant pinhole camera, as large as a room. Its invention was a milestone in the developing theory of how we see things.

4 If light came out of a person's eyes, what would the observer inside the camera obscura see when looking at the screen?

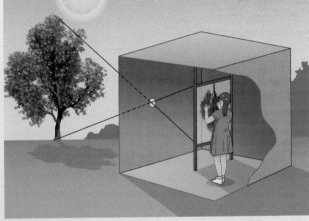

● **Figure 5** The camera obscura only works if light is travelling from the object to the camera

Publishing results and peer review

In any scientific enquiry, the results obtained are published so that other scientists can discuss them. This helps to decide whether the theories put forward are correct, or whether they can be improved.

Your final task is to decide which of the original ideas is the most correct. Write a paragraph describing why, and how it should be modified (if it needs it) in order to make it more correct. Report back to your class to see which ideas others have decided are the best. Has anyone, like a good scientist, modified their ideas based on the evidence you have seen?

10.1 Turning moments

To make a see-saw balance, you do not actually need to have the same weight on each side. You learn quite early in life that an adult can balance a child, as long as the adult sits closer to the middle of the see-saw.

● **Figure 1** What is wrong with this picture?

→ Moment of a force

The turning effect of a force is called its **moment**. You can work out the moment by multiplying the size of the force by the perpendicular distance of the force from the **pivot** (**fulcrum**). The unit for the moment of a force is the newton metre (Nm).

> moment   =   force   × perpendicular distance from pivot (fulcrum)
> (newton metre, Nm)   (newton, N)   (metre, m)

→ Increasing the moment

You can increase the moment you exert – or produce the same moment with a smaller force – by moving the force further away from the pivot.

- A door handle is placed as far from the hinge as possible so that a small force will open the door.
- Nut crackers have long handles so you need less force to crack open the nut.
- A bottle opener with a long handle makes it easier to remove the bottle top.
- It is easier to remove a nut from a bolt using a spanner than with your fingers.

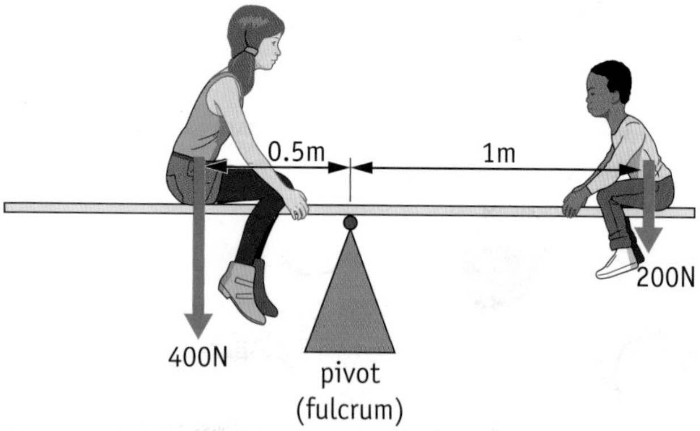

→ The principle of moments

For the see-saw in Figure 2 to be balanced (to be in **equilibrium**), the anticlockwise moment must equal the clockwise moment. This is called the **principle of moments**.

● **Figure 2** You can tell that the see-saw is balanced because the anticlockwise and clockwise moments are the same size

> force on left × perpendicular distance from pivot = force on right × perpendicular distance from pivot

150

If more than one person is sitting on each side, the see-saw can still balance. Work out the moment for each person separately, then add them together on each side. If the total of the anticlockwise moments equals the total of the clockwise moments, then the see-saw is in equilibrium.

On the see-saw in Figure 2 the girl has to sit nearer the pivot than the boy. To balance the moments, a larger force needs to be nearer to the pivot than a smaller force. If their weights were equal, then their distances from the pivot would also be equal.

? Questions

1 In the figure below, the distances are in centimetres and each block weighs 1 newton. For each ruler, state whether it is balanced or not. If it is not balanced, state which side will tip down.

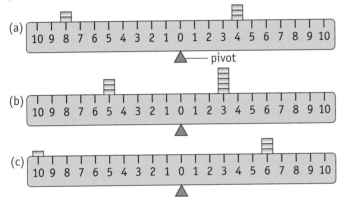

2 For each ruler in the figure below, state where would you place the weight shown on the right hand side to make it balance. The scale is in centimetres and each block weighs 1 N.

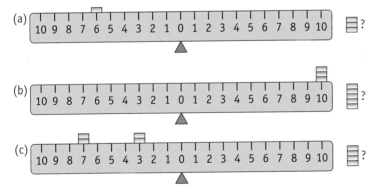
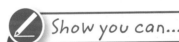

3 Sophie and Oscar sat on opposite sides of a see-saw. Sophie sat 3 m from the pivot. Oscar sat 2.5 m from the pivot. The see-saw balanced.
   a) If Oscar weighed 600 N, how much did Sophie weigh?
   b) Their father joined them and sat 4 m from the pivot. To balance him, Sophie and Oscar both sat on the opposite side to him, Sophie at 4 m and Oscar at 2 m. How much did their father weigh?
4 If you balanced a wooden spoon horizontally and then cut it in half at the balance point, which would be the heaviest piece, the head or handle?

Show you can...

Complete this task to show that you understand the concept of moments.

Two people stood either side of a door. One, a large, muscular rugby player, pushed as hard as he could to keep the door shut, but placed his hands next to the hinge. The other, a small young girl, opened the door easily by pushing on the handle. Explain why this happened.

10.2 Levers

Have you ever used a wheelbarrow or a bottle opener? Then you've used a lever. Levers can change the size or direction of the force you apply to do a job, or the distance your force moves.

● **Figure 1** Does a fishing rod increase or decrease the force you exert?

→ First-order levers

First-order levers put the pivot in the middle, with the **effort** and the **load** on either side, like a see-saw. A claw hammer is a first-order lever; so is a screwdriver being used to open a tin of paint.

In both of these the effort is further from the pivot than the load is, so it is smaller than the load. The principle of moments will give you the size of the effort:

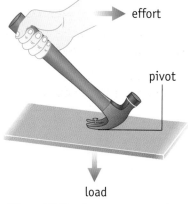

● **Figure 2** The force you exert is called the effort

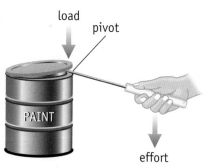

● **Figure 3** The force the lever pushes against is called the load

> large load × small distance = small effort × large distance

When used this way a first-order lever is a **force-multiplying lever**.

→ Second-order levers

Second-order levers put the load in the middle, with the pivot and effort at opposite ends. A wheelbarrow is a second-order lever.

In Figure 4, the centre of the rock is 0.3 m from the pivot (the wheel). The moment due to the rock's weight is 600 N × 0.3 m = 180 Nm. The man also produces a moment of 180 Nm to balance this.

A second-order lever is always a force-multiplying lever.

● **Figure 4** The force you exert on a wheelbarrow is less than the load

→ Third-order levers

Third-order levers put the load and the pivot at either end, with the effort in the middle.

Because the effort is closer to the pivot than the load, the effort is actually larger than the load. But third-order levers are useful because moving the effort a small amount moves the load a larger distance. This is why third-order levers are **distance-multiplying levers**.

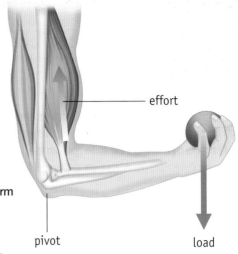

● **Figure 5** Your biceps muscle pulls on your forearm to raise your hand. This is a third-order lever

→ Bicycle gears

The gears on your bicycle alter the distance your feet pedal to make the rear wheel go around once. They also change the force you pedal with. The gear wheel on your pedals pulls the chain around, and the chain pulls the rear wheel round.

- A ratio of 1:1 would mean that your rear wheel goes around once every time your pedals go round.
- A ratio of 2:1 means that the rear wheel goes around twice every time your pedals go around once. You cover twice the distance for the same amount of pedalling, but you push twice as hard.

Bicycles are designed so that you travel a greater distance on the road than you pedal, but you have to push harder than you do when walking.

● **Figure 6** Are bicycle gears distance or force multipliers?

? Questions

1 Which order of lever are the following:
   a) a pair of scissors
   b) a door
   c) a pair of pliers?
2 Name the orders of lever that can be:
   a) distance multipliers
   b) force multipliers.
3 a) Explain why the force exerted on the walnut is greater than the force exerted on the handles.

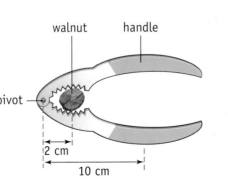

   b) If the force exerted on the handles is 10 N, what force is exerted on the walnut?
   c) The walnut needs 60 N to crack it. What two things could be done to crack the walnut?

✎ Show you can...

Complete this task to show that you understand how to apply the principle of moments to the action of levers.

Josh was carrying a heavy box with both hands. Connor suggested that he would find it easier to carry if he cradled it on his forearms, rather than holding it in his hands. Explain why Connor was right.

10.3 Pressure

The effect of a force depends on the area over which it acts. A big force will not have much effect spread over a large area, but a small force can have a great effect if the area is tiny.

● **Figure 1** Why can you not slice bread with a rolling pin?

→ Calculating pressure

Pressure is a measure of how effective a force is. We work out the pressure by dividing the magnitude of a force by the area it acts on.

- If the force is in newtons (N) and the area is in cm² then the unit is N/cm².
- If the area is in m² then the unit is N/m² or pascals (Pa).

> Remember – there are 10 000 cm² in 1 m²

$$\text{pressure (Pa)} = \frac{\text{force (N)}}{\text{area (m}^2)}$$

● **Figure 2** Why do football players' studs grip the ground so well?

→ High and low pressure

In some situations you want to exert a high pressure, whereas other times you need low pressure.

The studs on football boots have a small surface area. A small number on the bottom of a fraction gives a large value overall, so a small area results in a large pressure. This makes the studs sink into the ground, gripping far better than flat-soled trainers would.

Snow shoes have a larger surface area than normal shoes. In this case the number on the bottom of the fraction is large, so the pressure is small. A small pressure under snow shoes means they will not sink into the snow.

Tractors use wheels with a large surface area. Their smaller pressure means that they do not sink into soft ground. Camels have feet that spread wide on the sand for the same reason.

● **Figure 3** How do snow shoes let you walk on snow?

→ Finding your own pressure

You can find the pressure you exert on the floor by dividing your weight by the area of your feet. If you do not have scales calibrated in newtons, multiply your mass in kilograms by 10. Measure the area of your feet by drawing around them on some graph paper. (You can keep your shoes on.)

force = 628 N

area of both feet = 312.5 cm²

pressure = $\frac{force}{area}$

$= \frac{628}{312.5}$

$= 2.01 \text{ N/cm}^2$

$= 20100 \text{ Pa}$

scales calibrated in newtons

● **Figure 4** You can also find your pressure standing on one foot, or on tiptoe

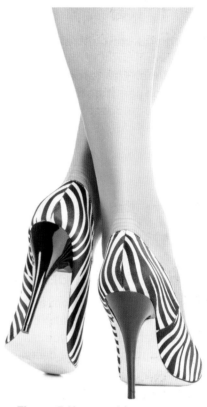

● **Figure 5** How would your pressure change if you wore stiletto heels?

❓ Questions

1 Use the pressure equation to explain your answers to the following.
  a) Why would you do more damage to a wooden floor by walking on it in football boots than in snow shoes?
  b) Would an elephant or a woman wearing stiletto shoes find it easier to walk in deep snow?
2 When you push a drawing pin into a notice board, the force of your thumb on the head of the pin is the same as the force of the pin on the notice board. Explain why the pin goes into the notice board, and not into your thumb.
3 The sole of one of Zoe's feet has an area of 200 cm². If Zoe weighs 700 N, what pressure does she exert on the ground?
4 A car spreads its 15 000 N weight over four tyres, each of which makes 170 cm² contact with the ground. A tank spreads its 700 kN weight over 7 m² of caterpillar tracks.
  a) Calculate the pressure:
    i) under the car
    ii) under the tank.
  b) Which one would drive more easily over soft ground? Explain.

✎ Show you can...

Complete this task to show that you understand how to control the pressure you exert by changing the area underneath you.

On a winter's day a friend of yours falls through the ice that covers a lake. Explain why you would rescue them by lying down on the ice instead of standing on it.

10.4 Pressure in fluids

When you concentrate a jet of water onto a very small area it can cut through materials like glass. The force may not be very large, but because it acts on a very small area its pressure becomes much higher.

● **Figure 1** How is the jet of water like a knife or a pin?

→ Feeling the pressure

If you dive to the bottom of a swimming pool you can feel the increased pressure on your ears. If you take a flight in an aeroplane, you may feel your ears 'pop' as the pressure around you decreases. These both happen because of changes in pressure in a **fluid** (a liquid or a gas).

The size of the pressure depends on the weight of the gas or liquid above you, so the pressure in a fluid increases with depth. **Atmospheric pressure** at the Earth's surface (due to the weight of the air above you) is around 100 kPa. This is sometimes called 'one atmosphere' of pressure. Ten metres depth of water will produce the same pressure as all the air above you, so water pressure increases by one atmosphere with every 10 metres you descend.

● **Figure 2** How can you tell that water pressure is greater lower down?

→ More pressure on the outside

Your teacher may have shown you a can-crushing experiment. When filled with air, the pressure inside a can equals the pressure on the outside. But with the air removed, there is nothing to balance the pressure from outside. This is what makes the can crush.

● **Figure 3** The polystyrene cup on the right was sent to the bottom of the ocean, but why did it crush?

The can and the cups were not squashed flat, like a pancake, even though the pressure on them was due to the weight of the air or water above them. Because fluids can flow, they exert pressure in every direction, at right angles to every surface, whichever way it is facing.

● **Figure 4** Pressure in a fluid acts in every direction, pushing inwards on every surface

→ More pressure on the inside

The **air pressure** acting on you is due to the weight of the air above you. The higher you go, climbing a mountain or flying in a plane, the less air remains above you, so the lower the air pressure you feel.

If the pressure outside an object decreases, it will be less than the pressure inside the object. If the object is made of a soft material, it will expand. You can see this happen by putting small marshmallows in a glass bottle and sucking the air out with a straw (you will need to seal the neck of the bottle with modelling clay).

● **Figure 5** What happens to the marshmallows when you let the air back in?

? Questions

1 Atmospheric pressure on the surface of Venus is 90 times larger than on Earth. On Mars it is 100 times smaller than on Earth. What else does this tell you about the atmospheres of these planets?

2 A jet pilot and a submarine captain each had a packet of marshmallows. The pilot opened hers while flying 10 km high. The captain opened his when 1 km under water.
   a) Whose marshmallows looked the biggest?
   b) Why?

3 A window ornament is held in place with a rubber sucker. The sucker has an area of 5 cm², and it takes a force of 50 N to pull it off the window. Calculate the atmospheric pressure.

Show you can...

Complete this task to show that you understand how pressure varies with depth.

Explain why a dam is thicker at the base than it is at the top.

10.5 Floating and sinking

When you get in the bath the water level rises. You displace the same volume of water as the volume of your body that is underwater. If this volume of water weighs more than you, then you float.

● **Figure 1** Why would you float higher than normal in the Dead Sea?

→ Density

To predict whether something will float in water you need to know its density. Objects that are less dense than water will float, whereas those that are more dense will sink.

The object's density is its mass per unit volume, and can be found by dividing its mass by its volume:

$$\text{density (kg/m}^3) = \frac{\text{mass (kg)}}{\text{volume (m}^3)}$$

If mass is measured in g and volume in cm³ then the units of density are g/cm³.

The density of water is 1000 kg/m³ or 1 g/cm³

● **Figure 2** A pound coin sinks in water, so why does it float in mercury?

→ Buoyancy

Archimedes' principle states that an object placed in water will feel an **upthrust** force (buoyancy), equal to the weight of the water it displaces. It is this upthrust, if it is large enough, that makes an object float.

What causes buoyancy?

The pressure of water pushing on an object increases with depth. Thus the water pressure on top of the fish, pushing it down, is less than the pressure beneath the fish, pushing it up (the pressures on each side are the same and cancel out).

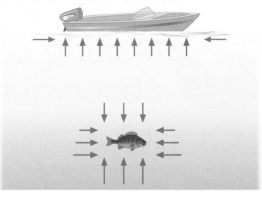

● **Figure 3** The boat and the fish both float because of water pressure

$$\underset{\text{(newtons, N)}}{\text{force}} \quad = \quad \underset{\text{(pascals, Pa)}}{\text{pressure}} \quad \times \quad \underset{\text{(square metres, m}^2\text{)}}{\text{area}}$$

The water pressure beneath the boat, multiplied by the area of the boat's hull, provides the force of the water pushing up on the boat. If this force (upthrust) is more than the weight of the boat, the boat floats.

→ Floating in air

A helium balloon floats in air because it is less dense than air. A volume of air is displaced equal to the helium's volume, and the balloon feels an upthrust equal to the weight of the air that it displaces.

? Questions

1 A 100 cm³ measuring cylinder is filled to the 50 cm³ mark and an object is dropped into it. Find the volume of the object if:
   a) a coin makes the water level rise to 52 cm³
   b) a pen makes the water level rise to 65 cm³
   c) a stone makes the water level rise to 59 cm³.
2 A block of wood and a metal boat both float. Both of them were cut in half. Explain which one now floats, which sinks, and why.
3 a) Calculate the density of:
      i)   a gold ring, of mass 135 g and volume 7 cm³
      ii) an iceberg, of mass 100 000 tonnes and volume 109 000 m³
      iii) a cork, of mass 7.2 g and volume 30 cm³.
   b) Which will float and which will sink?
4 a) Draw a diagram to show how the pressure acting on a submarine produces an upthrust force.
   b) What calculation would you have to do to find the size of the upthrust force?
   c) How big does this upthrust need to be for the submarine to float?

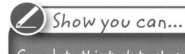

Show you can...

Complete this task to show that you understand how the density of an object determines whether or not it will float.

Explain why a solid block of steel will sink, whereas a boat made of the same steel would float.

Making and recording observations

→ Finding the density of air

We often think of air as having no mass. In fact it has a very small mass per unit volume (density). Did you know that the air above you (all the way to space) weighs the same as a column of water 10 m high?

Making observations

To find the density of air you need to measure a mass of air and its volume. There are several ways that you can do this, as shown below.

Measuring mass

To find the mass of your sample of air, you could weigh a balloon both with and without air inside. The mass of the air is the difference between these two measurements. Use electronic scales, with a **precision** of at least 0.1 g.

There are four ways that you could do this:

1 Weigh an empty balloon. Then inflate it by mouth and weigh it again.

2 Weigh an empty balloon. Then inflate it with a balloon pump and weigh it again.

3 Weigh a full balloon, inflated by mouth, then pop it and weigh it again.

4 Weigh a full balloon, inflated with a balloon pump, then pop it and weigh it again.

Measuring volume

There are three ways that you can find the volume of the air in your sample:

1 Measure the diameter of your balloon using three 50 cm rulers, as in Figure 3. Then calculate the volume of the balloon using the formula below:

$$\text{volume of air} = \frac{\pi \times d^3}{6} = \dots \text{ cm}^3$$

$$\left(\text{or } \frac{\pi \times d \times d \times d}{6}\right)$$

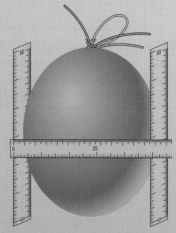

● **Figure 1** If you pumped more air into this ball, would it float higher or lower in the water?

● **Figure 2** What is the difference between a balloon inflated by mouth and one inflated with a pump?

● **Figure 3** You can find the diameter of a balloon using three rulers

2 Part fill a large bucket with water. Hold the balloon underwater and mark the level that the water has risen to on the side of the bucket. Now remove the balloon and pour in measured amounts of water until the level is where it was when the balloon was under water. The volume of water you have added is the same as the volume of the balloon.

3 Fill a measuring cylinder with water and invert it in a water trough, while keeping the top of the measuring cylinder underwater. Attach a tube to your balloon pump, long enough to reach the top of the measuring cylinder. Pump the balloon pump once and collect the air from it in the measuring cylinder. Read the volume of air in one pump from the measuring cylinder and multiply this by the number of pumps it took to inflate your balloon.

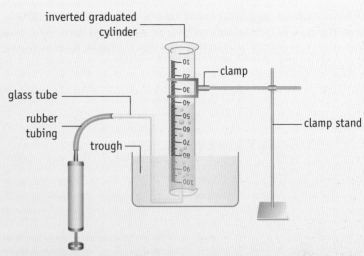

● **Figure 4** You can measure the air from several pumps and find the average volume for more accuracy

Recording observations

When you record your observations, it helps to do so in a way that makes your calculations straightforward.

For example, you might present your results like this:

> **Example**
> mass of balloon containing air: .... g
> mass of empty balloon: .... g
> mass of air: .... g
> volume of water: .... cm³
> (= volume of air)

In this way the mass of air can be found by a simple subtraction, and the figures for mass and volume are together, ready for the density calculation.

Planning your practical

Decide which methods you wish to use to find the density of air, and plan the steps that you will take. Draw up tables for any results you will take.

Evaluating your work

It is important as a scientist to evaluate the work that you do.

● Make a list of the pros (the good points) and cons (the bad points) of each method.
● State which you believe gave the best results and explain why.
● Suggest ways in which the experiment could be improved even further.

11.1 Conduction

People often say that heat rises. If this is true, what keeps a sunbather warm? Heat does not simply travel in one particular direction but flows from hot objects to cold ones.

● **Figure 1** If the butter is getting hotter, what is happening to the toast?

→ Heat

Substances hold energy in a **thermal store** due to the random movement of their molecules. 'Heat' is the name for the transfer of energy from one thermal store to another. In Figure 1, the butter on the toast is melting – it is changing state from solid to liquid. To do this, it must gain energy. The energy comes from the hot toast. Because heat is the transfer of energy from hot objects to cold ones, the toast must be hotter than the butter.

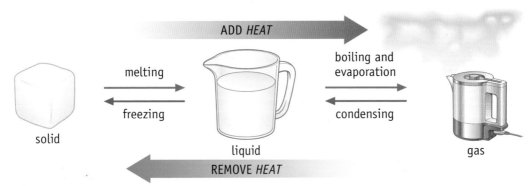

● **Figure 2** Gaining heat increases an object's temperature, or makes it melt or boil. Losing heat does the opposite

→ Conduction

Energy transfers from hot objects to cold objects in several ways. The main method of heat transfer in solids is **conduction**. In Figure 1 particles in the hot toast have more energy and vibrate more vigorously than particles in the butter. They collide with particles in the butter and transfer some of their energy.

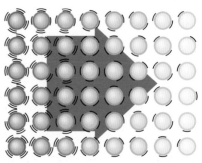

● **Figure 3** Conduction of heat is due to the transfer of energy by particle collisions

162

If the temperature is higher at one end of a solid, the particles at that end have more energy and vibrate more vigorously than particles at the other end. Because each particle is bonded to the next, the vibrations are passed on from particle to particle, so heat conducts along the solid. Eventually all of the particles in the solid end up vibrating at the same rate. This means that the temperature is the same throughout the solid.

→ Good and bad conductors

All metals, and some other materials, conduct heat well. Materials that do not, such as wood or plastic, are called **insulators**. You can investigate how well different metals conduct by heating some metal rods at one end in a Bunsen flame (see Figure 4). When the Vaseline on the other end melts, the pin falls.

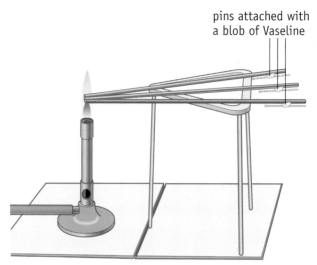

pins attached with a blob of Vaseline

● **Figure 4** The pin will fall off the best conductor first

→ Conduction in liquids and gases

- Particles in solids have stronger bonds between them than particles in liquids.
- Particles in gases have virtually no bonds between them and are extremely far apart compared with particles in solids or liquids.

These facts can explain why solids are generally the best **conductors** and gases the worst conductors of heat.

→ Conduction in metals

Metals are very good conductors because they have an extra process that conducts heat. Because electrons are free to move in metals (this is why metals conduct electricity so well) they are also free to carry heat. The **free electrons** at the hotter end have more energy than those at the colder end, so they move around faster and transfer this energy by bumping into atoms and electrons further down the metal. This is more effective than transferring heat through vibrations, so metals are better conductors of heat than non-metals.

❓ Questions

1 Which way does heat travel?
2 What are the names for materials that transfer heat quickly and those that transfer heat very slowly?
3 Name three things that can happen when an object:
   a) gains heat   b) loses heat.
4 Why would a good frying pan have a copper base and a plastic handle?
5 a) Why do metals conduct heat better than most non-metals?
   b) Why is diamond a better conductor of heat than even the best metal conductor?

✎ Show you can...

Complete this task to show that you understand how vibrations carry heat through a substance.

Explain why solids are usually the best conductors of heat, whereas gases are usually the worst.

11.2 Convection

When air is heated it expands. Its volume increases but its mass does not. So the density of hot air is less than the density of cold air. Less dense air floats, which is why a hot-air balloon rises.

● Figure 1 What must the pilots do to bring the hot-air balloons down again?

→ Heat transfer by particles

Convection is another way in which heat travels. Like conduction, in convection the energy is carried by particles. This time the particles actually move from one place to another, rather than transferring energy by collisions or through bonds between neighbouring particles.

Convection cannot happen in solids, but it is the main method of heat transfer in liquids and gases, which together are called fluids because they can flow. Convection makes hot air and hot water rise, which is why some people mistakenly say 'heat rises'.

→ Convection currents

A lava lamp has an old-fashioned filament bulb in its base, which produces a lot of heat as well as light. The heat from the bulb melts the coloured wax in the lamp and makes it expand (the particles in the wax get further apart – they do not increase in size). As it expands, the wax becomes less dense and floats upwards. Wax floating at the top of the lamp gradually cools down, becomes denser and sinks again. This type of flow is called a convection current.

● Figure 2 Would a lava lamp work with an energy-saving bulb?

→ Heating a room

Despite its name, a radiator actually heats a room by convection (not radiation), in a similar way to the lava lamp.

1 Cold air near the base of the radiator warms up and expands.
2 The warm air is less dense, so it floats upwards.
3 The warm air travels through the room, giving out heat and contracting.
4 Cooler air sinks down to the floor.
5 Cold air flows towards the radiator.

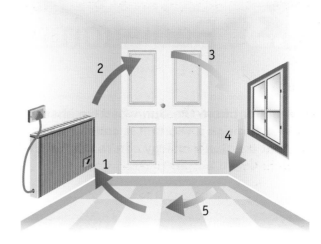

● **Figure 3** Air circulates throughout a room because of convection currents caused by a radiator

? Questions

1 Why can you get convection in water but not in ice?
2 Ice cubes float on the surface of a glass of water, yet they can still cool down the whole of the drink. Explain why.
3 a) Where is the coldest place in a room that is being heated by a radiator?
   b) Where is the warmest?
4 Use the idea of convection to explain how you could avoid breathing smoke in a house fire by crawling on the floor instead of walking.
5 During the daytime the land heats up more quickly than the sea. At night, the land cools down quicker than the sea. Use these facts and your knowledge of convection to explain why you feel a breeze blowing from the sea to the land during the day, and in the opposite direction at night.

Show you can...

Complete this task to show that you understand how a convection current is set up.

Before ventilation fans were invented, mines used to be ventilated by lighting a fire at the base of a mine shaft. Explain how this would ventilate the whole mine.

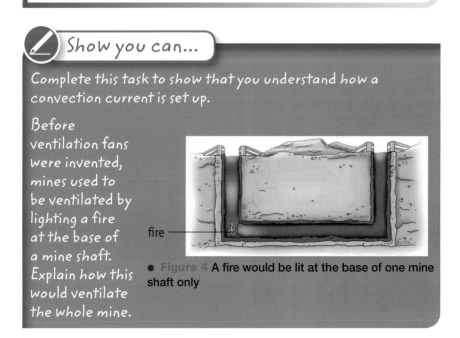

fire

● Figure 4 A fire would be lit at the base of one mine shaft only

11.3 Radiation

We feel heat from the Sun when we absorb its rays. Heat transfer through a vacuum is called radiation. It is the only way that the Sun's energy can reach the Earth through space.

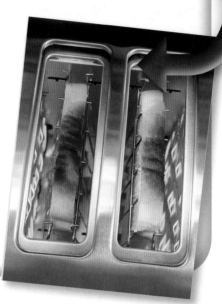

● **Figure 1** Would a toaster work in outer space?

→ No particles required

Heat radiation is the only form of heat transfer that does not involve particles. In fact, it is an electromagnetic wave with a wavelength only slightly longer than red light. It is given out by all hot objects – the hotter an object is, the more heat it radiates. This is also known as **infrared radiation** and can even be seen with a special camera.

Heat radiation can reach us from the Sun through the vacuum of space, but it can also pass to some extent through transparent substances, such as glass or water.

Most of the energy transferred to the toast from the toaster in Figure 1 is in the form of infrared radiation.

● **Figure 2** Heat radiation can pass through some materials. How might the police, the armed forces, or earthquake rescue teams use this fact?

166

→ Absorbing or reflecting radiation

Dark colours are better than light colours at absorbing heat radiation, and matt (dull) surfaces are better than shiny surfaces. You can show this by a simple demonstration.

Two metal squares are stuck onto corks with wax. One square has been painted matt black, the other is a shiny metallic colour. The corks are held in clamp stands so that the squares are next to each other, the same distance from a **radiant heater**. Soon the radiation absorbed by the squares makes the wax melt and the squares fall off.

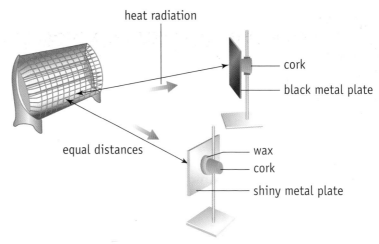

heat radiation

cork

black metal plate

equal distances

wax

cork

shiny metal plate

● **Figure 3** The square that falls first is the best absorber of heat radiation

→ Emitting radiation

As well as being the best at absorbing radiation, dark, matt colours are also the best **emitters** of radiation. If you take two empty food cans, leave one shiny and paint one black, and fill them with hot water, the water in the black can will cool down the quickest.

Mediterranean houses are painted white to reflect the Sun's heat in the summer. Because white is a poor emitter of heat, this would also keep British houses warmer in winter!

→ Other types of 'radiation'

The word 'radiation' is simply a term given to energy moving outwards from a central point. Nuclear radiation refers to dangerous rays that are given off by radioactive substances. Microwave radiation is given off by a microwave oven or your mobile phone. You are emitting heat radiation right now, because you are warm blooded, and even the sound that comes from your voice box could be called 'sound radiation'.

shiny metal can

matt black can

● **Figure 4** Fill the cans from a kettle, then read their temperatures after half an hour

? Questions

1 a) How do we know that heat radiation can travel through a vacuum?
  b) Can heat radiation pass through a gas? How can you tell?
  c) Are there any solids or liquids that heat radiation can pass through?

2 a) Why do cricketers and tennis players wear white clothing?
  b) Why are solar water heating panels painted black?

3 a) Why is a kettle often made of white plastic or shiny metal?
  b) How does a polar bear's white colouring help it keep warm?
  c) Why does an arctic fox have smaller ears than a European red fox?

Show you can...

Complete this task to show that you understand the absorption and reflection of radiation.

Fire-fighters tackling a blaze wear silver suits to prevent them from getting too hot, but after a marathon the runners are wrapped in shiny 'space blankets' to stop them cooling down too fast. Explain how a shiny material can serve both of these purposes.

11.4 Insulation

An insulator is a material that is a bad conductor of heat. We use insulators in clothing to keep ourselves warm, or in the loft to keep our houses warm. But they can also be used to keep things cold.

● **Figure 1** Will the snowman melt faster or slower when wearing a coat?

→ Trapping air

As we saw earlier in this topic, on page 163, air is a bad conductor because its particles are very far apart. However, as we found on page 165, it will transfer heat quite well by convection. So if we can trap air and stop it from flowing, we will have made a very good insulator.

Convection can also be reduced very simply by placing a lid on a saucepan, or covering a bowl of food with clingfilm. Installing **loft insulation** is like placing a lid on your house, while **cavity wall insulation** is like wrapping your home in a jumper. **Double glazing** also traps an insulating layer of air – this time between two panes of glass.

● **Figure 2** In what way is a knitted jumper . . .

● **Figure 3** . . . like bubble wrap?

● **Figure 4** Why is loft insulation so effective?

→ Insulation in nature

Many animals that live in cold regions trap air in a similar way to a jumper. Polar bears have hollow hairs, a sheep's woolly coat keeps it warm on cold hillsides and birds 'fluff up' their feathers in the winter.

→ The vacuum flask

A Thermos flask, or **vacuum flask**, is made of a double-layered glass bottle with a vacuum in between the two layers. Because there are no particles in a vacuum, it not only stops conduction, but convection as well. To reduce radiation, the outer and inner glass bottles are coated with a thin layer of silver or aluminium on the side facing the vacuum.

If the flask contains a hot drink, the inner bottle will be a bad emitter due to its silvered surface. The outer layer will reflect back most of the heat that does get emitted, and the vacuum will prevent conduction and convection between the two layers. This keeps the drink hot for a long time.

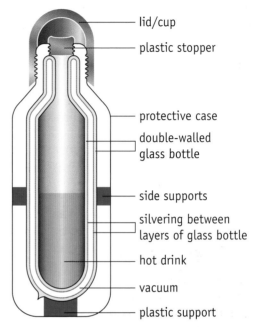

- lid/cup
- plastic stopper
- protective case
- double-walled glass bottle
- side supports
- silvering between layers of glass bottle
- hot drink
- vacuum
- plastic support

● **Figure 5** A vacuum flask can also be used to keep cold drinks cold

? Questions

1 Two identical saucepans were used to cook the same amount of pasta. Only one saucepan had its lid on, and both saucepans contained the same amount of boiling water. Which saucepan used the most energy to cook the pasta, and why?
2 Why does a robin 'fluff up' its feathers in the winter?
3 In the UK we insulate our houses to keep them warm in winter. Will a well-insulated house be hotter or cooler in the summer? Explain your answer.
4 Houses can be kept warmer in winter by closing the curtains at night, fitting draught excluders to doors and letterboxes, and by having a deep pile carpet with a foam rubber underlay beneath it. State the heat transfer method that each of these reduces, and explain how it is reduced.
5 Look at Figure 5. Explain how each of the following reduces heat loss:
   a) stopper
   b) glass bottle
   c) plastic supports
   d) vacuum
   e) silvering on outer layer of glass bottle.

✎ Show you can...

Complete this task to show that you understand how trapping air makes many insulating materials effective.

Explain how a woollen jumper resembles bubble wrap, and how it keeps you warm.

169

Asking questions and making predictions

→ Copious coffee cups

Hot stuff – prior knowledge and experience

1 You may have had a take-away hot drink recently. If so, what was the cup made from? Did it have a lid? Was there an insulating sleeve? Did your hand feel hot? How well did it keep your drink warm? Make notes about anything you can remember.

Not too hot, but not too cold

The owners of a coffee shop want their customers to enjoy every cup they drink. A customer will not be pleased if they burn their mouth on their coffee, nor if it goes stone cold before they can finish it.

● **Figure 1** Which is best at keeping drinks hot – the insulation or the lid?

Task 1: Asking questions

Imagine that you work for a coffee shop and have been given the task of deciding which type of cup to use. Your boss has told you to test some samples.

2 What would you like to test as you tried to find the optimum cup for keeping coffee hot? Which you would expect to insulate best, and which do you think would be worst? Make a list of the questions you might ask.

● **Figure 2** What could you use?

Task 2: Developing a line of enquiry

Think about how you would proceed from here, and what you could do to test the insulating properties of your cups, sleeves and lids.

Remember that a scientist changes only one independent variable at a time.

3 If you were testing cups made of different materials, how would you find which held heat the best? Describe the test that you would want to try.
4 If you were testing the same cups, but with or without lids or sleeves, what variables would you have to keep constant to make it a fair test?
5 How could you track the temperature throughout the experiment? How could you display your results to show how the temperatures changed?
6 Would you follow the cups' temperatures for a set period of time, or until they had cooled down by a certain amount? Explain the reason for your decision.

7 Make a table that gives space for all of your possible results, ensuring that the column headings have the correct labels, including units.

8 Gather the equipment that you need to take your readings.

9 Note down the values of the **control variables**, to show that you are ensuring they stay constant.

Task 3: Observations of the real world

You now need to carry out your experiment.

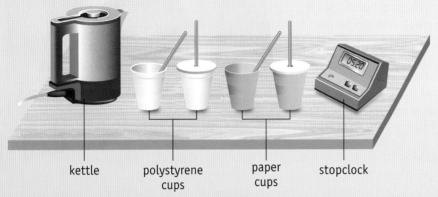

kettle   polystyrene cups   paper cups   stopclock

● **Figure 3** Ensure that the initial conditions are the same for each cup in your experiment. You can also stand each cup in a beaker so it doesn't fall over

Task 4: Presenting and interpreting data

10 Once you have gathered all of your results, you need to present them in a way that helps you explain your findings to the coffee shop owner.
   • Would your table of results alone demonstrate what you want to tell the owner clearly enough?
   • Do you need to draw a graph or chart? If so, decide on the type that best suits your needs – do you just want to show the overall temperature differences at the end of the experiment, or do you want to show how the temperatures changed from minute to minute?

11 Finally, write down in your book which type of cup you felt was the best, and whether you would also use a lid or a sleeve. Give a full explanation of your decision, using the results of your experiment to back up your decision.

A solar eclipse occurs when the Moon lies directly between the Sun and the Earth. The Sun's atmosphere, the corona, becomes visible because the Moon blocks out the Sun's intense light. Solar prominences can also be seen past the edge of the Moon.

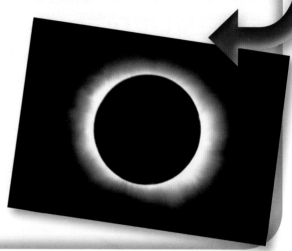

● **Figure 1** Why can you not see the Moon's face during an eclipse?

→ The seasons of the year

The Earth spins on its axis, which gives us day and night. But because the Earth's axis is tilted slightly, the days are longer in summer and shorter in winter. To see how this tilt is responsible for the seasons, we need to look at the Earth's **orbit** around the Sun.

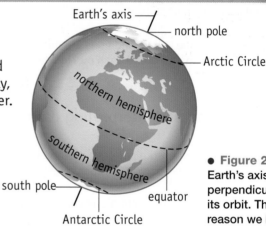

Earth's axis — / — north pole
— Arctic Circle
northern hemisphere
southern hemisphere
south pole
equator
Antarctic Circle

● **Figure 2** The Earth's axis is not perpendicular to its orbit. This is the reason we have seasons

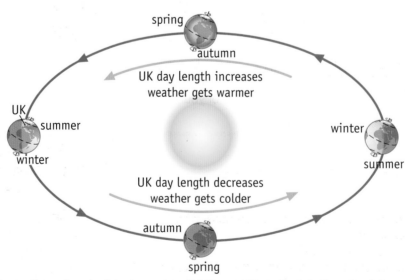

spring
autumn

In summer, the tilt of the Earth's axis means the UK spends more time in daylight than it does in shadow.

UK day length increases weather gets warmer

UK
summer
winter

winter
summer

In the UK's winter, due to the tilt of the Earth's axis, the nights are longer than the days.

UK day length decreases weather gets colder

autumn
spring

● **Figure 3** In the **northern hemisphere**'s summer, it is the **southern hemisphere**'s winter. Note that this is a side-on view – the Earth's orbit is really an almost perfect circle (diagram not to scale)

In the northern hemisphere's summer, the land near the north pole, north of the **Arctic Circle**, is in permanent daylight. This is sometimes referred to as the 'Land of the Midnight Sun'. The Sun still appears to move across the sky and dips towards the horizon at 'night-time', but it never actually sets.

→ The phases of the Moon

Just as the Earth orbits the Sun, the Moon orbits the Earth. The same side of the Moon faces us all the time, but it is not always lit by sunlight. When the Moon is between us and the Sun, the side we see is not lit up. This is called a **new moon**. When the Moon is on the far side of the Earth, the side we see is totally lit up. This is a **full moon**.

In Figure 5, the images of the Moon on the red circle show how sunlight illuminates it as it orbits the Earth. The larger images of the Moon show how it looks to us on the Earth in each position.

A **lunar eclipse** occurs when the Earth's shadow falls on the Moon. Eclipses do not occur every month because the Earth, Moon and Sun do not always line up exactly.

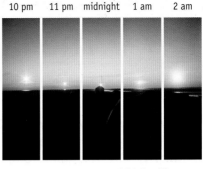

| 10 pm | 11 pm | midnight | 1 am | 2 am |

● **Figure 4** What would it be like to live in the Land of the Midnight Sun?

● **Figure 5** The Moon does not give off its own light. We see it because sunlight reflects off its surface towards us (diagram not to scale)

? Questions

1 What season is it in Australia when it is the following in the UK:
   a) summer
   b) spring?
2 When the north pole experiences 24 hours of darkness, what does the south pole experience?
3 Why can we not see a true new moon?
4 Find out why a solar eclipse can only be seen from certain points on the Earth's surface, whereas a lunar eclipse can be seen from the whole of the night side of the Earth.

✎ Show you can...

Complete this task to show that you understand how the tilt of the Earth produces the seasons.

Explain what the Earth's climate would be like if the planet was not tilted.

Our Sun is one of a hundred thousand million stars in the Milky Way galaxy, a vast collection of stars bound together by gravity. On average one star is 'born' in our galaxy every year.

● **Figure 1** The Whirlpool Galaxy, showing star birth regions in red. How many stars in our **galaxy** are younger than you?

Light travels at 300 000 km/s, and would travel almost 6 **trillion** miles in a year. This distance is called a **light year**. The **Milky Way** is 100 000 light years across.

→ Star birth

Figure 2 shows a **nebula** – a cloud of gas and dust – called the Pillars of Creation, where thousands of new stars are forming. The largest pillar is 60 trillion miles long.

A region inside a nebula that has a slightly higher density pulls gas towards it by gravity. As the gas falls inwards, this region heats up, becoming a **protostar** – a ball of hot gas glowing with infrared radiation.

● **Figure 2** How many years would it take light to travel the length of the longest pillar?

→ Switching on

When the protostar grows large enough, it 'switches on' – **nuclear fusion** begins in its core, turning hydrogen into helium and releasing vast amounts of energy. The radiation it gives off clears away the remnant of the nebula. In its place a **star cluster** shines brightly.

Nuclear fusion will power a star like our Sun for around ten billion years.

There are more galaxies in the Universe than there are stars in our galaxy.

Larger stars shine brighter because they use their hydrogen faster. The opposite is true for smaller stars. Stars 100 times the Sun's mass will only last a few million years, whereas a **red dwarf** star can live a thousand times longer than our Sun.

● **Figure 3** It takes light 180 years to cross this nebula. How large is it in light years?

→ Star death

When the Sun has used up all the hydrogen in its core, it will become a **red giant**, expanding until it engulfs the Earth. Its outer layers will drift off to become a **planetary nebula**, leaving only the core behind.

The core will become a **white dwarf**. It will not produce any energy and will very slowly cool down. A white dwarf has around half the original star's mass, but its gravity is so strong that this is pulled into a sphere the size of the Earth. One teaspoon of white dwarf matter would weigh several tonnes.

Stars much larger than our Sun end their days differently. They expand to become **supergiants**, then explode as **supernovae**. The star Betelgeuse will go supernova within a million years. When it does it will outshine the Moon and be visible during daylight for months.

In a supernova explosion all of the heavier elements in the Universe are formed and thrown into space – your body is made of dust from a supernova.

The remnant collapses to form a **neutron star**, only 7 miles across, or a **black hole** – a point of infinite density, from which even light cannot escape.

● **Figure 4** Planetary nebulae come in many different shapes. What might the nebula left behind by our Sun look like?

 ## ? Questions

1 What makes a nebula collapse to form a protostar?
2 **a)** What process powers a star?
  b) In this process, what does hydrogen become?
3 Why is it true to say that the Earth and everything on it is made of stardust?
4 A high mass star has more fuel than a low mass star. Which shines for longest, and why?

Show you can...

Complete this task to show that you understand how a star's mass determines its eventual fate.

Match the stars (a–c) with the final stage in their life span (i–iii):

a) small star, like our Sun
b) large star, much more massive than the Sun
c) extremely large star

i) black hole
ii) white dwarf
iii) neutron star

12.3 Changing ideas about the solar system

Around 250 BC a Greek scientist called Eratosthenes worked out the circumference of the Earth, simply by using shadows to find the angle of the Sun in the sky. His value was amazingly accurate.

● **Figure 1** If you had never seen a photograph of the Earth from space, would you think it was round like a ball or flat?

→ Ptolemy's geocentric model

In 150 AD Ptolemy published his ideas about the **solar system**. Ptolemy's model was **geocentric** – all heavenly objects orbited the Earth, which was fixed and did not rotate. This picture was similar to earlier Greek ideas, but Ptolemy turned them into a mathematical model. His calculations could also be used to predict planetary movements.

→ Copernicus's heliocentric model

In 1543 Nicolaus Copernicus revived an idea put forward by Aristarchus of Samos at the time that Eratosthenes had measured the size of the Earth. He suggested that the planets orbited the Sun instead of the Earth, and used a mathematical model to describe their motion. However, not many people believed that the Earth actually moved, which disagreed with religious ideas of the time.

● **Figure 2** Ptolemy's model was very complex, but data on the position and movement of the Sun, Moon and planets supported his theory

→ Kepler's elliptical orbits

Around 60 years later Johannes Kepler realised that Copernicus might be right. He worked for a famous astronomer called Tycho Brahe. Although telescopes had not yet been invented, Brahe had made very precise observations of the planets. Kepler tested Copernicus's theory and used mathematics to work out that the planets' orbits were elliptical (oval).

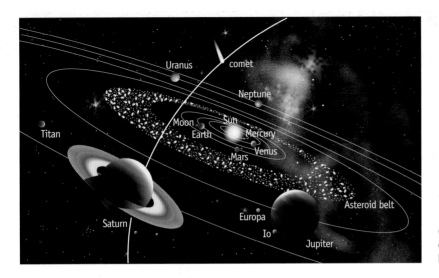

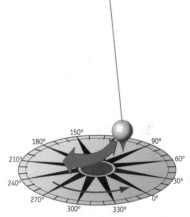

● **Figure 3** A comet's orbit is clearly elliptical, whereas a planet's orbit may be almost circular

→ Galileo's proof

Galileo studied the solar system with the newly invented telescope. He found further evidence that the Sun was at the centre, and in 1632 published a book supporting the **heliocentric** model.

→ Newton's theory of gravity

Published in 1687, this explained why the planets orbit the Sun rather than the Earth. The Sun is nearly 100 times heavier than all of the planets put together. Therefore the Sun's gravity is the dominant force. Newton showed mathematically that the inwards gravitational force from the Sun keeps the planets moving about it in elliptical orbits. Similarly a planet's gravity causes its moons to orbit the planet.

→ Foucault's pendulum

In 1851 Foucault used a pendulum to prove that the Earth is spinning on its axis, which explains why the Sun, Moon and stars appear to move across the sky.

→ Modern ideas

Since 1957 humans have put **satellites** into orbit around the Earth and sent probes to most of the other planets in our solar system. We have put people on the Moon and robotic **rovers** on Mars. These steps give us information that develops our ideas about the solar system.

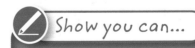

● **Figure 4** The line of the pendulum's swing on the Earth's surface changes slowly, as the Earth spins underneath

? Questions

1 How long have people known that the world is round?
2 **a)** What is the shape of a planet's orbit around the Sun?
   b) What causes the planets to orbit the Sun?
3 Why do the Sun, Moon and stars appear to move across the sky?
4 Do some research to find out ways that people could prove the Earth was a sphere before the age of space travel. Write a page of notes on what you find.

✎ Show you can...

Complete this task to show that you understand orbits.

Planets orbit the Sun, whereas moons orbit planets. Explain what causes both planets and moons to follow their orbits.

12.4 To the Moon and beyond

In 1969 Neil Armstrong became the first human to set foot on the Moon. In 1972 Gene Cernan became the last. Since then no one has been further than the International Space Station (ISS), which is in orbit around the Earth.

● **Figure 1** If you could set foot on another planet or moon, where would you go?

→ Getting into orbit

Giant rockets launched the first astronauts into space. These could only be used once, making space travel extremely expensive. The **Space Shuttle** was a partially reusable craft that would launch like a rocket and land like a plane, but it wasn't until the 21st century that the first totally reusable spacecraft was launched: **SpaceShipOne**.

→ Artificial satellites

Satellites are one of the main benefits of space travel. **Telecommunication satellites** transmit phone and television signals around the Earth. **GPS satellites** make navigation easier – you probably have a receiver in your phone. **Meteorologists** use satellites to monitor the weather across large areas of the Earth, and satellites like the Hubble Space Telescope provide astronomers with images that are not distorted by the Earth's atmosphere.

● **Figure 2** Virgin Galactic will soon be offering flights to space tourists aboard **SpaceShipTwo**. How much would you pay to travel into space?

→ International Space Station

The ISS is an orbiting research lab. Here astronauts can carry out experiments that are not possible on Earth, such as examining the effect of weightlessness on the growth of crystals or seedlings, and on the human body. Experience of living in the ISS also increases our understanding about how to live safely in space, which is vital if we are to send people on longer space voyages.

● **Figure 3** How could the ISS help to get humans to Mars?

→ Roving over Mars

Between 1997 and 2012 four rovers – wheeled robots that could carry out scientific experiments – have successfully landed on Mars. Images of the planet show evidence that liquid water once existed on the surface of Mars. If water is still present below the surface then it is just possible that there is microscopic life. No probes have discovered any water yet but future missions will test for water and for the presence of life using more sensitive techniques.

→ Going beyond

By sending robotic spacecraft to orbit or land on other planets and their moons it is possible to study their surface and atmosphere close up. The photo shown in Figure 5, taken by the Galileo spacecraft orbiting Jupiter, shows the plume of gas and dust from a volcano erupting on its moon Io.

● **Figure 4** Will future missions leave tyre tracks rather than footprints?

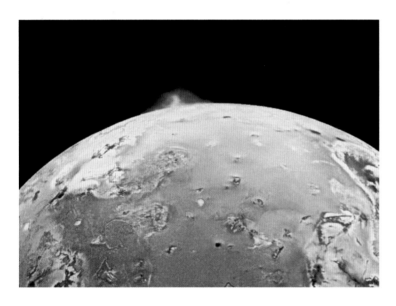

● **Figure 5** Do you think we will ever find life on another planet?

? Questions

1 How long is it since humans walked on the Moon?
2 **a)** Why is a weather satellite more useful than a ground-based weather station?
  b) What advantage does a space telescope have over a ground-based telescope?
3 Space travel used to be so expensive that only the world's superpowers could send people into space. What developments have made it potentially affordable to tourists?
4 Do some research into the life of an astronaut on the ISS. Write a short story about life on board the ISS, describing what the astronauts do and what it is like to live onboard.
5 Why do you think we have sent four robotic rovers to Mars, but not a single person? Discuss possible reasons in a group and report your ideas to the class.

✎ Show you can...

Complete this task to show that you know about space exploration.

Write a short paragraph explaining some of the challenges and advantages of exploring space (at least two of each).

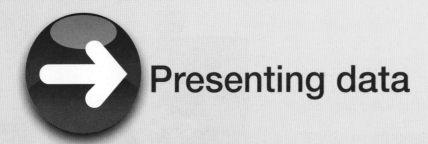

Presenting data

→ Solar system sorted

A solar system chart

Stan and Ella wanted to make a chart of the solar system on the 10 m-long wall of the Science Lab. Stan suggested drawing circles to scale to represent the planets. Ella wanted the planets' positions on the wall to reflect their positions in space.

They found out the following:

● **Figure 1** The objects in the solar system vary considerably

Name	Position, in order from the Sun	Average distance from the Sun (millions of km)	Diameter (thousands of km)
Sun	0	0	1400
Mercury	1	58	4.9
Venus	2	110	12
Earth	3	150	13
Mars	4	230	6.8
Jupiter	5	780	140
Saturn	6	1400	120
Uranus	7	2900	52
Neptune	8	4500	50
Pluto	9	5900	2.3

1 Stan suggested placing the Earth 1 m from the Sun. Ella argued that this would not work. Explain why Ella was correct, by working out the distances to the different planets at this scale, and entering them in a table. (Hint: divide every distance by the Earth's distance.)

2 Work out the distances Ella and Stan would actually have to use for the different planets if Pluto were to be placed 10 m from the Sun. Record these in your table. (Hint: divide every distance by Pluto's distance, then multiply by 10.)

3 Ella wanted to make the circle representing Earth 1 m wide, and to base the sizes of the other planets on this. Stan said that there was not enough space to do this. Explain why Stan was correct by working out the sizes of the planets at this scale, and enter them into your table. (Hint: divide every diameter by the Earth's diameter.)

4 A better scale would make Jupiter 1 m wide and base the diameters of the other planets on this. Work out the sizes of the planets using this scale and enter these into the next column in your table. (Hint: divide every diameter by Jupiter's diameter.)

5 Build a scale model of the solar system, based on the sizes you have worked out. Use sugar paper or card to make your planets, using a pencil on a piece of string to draw your larger circles (remember to use the radius, not the diameter). Sketch an arc to show the edge of the Sun.

If you are to display these on the wall, your teacher will provide the distances to each planet.

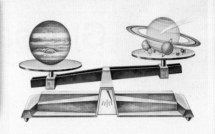

● **Figure 2** Jupiter has more mass than all of the other planets put together

More facts about the planets

Ella and Stan found out some other information on the solar system:

Name	Mass (billion trillion tonnes)	Gravitational field strength (N/kg)	Length of Year (years)	Surface temperature (°C)
Sun	1 990 000	274	–	5 800
Mercury	0.330	3.7	0.24	400
Venus	4.87	8.9	0.62	480
Earth	5.97	9.8	1.0	10
Mars	0.642	3.7	1.9	−100
Jupiter	1 900	24.8	11.9	−150
Saturn	568	10.4	29.7	−175
Uranus	86.8	8.9	84.3	−210
Neptune	102	11.2	165	−220
Pluto	0.0127	0.6	248	−240

6 Use the information provided in the tables to produce graphs that illustrate one of the following:
   • how **gravitational field strength** varies with mass
   • how length of year varies with distance from the Sun
   • how surface temperature varies with distance from the Sun.

Your graphs should be X–Y scatter graphs, and you should draw best-fit lines if the graph shows a pattern in the data.

If you are working as a group, divide the graphs among yourselves.

Comparing the graphs

7 When your group has finished drawing their graphs, discuss them together and identify any patterns that they show. Write a short paragraph about each in your book, explaining your ideas.

Glossary

Abdomen The lower part of the body separated from the thorax by the diaphragm

Absorbency A measure of how much liquid a material can hold

Acid A substance with a pH less than 7

Acid rain Rain that is more acidic than normal

Adaptation A feature that helps an organism to survive in its habitat

Aerobic respiration The type of respiration that uses oxygen

Agar plate A sterile dish containing nutrients in an agar jelly on which micro-organisms are grown

Air pressure The push of the air on all objects, due to the weight of the atmosphere above us

Alkali A substance with a pH above 7 which can dissolve in water

Alveoli The tiny air sacs that provide a large surface area for gas exchange in the lungs

Amplitude The distance from the equilibrium position to the peak of a wave

Anaerobic respiration The release of energy from glucose without oxygen

Angle of incidence The angle between the incoming wavefront and the reflector; also the angle between the incident ray and the normal

Angle of reflection The angle between the reflected wavefront and the reflector; also the angle between the reflected ray and the normal

Antagonistic muscle action Muscles work together in pairs: as one muscle contracts to pull a bone the other muscle relaxes

Antibiotic A medicine that either kills bacteria or stops them growing. Viruses are not affected by antibiotics, so antibiotics cannot be used to treat a viral infection

Antibody A chemical produced by white blood cells that kills microbes or makes them clump together

Antitoxin A chemical produced by white blood cells that neutralises a poison produced by a microbe

Archimedes' principle An object placed in water experiences an upthrust force that is equal to the weight of the water it displaces

Arctic circle The southernmost part of the northern hemisphere that will see 24 hours of daylight in midsummer. All areas north of this will see the midnight Sun at some point, while no areas south of this will do so

Artificial selection When people select organisms with useful characteristics to breed from them

Asthma A long-term condition that causes coughing, wheezing and shortness of breath

Atmosphere The layers of gases that surround the Earth

Atmospheric pressure The pressure at the Earth's surface, due to the weight of air above us; approximately 100 kPa

Atom A small particle that makes up all substances

Bacteria (singular 'bacterium') Unicellular organisms, some of which cause disease; others are useful to us, e.g. to make yoghurt

Ball and socket joint A joint where the ball-shaped end of one bone fits into a cup-shaped socket in the other bone; a type of synovial joint

Base A substance with a pH above 7 which cannot dissolve in water

Biceps The muscle at the front of the upper arm that contracts to bend the arm

Billion One thousand million

Biodegradable A substance that can be broken down by living organisms (bacteria)

Biodiversity The wide range of living organisms that exist

Biogas The gas produced when bacteria break down plant and animal wastes during anaerobic respiration

Biogas generator A vessel for producing and collecting biogas

Black hole One possible fate for a star that has gone supernova – all of its matter is compressed into an infinitesimal point, of infinite density. Not even light can escape a black hole

Breathing The movement of muscles to alter the volume of the chest cavity and bring air into the lungs

Breathing system An organ system made up of the lungs, trachea, rib cage and diaphragm. Its function is to move air in and out of the lungs

Bronchiole A narrow airway that branches off the bronchus and carries air to the alveoli

Bronchus A tube that leads off the trachea and carries air into a lung. There are two bronchi: one connected to the right lung and the other connected to the left lung

Camera obscura a darkened room with a small hole in one wall (possibly with a lens to project the image) and a screen, to make an image of the outside world

Carbon monoxide A poisonous gas that reduces the amount of oxygen carried in the blood

Cartilage The substance that forms the ears, nose and sternum. It is found on the ends of limb bones and between the vertebrae to reduce friction and act as a shock absorber

Cavity wall insulation A foam or fibre filling placed in between the outer and inner layers of a building's outer wall, reducing heat loss through the wall

Charging by friction Making electrons jump from one object to another by touching or rubbing them together

Charging by induction Separating out the charges in an uncharged object by bringing another, charged, object close to it, making one end of the original object positively charged and the other end negatively charged

Chromosomes The threadlike structures that carry genes found in the nuclei of cells

Cilia The tiny hair-like structures on the cells lining the trachea

Circulatory system The system made up of the heart, blood vessels and blood; it transports substances around the body

Cobalt chloride paper The paper used to detect water, which turns from blue to pink when water is present

Collide Hit, or bump into

Combustion A reaction where a substance burns and reacts with oxygen

Competition When resources are limited, organisms will compete with each other for them; organisms that are better adapted are more likely to survive

Compression A region in a longitudinal wave where the particles are closer together than normal

Concentrated Containing high amounts of a substance

Conduction (of heat) The transfer of heat by particle vibrations passing along through bonds, or by free electrons moving in metals (occurs mainly in solids)

Conductor (of heat) A substance that allows heat to pass through it easily, such as a metal

Cones Cells in the retina that detect colour

Control variables The things that must be kept the same in an experiment to make sure they do not affect the dependent variable

Convection current The pattern of circulation produced when hot, less dense fluid rises and colder, denser fluid sinks

Convection Heat transfer by movement of a liquid or gas due to changes in density

Convex lens A lens that is wider in the middle than it is at the edges, which brings light to a point called a focal point

Core The central part of the Earth

Cornea A transparent surface at the front of the eye

Corona The Sun's atmosphere, not normally visible because the surface of the Sun is so bright, but visible during an eclipse because the Moon blots out the Sun's surface

Coronary arteries The blood vessels that carry blood to the heart

Crust The solid outer layer of the Earth

Crystalline Made up of crystals

Decomposition When a compound is split into two or more simpler substances

Density The ratio of an object's mass to its volume; the greater the density, the more 'compact' an object is

Deoxygenated blood Blood that is low in oxygen and rich in carbon dioxide

Dependent variable The outcome of an experiment – the variable that changes, depending on the other variables in the experiment

Depressant A drug that slows down the nervous system and makes you feel more relaxed

Diaphragm In Biology: the sheet of muscle that separates the thorax from the abdomen. It is also involved in breathing

Diaphragm In Physics: a thin membrane in a microphone that detects sound by oscillating at the same frequency as the incoming sound

Diffuse reflection How light reflects from matt surfaces, in random directions, without forming an image; sometimes called scattering

Diffusion Movement of particles from an area where they are in high concentration to an area where they are less concentrated

Dilute Contains a lower proportion of particles of a substance as water has been added

Dispersion Splitting white light into all the colours of the spectrum

Displace To push aside, as in displacing a volume of water equal to the volume of the object immersed in water

Displacement A reaction in which a more reactive element replaces a less reactive one in a compound

Dissociation The separation of ions when a substance dissolves in water

Distance-multiplying lever A lever in which the effort is larger than the load, but the load moves through a greater distance than the effort

DNA The chemical that makes up genes and chromosomes

Double glazing A window made from two panes of glass separated by an air gap, thus reducing heat loss by convection and conduction

Drug A chemical that alters how the body works

Ductile Can be pulled into thin strands like wires

Eardrum A thin layer of skin within the ear that oscillates at the same frequency as incoming sound

Echo Reflected sound

Echolocation Finding the distance or position of objects (e.g. prey) by bouncing sound or ultrasound waves off them

Effervescence The escape of gas from a solution, recognised by bubbles rising

Effort The force exerted on a lever

Electric current (I) The rate of flow of electric charge

Electric field The area around a charged object where the electrostatic force can be felt

Electrodes A conductor that can be charged to pass electricity through an electrolyte

Electrolyte A non-metallic substance that can conduct electricity when in solution or when molten

Electrostatic force The force of repulsion or attraction between electric charges

Element A substance made from the same type of atoms

Emitter Something that gives off (heat) radiation

Endothermic A reaction where energy is taken into the reaction from the surroundings. A temperature drop would be measured

Environmental variation The variation in organisms caused by environmental factors such as diet

Equilibrium Balanced

Equilibrium position Where the medium would lie if there were no waves travelling through it

Erosion The removal and transportation of rocks, by wind or water

Ethanol The type of alcohol found in alcoholic drinks

Evolution A gradual change in a species, usually over a long period of time; it may result in a new species being formed

Exhaled air The air we breathe out

Exothermic A reaction where energy is given out from the reaction into the surroundings; a temperature rise would be measured

Extinct When the last member of a species has died

Extraction Separating a useful raw-material from other substances; may be physical or chemical

Fermentation Another name for anaerobic respiration

First-order lever A lever that has its pivot in the middle

Fixed joint A joint where bones are fused together; found in the skull

Fluid A liquid or a gas, i.e. matter that can flow

Focal point The point at which parallel rays of light cross after passing through a convex lens

Force An action that can stretch or compress an object, or cause it to speed up, slow down or change its direction of motion

Force-multiplying lever A lever in which the load is larger than the effort

Freak wave *See* rogue wave

Free electrons Electrons that are free to move within a metal, making them good conductors both of electricity and of heat

Frequency The number of oscillations, or waves, per second

Fulcrum *Same as* pivot

Full moon When the Earth lies between the Sun and the Moon, so the whole of the side of the Moon visible from the Earth is sunlit

Fungi Simple organisms, some of which cause disease; others are useful to us, e.g. to make bread

Galaxy A system of millions of stars bound together by gravity

Gas exchange The movement of oxygen from air into the blood, and the movement of carbon dioxide in the opposite direction

Gene The unit of inheritance that codes for a specific protein; made of DNA

Gene banks Stores of tissue or cell samples from endangered species

Geocentric A model of the solar system that puts the Earth at the centre, with the Sun and the other planets orbiting it

Global warming The increase in average temperature of planet Earth

Goblet cell A cell found in the trachea that produces mucus

GPS satellites Artificial satellites that provide the signals for the Global Positioning System network, which is used by many devices, including satnavs and smartphones

Greenhouse effect When heat from the Sun is trapped by the atmosphere

Hallucinogen A drug that makes you imagine things that are not there

Heart The muscular organ that pumps blood around the lungs and body

Heat The transfer of energy from a hot object to a cold one

Heat radiation *Same as* infra-red radiation

Heliocentric A model of the solar system that puts the Sun at the centre, with the planets orbiting it

Hertz (Hz) The unit of measurement of frequency; 1 Hz = one oscillation per second

Hinge joint A joint that allows movement in one direction, e.g. the elbow or the knee; a type of synovial joint

Humidity The amount of moisture in the air

Infrared radiation Heat transfer by an electromagnetic wave with a wavelength just outside the visible spectrum, beyond red

Infrasound Sounds with frequencies below the range of human hearing, i.e. below 20 Hz

Inhaled air The air we breathe in

Inherited variation The variation in organisms that is controlled by genes. Also called 'genetic variation'

Insulator A substance that does not let heat pass through it very easily, such as wool

International Space Station The largest spacecraft ever made, placed in orbit around the Earth, and used as an orbital science laboratory and living quarters for up to six astronauts at a time

Immunity When the body can quickly produce antibodies against a particular microbe to kill it, so you do not become ill; you are immune to it

Independent variable The variable that is changed in an experiment to see what effect it has

Ion A charged particle

Iris The coloured part of the eye, a ring of muscle that changes size to control the amount of light entering through the pupil

Joint Where bones meet

Kilowatt-hour (kWh) A measure of energy used by electricity companies; it is the amount of energy used when a 1 kW appliance is run for 1 hour; also called a 'unit' of electricity

Lactic acid The chemical produced during anaerobic respiration that can cause cramp

Larynx The part of your throat containing your vocal cords, where sounds are made

Lens A shaped piece of glass made to focus light rays, or a part of the eye, made of cells, which focuses light on the retina

Lever A device using a long rod and a pivot, used to increase the force on an object, change the force's direction, or increase the distance over which the force acts

Ligament The structure that attaches one bone to another and holds them together in a joint

Light year The distance that light will travel in a year

Limewater A liquid used to test for carbon dioxide; it changes from clear to cloudy with carbon dioxide

Load The force that a lever acts against

Loft insulation A thick layer of loose fibre that traps air pockets that is laid on the floor of a loft to reduce heat loss from the house below; often made of mineral wool, but also sometimes made from sheep's wool or recycled plastic.

Longitudinal wave A wave in which the medium oscillates parallel to the direction that the wave travels

Lunar eclipse When the Earth casts its shadow on the Moon's surface, often turning it a blood red colour

Lungs The organs in the thorax where gas exchange occurs

Lymphocyte The type of white blood cell that makes chemicals called antibodies

Magma Molten rock under the Earth's surface

Malleable Easy to shape

Mantle The layer of the Earth between the core and the crust

Medicine A drug that is used to treat or prevent disease

Medium The substance that a wave travels through

Meteorologists Scientists who study weather and climate

Methane A flammable gas used as a fuel. It is produced when bacteria break down plant and animal wastes

Microbiology The study of micro-organisms such as bacteria, fungi and viruses

Micro-organisms Small organisms that can only be seen with a microscope; also called 'microbes'

Milky way Our galaxy; a collection of 100 000 million stars which orbit their common centre due to gravity. Visible as a band of light lying across the night sky

MMR vaccine The vaccine given to provide immunity against measles, mumps and rubella (rubella is also called 'German measles')

Moment The turning effect of a force, equal to the force multiplied by the perpendicular distance from the pivot

Mucus The thick, sticky secretion produced in the lungs, for example

Multicellular organism An organism that is made up of many cells

Nanotechnology Using nanoparticles, which are 1 millionth of the size of a millimetre

Natural selection When the organisms best suited to their environment survive and breed, and those less suited to their environment die out

Nebula A vast cloud of gas and dust in space

Neutralisation A reaction between an acid and an alkali where a salt and water are formed

Neutron star One possible fate for a star that has gone supernova – all of its matter has been compressed until it is made of nothing but neutrons

New Moon When the Moon is between the Sun and the Earth, so the sunlit side of the Moon is not visible from the Earth

Nicotine The very addictive drug in cigarette smoke

Non-contact force A force between objects that can be felt at a distance, without them actually touching

Normal (normal line) An imaginary line used to measure angles with rays, which is drawn at right angles to a surface

Northern hemisphere The half of the Earth that lies north of the equator, including the land masses of Europe, North America and Asia

Nuclear fusion Hydrogen nuclei combining under immense temperatures and pressures to form helium nuclei, and releasing heat

Nucleus (plural 'nuclei') The structure that contains instructions to control the cell

Octave A scale or separation of eight musical notes, between two notes of the same name; notes that are one octave apart are the same note, but higher or lower in pitch

Opaque Something that blocks all light that falls on it

Optic nerve The nerve that takes the signal from the retina to the brain

Orbit The path that one object takes travelling around another object, for example the Earth travelling around the Sun during the course of a year, or the Moon travelling around the Earth over a month

Ore A rock from which a metal can be extracted

Oscillation Vibration about a fixed position

Oscilloscope A device for displaying waveforms on a screen, in order to make measurements of them

Oxidation A reaction where oxygen is gained

Oxygenated blood Blood rich in oxygen and low in carbon dioxide

Oxygen debt The amount of oxygen needed to break down the lactic acid built up during anaerobic respiration in the muscles

Ozone A chemical substance, O_3, which absorbs ultraviolet radiation from the Sun

Pathogen An organism that causes a disease, for example a virus

Peak The highest point on a transverse wave

Penicillin The first antibiotic that was developed

Phagocyte The type of white blood cell that engulfs and digests microbes

Phagocytosis The process of engulfing and digesting microbes carried out by a phagocyte

Pinhole camera An optical device made from a small hole opposite a translucent screen, on which an image can be projected

Pitch How high or low a musical tone is

Pivot The point or line about which an object turns, e.g. the point that a see-saw balances on; also called the fulcrum

Pivot joint A joint that allows side-to-side movement, e.g. in the neck

Plane Flat, as in a plane mirror

Planetary nebula A nebula formed when a red giant star loses its outer layers, which float off into space

Plasma The liquid part of the blood that carries dissolved chemicals such as carbon dioxide and food

Platelets Fragments of cells in the blood that help to form a scab if you get cut

Population The number of organisms of one species in an area

Potential difference (p.d.) The energy that a battery or cell gives to the charge that it pushes around a circuit

Precision How small a reading you are able to take in a particular measurement

Pressure The ratio of force applied to the area it is applied over, or the force per unit area

Primary colours (of light) The three colours that can be mixed in different proportions to make all of the colours that we see: red, green and blue

Principle of moments An object is balanced when the sum of all anticlockwise moments equals the sum of all clockwise moments

Prism Any 3D solid shape; these are often made of glass for optics experiments

Products The new chemical substances formed in a chemical reaction

Properties Features of a chemical substance or material

Prosthesis An artificial replacement for a body part

Proton acceptor A substance that can combine with a proton (also called a hydrogen ion, H^+)

Proton donor A substance that can give up a proton (also called a hydrogen ion, H^+)

Protostar A sphere of gas formed from the gravitational collapse of a nebula. It glows due to its high temperature, but is not yet hot enough for nuclear fusion to make it shine as a star

Pupil A hole of varying size (controlled by the iris) which allows light to enter the eye

Pure Made up of a single substance

Radiant heater A heater that gives off heat as radiation, rather than by convection

Radiation The transfer of energy from a central point. This word is usually used when referring to infra-red rays, microwaves, or emissions from the nucleus of an unstable atom

Rarefaction A region in a longitudinal wave where the particles are further apart than normal

Reaction rate How quickly a reaction occurs. Monitored as change over time

Reactivity series A list of elements in order of how reactive they are

Real image An image that light actually reaches or passes through, e.g. the image on a cinema screen

Recovery time The time it takes for the heart and breathing rates to return to normal after exercise

Recreational drug A drug that people take because they like the effect it has on them, e.g. alcohol and tobacco

Recycling Changing waste materials into new products

Red blood cells The cells that transport oxygen around the body

Red dwarf A star much smaller and cooler than our Sun, which glows red

Red giant What a star becomes when it has used up all of its hydrogen fuel. A red giant is hundreds of times larger than a normal star

Reducing agent A substance that can remove oxygen from another compound

Reduction A reaction where a substance loses oxygen

Refraction How a wave bends as it passes from one medium to another, due to a change in wave speed

Respiration The release of energy from food

Retina A light-sensitive layer of cells at the back of the eye

Rogue wave An extremely tall wave that appears as if from nowhere on otherwise normal seas

Rover A wheeled robot that can explore an alien planet's surface, as on Mars

Salt A compound formed in the reactions of acids

Satellite An object that orbits a planet, either a natural satellite (a moon) or an artificial satellite – put there by humans

Secondary colours Colours made from mixing two primary colours in equal proportions: cyan, magenta and yellow

Second-order lever A lever that has the load in the middle

Selective breeding When humans select organisms with desirable characteristics and breed from them

Semen The liquid that a man ejaculates from his penis during sex, which contains the sperm cells

Skeletal system The organ system made up of bones and muscles. Its functions include support of the body, movement, protection of body organs and production of blood cells

Smart material Material where the properties change depending on the environment it is in

Snow shoe A shoe with a large area of sole that keeps the pressure underneath it low, so that it does not sink in snow

Solar eclipse When the Moon comes directly between the Sun and the Earth, so that observers standing on the Earth under the Moon's shadow see the Moon blocking out the Sun

Solar prominences Loops of hot gases, hundreds of times the volume of the Earth, thrown out from the Sun's surface into the corona, and visible during a solar eclipse

Solar system The Sun, and all of the planets and other objects (such as asteroids and comets) that orbit it

SONAR A device used by ships and submarines to find the depth of the sea, or the distance to underwater objects

Sound wave A longitudinal wave that is passed along by vibrating particles in a solid, liquid or gas

Source (of light) An object that gives off its own light

Southern hemisphere The half of the Earth that lies south of the equator, including the southern half of Africa, most of South America and Australia

Space Shuttle A partially reusable spacecraft, built by the USA's National Aeronautics and Space Administration (NASA). No longer in service

SpaceShipOne The first truly reusable spacecraft, built by Scaled Composites of California, USA. It launches from an aeroplane and lands like a glider

SpaceShipTwo A larger spacecraft, based upon spaceshipone, to be flown by Virgin Galactic, providing flights for space tourists

Species One particular type of organism; organisms of the same species can interbreed to produce fertile offspring

Spectrum White light split into the colours of the rainbow, from red to violet

Specular reflection How light reflects from very smooth or shiny surfaces, forming an image

Star cluster A group of stars that formed from the same nebula

Static electricity When electrons jump from one object to another, making one positive and the other negative. This causes them to attract or repel other charged objects

Stimulant A drug that speeds up the nervous system and makes you feel more alert

Stoma (plural 'stomata') A hole/pore found mainly in the lower surface of a leaf

Strain A variety of bacterium

Subatomic Smaller in size than the atom

Supergiant The penultimate stage in a giant star's life, like a red giant but much larger

Supernova The explosion of a supergiant star, which creates all of the heavy elements and throws them out into space

Survival of the fittest When organisms with adaptations most suited to the environment survive and breed

Synovial fluid The fluid found in freely movable joints that reduces friction between bones

Synovial joint A freely movable joint that contains synovial fluid

Synthetic Manufactured or human-made

Tar A thick, poisonous substance that contains chemicals that cause cancer

Telecommunication satellites Artificial satellites used to relay telephone or television signals between countries

Tendon The tough cord that attaches muscle to bone

Thermal decomposition A reaction where a substance is broken down using heat

Thermal store The energy of a substance due to the random motion of its particles

Third-order lever A lever that has the effort in the middle

Thorax The upper part of the body, separated from the abdomen by the diaphragm

Trachea The tube that carries air from the mouth and nose into the lungs

Transducer A device, such as a microphone or a loudspeaker, that shifts energy from one store into another

Translucent Something that allows light to pass through it, but does not transmit an image

Transparent Something that allows light to pass so easily that an image can be seen through it

Transverse wave A wave in which the oscillations are at right angles to the direction of wave motion

Triceps The muscle at the back of the upper arm that contracts to straighten the arm

Trillion One thousand billion, or one million million (*see also* billion)

Trough The lowest point on a transverse wave

Ultrasound Sound with a frequency above 20 kHz, or above the range of human hearing

Ultraviolet light Electromagnetic radiation just outside the visible spectrum, beyond violet

Unicellular organism An organism that is made up of one cell only

Upthrust The upwards force of a fluid on an object within it, e.g. the force that keeps a boat afloat in water or a helium balloon afloat in air

Vaccination The process that involves taking or injecting a vaccine to make you immune to a particular disease

Vaccine The substance containing either a weakened or dead form of a micro-organism

Vacuum flask A flask made of a double-walled bottle with a vacuum between its walls, which will keep hot drinks hot and cold drinks cold

Virtual image An image that the light rays do not actually pass through, e.g. the image in a mirror

Virus A very small organism that causes disease

Voicebox *See* larynx

Weathering The breaking down of rocks into smaller pieces

Wave A regular oscillation passing through a medium that transfers energy without moving the medium

Wavefront A line drawn along the top of a water wave, showing where the crest or peak is

Wavelength The distance between two adjacent peaks, or between two adjacent troughs in a wave

White blood cells The cells that defend the body against disease

White dwarf The core of a star, left behind when a red giant has lost its outer layers as a planetary nebula

Withdrawal symptoms The symptoms such as sickness and anxiety experienced by people who stop using a drug that they are addicted to

Index

Free online extras

For each topic you can find free:

- Hints for all the Questions in this book
- Extended glossary definitions and examples
- Quick quizzes to test your knowledge.

Scan the QR codes below for each topic.

Alternatively, you can browse to www.hodderplus.co.uk/scienceprogress.

→ How to use the QR codes

To use the QR codes you will need a QR code reader for your smartphone/tablet. There are many free readers available, depending on the smartphone/tablet you are using. We have supplied some suggestions below, but this is not an exhaustive list and you should only download software compatible with your device and operating system. We do not endorse any of the third-party products listed below and downloading them is at your own risk.

- for iPhone/iPad, search iTunes for Qrafter

- for Android, search the Play Store for QR Droid

- for Blackberry, search Blackberry World for QR Scanner Pro

- for Windows/Symbian, search the Store for Upcode

Once you have downloaded a QR code reader, simply open the reader app and use it to take a photo of the code. You will then see a menu of the free resources available for that topic.

→ Biology

Topic 7 Lungs and gas exchange

Topic 8 Respiration

Topic 9 Muscles and bones

Topic 10 Inheritance and evolution

Topic 11 Drugs and health

Topic 12 Microbes

→ Chemistry

Topic 7 Periodic Table

Topic 8 Extracting metals

Topic 9 Reactions of acids

Topic 10 Describing reactions

Topic 11 Earth and atmosphere

Topic 12 Innovative materials

→ Physics

Topic 7 Domestic and static electricity

Topic 8 Waves and sound

Topic 9 Light

Topic 10 Application of forces

Topic 11 Heat transfer

Topic 12 Exploring space

# Acknowledgements

The Publisher would like to thank the following for permission to reproduce copyright material:

p.6 © Dept. of Nuclear Medicine, Charing Cross Hospital/Science Photo Library; **p.7** © Steve Gschmeissner/Science Photo Library; **p.8** © Alexis Rosenfeld/Science Photo Library; **p.9** *t* © Dr Jeremy Burgess/Science Photo Library, *b* © Pavel Chernobrivets – Fotolia.com; **p.10** © Science Source/Science Photo Library; p.12 © Steve Allen/Science Photo Library; p.13 both © Science Photo Library; **p.14** © Arkady Chubykin – Fotolia.com; **p.15** © age fotostock/Alamy; **p.16** © Sergej Razvodovskij – Fotolia.com; **p.18** *t* © Philippe Plailly/Science Photo Library, *m* © Pete Saloutos – Fotolia.com, *b* © Jean-Yves Ruszniewski/TempSport/Corbis; **p.20** *from t to b* © ematon – Fotolia.com, © vladi59 – Fotolia.com, © Ellen Isaacs/Alamy, © Paul Walters Worldwide Photography ltd/HIP/TopFoto; **p.21** © S Teutsch/Look at Sciences/Science Photo Library; **p.23** *l* © Pete Saloutos – Fotolia.com, *r* © Steve Gschmeissner/Science Photo Library; **p.24** © Simon Fraser/Science Photo Library; **p.26** *t* © Jale Ibrak – Fotolia.com, *bl* © Eric Grave/Science Photo Library, *bm* © Innerspace Imaging/Science Photo Library, *br* © Science Photo Library; **p.28** *t* © Mehau Kulyk/Science Photo Library, *m* © B Christopher/Alamy, *b* © Zephyr/Science Photo Library; **p.29** © Philippe Psaila/Science Photo Library; **p.30** *t* © maurice joseph/Alamy, *b* © Martyn F. Chillmaid/Science Photo Library; **p.32** *tl* © lichtreflexe – Fotolia.com, *tm* © flucas – Fotolia.com, *tr* © Terrill White/Wikipedia Commons (http://commons.wikimedia.org/wiki/File:Mule_(1).jpg), *b clockwise from tl* © MAK – Fotolia.com; © sp4764 – Fotolia.com, © martyns2011 – Fotolia.com, © Laguna Design/Science Photo Library, © fotoliaxrender – Fotolia.com; **p.33** *t* © biglama - Fotolia.com, *bl* © M & J Bloomfield/Natural Visions, *br* © Stephan Morris; **p.34** *t* © Studiotouch – Fotolia.com, *b* © Pr. G Gimenez-Martin/Science Photo Library; **p.35** *both* © CNRI/Science Photo Library; **p.36** *t* © Diyanski – Fotolia.com, *m* © Paul D Stewart/Science Photo Library, *b* © andreanita – Fotolia.com; **p.37** *t* © Steve Gschmeissner/Science Photo Library, *b* © Bon Appetit/Alamy; **p.38** *t* © David Davis/Science Photo Library, *b* © The Natural History Museum/Alamy; **p.39** *tl* © Gail Johnson – Fotolia.com, *tm* © lunamarina – Fotolia.com, *tr* © jptenor – Fotolia.com, *br* © Oli Scarff/Getty Images; **p.40** © Xavier – Fotolia.com; **p.42** *t* © yogesh more/Alamy, *in table from t to b* © volff – Fotolia.com, © volff – Fotolia.com, © Marc Dietrich – Fotolia.com, © Opra – Fotolia.com; **p.44** *t* © Medical-on-Line/Alamy, *m* © mgp – Fotolia.com, *b* Ian Boddy/Science Photo Library; **p.45** © Image Source/SuperStock; **p.46** *t* © Lee Powers/Science Photo Library, *m* © Stuart Boulton/Alamy, *b* © MARKA/Alamy; **p.47** © Janine Wiedel Photolibrary/Alamy; **p.50** *both* © Science Photo Library; **p.51** *t* © David Scharf/Science Photo Library, *b* © NIBSC/Science Photo Library; **p.52** *t* © Tyler Olson – Fotolia.com, *b* © Eric Grave/Science Photo Library; **p.54** © Dr Kari Lounatmaa/Science Photo Library; **p.55** © St Mary's Hospital Medical School/Science Photo Library; **p.56** *t* © John Durham/Science Photo Library, *b* © Wellcome Library, London; **p.57** © Medical-on-Line/Alamy; **p.58** © Serghei Veluşceac – Fotolia.com; **p.60** © si evans – Fotolia.com; **p.62** *l* Used with permission of Popular Science. Copyright © 2013. All rights reserved. *r* © Mike Walker Photography; **p.64** © W. Oelen/Wikipedia Commons (http://en.wikipedia.org/wiki/File:Halogens.jpg); **p.68** *tl* © Andrew Lambert Photography/Science Photo Library, *tr* © Andrew Lambert Photography/Science Photo Library, *b* © Mike Walker Photography; **p.69** © Leslie Garland Picture Library/Alamy; **p.70** *t* © vvoe – Fotolia.com, *b* © Greenshoots Communications/Alamy; **p.71** © RGB Ventures LLC dba SuperStock/Alamy; **p.72** © Science Photo Library; **p.74** © PHB.cz – Fotolia.com; **p.78** © Cristina Pedrazzini/Science Photo Library; **p.80** *t* © Stefan Redel – Fotolia.com, *b* © novotnyms – Fotolia.com; **p.82** © Phil Degginger/Alamy; **p.84** © Viktor Fischer/Alamy; **p.86** © Martyn F. Chillmaid/Science Photo Library; **p.87** © Andrew Lambert Photography/Science Photo Library; **p.88** © Jerry Mason/Science Photo Library; **p.90** © Charles D. Winters/Science Photo Library; **p.92** © Julian Finney/Getty Images; **p.94** © Crown Copyright/Health & Safety Laboratory /Science Photo Library; **p.98** © Yasar Simit – Fotolia.com; **p.100** *t* © Joe Gough – Fotolia.com, *bl* © Michael Szoenyi/Science Photo Library, *br* © Geoscience Features Picture Library; **p.101** © John Cancalosi/Getty Images; **p.102** *t* © siimsepp – Fotolia.com, *b* © Natural History Museum, London/Science Photo Library, **p.104** © University Corporation for Atmospheric Research/ Science Photo Library; **p.106** © trgowanlock – Fotolia.com; **p.107** *l* © Remarkable Ltd, *r* © M. Schuppich – Fotolia.com; **p.110** *t* © Nick Veasey/Getty Images, *b* © Martyn F Chillmaid; **p.112** © Pink Badger – Fotolia.com; **p.113** © Eye of Science/Science Photo Library; **p.114** *t* © Andy Crump/Science Photo Library, *b* © David J. Green - financial/Alamy; **p.115** © US Air Force/Science Photo Library; **p.116** *t* © Marek Cech – Fotolia.com; **p.117** © Cordelia Molloy/Science Photo Library; **p.120** © demarfa – Fotolia.com; **p.121** © shishiga – Fotolia.com; **p.122** *t* © Tim Jones/Alamy, *b* © Sheila Terry/Science Photo Library; **p.124** *t* © EggImages/Alamy, *b* © Lourens Smak/Alamy; **p.125** © Martyn F Chillmaid; **p.126** *t* © Alexandr Chubarov – Fotolia.com, *b* © the food passionates/Corbis; **p.127** © Peter Menzel/Science Photo Library; **p.128** © Digital Vision/Thinkstock; **p.130** *t* © 2011 Rex Features, *b* Image used courtesy of Pelamis Wave Power; **p.131** © Mikkel Juul Jensen/Science Photo Library; **p.132** *t* © ZUMA Press, Inc./Alamy, *b* © lausher – Fotolia.com; **p.134** © Ashley Cooper/Alamy; **p.135** © Kumar Sriskandan/Alamy; **p.136** © Action Plus Sports Images/Alamy; **p.137** © GE Medical Systems/Science Photo Library; **p.138** © auremar – Fotolia.com; **p.139** © Anna Omelchenko – Fotolia.com; **p.140** *t* © Jerry Lodriguss/Science Photo Library, © Soho-Eit/Nasa/Esa/Science Photo Library; **p.142** *t* Courtesy Rachel Pfleger, *b* © Alen MacWeeney/Corbis; **p.144** © Darius Kuzmickas (http://www.flickr.com/photos/kudaphoto); **p.146** *t* © David Parker/Science Photo Library, *b* © Alexander Kaludov/Alamy; **p.148** © Cordelia Molloy/Science Photo Library; **p.149** *l* © xalanx – Fotolia.com, *r* © bellan – Fotolia.com; **p.150** © RTimages – Fotolia.com; **p.152** © lunamarina – Fotolia.com; **p.153** © steamroller – Fotolia.com; **p.154** *t* © razmarinka – Fotolia.com, *m* © Lucky Dragon USA – Fotolia.com, *b* © Gina Sanders – Fotolia.com; **p.155** © Alexandr Makarov – Fotolia.com; **p.156** *t* © Pascal Goetgheluck/Science Photo Library, *m* © GIPhotostock/Science Photo Library, *b* USS TUNNY (SSN 682) http://www.ssn682.com; **p.157** © Martyn F Chillmaid; **p.158** *t* © Rafael Ben-Ari/Alamy, *b* © Alby (talk) (http://en.wikipedia.org/wiki/File:Pound-coin-floating-in-mercury.jpg) (This file is licensed under the Creative Commons Attribution-Share Alike 3.0 Unported license.); **p.160** *t* © astronoman – Fotolia.com, *m* © Aaron Amat – Fotolia.com; **p.162** © Lee Hacker/Alamy; **p.164** *t* © marytmoore – Fotolia.com, *b* © veneratio – Fotolia.com; **p.166** *t* © Mark Sykes/Science Photo Library, *bl & br* © Paul Das/Cotswold Thermal Imaging (www. cotswoldthermalimaging.co.uk); **p.168** *t* © Marc Trigalou/Getty Images, *ml* © Ekaterina Garyuk – Fotolia.com, *mr* © Ilya Akinshin – Fotolia.com, *b* © DWImages/Alamy; **p.170** *t* © Fotofermer – Fotolia.com, *b* © Food Features/Alamy, © Ingus Evertovskis – Fotolia.com, © nadezhda1906 – Fotolia.com, © babimu – Fotolia.com; **p.172** © Rev. Ronald Royer/Science Photo Library; **p.173** *t* © Doug Plummer/Science Photo Library; **p.174** *t* © NASA/ESA/STScI/Hubble Heritage Team/Science Photo Library, *m* © NASA, *b* © Hubble Heritage Team/NASA/ESA/STScI/AURA/Science Photo Library; **p.175** © NASA/ESA/STScI/Science Photo Library; **p.176** *t* © NASA, *b* © Sheila Terry/Science Photo Library; **p.178** *t* © Detlev Van Ravenswaay/Science Photo Library, *m* © MARS Scientific and Clay Center Observatory, *b* © B.A.E. Inc./Alamy; **p.179** *t* © NASA/Science Photo Library, *b* © NASA/Science Photo Library; **p.180** © Tristan3D – Fotolia.com; **p.181** © Lynette Cook/Science Photo Library.

t = top, *b* = bottom, *l* = left, *r* = right, *m* = middle

Every effort has been made to trace all copyright holders, but if any have been inadvertently overlooked, the Publisher will be pleased to make the necessary arrangements at the first opportunity.

Contents

Contents

iv Contents

Revision Guide

Cambridge International AS and A Level

Economics

Susan Grant

CAMBRIDGE

Windsor and Maidenhead

CAMBRIDGE UNIVERSITY PRESS
Cambridge, New York, Melbourne, Madrid, Cape Town,
Singapore, São Paulo, Delhi, Mexico City

Cambridge University Press
4381/4 Ansari Road, Daryaganj, Delhi 110002, India

www.cambridge.org
Information on this title: www.cambridge.org/9781107661783

© Cambridge University Press 2013

First Published 2013
Reprinted 2013

Printed in India by Sanat Printers, Kundli

A catalogue for this publication is available from the British Library

ISBN 978-1-107-66178-3 Paperback

Introduction

The structure of the book

This book is designed to help students revise both AS Level and A Level Economics. The first section provides guidance by discussing revision techniques and the techniques used in answering the three different types of questions students will face in examinations.

Section 2 consisting of Chapters 5 to 11 is devoted to AS Level Economics (Core), while chapters 12 to 17 of section 3 cater to A Level (Supplement). Each of these chapters contain a number of common features. They start by a recapitulation of the key economic content which would help students to recall the topics covered during the course. The mind maps are designed to show links within certain topics encouraging students to draw more mind maps as part of their revision. They may even wish to place some of these on their walls in school or at home.

The short questions at the end of each chapter provide a quick check on the reader's knowledge and understanding. The short questions are followed by a number of revision activities which seek to allow students to apply their skills built up to a range of different situations and tasks. The multiple choice questions are designed to give them practice of the types of questions one will face in the examination. The data response and essay questions also provide examination practice. In addition, they assess the readers' ability to explore topics in more detail and to demonstrate their analytical and evaluative skills.

Answers for all the questions are provided at the end for the benefit of the students. Students are instructed not to look at the answers to one section until they have attempted all the questions in that section. Also, bear in mind that the answers to the essay questions are suggested answers. No answer can cover all the possible points and equally valid answers can be structured in a slightly different way.

Cambridge International Examinations bears no responsibility for the example answers to questions taken from its past question papers which are contained in this publication.

The difference between IGCSE and AS/A Level

Some AS/A Level candidates have previously studied IGCSE Economics whilst others are new to the subject. In either case, it is important to remember that AS Level is a step up from IGCSE and, in turn, A Level (Supplement) is a step up from AS Level (Core). As one progresses up the levels, they are required to demonstrate higher order skills. At IGCSE Level there is, for instance, more emphasis on knowledge and understanding than at AS/A Level. The skills that become more important at AS/A Level are analysis, evaluation and judgement making.

At AS/A Level one is beginning to work as an economist. Economists working for a government, an NGO, bank or multinational company, for instance, analyse economic data, make judgements about the best strategies to follow and write reports which analyse the significance of events and government policy changes for their organisations. They will be able to write clearly, carry out numerical calculations and interpret and use statistics, graphs and diagrams. Through the examination questions one will be able to exhibit all of these skills.

At AS Level, the multiple choice questions have a 40% weighting, the data response question a 30% weighting and the structured essay a weighting of 30%. The AS Level papers provide half the weighting for A Level with remaining 50% being made up of 15% for the multiple choice questions, 10% for the data response question and 25% for the essay question.

When revising the A Level (Supplement) part of the syllabus, it is important not to forget the AS Level (Core) part of the syllabus. For instance, in considering sources of market failure, the student also has to consider externalities, public goods, merit and demerit goods which are first encountered at AS Level.

Coverage

Chapter 5 covers basic ideas including scarcity, opportunity cost, different economic systems, production possibility curves, factors of production as well as division of labour and money. Chapter 6 focuses on demand and supply. It covers the nature and determinants of demand and supply, elasticities and changes in market conditions. Whilst Chapter 6 concentrates on the free workings of the price system, Chapter 7 examines the reasons why governments intervene in the price system including externalities, public goods, merit goods and demerit goods. It also examines some of the ways in which governments intervene – maximum price control, price stabilisation, taxes, subsidies and the direct provision of goods and services.

In Chapter 8, the focus switches from the national to the international economy. The basis of international trade, arguments for and against protectionism, trade blocs, the terms of trade and the composition of the balance of payments are considered. Chapter 9 covers the labour market, unemployment, the general price level and aggregate demand and supply. In Chapter 10, the causes and consequences of inflation and balance of payments problems are explored and the factors influencing the value of exchange rates and the effects they have on the economy are discussed. Chapter 11 is the last chapter in the AS Level (Core) part of the book. It covers macroeconomic policies that can be used to correct balance of payments disequilibrium and to influence the exchange rate as well as some policy conflicts.

Chapter 12 is the first of the A Level (Supplement) chapters. It is a short chapter concentrating on efficiency. It is followed by Chapter 13 which explores demand and supply in more depth than in Chapter 6. It examines marginal utility theory and costs of production. It examines the labour market in some depth, considers the benefits that large firms may gain from economies of scale and why, despite these benefits, so many small firms exist. The different market structures of monopoly, oligopoly, monopolistic competition and perfect competition are assessed and compared.

Chapter 14 returns to the reasons why governments intervene in the price system and the microeconomic policy measures they employ. Chapter 15 is another relatively short chapter covering measures of national output and living standards, the money supply, Keynesian and Monetarist schools of thought, aggregate expenditure, the multiplier, the accelerator and theories of interest rate determination.

Chapter 16 examines economic growth and development, the causes and consequences of unemployment and the links between the macroeconomic problems of balance of payments deficits, inflation and unemployment. The final chapter, Chapter 17, explores fiscal, monetary and supply side policies which can be used to tackle macroeconomic problems.

Study skills

This book aims both to strengthen the readers' understanding of the economics they have covered during their course and to develop their study skills. It is designed to improve their ability to interpret and draw diagrams, interpret other forms of data, undertake numerical calculations and to write lucid and well structured answers.

This book is designed to make revision more effective and hopefully increase the students' enjoyment of the subject enabling them to excel in the examinations.

Acknowledgement

The following is reproduced by permission of Cambridge International Examinations.

Syllabus Name and Code	Paper and Question Number	Month/Year	Page/chapter in book
Cambridge International AS & A Level Economics 9708	Q4 Paper 9708/02	May/June 2005	14
Cambridge International AS & A Level Economics 9708	Q2 Paper 9708/21	May/June 2009	27
Cambridge International AS & A Level Economics 9708	Q2(a) Paper 9708/02	Oct/Nov 2008	27
Cambridge International AS & A Level Economics 9708	Q2 Paper 9708/02	May/June 2008	27
Cambridge International AS & A Level Economics 9708	Q1 Paper 9708/21	Oct/Nov 2009	44
Cambridge International AS & A Level Economics 9708	Q1 Paper 9708/02	Oct/Nov 2008	45
Cambridge International AS & A Level Economics 9708	Q2 Paper 9708/02	May/June 2006	45
Cambridge International AS & A Level Economics 9708	Q1 paper 9708/21	May/June 2007	57/58
Cambridge International AS & A Level Economics 9708	Q4 Paper 9708/22	May/June 2010	59
Cambridge International AS & A Level Economics 9708	Q3 Paper 9708/02	Oct/Nov 2010	59

Part 1   Techniques

Revision Techniques 1

Revision Objectives

After you have studied this chapter, you should be able to:

☞ understand the techniques used in revision
☞ know the purpose of revision in economics
☞ outline the appropriate place to revise

☞ distinguish between ways to and not to revise
☞ identify the various methods used to revise.

1.1 What is revision

Many students think that revision is cramming facts just before an examination. The word 'revision' actually means to review work. This is something that you should do throughout your course. If you spend at least ten minutes after each lesson checking on what you have just learned, this will increase your understanding of the topics and help you remember the key features. It may also encourage you to add notes on topics you feel uncertain about or are particularly interested in, and to ask your teacher for clarification on some aspects. You should also review all your work on a regular basis, for example, every two weeks. By doing this, you should already be relatively familiar with the work you understand by the time you come to your final revision.

1.2 The purpose of revision in economics

The main purpose of revision in economics is not to learn facts. It is to develop your skills including your ability to make links between causes and effects, problems and policies, theory and real world examples. For instance, you should recognise that a rise in aggregate demand would be expected to raise real GDP and lower unemployment. You also need to develop the ability to think critically about the sequence of events and to question the extent to which the events will happen and, indeed, whether

they will happen at all. If you start to think like an economist, you will enjoy the subject more and you will perform better in examinations.

1.3 Where to revise

The best place to revise is influenced by how you are revising and your level of concentration. For instance, one of your revision activities may be to write an answer to an essay question. In this case, it would be useful to write in a room where you have access to a desk. On another occasion, you may be getting a fellow student to test you with oral questions. In this case, it would be sensible to be in a room where the two of you cannot be disturbed. Besides, different people have different levels of concentration.

1.4 Ways not to revise

One of the most common ways that students try to revise is also one of the least effective. This way is just to re-read through class notes. Students who do this are trying to remember a mass of information which may have little meaning for them.

You must also not leave your revision to the last minute. If you do this you will feel very pressurised and you are unlikely to do well.

In addition, it is important that you do not revise for long periods of time or when you are tired. Diminishing returns can set in relatively quickly.

1.5 Ways to revise

Throughout your course, regularly check that you have covered all aspects of the syllabus. Use the syllabus to inform your revision activities.

The best way to revise is to engage in active revision. This means to process the information you have learned, in order to develop your understanding, not only of the topics but also the links between topics. It also means developing your skills of application, analysis and evaluation.

Undertaking a variety of revision methods will be both more rewarding and interesting. There are a number of ways you can revise and below are a number of suggestions.

1.6 Revision methods

Adding to your notes

As you progress through the course and check through your notes, you should get used to adding to them. You should do this for a number of reasons including:

✓ to fill in any gaps
✓ to provide relevant examples
✓ to show links to other topics
✓ to take into account recent developments.

Revision cards

These do not have to be left to just before the examination. After you have covered a topic you can draw up revision cards which you can look at later. Keep the card to a size that is easy to carry around. Do not put too much information on one card and leave space in case you want to add to it later. You may also want to colour code it – for example, green cards for Core Section 2 etc.

Following is an example of a revision card on elastic demand.

Elastic Demand

Definition: a % change in price causes a greater percentage change in demand.

PED > 1

Characteristics of products with elastic demand.

✓ Have close substitutes – key influence
✓ Luxuries
✓ Take up a large % of income
✓ Purchase can be postponed
✓ Not habit forming.

Tables

It is useful to draw up tables as these can enable you to process information and make comparisons. For example, you might find it useful to compare demand-pull and cost-push inflation.

Try filling in the table below in Activity 1.1.

Activity 1.1	Demand-pull inflation	Cost-push inflation
Definition		
Illustrated by		
Examples of causes		
Impact on real GDP		
Policy to reduce		

Mind maps

Mind maps can also be called spider diagrams. You can put a topic in the centre and then move out to show connected points. You may find it useful to draw up mind maps with fellow students and you may want to put some up on your wall.

Try completing the following mind map on Cambridge Economics A Level in Activity 1.2.

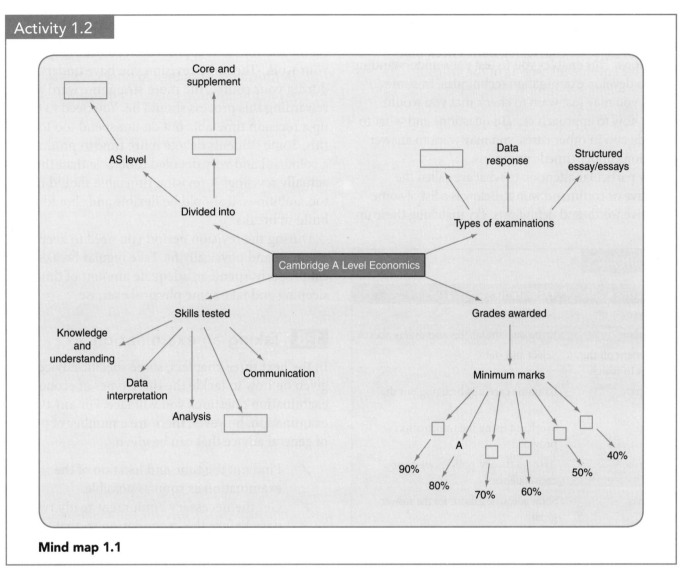

Mind map 1.1

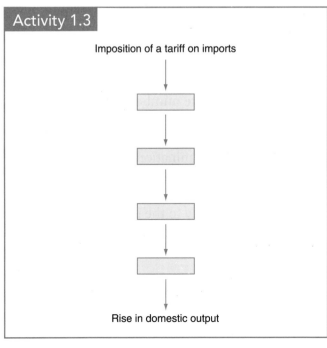

Flow charts

Flow charts are useful to show the links between events. Complete the flow chart in Activity 1.3.

Diagrams

It would be useful to make a list of the key diagrams you need to know for each section of the syllabus. For example, you could write a list of diagrams for Section 3 of the Core syllabus (Activity 1.4).

You may want to put some correct versions on your wall and to include some on your revision cards.

Question and answer sessions

It can be rewarding to revise with a fellow student or students. You can meet to ask each other questions on particular topics. This activity may involve you explaining points and this will really challenge your understanding.

Past examination papers

It is very useful to work through past examination questions. This enables you to test your understanding and to develop examination techniques. In some cases, you may just want to check that you would know how to approach certain questions and what to include and in other cases, you may want to answer questions under timed conditions.

Pay particular attention to what are called the directive or command words. Below is a list of some directive words and definitions. Try matching these up.

Activity 1.5

Directive words	Definitions
Analyse	Make clear
Calculate	Bring out similarities and differences
Comment on the extent to which	Select and state
Compare	Examine carefully bringing out the links
Define	Work out using the information provided
Describe	Assess a theory, policy, causes or consequences
Discuss	Show adequate reasons for the answer given
Explain	Consider likelihood of something happening, size of a change, significance of a change
Evaluate	Briefly describe the main features
Identify	Select the main points
Justify	Give an account of the main characteristics
Outline	Give the precise meaning
Summarise	Examine advantages and disadvantages, reasons for and against, qualifying factors in a critical way

Using a revision guide

Working through a revision guide should help consolidate your understanding and develop your examination techniques.

1.7 Preparing for an examination

As the examination approaches you need to review your work. The more revision you have undertaken during your course, the more straightforward and rewarding this process should be. You need to draw up a revision timetable but do not spend too long on this. Some students devote more time to producing a colourful and very detailed timetable than they do actually revising! A revision timetable should not be too ambitious; it should be flexible and should have build in breaks.

During the revision period you need to keep mentally and physically fit. Take regular breaks, eat properly, spend an adequate amount of time sleeping and take some physical exercise.

1.8 Taking an examination

In the next three chapters, some specific advice is given on how to tackle the three types of economics examination questions you will face. For any type of examination, however, there are a number of points of general advice that can be given.

✓ Find out the time and location of the examination as soon as possible.

✓ Get the necessary equipment ready two days before the examination so that, if required, you would have time to find any items you are missing or need replacing. The equipment you require for economics examinations is: two black pens (in case one runs out), a pencil, pencil sharpener, ruler, eraser (for diagrams) and a calculator. You may also have to take your candidate number – check with your teacher on this.

✓ Ensure you get sufficient sleep the night before the examination.

✓ Arrive at the examination room in plenty of time.

✓ Listen carefully to the instructions given to you by the invigilator and read carefully the instructions on the front of the paper.

Multiple Choice Techniques

2

Revision Objectives

After you have studied this chapter, you should be able to:

☞ understand what a multiple choice question is
☞ know the ways of answering multiple choice questions

☞ outline the amount of time required to answer a question
☞ draw diagrams to help in answering questions.

2.1 What is a multiple choice question

A multiple choice question contains a question and four alternative answers. The question is known as the stem, the correct answer is called the key, and the three incorrect answers are referred to as distracters.

2.2 Ways of answering multiple choice questions

With some multiple choice questions, it is possible to consider the answer before you look at the four options. For example, in the case of the following question:

The supply of a product rises from 400 to 500 as a result of its price rising from $10 to $12. What is the price elasticity of supply of the product?

You could work out the answer before you look at the options. These may be:

A   0.8   B   0.9   C   1.20   D   1.25

In other cases, you will have to consider the stem in connection with the options as in the case below.

What is the difference between a firm producing under conditions of perfect competition and a firm producing under conditions of monopoly?

A   A monopolist can earn supernormal profit in the short run whereas a perfect competitor cannot

B   A perfect competitor can sell any quantity at the market price whereas a monopolist can only sell more by lowering price

C   The average cost will equal marginal cost under conditions of monopoly but not under conditions of perfect competition

D   To maximise profits, the perfect competitor will produce where MC = MR whereas the monopolist will produce where AC = AR

In such a case, it is useful to consider not only why you think the option you have selected is correct but also why the three options you have rejected are incorrect. Indeed, sometimes a way of arriving at the correct answer is by eliminating the incorrect answers.

Always take care when answering questions. In the case of question such as the one above, you might look at option A and consider that it is right. Nevertheless, check the other options to decide whether you can reject them.

Activity 2.1

Answer the two multiple choice questions above and explain why you think the option you have selected is correct and why the other three options are incorrect in the case of the second question.

8   Multiple Choice Techniques

Time allocation

Check how long you have to answer each question
in the paper (usually two minutes per question).
Each question is worth the same number of marks.
The level of difficulty of the questions, however,
varies. Some will be very straightforward and these
you will be able to answer in about half a minute.
In contrast, others are more challenging and will
take longer than two minutes, but do not spend
too long on any one question. If you are struggling
with it, leave it and come back to it when you have
completed the other questions.

Diagrams

To answer some questions it may be useful to draw a
diagram. You can do this on the question booklet.

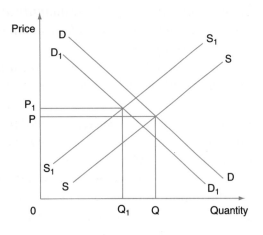

Unanswered questions

Never leave a question unanswered. If after you have
considered all the questions, there is a question you are
uncertain about, still attempt it. You do not lose marks
for getting a question wrong and you have a one in four
chances of getting it right. If you have prepared well for
the examination, it will not be a complete guess.

Data Response Techniques

3

After you have studied this chapter, you should be able to:

☞ understand what a data response question is
☞ outline the ways to approach a data response question

☞ know how to avoid reproducing the data in the answer
☞ examine the data provided critically in order to come to an overall judgement.

3.1 What is a data response question

A data response question is actually a series of connected questions based on real world information. The information may be in prose form, a table of figures, a diagram or more commonly a combination of two or three of these.

3.2 Approach

It is useful to read through the questions first, then examine the data and then to go back to the questions. Pay careful attention to the command words. It is advisable to answer the questions in order, as there is usually a logical structure to the questions.

Take into account the number of marks given to each question as this will indicate how much time you should spend on them. The quality of your answers is more important that the number of words you write, so think through your answers carefully.

There are usually between four to six questions. The first question is often a straightforward one. It may be a definition or one requiring you to draw on the data in a relatively uncomplicated way. The last question will ask you to discuss, which may involve examining two sides of an issue and making a judgement. The questions in between will draw both on your knowledge and understanding of economics and your ability to interpret economic data.

Critical approach

You need to examine the data in a critical manner. You may, for instance, be asked whether the data supports the view, for instance, that China should raise the value of its currency. In this case, you should identify and explain the information that does support the view, the evidence that does not and come to an overall judgement as to whether the information does largely support the view or whether it does not. You might also mention some of the other information it would be useful to examine. It might also be appropriate to consider the reliability of the data.

Activity 3.1

India is the world's biggest sugar market and is set to grow even larger. In 2011, the country consumed 1 million tonnes of sugar. This compares with 21 million tonnes in the US. Some people in the sugar industry predict that India will consume twice as much sugar as the US in 2030.

Demand for sugar in India is growing by one million tonnes a year as population and incomes grow. More sweets and fizzy drinks are being purchased every year although some Indians are becoming more health conscious.

Comment on whether the information provided supports the view that India will consume more sugar than the US in 2030.

Diagrams and statistical tables

Diagrams may take a number of forms. They may be time series graphs, cross-sectional graphs, bar charts or pie charts. Check what the axes are showing and the scales used.

Statistical tables may show a range of information. In the case of both diagrams and statistical tables, it is useful to consider the following.

✓ The reliability of the source or sources (as with prose data)

✓ Whether the data has been adjusted for inflation or is in current figures

✓ Is the information in absolute figures, percentages or index numbers

✓ What denominations the figure are in, for instance, billions or millions

✓ If average figures are shown, are they using the arithmetic mean, the median or the mode

✓ What trend or trends are being shown and are there any exceptions

✓ What relationships are being shown and are there lagged relationships

✓ Is the data for different countries measured in the same way, example are the countries using the same method of measuring unemployment

✓ Is there any missing information.

3.3 Avoid reproducing the data

You may want to quote from the data but if you do, quote briefly and selectively. You should avoid

Table 3.1

Returns for a man attending tertiary education (2010)000s.		
Country	Return to individual	Return to society
Australia	100	185
Canada	155	210
Germany	130	310
Hungary	210	385
South Korea	110	130
UK	210	300

a pedestrian approach to the data. For instance, you might be asked to compare the return from education in the countries shown in Table 3.1.

Just stating that men in Hungary had the highest return, Germany the second highest, the UK the third, followed by Canada, then Australia and then South Korea, is not very revealing.

You should interpret the data using your knowledge and understanding of economics to assess the data. For instance, you might mention the following.

✓ The social benefit was highest in Hungary and lowest in South Korea

✓ The private benefit was highest in Hungary and the UK and lowest in Australia

✓ External benefit was highest in Germany in absolute and percentage terms and lowest in South Korea

✓ Private benefit as a percentage of social benefit was highest in South Korea

✓ The return for a woman is not shown.

Similarly you might be asked to compare what happened to unemployment in Egypt and South Africa over the period 2008–2011 using Table 3.2.

Table 3.2

Unemployment rate (%)		
Year	Egypt	South Africa
2008	9.6	25.5
2009	9.4	23.5
2010	9.0	25.3
2011	8.9	24.0

It would not be productive to write that between 2008 and 2009, the unemployment rate fell in both countries; it fell in Egypt in 2010 and rose in South Africa, and then in 2011 it fell in both countries. Such an approach does not show any knowledge and understanding of economics nor any data handling skills.

Using the information in Table 3.2, provide four comparison points.

Points to note

Do not confuse percentage changes and percentage point changes. For instance, a change from 5% to 7.5% is a rise of 2.5% points or 50%. Also remember that whilst a country may have, for instance, a higher unemployment percentage than another country, it may have fewer people unemployed if it has a smaller labour force. Similarly, a country may only export a small percentage of its exports but it might account for a large percentage of global trade if it is a very large economy.

Take care when examining data on inflation, economic growth and changes in exchange rates. A fall in the inflation rate, for instance, from 8% to 5%, means that the price level is rising but rising more slowly. Similarly, a decline in an economic growth rate from 4% to 2% means that real GDP is still increasing but at a slower rate.

A question might be set on Figure 3.1.

Discuss whether Figure 3.1 suggests that Japan would have experienced an increase in its current account surplus over the period 2007–2011. Outline how you would approach this question.

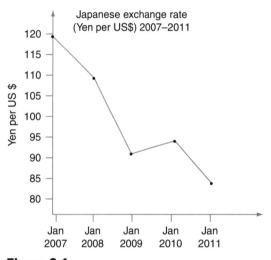

Figure 3.1

Essay Writing Techniques

4

Revision Objectives

After you have studied this chapter, you should be able to:

☞ understand what an essay question is
☞ know how to structure the essay and plan the answer

☞ understand how to use and explain economic terminology in essay writing
☞ outline how to select an appropriate question from the choices available.

4.1 What is an essay

An essay is an extended piece of writing. It enables you to explore a topic in depth and to show the connections between causes, consequences and policies and to apply economic theory to examine current issues.

Essay style

It is important for an essay to have a logical structure with an introduction, main body and a conclusion. The introduction should be relatively brief and direct. It is often useful to define key terms and to provide a brief outline of your answer or an outline of how you are approaching the question. The main body of the essay should examine the topic in depth, with each point following on logically from the previous one. The conclusion should be a summary of the main points you have made and may involve an overall judgement.

You should write essays in paragraphs as this will help you to develop a logical structure and enable the different points to stand out. The introduction should be written as one paragraph and a paragraph should be devoted to the conclusion. Within the main body of the answer, each major point should have a paragraph devoted to it.

Activity 4.1

The following is part of an answer to the question:

Assess how useful the accelerator theory is in explaining net investment.

Divide the answer into 7 paragraphs and then consider whether the answer is now easier to read and whether the main points stand out more.

The accelerator theory is useful in explaining net investment. It concentrates on the main influence on investment which is changes in demand for consumer goods. Firms will invest when the expected yield exceeds the cost of investment. If demand is increasing, firms will expect to sell more goods and hence receive more revenue. However, the accelerator theory does not provide a complete explanation of the behaviour of net investment. Demand for consumer goods may rise without a greater percentage rise in demand for capital goods. Indeed, there may be no change in investment. Firms will not invest to expand capacity if they do not believe that the increase in demand for consumer goods will last. Expectations are a significant influence on investment. If firms are pessimistic about the future they may not even replace machines as they

(continued)

wear out. Firms may also not buy new capital goods if they have spare capacity. They will be able to respond to the rise in demand by making use of previously unused or underused capital equipment and plant. So spare capacity in consumer goods may result in no or a smaller change in demand for capital goods. In contrast, it may be an absence of spare capacity in the capital goods industry which may prevent firms from being able to purchase capital equipment. The consumer goods industries may want to buy, example, more machines but the capital goods industries may not have the resources to produce them. Changes in technology may also mean that an increase in demand for consumer goods may bring about a smaller percentage increase in demand for capital goods. A new machine, embodying advanced technology, will be able to produce more goods than the machine or machines which it replaces. Other influences on investment may change. For example, there may be changes in the cost of machinery, corporation tax or government subsidies. Investment is also influenced by changes in the rate of interest. An increase in the rate of interest will increase the opportunity cost of investment.

Try to ensure that your writing is legible and also, write your answer in good English with accurate spelling, grammar and punctuation. Do not abbreviate ordinary words. It is acceptable to do so for abbreviate certain economic terms if you first write out the word or words in full and put the abbreviation in brackets after it. For instance, you could write 'aggregate demand (AD)' and thereafter write 'AD'.

Answer the specific question set directly. Do not write around the topic including irrelevant information. For instance, if a question asks you to '*Explain the effect of a rise in the rate of interest on consumer expenditure*' you should not discuss the effect of a fall in the rate of interest on consumer expenditure. This is not what the question is asking and would, anyway, be repetitive.

Write in black ink and use pencil and ruler for diagrams.

Economic terminology

Your answer should make use of relevant economic terminology. For example, if you are writing about the value of a currency due to market forces use the term 'appreciation'. Bring out the meaning in your answer and try to write in a clear and straightforward manner.

Explanation

It is important to explain what you write. Do not just assert points. For instance, it is not sufficient to write 'a cut in government spending will cause unemployment'. An examiner would want to know why and whether this would always be the case.

Diagrams

Diagrams are a good analytical tool and can be usefully included in a relatively high proportion of essays. It is, however, important not just to drop them into your answer. You must explain in the text of your answer what they are showing. Make sure your diagrams are large enough to be clear – approximately a one-fourth to one-third of a page. Also ensure they are accurately labelled.

4.2 Selecting a question

In Paper 2 (AS Level) you have to select one question out of three and in Paper 4 (A Level) you have to choose two questions out of six. In Paper 2, the questions are divided into two parts (a) and (b). This is also the case for most, but not all, the questions on Paper 4 where some are just in one part. Where there is two parts, part (a) usually asks you to explain and part (b) to discuss.

Read through all the questions carefully. You may want to highlight key words. Then select your question or questions carefully. It is important that you think you will be able to answer both parts well. You should also be confident that you understand the two parts and that you can apply relevant economics in both parts. Do not attempt more questions than required.

Clearly indicate which question you are answering by putting the number at the start and the question part. Do not waste time by writing out the question.

Planning your answer

There are a number of advantages of planning an answer which includes the following.

- ✓ It makes sure you cover all the points you first think of
- ✓ It may stimulate you to think of additional points
- ✓ It helps you develop a logical structure
- ✓ It prevents repetition.

The plan can be in any format – whichever style suits you. For instance, it may be in note form or a mind map.

Activity 4.2

Produce a plan to answer the following question which appeared on Paper 2 of a Cambridge examination.

(a) Explain how a country's balance of payments is organised to account for all of its international transactions. [8]

(b) A country has a deficit on the current account of its balance of payments. Discuss whether this is necessarily harmful to the country. [12]

Cambridge 9708/02 Q4 May/June 2005

Time allocation

It is obviously important to allocate your time appropriately. You should plan how long you have to spend on each question at the start of the examination. Not all of the time allocated to an essay question, however, should be spent writing. As with the data response paper, the quality of answers is more significant than the quantity of words. This is why planning your answer is important. Spend time thinking through the question and organising your ideas before you start writing your answer.

You should not write in note form unless you are running out of time and this is something you should avoid doing. An answer written in note form does not enable the writer to develop links and answer in sufficient depth and therefore is not likely to gain high marks.

Checking your answer

It is useful to allow a few minutes at the end of the examination to check through your answer. This is because when you are writing under pressure it is easy to make mistakes or leave words out.

Part 2   Core

Basic Economic Ideas

5

Revision Objectives

After you have studied this chapter, you should be able to:

- ☞ explain the meaning of scarcity, opportunity cost and the basic economic questions
- ☞ compare different economic systems
- ☞ draw and interpret production possibility curves
- ☞ recognise the importance of decision making at the margin
- ☞ distinguish between positive and normative statements

- ☞ define ceteris paribus
- ☞ outline the characteristics of factors of production
- ☞ assess the advantages and disadvantages of division of labour
- ☞ explain the functions and characteristics of money.

5.1 Meaning of scarcity and the inevitability of choices at all levels

Resources are limited in supply (finite) whilst wants are unlimited (infinite). As there is scarcity of resources, choices have to be made. Consumers have to decide what to buy, workers – which jobs to do, firm – what to produce, governments – what to spend tax revenue on.

5.2 Opportunity cost

Having to select one option involves an opportunity cost. Opportunity cost is the best alternative forgone.

Due to the economic problem of wants exceeding resources, economies have to decide what to produce, how to produce it and who will receive what is produced.

5.3 Different allocative mechanisms

An economic system is a way of allocating resources to answer the three fundamental questions of what to produce, how to produce it and for whom. There are three main types of economic systems – a market economy, a planned economy and a mixed economy.

Market economies

A market economy is one in which resources are allocated by means of the price mechanism. Consumers indicate what they are willing and able to buy through the prices they are prepared to pay. Private sector firms respond to changes in consumer tastes by altering what they produce. Property is privately owned and the government's role in the economy is minimal.

Among the advantages claimed for a market economy are consumer sovereignty, incentives for workers and firms to be efficient and innovative, and a lack of bureaucracy.

The possible disadvantages of a market economy include an inequitable distribution of income, a risk of unemployment of resources, under-consumption of merit goods, over-consumption of demerit goods, lack of provision of public goods, information failure, and abuse of market power.

Planned economies

A planned economy is one in which the government makes most of the decisions on how resources are allocated. Property is largely state owned and most workers are employed in state owned enterprises (SOEs). The private sector's role in the economy is minimal.

The advantages of a planned economy includes full employment of resources, avoidance of wasteful duplication, an equitable distribution of resources, consideration of externalities, provision of merit goods and public goods, discouragement of demerit goods, long term planning and support for vulnerable goods.

Among the possible disadvantages of a planned economy are slow responses to changes in consumer demand, too much bureaucracy, a lack of incentives, and too much concentration on capital goods.

Mixed economies

In a mixed economy, both the private and public sector play a key role. Resources are allocated using both the price mechanism and state planning.

A mixed economy seeks to gain the advantages of both a market and a planned economy whilst seeking to avoid the disadvantages. How successful it is depends on the effectiveness of government policies and how efficient the private sector is.

In the late twentieth and early twenty-first century, a number of economies moved from a planned towards a market economy.

There are a number of problems that can arise when central planning in an economy is reduced. Inflation may rise when price controls are removed. It may take time to build up entrepreneurial skills, to develop a financial sector including a stock exchange and a social welfare network. The removal of government support and trade restrictions can result in some enterprises going out of business and can cause unemployment.

5.4 Production possibility curve: shape and shifts

A production possibility curve (PPC) shows the maximum output of two types of products that can be produced with existing resources and technology.

A production point on the curve represents full use of resources, a production point inside the curve indicates unemployed resources and a production point outside the curve is currently unattainable.

A shift to the right of a PPC is caused by an increase in the quantity or quality of resources. A change in the slope of a PPC will occur if the ability to produce only one of the two products alters. A straight line PPC indicates a constant opportunity cost.

Decision making at the margin

Individuals, households, firms and governments often have to make marginal decisions. These involve considering whether to make slight changes. For instance, whether to buy one more apple, produce one more car and whether to reduce the number of teachers employed in state schools.

5.5 Positive and normative statements

A positive statement is a statement of fact. It can be tested to assess whether it is right or wrong. A normative statement is a statement based on opinion. It is a value judgement and, as such, cannot be proved right or wrong.

There are both positive and normative statements in economics. 'The unemployment rate in a country is 6%' is a positive statement. In contrast, 'the government's key priority should be reducing unemployment' is a normative statement.

Ceteris paribus

Ceteris paribus means other things being equal. Economists often make use of ceteris paribus to consider the possible effects of a change in one variable on another variable. For instance, an increase in real disposable income would be expected to lead to an increase in demand for gold watches, on the assumption that the other influences on demand for gold watches are not changing.

5.6 Factors of production: land, labour, capital, enterprise

Factors of production are resources used to produce goods and services.

Land covers all natural resources, example, the surface of the earth, the sea, rivers, minerals below the earth etc. Most land is geographically immobile but occupationally mobile. The reward to land is rent.

Labour is human effort, mental or physical, used in the production of goods and services. Labour may be geographically immobile due to differences in housing costs and because of family ties. It may be occupationally immobile if workers lack education and training. Spending on education and training increases human capital. Wages are the reward to labour.

Capital is human made goods used to produce other goods and services. Investment is spending on capital goods. Net investment occurs when firms purchase more capital goods than are needed to replace those capital goods which have become obsolete – gross investment exceeds depreciation. Capital varies in its occupational and geographical mobility. A photocopier, for instance, can be used in most types of industries and can be moved from one part of the country to another. In contrast, an operating theatre is likely to be occupationally immobile and a gold mine is geographically immobile. The reward for capital is interest.

Enterprise is the willingness and initiative to organise the other factors of production and, crucially, to bear the uncertain risks of producing a product. Entrepreneurs are the people who have the willingness and initiative to make decisions and to take the risks involved in production. In a public limited company, the role of the entrepreneur is divided between the managers (who make the business decisions) and shareholders (who bear the risks). Entrepreneurs tend to be relatively, occupationally and geographically mobile. The reward for enterprise is profit.

5.7 Division of labour

Division of labour involves breaking down the production into separate tasks and having each worker concentrating on a particular task.

One of the first economists to describe division of labour was Adam Smith. In his book *An Enquiry into the Nature and Causes of the Wealth of Nations* (often shortened to *The Wealth of Nations*), he described the eighteen separate processes involved in producing a pin.

Advocates of division of labour claim that it increases output and reduces the average cost of production. This is because it enables workers to concentrate on what they are best at, increases their skill ('practice makes perfect'), reduces the time it takes to train them, reduces the equipment needed, cuts back on the time involved in moving from one

activity to another and makes it easier to mechanise the process.

Critics of division of labour, in contrast, argue that it may reduce output and increase the average cost of production. They claim that workers can get bored, doing the same task time after time. Boredom can lead to workers making mistakes and leaving the firm after a short time. In addition, division of labour may mean that a firm does not find out what task a worker is best at and may mean that a firm will find it difficult to cover for workers who are absent from work due to illness or because they are undergoing training.

5.8 Money: its functions and characteristics

Money covers any item which carries out the functions of money. The four functions of money are a medium of exchange, a store of value, a unit of account and a standard of deferred payments.

Probably the best known function of money is as a medium of exchange. Money makes it easy for people to buy and sell products. In the absence of money, people would have to engage in barter. A store of value means that money enables people to save. Money can be saved in a range of financial institutions to be used in the future.

Money acts as a unit of account, or a measure of value, as it permits the value of goods, services and assets to be compared. A standard of deferred payments allows people to agree prices of future payments and receipts. This enables payments to be made and received in the future and allows people, firms and governments to lend and borrow.

To act as money, an item has to be generally acceptable. An item may have all the other characteristics needed for it to act as money but if people are not prepared to accept it in exchange for products and in its other capacities, it will not act as money. The other characteristics money should possess are durability, recognisability, divisibility, portability, limited in supply, stability in value and uniformity. There are links between the characteristics. For instance, to be stable in value it should be limited in supply.

5.9 Mind maps

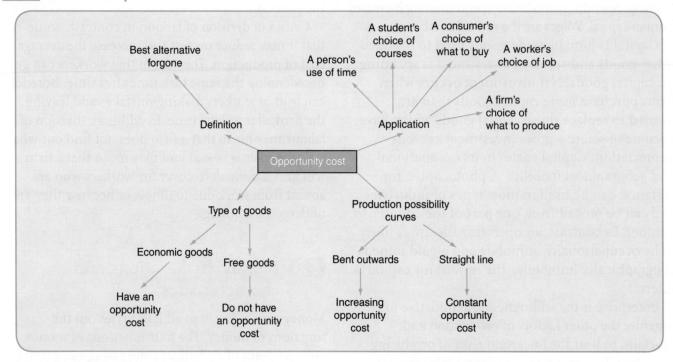

Mind map 5.1 Opportunity cost

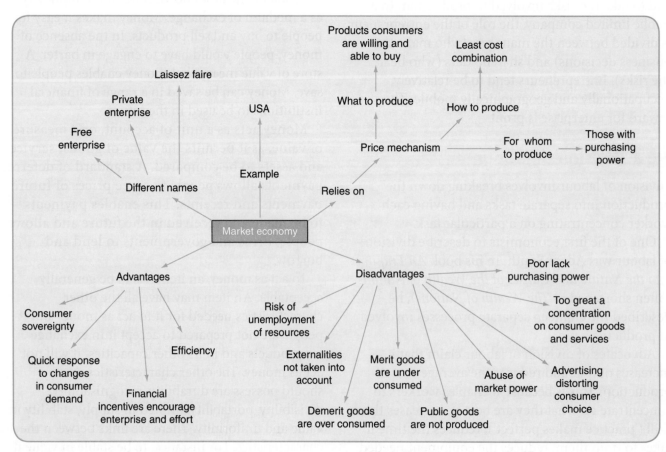

Mind map 5.2 Market economy

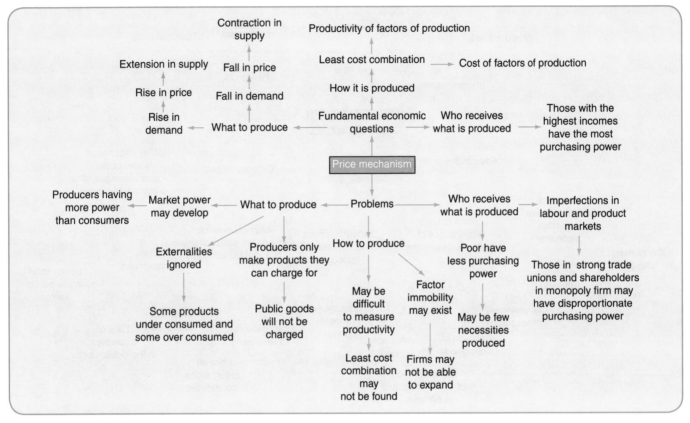

Mind map 5.3 Price mechanism

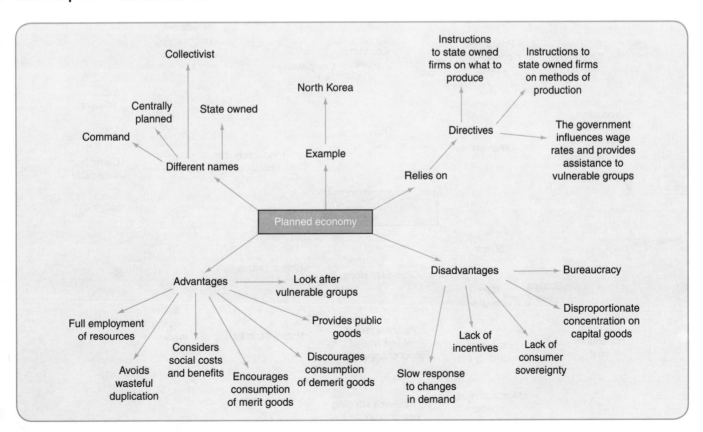

Mind map 5.4 Planned economy

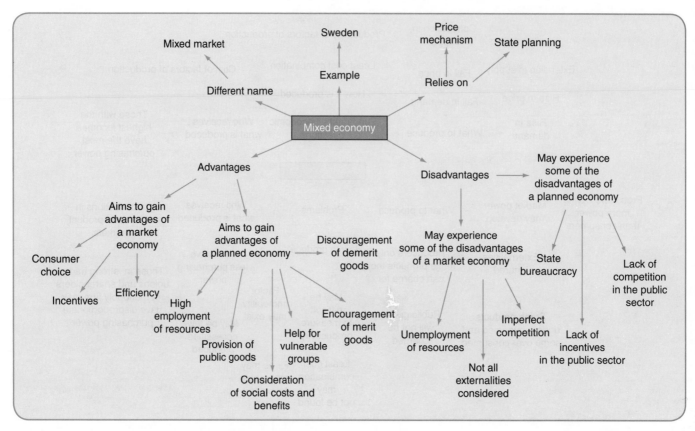

Mind map 5.5 Mixed economy

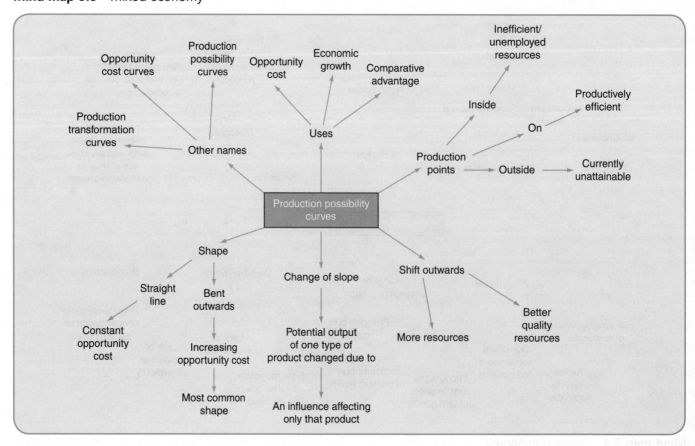

Mind map 5.6 Production possibility curves

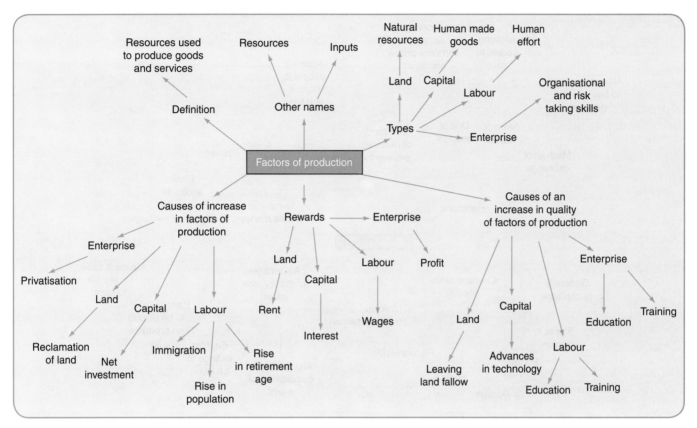

Mind map 5.7   Factors of production

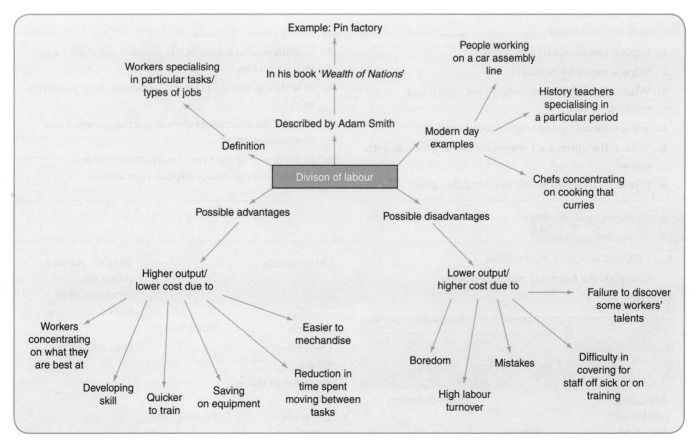

Mind map 5.8   Division of labour

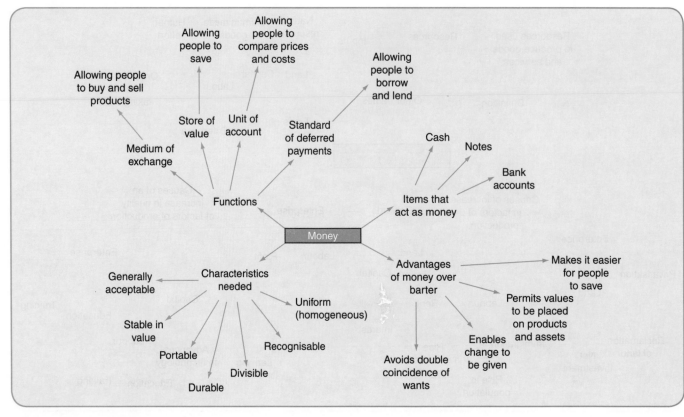

Mind map 5.9   Money

Short Questions

1. Explain two disadvantages of barter.
2. What is meant by liquidity?
3. What is the difference between fixed capital and working capital?
4. What is meant by fixed capital formation?
5. What is the difference between economic goods and free goods?
6. Is health care an economic good or a free good?
7. Explain what is meant by the primary, secondary and tertiary sectors.
8. In which sectors do teachers, farmers and car producers work?
9. What is the difference between microeconomics and macroeconomics?
10. Is the theory of the firm, a microeconomic or a macroeconomic topic? Explain your answer.

Revision Activities

1. Different allocative mechanisms

   Complete the following table given below.

Table 1

A companison of a market economy and planned economy		
Features	Market economy	Planned economy
Allocative mechanism		State directives
Key sector	Private	
Key decision makers	Consumers	

Other names		Centrally planned, collectivist, command, state owned
Example	Hong Kong	
Ownership of means		State owned
Provision of public goods		
The profit motive	Present	

2. Production possibility curves

Look at Figure 5.1 and answer the questions which follow.

(a) What does a straight line production possibility curve (PPC) such as AB indicate?

(b) What does the movement of the PPC from AB to AC show?

(c) Why may the PPC have shifted from AB to DE?

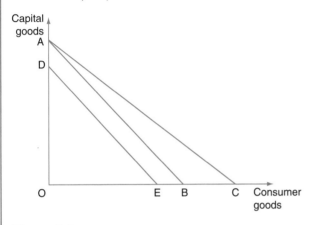

Figure 5.1

3. Factors of production

(a) Identify an example of each factor of production that is employed in the film industry.

(b) Give an example of a capital intensive industry and a labour intensive industry.

(c) What factors influence the supply of labour to a particular occupation?

(d) Explain the link between enterprise and opportunity cost.

(e) Why is the rent on land in city centres usually higher than that on land in rural areas?

4. Money

Fit the following terms into the sentences, using each term only once:

1. medium of exchange   2. store of value   3. unit of account   4. standard of deferred payments   5. general acceptability   6. durability   7. liquid   8. Cheques   9. divisibility

(a) .... are not money. They are a means of transferring a bank deposit from one person to another.

(b) The function of money which allows products to be bought on credit is a ....

(c) To act as a .... money has to be .... in order that payments of different values can be made and change can be given.

(d) A sight deposit (current account) is more .... than a time deposit (deposit account).

(e) The .... and .... of money allows it to act as a ....

(f) Money acts as a .... when the value of products is compared.

Multiple Choice Questions

1. A young woman decides to go to university to study full time. What is the opportunity cost of this decision?

   A The income she could have earned in a job

   B The income she earns when she leaves university

   C The qualifications she gains at university

   D The tuition fees she has to pay

2. Which is a normative statement?

   A A mixed economy includes both a private sector and a public sector

   B Public goods are likely to be provided in a planned economy

   C The government should play a larger role in a mixed economy than it does

   D The price mechanism plays a key role in a market economy

3. Which point represents complete specialisation?

   A A

   B B

   C C

   D D

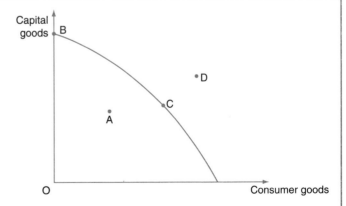

Figure 5.2

4. What is the fundamental economic problem which faces all economies at all times?

   A The need to increase the money supply

   B The need to reduce inflation

   C What, how and for whom to produce

   D When and how to increase the role of the public sector

5. What is the key characteristic of a mixed economy?
   A Some of the products produced are domestically produced and some are imported
   B The economy has both declining and expanding industries
   C There is both a private sector and a public sector
   D The size of the primary, secondary and tertiary sectors are equally balanced

6. Who determines how resources are allocated in a planned economy?
   A Consumers
   B Managers
   C Shareholders
   D The government

7. What does the existence of scarcity mean?
   A Economies are inefficient
   B Economies are developing
   C Households, firms and governments have to make choices
   D It is not possible to increase the quantity of resources

8. Which function of money enables people to save for the future?
   A A measure of value
   B A medium of exchange
   C A standard of deferred payments
   D A store of value

9. The tourist industry makes use of a range of factors of production. Which of the following is an example of land used in the tourist industry?

   A A hotel
   B A beach near the hotel
   C An air ventilation system in the hotel
   D An indoor swimming pool in the hotel

10. Figure 5.3 shows the production possibility curve of an economy. What is the opportunity cost of the economy increasing the output of capital goods from 20 m to 50 m?

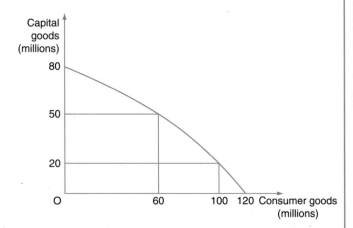

Figure 5.3

   A 30m capital goods
   B 40m consumer goods
   C 70m consumer goods
   D 120m consumer goods

Data Response Questions

1. Since 1989 Eastern European countries have moved from operating planned economies towards operating market economies. This rate of change has varied. Russia and Kazakhstan, for instance, sold off its state owned enterprises (SOEs) and generally reduced government intervention at a quick rate. In contrast, Croatia, Poland and Slovenia removed price controls and subsidies, and privatised relatively slow.

   Some economists claim that the 'shock therapy' swept away safety nets with state ownership. In a number of Russian towns, the SOEs provided not only jobs but also healthcare, childcare and pensions. When the SOEs were sold off from the early 1990s to the mid 1990s, unemployment rose and life expectancy fell to below 60. The lives of men were particularly badly affected with their lifespan falling below 58.

   The questions of how fast to make the transition from a planned to a market economy is now facing China.

The Chinese economy has grown rapidly in recent years boosted by high levels of investment and by taking advantage of a low exchange rate to boost exports. As the economy is developing, the role of the secondary and tertiary sectors is increasing.

Key social and economic data			
Country	Life expectancy 2007	Average annual growth	Unemployment %
	Male Female	In real GDP 2002–07	2007
China	71.3 74.8	13.7	4.0
Poland	71.3 79.8	5.7	9.6
Russia	60.3 73.1	8.5	6.1
Slovenia	74.6 81.9	5.3	4.6

(a) Identify from the extract three indicators of an economy moving from a planned to a market economy.   [3]

(b) Which stage of production is not referred to in the extract?   [1]

(c) Explain, using a production possibility curve, what happened to the Russian economy from the early 1990s to the mid 1990s.   [5]

(d) Using the table, comment on whether the Polish economy performed better in 2007 than Russia.   [5]

(e) Discuss whether the sale of state owned enterprises will benefit an economy.   [6]

Essay Questions

1. (a) An economy can produce agricultural and industrial goods. Explain the possible effects on its production possibility curve if there is an increase in the productivity of its agricultural workers.   [8]

   (b) Discuss whether a market economy can solve the problem of scarcity more effectively than a command economy.   [12]

   Cambridge 9708 Paper 21 Q2 May/June 2009

2. (a) Explain the characteristics required by money if it is to carry out its functions effectively.   [8]

   Cambridge 9708 Paper 2 Q2(a) Oct/Nov 2008

3. (a) Explain the three economic questions that all economies face because of the basic economic problem.   [8]

   (b) Discuss whether the price mechanism is an effective way to solve the basic economic problem.   [12]

   Cambridge 9708 Paper 2 Q2 May/June 2008

6.1 Individual demand curves

Demand is the willingness and ability to buy a product. A demand curve is drawn up from a demand schedule. It shows the quantity demanded at different prices. Price and demand are inversely related.

A market demand curve shows the total demand for a product. The market demand is found by adding up the amount demanded by individual consumers at different prices. Demand for a product is influenced by changes in its price, disposable income, the price of related products, tastes and fashion, population size and age structure, advertising and a number of other factors.

A movement along a demand curve can only be caused by a change in the price of the product itself. A rise in price will cause a contraction in demand. This movement can also be referred to as a decrease in quantity demanded. A fall in price will cause an extension in demand. This movement can also be called an increase in quantity demanded (see Figure 6.1).

A change in any influence on demand for a product, other than its own price, will result in new quantities being demanded at each and every price. As a result,

the demand curve will shift its position. A shift to the right of the demand curve is called an increase in demand, with higher quantities being demanded at each and every price.

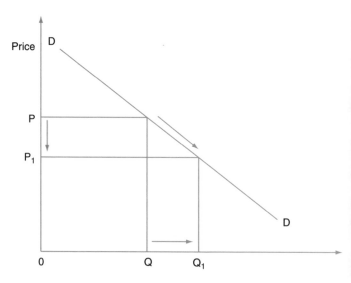

Figure 6.1

A decrease in demand will lower quantities being demanded at each and every price and is illustrated by a shift to the left of the demand curve (as shown in Figure 6.2).

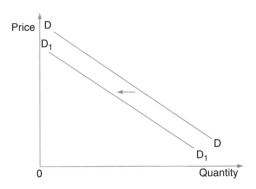

Figure 6.2

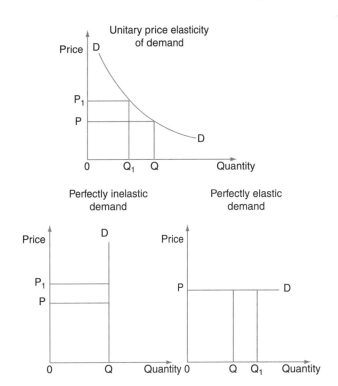

Figure 6.3

6.2 Price elasticity of demand

Price elasticity of demand (PED) is a measure of the responsiveness of demand to a change in price. It is calculated by using the following formula.

$$PED = \frac{\text{percentage change in quantity demanded}}{\text{percentage change in price}}$$

As price and demand are usually inversely related, PED is normally a minus figure. A PED figure (ignoring the sign) greater than one and less than infinity means that demand is elastic, with a change in price causing a greater percentage change in quantity demanded. A PED figure (ignoring the sign) of less than one and greater than zero means that demand is inelastic. In this case, demand is not very responsive to a change in price, with a change in price resulting in a smaller percentage change in quantity demanded.

The three other degrees of PED are perfectly inelastic demand, unitary elasticity of demand and perfectly elastic demand. Perfectly inelastic demand means that demand remains unchanged when price alters. In this case, PED is zero. Unitary elasticity of demand occurs when a change in price causes an equal percentage change in quantity demanded, giving a PED figure of –1. Perfectly elastic demand means that a change in price will cause an infinite change in quantity demanded, giving a PED of infinity.

Unitary price elasticity of demand is illustrated by a rectangular hyperbola, perfectly inelastic demand by a straight vertical line and perfectly elastic demand by a straight horizontal line (see Figure 6.3).

Factors affecting PED

The main factor which influences the degree of PED is the extent to which the product has close substitutes.

The existence of very similar products at a similar price would make demand elastic. Other factors that influence PED are the proportion of disposable income spent on the product; whether the product is habit forming, a luxury or a necessity, and whether its purchase can be postponed or not. Demand tends to be more elastic over time as consumers have longer to recognise the price change and to find alternatives.

If the price of a product takes up a small proportion of income, demand is likely to be price inelastic. A relatively large rise in price would not have much of an impact and so demand would be expected to fall by a smaller percentage.

If a product is habit forming, it will be relatively insensitive to price changes. Similarly, if a product is considered to be a necessity and its purchase cannot be delayed, demand will be inelastic PED for the same products can vary between countries with, for instance, what is viewed as a necessity in one country being viewed as a luxury in another country.

Implications of PED for revenue and business decisions

If demand is inelastic, a rise in price will cause a rise in revenue and a fall in price will cause a fall in revenue. In contrast, if demand is elastic, price and revenue will move in opposite directions.

Discovering that demand for its product is elastic, would usually indicate to a firm that close substitutes are available. This knowledge may make the firm reluctant to raise its price, as it will expect to lose a significant proportion of its sales. It may, however, be tempted to lower its price, as it may be able to capture more of the market. Firms may try to make their products seem unique through, for instance, advertising and brand names. If successful, this would make demand more inelastic, giving firms greater market power.

Firms may estimate PED figures by examining past changes in price and demand and by carrying out market research. In basing their business decisions on PED estimates, however, firms have to take care. This is because the figures are only estimates and PED can change over time.

6.3 Income elasticity of demand

Income elasticity of demand (YED) is a measure of the responsiveness of demand to a change in income. It is calculated using the following formula.

$$YED = \frac{\text{percentage change in quantity demanded}}{\text{percentage change in income}}$$

If YED is greater than one, demand is income elastic. This means that a change in income will cause a greater percentage change in demand. A YED of less than one, in contrast, means demand is income inelastic. In this case, a change in income will result in a smaller percentage change in demand.

Most products have positive income elasticity of demand. This means that a rise in income will result in an increase in demand and a fall in income will cause a fall in demand (see Figure 6.4). Products which have positive YED greater than one may be referred to as superior or luxury goods.

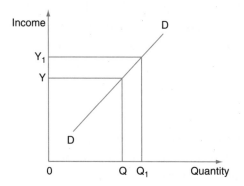

Figure 6.4

Products which have negative YED are called inferior goods. In the case of these products, income and demand are inversely related – an increase in income will cause a decrease in demand and a decrease in income will cause an increase in demand (see Figure 6.5) as shown below.

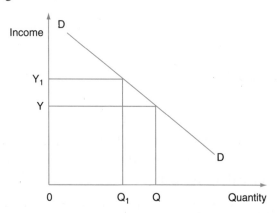

Figure 6.5

Factors affecting YED

Products which have substitutes of a higher quality and a higher price are likely to be inferior goods. Expensive and desirable products may have positive income elasticity greater than one.

Implications of YED for revenue and business decisions

As income usually rises, firms are likely to want to produce mainly normal goods. Producing a few inferior goods may protect firms from the risk of a recession; but usually the fear of the consumer finding out that they are making inferior goods will either stop the firms from producing them or will result in them trying to change their nature. To achieve the later, firms may try to convince consumers that the products have, for instance, health benefits.

6.4 Cross elasticity of demand

Cross elasticity of demand (XED) is a measure of the responsiveness of demand for one product to a change in the price of another product. It is calculated using the following formula.

$$XED = \frac{\substack{\text{percentage change in the quantity demanded} \\ \text{for another product}}}{\text{percentage change in the price of one product}}$$

Positive cross elasticity of demand means that a rise in the price of one product will result in an increase in demand for the other product. Similarly, a fall in the price of one product will cause demand for the other product to decrease. This is the case with substitutes (see Figure 6.6).

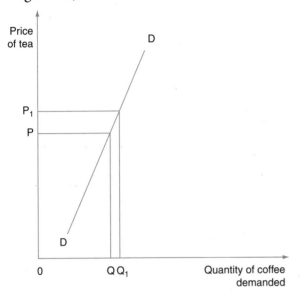

Figure 6.6

Negative cross elasticity of demand occurs when the change in the price of one product results in a change of the opposite direction in demand for the other product. Complements have negative XED. For instance, a rise in the price of PCs may result in a decrease in demand for printers, which are bought to be used with PCs (see Figure 6.7).

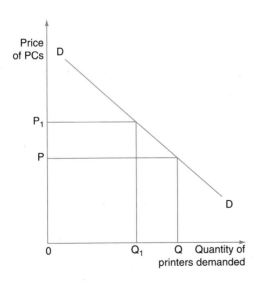

Figure 6.7

Factors affecting XED

The closer two products are as substitutes, the higher the positive XED figure will be. The higher the negative XED figure is, the closer the two products are as complements.

Implications of XED for revenue and business decisions

Firms have to be aware of the extent to which their products have close substitutes. The existence of close substitutes provides both a threat and a challenge for a firm. It has to be aware that raising its price may lose some of it customers. It may, however, be able to attract customers away from rivals by lowering price.

Knowledge about the existence of complements can help a firm to increase its revenue. A firm may offer one product at a lower price if it is purchased with a more expensive complement. For instance, a firm may seek to sell more TVs by offering to sell a CD player at a reduced price with every TV purchased.

6.5 Firms' supply curves

Supply is the ability and willingness to sell a product. A supply curve is drawn from a supply schedule. Price and supply are directly related. Market supply is the total supply of a product. It is found by adding up the supply curves of the individual firms in the market.

Supply is influenced by a range of factors including price, costs of production, advances in technology, indirect taxes, subsidies, and in the case of primary products, weather conditions and diseases.

A movement along a supply curve can be caused only by a change in the price of the product itself. A rise in price will cause an extension is supply. This movement can also be referred to as a decrease in quantity supplied (see Figure 6.8).

A change in any influence on the supply of a product other than its own price will result in new quantities being supplied at each and every price. As a result, the supply curve will shift its position. A shift to the right of the supply curve is called an increase in supply. Higher quantities are supplied at each and every price (see Figure 6.9).

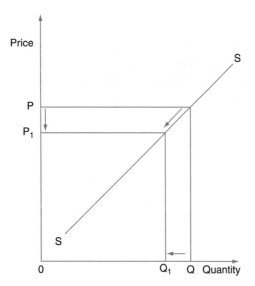

Figure 6.8

Figure 6.9

A decrease in supply with lower quantities being supplied at each and every price is illustrated by a shift to the right of the supply curve.

6.6 Price elasticity of supply

Price elasticity of supply (PES) is a measure of the responsiveness of supply to a change in the price of the product. It is calculated by using the formula below.

$$PES = \frac{\text{percentage change in quantity supplied}}{\text{percentage change in price}}$$

As price and supply are directly related, PES is a positive figure. A PES figure greater than one and less than infinity means that supply is elastic, with a change in price causing a greater percentage change in quantity supplied. A PES figure of less than one and greater than zero indicates inelastic supply. In this case, supply is not very responsive to a change in price, with a change in price resulting in a smaller percentage change in quantity supplied.

Supply may also be perfectly inelastic, unitary elastic and perfectly elastic. Perfectly inelastic supply means that a change in price has no effect on the quantity supplied. Unitary elasticity occurs when a change in price causes an equal percentage change in supply. Perfectly elastic supply means that a change in price will cause an infinite change in quantity supplied (see Figure 6.10).

The main factors that influence PES are whether the products can be stored, the time it takes to produce it and time itself. A product will have an elastic supply if it can be stored and is quick to produce. In this case, supply can be easily adjusted to changes in price. Over time the supply of most products become more elastic as there

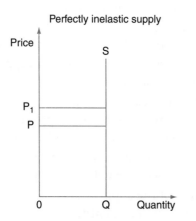
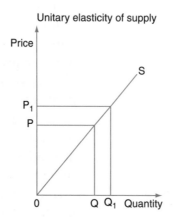
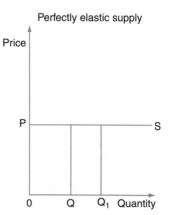

Figure 6.10

is more opportunity to make adjustments to the factors of production employed.

The supply of manufactured products is usually more elastic than agricultural products. Plants take time to grow and livestock to mature. Firms benefit from making the supply more elastic. Their profits are likely to be higher if they can easily raise the quality supplied when the price rises and can easily withdraw supplies from the market when price falls.

6.7 Interaction of demand and supply: equilibrium price and quantity

A market is in equilibrium when demand and supply are equal. In this situation, there will be no reason for price or quantity to change. The market will clear, with no surplus or shortage as shown in Figure 6.11.

Disequilibrium occurs when demand and supply are not equal. If demand exceeds supply, a shortage will occur. This shortage is likely to result in price rising until demand and supply are again equal (see Figure 6.12).

In contrast, if supply is greater than demand, a surplus will result. The surplus will put downward pressure on price, until equilibrium is restored as shown in Figure 6.13.

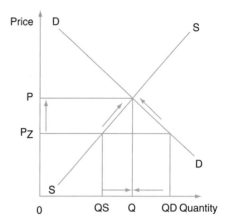

Figure 6.12

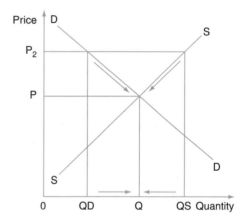

Figure 6.13

Applications of demand and supply

Demand and supply analysis enables economists to make predictions about changes in market conditions.

Governments apply demand and supply analysis to consider the effects of taxes and subsidies. For instance, the imposition of an indirect tax effectively increases firms' costs of production. This will cause supply to decrease. The decrease in supply will cause price to rise and demand to contract. How much of the price rises and the quantity bought and sold falls, will be influenced by the size of the tax and PED and PES.

Firms use demand and supply analysis to decide on their production and prices.

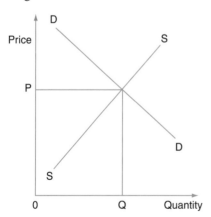

Figure 6.11

Effects of changes in supply and demand on equilibrium price and quantity

An increase in demand will result in a rise in price and an extension in supply. A decrease in demand will cause a fall in price and a contraction in supply. An increase in supply will result in a fall in price and an extension in demand. A decrease in supply will push up the price which, in turn, will cause a contraction in demand.

6.8 Consumer surplus and producer surplus

Consumer surplus is the difference between what consumers are willing to pay for a product and the amount they actually do. For instance, one person may be prepared to pay $20, another person $18, a third $15 and a fourth $11. If the actual price charged is $11, the first person would enjoy $9 consumer surplus, the second person $7, the third $4 and the fourth zero. On a diagram, consumer surplus is the area above the price line and below the demand curve (see Figure 6.14).

A fall in price will increase consumer surplus while a rise in price will cause a decrease in consumer surplus.

Producer surplus is the difference between what firms are willing and able to sell a product for and what they are actually paid. For instance, a firm may have been prepared to accept $9 for a product but if it is paid $12, it will have received $3 in producer surplus. On a diagram, producer surplus is the area above the supply curve and below the price line (see Figure 6.15).

A fall in price will reduce producer surplus whilst a rise in price will increase producer surplus.

Prices as rationing and allocative mechanisms

Prices signal to producers changes in consumers' demand. If consumers want more of a product and are prepared to pay for it, the price of the product will rise. Higher price will provide an incentive for firms to respond by supplying more of the product.

If demand exceeds supply, price will rise and the products will be sold to those who can afford the higher price. Changes in price should ensure that the market clears with no products being unsold

and no consumers who are willing and able to pay the market price, being unable to purchase it.

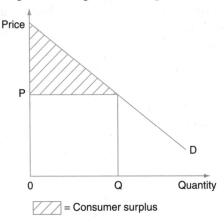

Figure 6.14

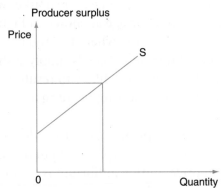

Figure 6.15

Mind maps

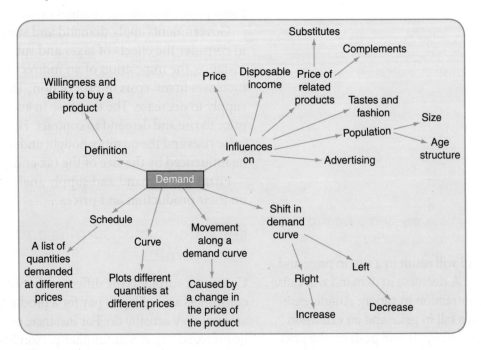

Mind map 6.1   Demand

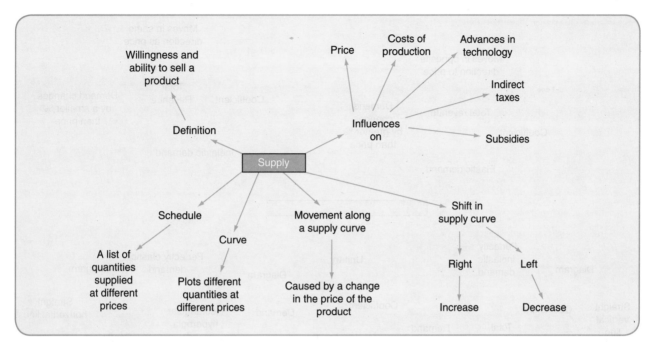

Mind map 6.2   Supply

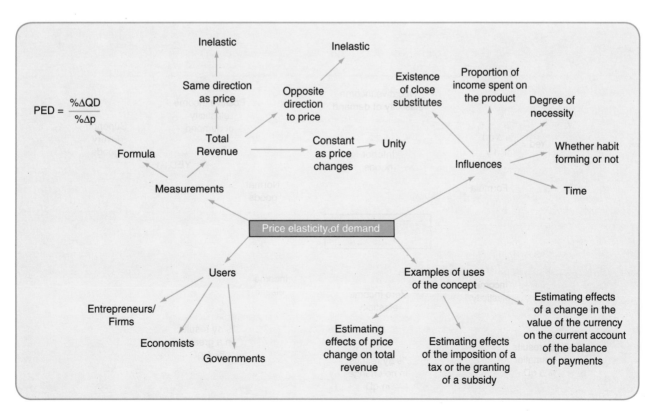

Mind map 6.3   Price elasticity of demand 1

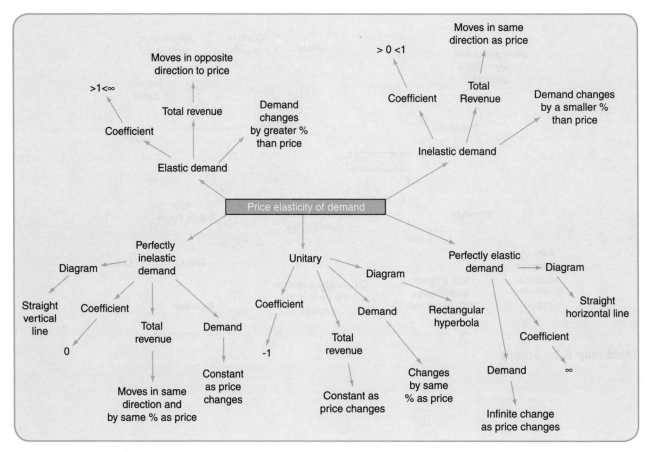

Mind map 6.4 Price elasticity of demand 2

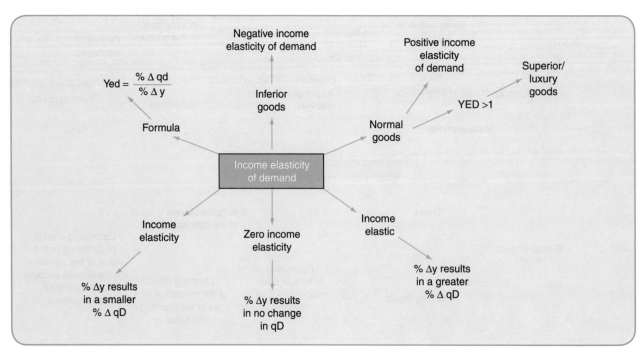

Mind map 6.5 Income elasticity of demand

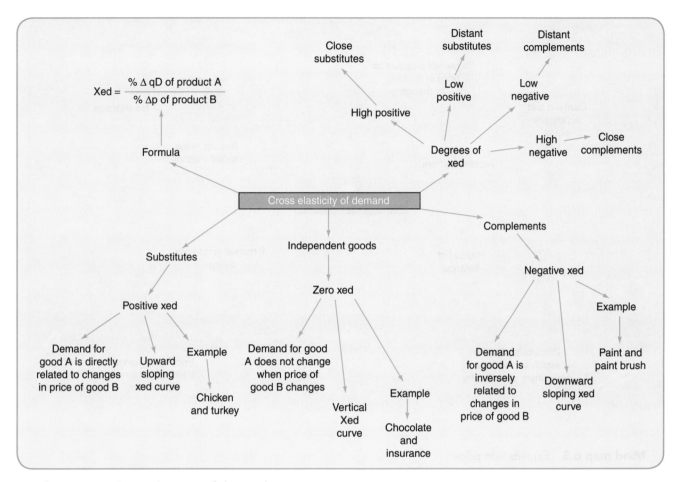

Mind map 6.6   Cross elasticity of demand

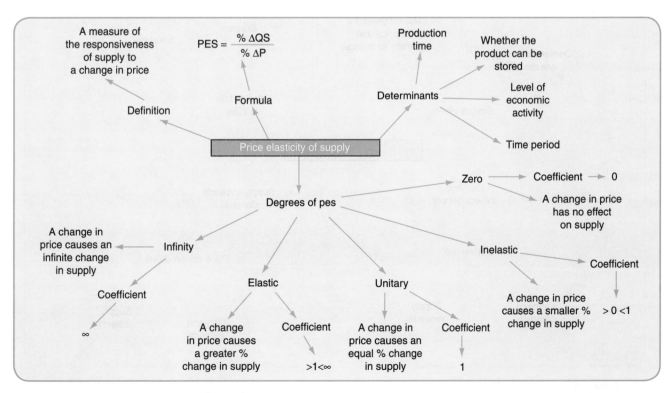

Mind map 6.7   Price elasticity of supply

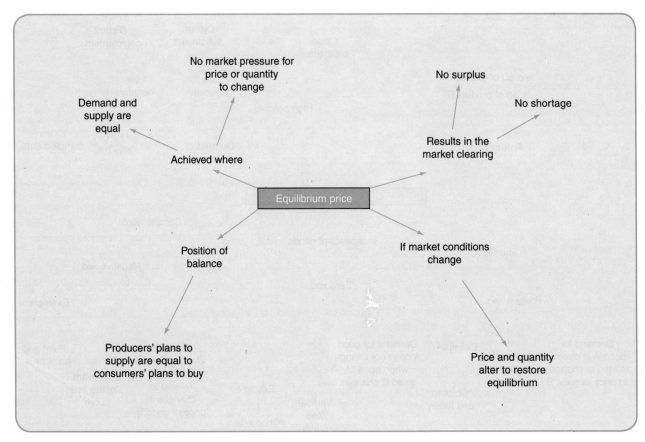

Mind map 6.8   Equilibrium price

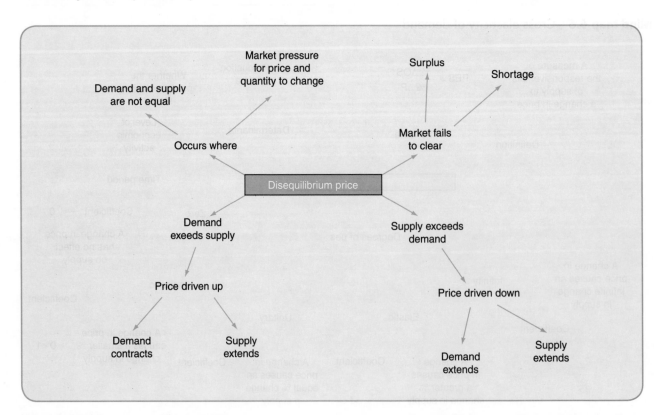

Mind map 6.9   Disequilibrium price

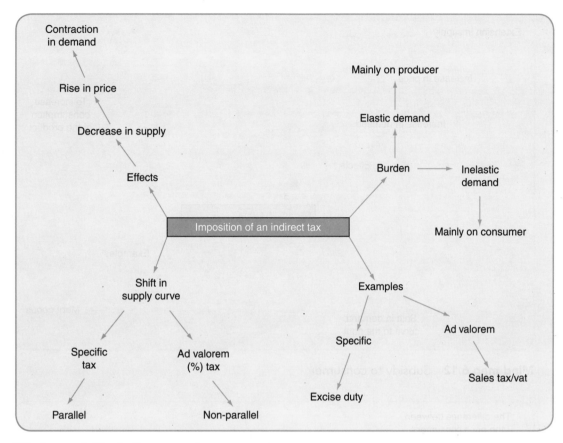

Mind map 6.10   Indirect taxation

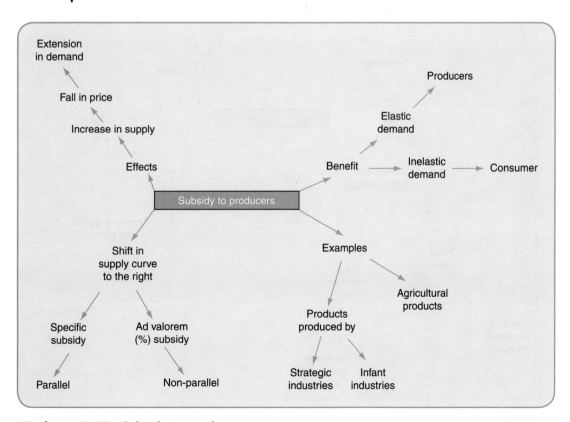

Mind map 6.11   Subsidy to producer

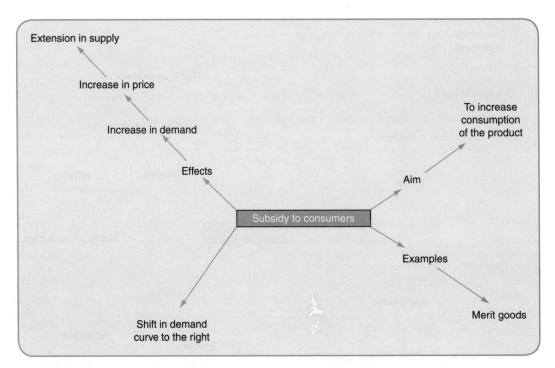

Mind map 6.12   Subsidy to consumer

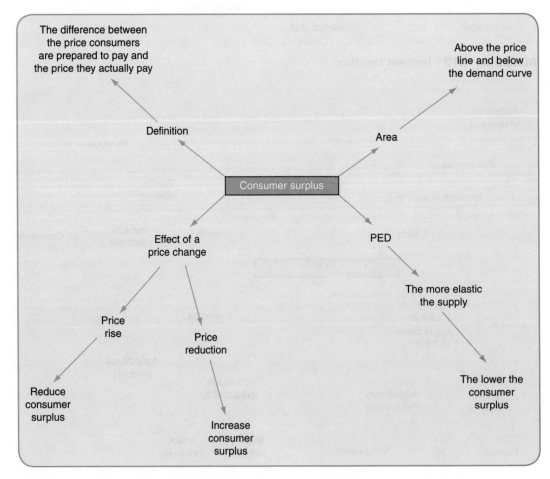

Mind map 6.13   Consumer surplus

Short Questions

1. Why do demand curves slope down from left to right?

2. What is meant by derived demand?

3. Explain three factors that influence demand for air travel.

4. What is the difference between a contraction in demand for ice cream and a decrease in demand for ice cream?

5. Explain what type of price elasticity of demand a product possesses if it has a perfect substitute.

6. Explain two factors that could make demand for a product more price inelastic.

7. How does the PED vary over a straight, downward sloping demand curve?

8. In most countries, services have an income elasticity of demand which is both positive and greater than one. Explain what this means.

9. How does the cross elasticity of demand between one model of car and petrol differ from the cross elasticity of demand between the model and other model of cars?

10. Explain three factors that could increase the supply of rice in a country.

11. A firm produces both blankets and duvets. Explain what effect a fall in the price of blankets is likely to have on its supply of blankets and duvets.

12. How does perishability influence the price elasticity of supply of a product?

13. Using a demand and supply diagram, analyse the impact of an indirect tax on a product which has inelastic demand and elastic supply.

14. What moves a market from disequilibrium to equilibrium?

15. What is meant by the laws of demand and supply?

16. What effect would an increase in supply and a decrease in demand have on the price and the quantity bought and sold?

17. What is the difference between joint demand and composite demand?

18. Why does the amount of consumer surplus received from the purchase of a product differ between countries?

19. How does the price mechanism ration products?

20. Why may the market price differ from the equilibrium price?

Revision Activities

1. In each of the following four cases, draw a demand and supply diagram and explain the effect on the market for newspapers of:
   (a) an increase in the cost of paper
   (b) a decrease in the quality of internet news websites
   (c) free gifts provided by newspapers
   (d) the introduction of a more efficient printing process.

2. A rise in the price of a particular model of car from $12,000 to $15,000 results in an extension in supply from 500 to 600 a week.
   (a) Calculate the price elasticity of supply.
   (b) Is supply, elastic or inelastic?
   (c) Would the car firm want to change its price elasticity supply?
   (d) If the car firm does want to change its PES, how could it achieve this?

3.

Data on a firm producing wrapping paper	
Luxury wrapping paper	Standard wrapping paper
PED = −0.5	PED = −2.0
YED = 2.5	YED = 0.5
XED in relation to rival brand = 0.2	XED in relation to rival brand = 1.5
XED in relation to own brand of luxury gift tags = −0.1	XED in relation to own brand of standard gift tags = −2.0

   (a) Why might the PED figures seem to be wrong but why might they be right?
   (b) What should the firm do to the price of each of its products to raise revenue?
   (c) What should the firm do to the price of each type of gift tags?
   (d) In the long run, which type of wrapping paper should the firm specialise in?

Multiple Choice Questions

1. What causes a movement along a demand curve?
   A A change in income
   B A change in population size
   C A change in price
   D A change in tastes

2. What would cause the demand curve for cinema tickets to shift to the right?
   A A fall in the price of cinema tickets
   B A rise in the price of transport to cinemas
   C The removal of a tax on cinema tickets
   D The release of a number of very popular films

3. A product falls in price by 8% and total expenditure remains unchanged. What type of price elasticity of demand does the product possess?
   A Elastic
   B Perfectly elastic
   C Perfectly inelastic
   D Unitary

4. The price of a product is initially $10 and 200 units are sold per day. The price elasticity of demand for the product is −0.8. By how much would the price have to fall to raise sales by 80 units?
   A $1
   B $3.2
   C $5
   D $8

5. Demand for a product is price inelastic. What effect will a fall in price have on demand and total revenue?

	Demand	Total revenue
A	Contract	Decrease
B	Contract	Increase
C	Remain unchanged	Decrease
D	Remain unchanged	Increase

6.

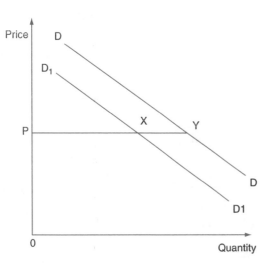

Figure 6.16

Figure 6.16 shows two demand curves. How does price elasticity at point X compare with the price elasticity of demand at point Y?
   A Demand is more elastic at point X than at point Y as consumers are less sensitive to price changes
   B Demand is more elastic at point X than at point Y as consumers are more sensitive to price changes
   C Demand is more inelastic at point X than at point Y as consumers are less sensitive to price changes
   D Demand is more inelastic at point X than at point Y as consumers are more sensitive to price changes

7. What is likely to cause a high price elasticity of demand for a product?
   A The product is habit forming
   B Expenditure on the product forms a small proportion of total spending
   C The product is considered to be a necessity
   D There are close substitutes to the product

8. A demand curve for a product is a downward sloping demand curve. What can be deduced about the price elasticity of demand for the product?
   A It decreases as the quantity demanded decreases
   B It increases as the quantity demanded decreases
   C It is infinity along the curve
   D It is unitary along the curve

9.

Price of Product X ($)	Quantity demanded of Product X	Quantity demanded of Product Y
10	200	500
15	180	600
20	120	800

What is the cross elasticity of demand for Product Y when the price of Product X rises from $10 to $15?
   A 0.2
   B 0.4
   C 0.6
   D 0.8

10. What does a cross elasticity of demand figure of −0.2 indicate about the relationship between two products?
   A Close complements
   B Distant complements
   C Close substitutes
   D Distant substitutes

11. Which types of products have negative income elasticity of demand?
   A Complements
   B Demerit goods
   C Goods in joint supply
   D Inferior goods

12. What would make the supply of a product more elastic?
    A   An increase in the number of substitute products
    B   An increase in the speed of production
    C   A reduction in the durability of the product
    D   A reduction in the spare capacity in the industry

13. The price elasticity of supply of a product is 0.8. Initially the price of the product is $20 and the quantity sold is 2,000. The price then rises to $30. What will now be the firm's revenue?
    A $36,000
    B $40,000
    C $84,000
    D $108,000

14. Supply of a product exceeds demand. What does this indicate about the market?
    A   There is a shortage with the market price exceeding the equilibrium price
    B   There is a shortage with the market price being below the equilibrium price
    C   There is a surplus with the market price exceeding the equilibrium price
    D   There is a surplus with the market price being below the equilibrium price

15. The market price of a product rises but the quantity bought and sold remains the same. What change in market conditions could explain this impact on price and quantity?
    A   A decrease in demand and an increase in supply
    B   A decrease in demand with supply conditions remaining unchanged
    C   An increase in demand and a decrease in supply
    D   An increase in demand with supply conditions remaining unchanged

16. Figure 6.17 shows the demand and supply curves of laptops. The initial equilibrium position is X.

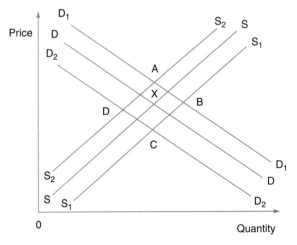

Figure 6.17

What will be the new equilibrium position if incomes rise and advances in technology reduce the cost of producing laptops?
    A   A
    B   B
    C   C
    D   D

17. Figure 6.18 shows the effect of imposing a tax on a product.

Figure 6.18

What is the fall in producers' revenue?
    A $50
    B $100
    C $150
    D $190

18. What effect would the granting of a specific subsidy to producers have on a market?
    A   It would cause a parallel shift to the right of the demand curve
    B   It would cause a non-parallel shift to the right of the demand curve
    C   It would cause a parallel shift to the right of the supply curve
    D   It would cause a non-parallel shift to the right of the supply curve

19. Figure 6.19 shows the initial demand and supply curves of a product. The supply curve then shifts to S1. What is the change in consumer surplus?
    A   XYZ
    B   PWZ
    C   PP1XZ
    D   Q1YZ

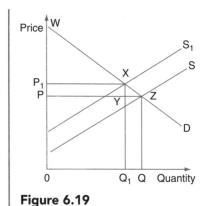

Figure 6.19

20. In which circumstances would consumers bear most but not all of an indirect tax?

	Price elasticity of demand	Price elasticity of supply
A	0	0.2
B	0.2	2.0
C	1.2	0.2
D	2.0	2.0

Data Response Questions

1. The Department of Finance of Canada examined twenty one studies of demand for air travel. These were mainly based on behaviour in the US. It produced a summary of what it thought were the most accurate estimates of elasticities for different segments of the market. Some of these findings are given in Table 1 and 2 below.

Table 6.1

Income elasticity of demand (YED) for air travel	
	YED value
Total market	1.1

Table 6.2

Price elasticity of demand (PED) for air travel	
Market segment	PED value
Long-distance international business flights	−0.3
Long-distance international leisure flights	−1.0
Short-distance business flights	−0.7
Short-distance leisure flights	1.5

(a) (i) State the formula used to calculate income elasticity of demand. [2]

(ii) What can be concluded about air travel from Table 1? [2]

(b) Using Table 2, explain a likely reason for the different price elasticity values for:

(i) business flights compared with leisure flights. [3]

(ii) long-distance flights compared with short-distance flights. [3]

(c) Explain the significance of the price elasticity values in Table 2 for an airline considering a policy of fare cutting. [4]

(d) Discuss the costs and benefits of an increased demand for air travel. [6]

Cambridge 9708 Paper 21 Q1 Oct/Nov 2009

2. Prices of agricultural products can fluctuate significantly. Two recent examples are changes in the prices of cotton and spices.

After a fall in prices between 2008 and 2009, global cotton prices trebled between 2009 and 2010. This reversal was, in part, caused by a fall in the global output of cotton from 107m bales in 2009 to 101m bales in 2010. While output rose in India and Brazil, it fell in China, Brazil and Pakistan. Supply in Pakistan was particularly badly affected by devastating floods. The following table shows the output of the major cotton producers.

Table 6.3

Output of cotton in 2010 (millions of bales)	
China	31.5
India	26.0
USA	18.9
Pakistan	9.3
Brazil	7.0

Unusual weather conditions and flooding also contributed to a trebling of the price of spices produced in Asia a year later. The price also rose because people not only in Asia but also in the US and Europe are using spices in their cooking; and spices are increasingly being used for other purposes. These include using them in health products and in natural colouring.

The higher prices were encouraging more farmers in 2011 to devote more of their land to spices but it was expected that it would take some time for this to increase the quantity supplied. For instance, it takes five years to grow nutmeg.

(a) What was Pakistan's share of the global cotton market in 2010? **[1]**

(b) What might cause a fall in the price of cotton? **[2]**

(c) What effect would the rise in the price of spices have on the price of natural colouring? **[2]**

(d) Does the information suggest that the supply of spices is elastic or inelastic? Explain your answer. **[3]**

(e) Analyse with a demand and supply diagram, why the price of spices rose in 2011. **[6]**

(f) Discuss whether farmers always benefit from a rise in the price of their products. **[6]**

3. Tyre problems for the mining industry.
In 2005, large-scale economic growth in China had increased the demand for minerals to record levels. As a result, 2005 was a boom year for mining worldwide, and output expanded rapidly.

The huge earth-moving trucks used in mining need massive off-the-road (OTR) tyres. These tyres cost $20,000 each and take a day to manufacture. By 2006, a major tyre producer, Bridgestone, estimated that

mining companies required 50% more OTR tyres than in the previous year, but the tyre producers struggled to meet this demand from the mining companies. Tyre producers' stocks of tyres were very low, while fixed production capacity meant that output had remained steady since 1999. It was not technically possible to switch from car tyre production to making OTR tyres.

Total planned production for all of 2006 had already been sold and no new factories were due to start producing before the end of 2007. Another tyre producer, Michelin, intended to spend $85m in its factory in the US, and Bridgestone intended to raise its factory capacity in Japan in 2008.

Rio Tinto, one of the world's largest mining companies, usually spent $100m on 5,000 tyres each year but instead faced the prospect of having to stop trucks working while they had expensive tyre checks and tyre repair. The company predicted that this would limit future exploration for new sources of minerals and cause a rise in mineral prices.

(a) Why can the demand for OTR tyres be described as a derived demand? **[2]**

(b) Explain why the supply of OTR tyres was highly inelastic in 2006. **[4]**

(c) Analyse with a demand and supply diagram, the change in the market for OTR tyres during 2006. **[5]**

(d) How might a shortage of OTR tyres affect the productivity of mining companies? **[3]**

(e) Discuss whether the shortage of OTR tyres required government intervention. **[6]**

Cambridge 9708 Paper 2 Q1 Oct/Nov 2008

Essay Questions

1. (a) Explain with examples, the significance of the value of a good's cross elasticity of demand in relation to its substitutes and complements. **[8]**

   (b) Discuss whether the demand for mobile phones (cell phones) is likely to be price-elastic or price-inelastic. **[12]**

   Cambridge 9708 Paper 2 Q2 May/June 2006

2. (a) With the aid of a diagram, explain how consumer surplus will be affected by the granting of a subsidy to producers. **[8]**

   (b) Discuss whether a fall in the price of a product will always be accompanied by a reduction in the quantity traded of that product. **[12]**

3. (a) Explain why price moves towards equilibrium in a free market and why the equilibrium price may change over time. **[8]**

   (b) Discuss whether all firms in a country will welcome a change in people's income. **[12]**

Government Intervention in the Price System

7

Revision Objectives

After you have studied this chapter, you should be able to:

☞ explain externalities
☞ distinguish between social, private and external costs and between social, private and external benefits
☞ evaluate cost-benefit analysis as a decision making tool
☞ distinguish between private goods and public goods

☞ explain the natures of merit goods and demerit goods
☞ assess the different forms of government intervention i.e. maximum price controls, price stabilisation, taxes, subsidies and direct provision of goods and services.

7.1 Externalities

Externalities are effects on third parties, i.e., people who are not directly involved in the production of a product. These effects are sometimes called spillover effects. Positive externalities (also referred to as external benefits) are beneficial effects that third parties receive without paying for them.

When making their consumption and production decisions, consumers and producers do not usually take into account externalities.

Negative externalities (also called external costs) are harmful effects imposed on third parties who do not receive financial compensation. First and second parties are the producers and consumers of the product.

7.2 Social costs and benefits

Social costs are the total costs of an economic activity. They consist of both private costs and external costs. Private costs are the costs incurred by consumers and producers of the product. Social costs minus private costs equal external costs.

Social benefits are the total benefits of an economic activity. They consist of both private

benefits and external benefits. Social benefits minus private benefits equal external benefits.

The socially optimum output (allocatively efficient output) is achieved where marginal social cost equals marginal social benefit.

7.3 Decision making using cost-benefit analysis

Cost-benefit analysis (CBA) is a method of appraising a major investment project such as a railway line, airport and a main road. A CBA takes into account social costs and benefits whereas a private investment appraisal considers private costs and private benefits. So, whereas a CBA includes external costs and benefits, a private investment appraisal does not.

The first stage of a CBA is to identify the costs and benefits involved. Decisions will have to be made on the depth and width of the scope of the CBA. Monetary values are assigned to the present and future costs and benefits. In some cases, shadow prices are used. These are estimated prices. Calculating monetary values is easier in the case of private costs and benefits than in the case of external costs and benefits. In calculating private benefits, an estimate has to be made of consumer surplus.

Externalities may be estimated by using questionnaires, considering how much those who suffer would have to be compensated and how much people would be prepared to pay for the benefits, valuing time saved when, example, a new road is built by using the average wage and examining the effects of similar projects.

Future costs and benefits are discounted (given a lower value). This is because money paid out in the future is less of a sacrifice than money paid out now, and money received now is more valuable than money received in the future as it can be put to use or can earn interest.

The costs may be weighted to take into account how they are distributed and the income of those incurring the costs. For example, a few people experiencing very high costs may be thought to be more significant than many people being only slightly affected. In addition, loss of income incurred by the poor might be considered to have more of an impact than the equivalent loss for the rich.

Calculations have to be adjusted for risk and uncertainty. When all the costs and benefits have been added up and compared, the net present value is calculated. Then a decision is made. The project will not be recommended if social costs exceed social benefits. Even if there is a positive net value (also referred to as a net social benefit), the project may not be approved if the net social benefit is less than that of a rival project or if it will be politically unpopular.

According to what is known as the Pareto criterion, a project is desirable only if there are gainers and no one is worse off. According to this view, a project would be approved only if the gainers fully compensate the losers, with the gainers still being better off after doing so. The Hicks-Kaldor criterion is more lenient. It suggests a project should be approved if the gainers could, in principle, compensate those who lose and still enjoy a net increase in welfare.

A CBA has the advantage that it seeks to make a decision based on the full costs and benefits of a project. This should make it more likely that an allocatively efficient decision will be made. A CBA may be limited in its scope and may not include all those likely to be affected by the project. In practice, it is difficult to place a monetary value on external costs and external benefits. For instance, how should the loss of wildlife habitats be valued?

People likely to be affected by the investment project tend to exaggerate costs (in the hope that they may be compensated) and underestimate benefits (for fear they may have to pay for them). It can be very expensive and time consuming to carry out a CBA. Political pressure may influence the recommendation of a CBA. Pressure groups may try to influence the outcome and a government may reject the recommendation of a CBA if it thinks it will lose votes.

7.4 Private goods and public goods

Most products are private goods. A private good is both excludable and rival. It is excludable in the sense that someone who is not prepared to pay for it can be prevented from consuming the product. It is rival in that if one person consumes the product, someone else cannot consume it. As private goods are excludable, they can be sold through the market. Private sector firms have an incentive to produce them as they can charge directly for them.

The two key characteristics of a public good are non-excludability and non-rivalry. It is not possible to stop non-payers from enjoying the product. As a result, people have no incentive to pay for a public good. Once provided, a public good is available for everyone including non-payers. So people can act as free riders, consuming the product without paying for it. When people consume a public good, they also do not reduce other people's ability to consume the product. For instance, one more person walking down a street will not reduce the benefit other people receive from the street lighting.

Two other characteristics of a public good are non-rejectability and zero marginal cost. It is not possible for people to reject public goods such as defence. It is also often the case that once provided, it will not cost any more to extend the benefit of a public good such as sea defence to another person.

As it is not possible to charge people directly for public goods, private sector firms lack the financial incentive to provide them. As a result, the provision of public goods has to be financed out of taxation. The government can produce them or pay private sector firms to provide them.

A quasi-public good is a product which possesses some of the features of a public good. For instance, it may be difficult to restrict entry to a beach, making it non-excludable but if it is crowded it may be rival.

It can be difficult for a government to determine the quantity of a public good to provide. This is because preferences are not revealed via the price mechanism. The optimum quantity will be where marginal social benefit (MSB) equals marginal social cost (MSC). To assess MSB, the individual marginal benefits can be estimated using questionnaires and then added up at each quantity of provision. A cost benefit analysis might also be undertaken to assess MSB and MSC.

7.5 Merit and demerit goods

Merit and demerit goods are special categories of private goods. A merit good is a product that a government considers people undervalue. It has two key characteristics. As well as people underestimating the benefit they receive from consuming the product, the consumption also provides external benefits. As a merit good is undervalued, it will be under-consumed and so under-produced if left to market forces. Output will be the allocatively efficient (socially optimum) level. The existence of information failure and external benefits results in market failure.

To encourage greater consumption of a merit good, a government may provide it free, may subsidise it, set a maximum price or provide some information about its benefits. If the government thinks it is very important for people to consume the product, it may make its consumption compulsory.

A demerit good is a product that the government considers people overvalue. As with a merit good, a demerit good has two key characteristics. People fail to appreciate the harmful effects they experience from consuming the product and consumption of the product generates external costs. As a demerit good is overvalued, it will be over-consumed and over-produced if left to market forces. Output will be above the allocatively efficient level. The existence of information failure and external costs results in market failure.

To discourage consumption of a demerit good, a government may tax it, set a minimum price or provide information about its harmful effects. If the government thinks the product is very harmful, it may ban its consumption.

Governments differ as to what they consider to be merit and demerit goods. The US government, for instance, believes that people are fully and accurately informed about the benefits and risks of owning guns and so does not impose many restrictions on gun ownership. In contrast, the UK government makes it more difficult to own a gun as it thinks information failure and the negative externalities involved are more significant.

7.6 Government intervention

Maximum price controls

To have any effect, a maximum price has to be set below the equilibrium price. A government may impose a maximum price on a product for a number of reasons. These include encouraging consumption, to make it more affordable for the poor and to counterbalance the power of monopolies.

A maximum price can result in shortages as shown in Figure 7.1. Demand exceeds supply, giving

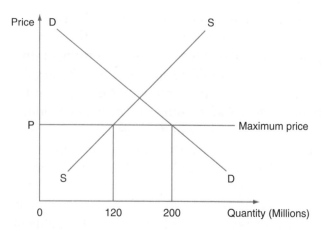

Figure 7.1

rise to a shortage. Some willing buyers will be unable to purchase the product.

A rationing system may have to be introduced with a maximum price and there is a risk that a maximum price will result in a shadow market developing with producers selling the product illegally at higher prices.

Minimum price controls

A government may set a minimum price to discourage consumption of a product or to increase the income of producers. To have any effect, a minimum price must be set above the equilibrium price. Among the disadvantages of a minimum price are that a surplus will be created and higher price will make it less accessible to the poor.

Price stabilisation

A government may operate a buffer stock to stabilise the price of a metal, mineral or an agricultural product that is non-perishable. A buffer stock is a store of a commodity.

Buffer stock managers buy the commodity when market prices are threatening to push the price below the lower limit and sell it when market prices are threatening to raise the price above the upper limit. To ensure that buffer stock managers do not have to spend large amounts on storing the product or run out of money buying the product, the price limits have to be set close to the long run equilibrium price.

Price stabilisation can encourage producers to plan ahead and invest in new production methods.

Taxes

Governments impose indirect taxes to raise revenue and to influence the products consumed and produced. A tax may be used to reduce consumption of a demerit good and to internalise external costs.

The imposition of an indirect tax on a product will add an additional cost to producers. Higher cost of production will cause a shift of the supply curve to the left. The decrease in supply will cause a rise in price and a contraction in demand.

A specific tax is the same charge whatever the price of the product. The imposition of a specific tax will cause a parallel shift of the supply curve to the left. An ad valorem, that is percentage, tax, will mean that higher total amount is charged in tax the higher the price. An ad valorem tax will cause a non-parallel shift in the supply curve as shown in Figure 7.2.

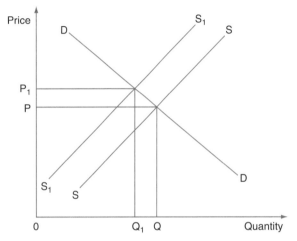

Figure 7.2

A tax which is equal to marginal external cost can move the equilibrium price, and the quantity bought and sold to the allocatively efficient level. This is sometimes referred to as internalising external costs. Producers and consumers would base their economic decisions on the full costs of the product. A tax has a number of other advantages in correcting market failure. These include that it will raise revenue and it works with the market, allowing the market to clear.

As external costs are difficult to measure, a tax may be set too high or too low. Among the other possible disadvantages of a tax are that it may make a domestic industry less competitive, may fall more heavily on the poor, may be politically unpopular, will be less effective when demand is inelastic, there is cost involved in collecting taxes and they may be inflationary.

Consumers bear most of the tax when demand is inelastic as producers can pass most of the tax on in the form of a higher price, knowing that demand will not contract significantly. Producers bear most of the tax when demand is elastic.

Subsidies

A subsidy is a payment by the government to consumers or producers to encourage the consumption and production of a product. A subsidy to consumers would be expected to increase demand, shifting the demand curve to the right. Subsidies to producers are more common than subsidies to consumers.

A subsidy will shift the supply curve to the right, lower price and cause supply to extend. A specific subsidy will cause a parallel shift of the supply curve to the right whereas an ad valorem subsidy will cause a non-parallel shift. Figure 7.3 shows the effect of a specific subsidy.

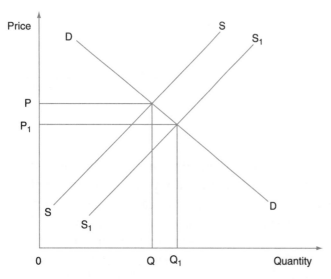

Figure 7.3

A subsidy which is equal to marginal external benefit can move price and the quantity bought and sold to the allocatively efficient levels. A subsidy works with the market and should lead to market clearing.

There is a risk that a subsidy may be set too low or too high. There is also a chance that not much of the subsidy may be passed on to consumers in the form of a lower price. The subsidy will be less effective if demand is inelastic, there may be a time lag involved before producers can alter supply and it will involve an opportunity cost as the government spending could have been used for another purpose.

Direct provision of goods and services

A government may provide both public and merit goods. In the case of public goods, a government has to finance their production. It may produce them through state owned enterprises, also called nationalised industries. It may also pay private sector firms to produce them. Merit goods may be provided free or at low prices to consumers to encourage their consumption and to make them accessible to all.

The quantity of goods and services which a government provides is influenced by two key factors. One is the extent to which it considers market failure occurs. The other is its ability to raise tax revenue.

Regulation

A government may seek to correct market failure by using rules and regulation. A government may make the consumption of some products illegal whilst making the consumption of other products compulsory. Regulation may also take the form of setting limits, for instance on pollution, or standards, for example, on the cleanliness of hotel kitchens. In addition, price controls may be regarded as a form of regulation.

There are a number of advantages to regulation. It is backed up by the force of law, is easy to understand and can influence people's behaviour.

Regulation does, however, have disadvantages. It can take time to introduce, maybe costly to enforce, may not be effective if other countries do not introduce similar measures and illegal (shadow) markets may develop.

Government failure

Government intervention may not increase efficiency in markets. The government may lack information about the extent of the externalities and so may set taxes or subsidies at the wrong rates and there may be time lags between deciding on the policy measure and the policy measure having an effect. Government policy measures may also have unexpected effects. In addition, government intervention may be relatively expensive and may be motivated more by a desire to be politically popular rather than to increase economic efficiency.

7.7 Mind maps

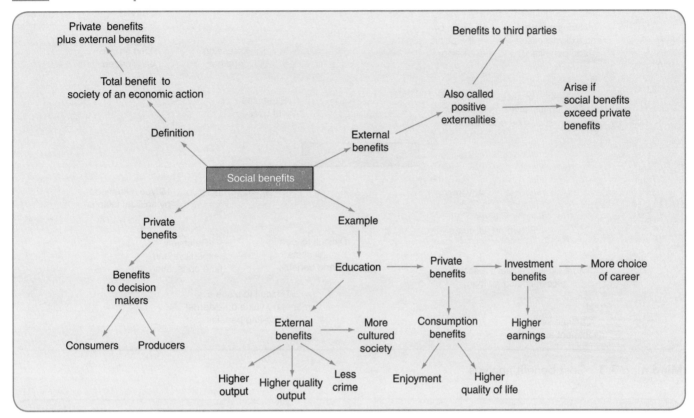

Mind map 7.1   Social benefits

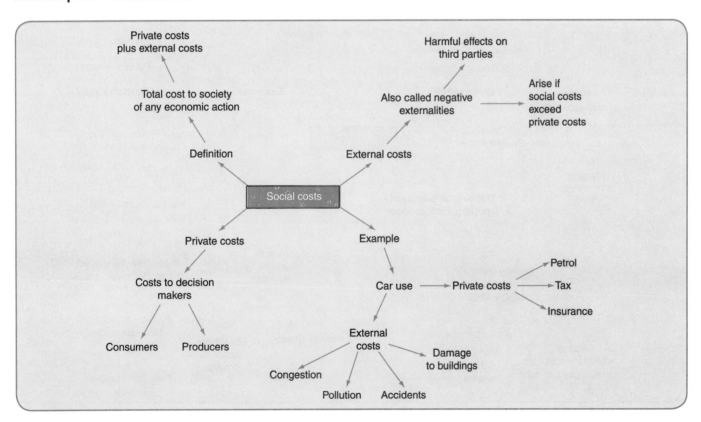

Mind map 7.2   Social costs

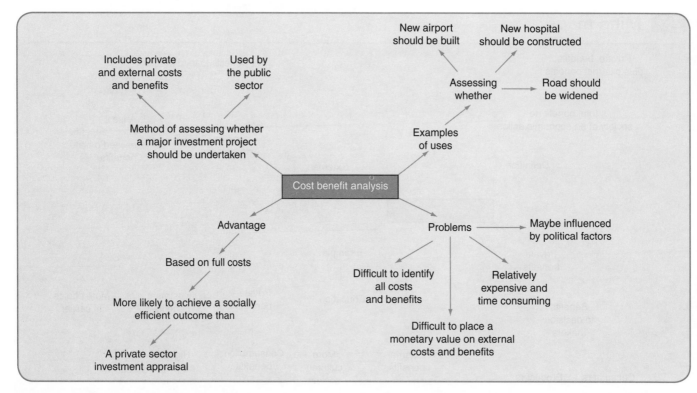

Mind map 7.3   Cost benefit analysis

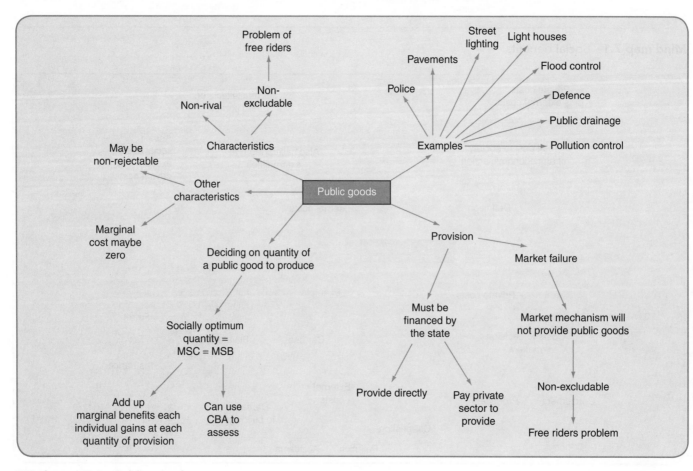

Mind map 7.4   Public goods

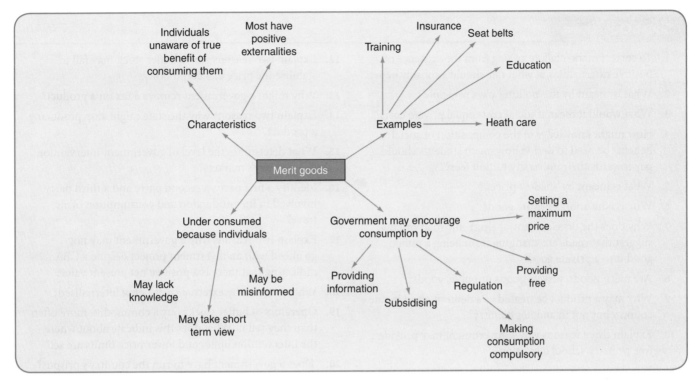

Mind map 7.5   Merit goods

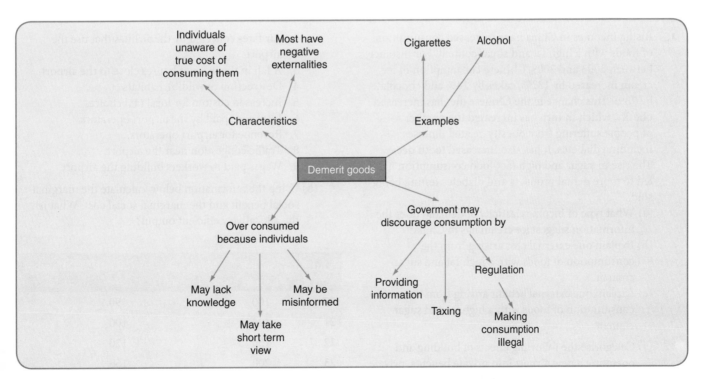

Mind map 7.6   Demerit goods

Short Questions

1. To correct market failure arising from the existence of positive externalities, at what rate should a subsidy be set?

2. What is meant by the 'polluter pays principle'?

3. What would it mean if social costs equal private costs?

4. How might knowledge of the composition of social benefits be used to decide how much students should pay towards their university tuition fees?

5. What is meant by 'shadow prices'?

6. Why is education a private good?

7. Why does the development of electronic road pricing suggest that roads are changing from being a public good into a private good?

8. Are merit goods, private goods or public goods?

9. Why may a product be treated as a demerit good in one country but not in another country?

10. Explain three reasons why a government may provide free primary school education.

11. Why may a maximum price result in an illegal market developing?

12. Explain two reasons why a buffer stock may fail to stabilise the price of a commodity.

13. Why might a government remove a tax on a product?

14. Explain two reasons why the state might stop producing a product.

15. What determines the level of government intervention in a country's markets?

16. Identify a first party, a second party and a third party involved in the production and consumption of air travel.

17. Explain two reasons why a government may not go ahead with an investment project despite a CBA indicating that there is a positive net present value.

18. What is meant by external costs being internalised?

19. Operators of buffer stocks buy a commodity more often than they sell it. What does this indicate about where the intervention upper and lower price limits are set?

20. Does a government have to run the country's prisons?

Revision Activities

1. Rising incomes in China have increased consumption of foods with a high fat and sugar content. For instance, between 2003 and 2009, Chinese consumption of ice cream increased by 132%, cakes by 24% and chocolate by 78%. This change in the Chinese diet has increased obesity, which in turn has increased the number of people suffering from obesity-related illnesses including diabetes. It has also increased tooth decay. The rise in sugar and high fact food consumption has led to more dental products and diabetes testing being sold.

   (a) What type of income elasticity of demand does the information suggest ice cream has in China?

   (b) Explain one external cost arising from the consumption of foods with a high fat and sugar content.

   (c) Explain one external benefit arising from the consumption of foods with a high fat and sugar content.

2. (a) Categorise the following effects of building and operating a new airport into private benefits, private costs, external benefits and external costs.

   1. Air and noise pollution generated by flights to and from the airport.

   2. Air fares collected by the airlines that use the airport.
   3. A fall in the price of houses close to the airport.
   4. Destruction of wildlife habitats.
   5. Increased custom for local taxi choice.
   6. Insurance paid by the airport operators.
   7. Revenue for airport operators.
   8. Traffic congestion near the airport.
   9. Wages paid to workers building the airport.

   (b) Using the information below, calculate the marginal social benefit and the marginal social cost. What is the allocatively efficient output?

Output	Social benefit	Marginal social benefit	Social cost	Marginal social cost
20	100		90	
21	120		100	
22	150		120	
23	200		150	
24	240		190	
25	260		220	

3. Identify, in each case, a government policy measure that could be used to correct the following examples of market failure.
   (a) Price instability in the rice market.
   (b) The smoking of cigarettes causing health problems for non-smokers.
   (c) Firms polluting rivers with chemical waste.
   (d) A lack of provision of street lighting.
   (e) Under-consumption of dental treatment.

Multiple Choice Questions

1. Which of the following is an example of an external benefit of increased train travel?
   A   Increased overcrowding on trains
   B   Reduced fares for train passengers
   C   Reduced congestion on roads
   D   Higher profits for train operating firms

2. What does the existence of external costs suggest?
   A   External benefits also exist
   B   Output is higher than the socially optimum level
   C   Social costs are less than private costs
   D   The product should be subsidised

3. After carrying out a cost-benefit analysis, a government decides to go ahead with a hospital building scheme as there is a net social benefit. Private costs were calculated to be $500m, private benefits at $600m and external benefits at $700m. What does this information indicate about the external costs of the scheme?
   A   External costs were equal to private costs
   B   External costs were equal to social costs
   C   External costs were less than external benefits
   D   External costs were less than $800m

4. What is a pure public good?
   A   A product that confers benefits on consumers which are greater than they themselves realise
   B   A product that involves no external benefits or external costs
   C   A product that involves no opportunity cost to produce it
   D   A product which when consumed by one person does not reduce the amount of the product available for others and which can be charged for directly

5. What causes the free rider problem?
   A   People avoiding paying fares on public transport
   B   People who would have paid higher fares benefiting from government subsidies to producers of public transport
   C   Public goods being non-excludable
   D   Public goods being non-rival

6. What is a defining characteristic of a private good?
   A   It is non-rival so that when a unit is consumed by one individual, it is available to others
   B   It is possible to exclude non-payers from consuming it

C   It is produced only by private sector firms
D   It provides only private benefits

7. Why might a government decide to provide a private good?
   A   It considers that people overvalue the benefits to themselves of consuming the product
   B   It considers that the product should be available to all on grounds of equity
   C   Private sector firms base their production decisions on social costs rather than private costs
   D   Private sector firms will not be able to charge for a private good

8. The demand and supply schedule for a product is shown below.

Price ($)	Quantity demanded per week	Quantity supplied per week
1	3,000	1,800
2	2,800	2,100
3	2,500	2,500
4	2,100	3,200
5	1,500	4,000
6	700	5,000

The government sets a maximum price of $4 per unit. What will be the shortage of the product after four weeks?
   A 0   B 1,100   C 2,500   D 4,400

9. Figure 7.4 shows a maximum price set at PX.

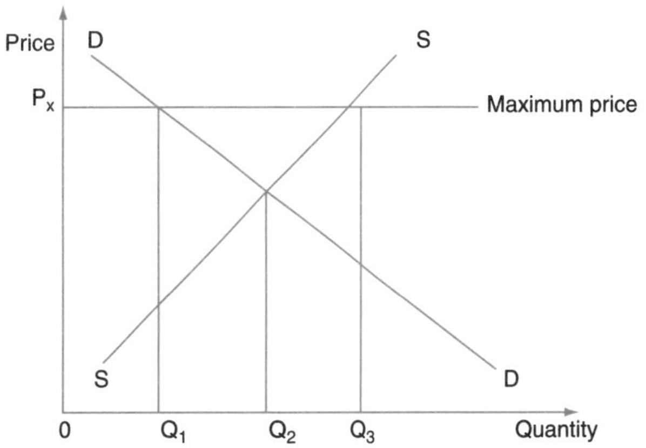

Figure 7.4

What will be the quantity demanded and the quantity supplied?

	Quantity demanded	Quantity supplied
A	Q1	Q2
B	Q2	Q3
C	Q2	Q2
D	Q3	Q1

10. A government sets a maximum price of PT which is below the equilibrium price of P as shown in Figure 7.5.

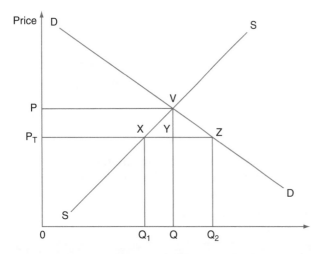

Figure 7.5

Which area shows consumers' expenditure after the imposition of the maximum price?

A OPTXQ1
B OPVQ
C OPTZQ2
D OPTYQ

11. If a government does not intervene, how will the output and price of a demerit good, compare with the socially optimum level?

	Price	Output
A	Too high	Too low
B	Too high	Too high
C	Too low	Too high
D	Too low	Too low

12. A government decides that product X is a demerit good and product Y is a public good. Which policy measures is it likely to adopt in relation to the two products?

	Product X	Product Y
A	Tax	Directly provides
B	Subsidise	Impose a maximum price
C	Directly provides	Subsidise
D	Impose a maximum price	Tax

13. Figure 7.6 shows the market for an agricultural product. The government maintains a price of PF by intervention buying. Which area shows the amount that government spends?

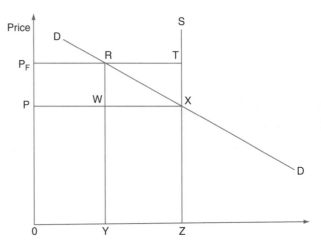

Figure 7.6

A RTX
B YRTZ
C OPFRY
D PPFTX

14. The imposition of a tax on a product causes a non-parallel shift of the supply curve to the left? What type of tax is this?

A Direct and ad valorem
B Direct and percentage
C Indirect and ad valorem
D Indirect and percentage

15. Figure 7.7 shows the effect of a subsidy on a market.

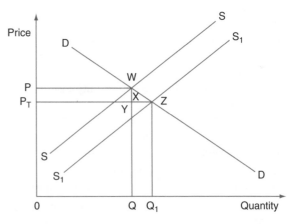

Figure 7.7

What is the subsidy per unit?

A WX
B XY
C WY
D WZ

16. Figure 7.8 shows that the imposition of an indirect tax shifts the supply curve to the left. What is the total tax revenue received by the government?

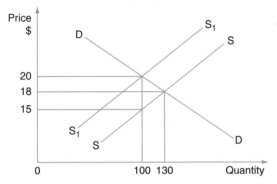

Figure 7.8

    A $200   B $300   C $500   D $650

17. A government intervenes in a market and as a result the demand curve shifts to the right. Which government measure could cause this effect?
  A A subsidy granted to consumers of the product
  B A subsidy granted to producers of the product
  C The imposition of a direct tax
  D The imposition of an indirect tax

18. A government discovers that production of a product gives rise to significant negative externalities. It decides to impose an indirect tax on the product. What would increase the effectiveness of this policy measure?
  A The tax is set at a rate equal to marginal private cost
  B The tax is set at a rate equal to marginal social cost
  C Demand for the product is elastic
  D Demand for the product is inelastic

19. Figure 7.9 shows that the market demand for a product DD is below the demand which would reflect the social benefit D1D1.

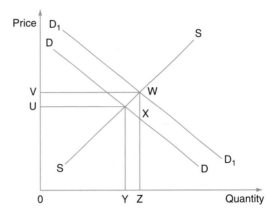

Figure 7.9

To achieve the allocatively efficient level of output, what size of subsidy should be given?
  A UV   B WX   C OU   D YZ

20. Which policy measure would be most likely to result in government failure?
  A A ban on consumption of a product which generates significant external costs
  B The imposition of a tax on a product which has greater social benefits than private benefits
  C The granting of a subsidy to producers of a product that gives rise to positive externalities
  D The provision of information about the beneficial effects of a merit good

Data Response Questions

1. The British Broadcasting Corporation (BBC) is a world-famous radio and television broadcaster. The BBC has an obligation to provide programmes that are educational and cultural, as well as the more popular programmes such as comedy and sport.

In 2005, the services the BBC provided to the UK were part of the public sector and financed mainly by a compulsory annual licence fee of £121 on those households owning a television. There were heavy fines for those who did not pay.

One economist has argued that the BBC should receive more finance, that BBC broadcasting is a public good and that it has greater value for viewers than they have to pay for. To support this view he produced two diagrams. Figure 7.10a shows the demand for BBC services among the 23 million television owning households in the UK and Figure 7.10b shows where this economist placed BBC broadcasting among a selection of public goods in 2005.

(a) (i) Draw a production possibility curve to show the trade-off between the provision of educational and popular television programmes. Explain the possible effects if more money is available to the programme makers from increased licence fees.
               [3]

     (ii) Explain how an increase in the licence fee to receive television programmes may affect the market for television sets. **[4]**

(b)   How does Figure 7.10a support the view that the BBC creates more value than viewers pay for? **[4]**

(c) (i) Comment on the relative positions in Figure 7.10b of national defence and police services.
               [3]

     (ii) Discuss whether it is correct to include education and health as public goods in Figure 7.10b. **[6]**

Cambridge 9708 Paper 2 Q1May/June 2007

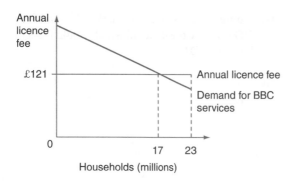

Figure 7.10a

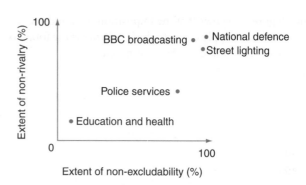

Figure 7.10b

2. Changing prices in Cuba.

Some Cubans had been buying soap and toothpaste in shadow markets. In part, because of this but also to save money, at the start of 2011, the Cuban Government removed its subsidies on the products and the maximum prices which had been imposed on them. Subsidies on cigarettes, salt, peas and potatoes had been removed in 2010.

The price of a bar of soap was expected to rise from $0.03 to $0.21 and the price of toothpaste to increase from $0.08 to $0.32.

The removal of the subsidies and maximum prices were expected to have a significant effect on the poor. Cuban consumers, in general, were not happy with the measures but the Government stated that the measures were necessary because of the shortage of government revenue.

(a) Using a demand and supply diagram, analyse the effect of removing a subsidy on cigarettes. [6]

(b) Compare the expected change in the price of soap and in the price of toothpaste. [2]

(c) Does the information suggest that the subsidy given to soap and toothpaste was sufficient to achieve the price the government wanted? [4]

(d) Discuss whether it is more justifiable to subsidise soap than to subsidise potatoes. [8]

3. Copper mining

Figure 7.11 shows demand for copper and the price of copper over the period 2004–2011, as well as copper consumption for 2004¬¬–2011.

Chile is the world's largest copper mining country but copper is also mined in a number of other countries. One of these is Zambia. One foreign multinational company operating in Zambia has been found to be emitting dangerous levels of sulphur dioxide. Indeed, the emissions are

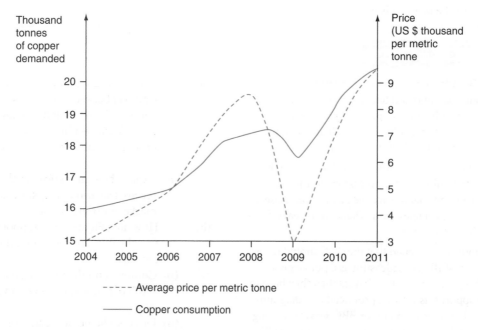

Figure 7.11

seventy times the maximum healthy limit set by the World Health Organisation. Such high levels of pollution are causing respiratory diseases among people living near the mine and destroying the crops of local farmers.

The company claims that it is seeking to reduce the sulphur emissions. It also points out that it has built a number of local roads and has financed the building and running of a local school.

(a)   What evidence is there in Figure7.11a that changes in the demand for copper affected the price of copper over the period shown?   [3]

(b) (i) Distinguish between an external benefit and an external cost.   [3]
     (ii) Explain two external costs created by the operation of the multinational company's copper mine.   [4]

(c) Explain two possible ways the Zambian government could reduce the pollution caused by the multinational company's copper mine.   [4]

(d) Discuss whether the financing of the building and running of a local school by the multinational copper company generates private benefits or external benefits.   [6]

Essay Questions

1. (a) Using a normal demand curve, explain how consumer surplus occurs.   [8]

    (b) With the help of diagrams, discuss whether consumers will benefit from the introduction of (i) an indirect tax and (ii) an effective maximum price.   [12]

    Cambridge 9708 Paper 22 Q4 May/June 2010

2. (a) Explain how information failure can cause market failure.   [8]

    (b) Discuss the effectiveness of government policy measures to correct market failure in the case of merit goods and demerit goods.   [12]

3. (a) Explain the market failure which arises from the characteristics of public goods.   [8]

    (b) Discuss whether the use of cost-benefit analysis helps improve economic decision making.   [12]

    Cambridge 9708 Paper 2 Q3 Oct/Nov 2010

International Trade

Revision Objectives

After you have studied this chapter, you should be able to:

☞ explain absolute and comparative advantage
☞ use opportunity cost to explore the nature of comparative advantage
☞ the arguments for free trade and the motives for protection
☞ distinguish between a free trade area, a customs union and an economic union

☞ the causes and consequences of changes in the terms of trade
☞ describe the components of the balance of payments.

8.1 Principles of absolute and comparative advantage

Absolute advantage is the ability of a country to produce a product using fewer resources than another country. Comparative advantage is the ability of a country to produce a product at a lower opportunity than another country.

In theory, trade should benefit two countries if there is a difference in their opportunity cost ratios and if the exchange rate lies between the opportunity cost ratios. International trade enables countries to specialise and enjoy higher output.

A trading possibility curve shows an economy's post trade consumption possibilities. Figure 8.1 shows that before trade, for instance, an economy can produce any combination on the line joining 500 cars and 2,000 computers. In this case, the opportunity cost of one car is four computers. When international trade is considered, one car can be exchanged for six computers. The economy has a comparative advantage in producing cars. The economy can increase its consumption from X to Y by concentrating on car production and exporting cars in return for computers.

In practice, specialisation and trade may not be as straightforward and beneficial as the law of comparative advantage suggests. With so many countries and products, it can be difficult for economies to determine where their comparative advantage lies. Advances in technology, discoveries of minerals and changes in labour productivity, for instance, can change comparative advantage over time. When economies change their output, their

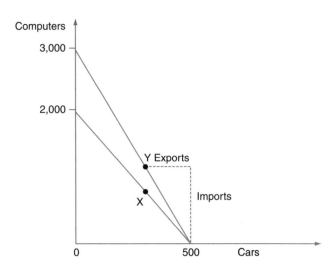

Figure 8.1

opportunity cost ratios may change as resources are unlikely to be equally efficient in different uses. So, for example, if an economy can produce 5,000 chairs with half of its resources, it does not necessarily mean that devoting all its resources to their production will increase output to 10,000. It may increase by more if advantage can be taken of the economies of scale, and by less if diseconomies of scale are experienced. The benefits from trade may be offset by high transport costs or import restrictions. In addition, the government may be concerned about depending on a narrow range of products. Overspecialisation can make an economy vulnerable to changes in costs of production and demand.

Other explanations/determinant of trade flows

Trade flows may be influenced by factors other than comparative advantage. If transport costs are high, a country may trade mainly with countries nearby. The existence of trading blocs may also influence trade flows. Whilst members are more likely to trade with each other, non-members may find it more difficult to export to member countries. A country may have a comparative advantage in a product but may still import it because demand for the product is so high.

8.2 Arguments for free trade and motives for protection

Free trade is the exchange of products between countries without any restrictions. Free trade allows countries to exploit their comparative advantages and so increase output. The competition created can lower prices and raise quality for consumers. It can also enable people to access a greater range of products. Firms will have a wider choice of raw materials which may reduce their costs of production. In addition, they will have a larger market which will enable them to take greater advantage of economies of scale.

Protection involves protecting domestic industries from foreign competition by placing restrictions on international trade. Arguments for protection include – to protect infant (sunrise) industries, to protect declining (sunset) industries, to protect strategic industries, to prevent dumping, to improve the balance of payments position, to protect domestic employment and to protect industries against cheap foreign labour.

Sunrise industries are also known as infant industries. The infant industry argument is that a new industry in a country, which may go on to develop a comparative advantage, may initially need help to compete with well established industries. This is because, before they have grown to a certain size, its unit costs may be higher as they cannot take full advantage of economies of scale. There is a choice that the social benefit of developing a new industry may outweigh the private cost of higher priced imports. It is, however, difficult to identify which industries will be successful and there is a risk that an industry may become dependent on protection.

Sunset industries are declining industries. A government may seek to protect sunset industries to allow them to decline gradually and so avoid a rise in unemployment. The intention would be to remove the protection gradually, but industries may try to resist the protection being removed. It can also be argued that any subsidies used to protect sunset industries might be better used to protect sunrise industries.

Some governments seek to protect agriculture in order to achieve food security. Governments may also want to ensure that it has its own supplies of weapons and may want to see other important industries to be under national control. Such action may, however, result in retaliation.

Taking action against dumping is generally regarded as a valid reason for imposing import restrictions. Dumping involves selling products in a foreign market at below cost price. A firm may engage in dumping to get rid of surplus stock to prevent driving down price in the home market. It may also seek to drive out domestic firms by selling at low prices. If it is trying to gain a monopoly share of the market, consumers may lose out in the long run. In practice, however, it can be difficult to determine whether foreign firms are engaging in dumping.

Import restrictions may reduce a current account deficit in the short run. If they lead to retaliation, however, lower imports may be matched by lower exports. Import restrictions also do not tackle the causes of balance of payments' problems.

A government may think that import restrictions will increase employment or at least prevent unemployment in declining industries. Again there is a risk of retaliation. In addition, if industries that are protected produce raw materials and components, other domestic industries will be disadvantaged. For instance, if steel is protected, domestic car firms will have higher costs of production. This will reduce their international competitiveness and so may lower the employment. Protecting domestic industries that produce finished products can also create disadvantages for other domestic industries. This is because if consumers have to pay higher prices for the protected products, they will have less to spend on other products.

Some argue that domestic industries should be protected from competition from low wage countries. They claim that such competition may drive down wages at home. This argument, however, confuses low wages with low costs of production. A country may have low wages but if it also has low productivity, its unit costs will be high. The US has high wages but relatively low unit costs in many industries because its labour force is well educated and trained and work with advanced technology.

8.3 Types of protection and their effects

Methods of protection include a tariff (a tax on imports), a quota (a limit on the quantity of imports), an embargo (a ban on imports of particular products or from particular countries), excessive paperwork, exchange control (a limit on the amount of foreign currency that can be bought) and a voluntary export restraint (an agreement between two countries to limit imports). A government may also seek to give domestic producers a competitive advantage by subsidising their output.

A tariff will, in effect, increase costs of production for importers. As a result it may raise price and cause demand to contract. The effect a tariff will have will depend on its size, how much of the tax is passed on to consumers and the price elasticity of demand.

A quota will restrict supply, push up price and, again, reduce the quantity demanded. Whilst a tariff will generate government revenue, a quota will not unless a government enforces the quota by selling licenses. A quota increases the price foreign producers receive, whereas a tariff may force foreign producers to lower the price in a bid to remain competitive. By restricting the availability of foreign currency, demand for imports may fall.

Voluntary export restraints create shortages of imports which can push up their price and reduce the quantity demanded. Excessive paper work, 'red tape', will raise importers' costs of production, raise price and lower the quantity demanded.

A subsidy given to domestic producers will encourage domestic firms to increase their output and lower price. This may encourage domestic consumers and foreigners to buy more of the country's products.

8.4 Economic integration

A trading bloc, also sometimes called a trade bloc, is a group of countries that have agreed to reduce or remove some trade restrictions between themselves. Some also have other economic links. Trading blocs have different levels of economic integration. A free trade area has the lowest degree of economic integration whilst an economic union has the highest degree. The more integrated a trade bloc is, the more the member countries act as one economy.

A free trade area is a group of countries that agree to remove barriers to the movement of products between each country. One of the most well known examples of a free trade area is the North American Free Trade Area (NAFTA) which consists of the US, Canada and Mexico.

A customs union goes further in terms of economic integration. As well as removing trade barriers between each other, member countries have to agree to impose a common external tariff that is the same tariff on non-members. An example of a customs union is Mercosur. This has Argentina, Brazil, Paraguay and Uruguay as full members and Venezuela about to become a full member.

Economic union involves a number of features. In addition to removing trade restrictions between each other and imposing a common external tariff, members operate a common market. This involves free movement between member countries not only of products but also of labour, capital and enterprise. Another feature is the operation of some common policies, for example, a common competition policy and a common agricultural policy. There is also greater co-ordination of some policies, for instance, setting similar tax rates. An economics union also involves monetary union, with member states operating the same currency and the central bank of the trading bloc setting the rate of interest for the member countries. The Caricom Single Market and Economy (CSME) is getting close to economic union. It has a common market, common policies in agriculture and tourism, co-ordination of national indirect taxes and budget deficits. Eight of the members of this trading bloc operate the same currency, the Eastern Caribbean dollar, and have their interest rate set by the Eastern Caribbean Central Bank.

Membership of a trading bloc which involves free trade between members and a common external tariff, can involve both trade creation and trade diversion. Trade creation, in this case, occurs when a country moves from buying products from higher cost non-members to lower costs member countries. In contrast, trade diversion takes place when trade is diverted away from more efficient non-members towards less efficient member countries.

8.5 Terms of trade

The terms of trade can refer to the ratio of products exchanged between countries. For example, Country A may exchange 1 unit of tin for 8 units of wheat from Country B. The terms of trade can also refer to the ratio of the index of average export prices to the index of average import prices.

An improvement or favourable movement in the terms of trade occurs when the ratio gets larger – export prices rise relative to import prices. It means that the country can gain more imports in exchange for a given volume of exports.

Whether an improvement in the terms of trade will have a beneficial effect on the current account of the balance of payments will depend on its cause. Higher export prices due to higher demand will increase export revenue. In contrast, higher export prices due to a rise in the costs of production may lower export revenue. Deterioration or an unfavourable movement in the terms of trade takes place when the ratio declines – export prices fall relative to import prices.

Again, the effect deterioration in the terms of trade has on the current account of the balance of payments will depend on its cause. A fall in the exchange rate, for instance, will cause deterioration in the terms of trade but its effect on the current account position will depend on the price elasticity of demand for exports and imports.

Countries which export mainly agricultural products may experience significant changes in their terms of trade. A bumper harvest of a country's main crop, for instance, may reduce the price of the crop which will lower the terms of trade. In contrast, a rise in global economic activity may increase demand for oil which, in turn, may raise the price of oil and improve the terms of trade of oil producing countries.

8.6 Components of the balance of payments

The balance of payments is a record of the economic transactions between a country's residents and residents in other countries over the period of a year. Flows of money into the country are given a positive sign and are referred to as credit items. In contrast, flows of money out of the country are given a negative sign and are called debit items.

The main sections of the balance of payments are the current account, the capital account, the financial account and net errors and omissions.

The current account is probably the section of the balance of payments which receives the most media attention. It consists of trade in goods (exports and imports of visible goods), trade in services (exports and imports of invisible items), income (investment income in the form of profits, interest and dividends) and current transfers (includes government contributions to international organisations, bilateral aid and workers' remittances). If credit items exceed debit items, the country has a current account surplus whereas if debit items are greater than credit items, there is a current account deficit.

The capital account is relatively small. It includes sales of embassy buildings and land and debt forgiveness.

The financial account can cover very significant sums of money. It records transactions in assets and liabilities. It includes direct investment, portfolio investment, loans and changes in reserves.

The sum of debit items and the sum of credit items should be equal and the balance of payments should balance. In practice, however, there are so many transactions and there can be a time delay in reporting these transactions. To ensure that debit and credit are equal, a net errors and omissions item is included. If, for instance, debit items exceed credit items by $10 billion, it means that there has either been a mistake or some credit items have been left out. A net errors and omissions figure of plus $10 billion would be added to balance the balance of payments. Compilers of the balance of payments continue to find where the discrepancy lies.

8.7 Mind maps

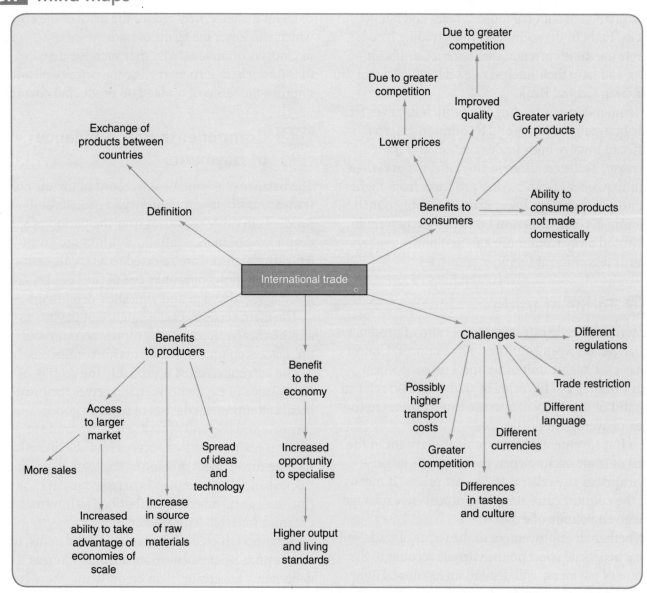

Mind map 8.1   International trade

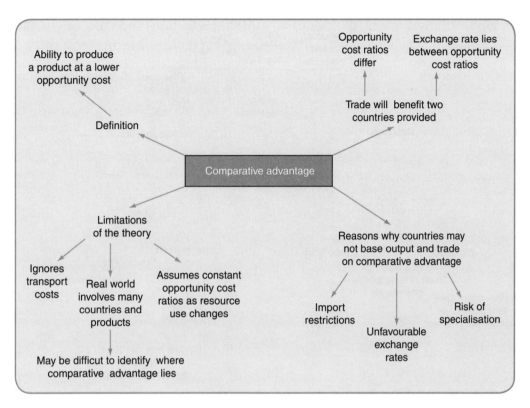

Mind map 8.2   Comparative advantage

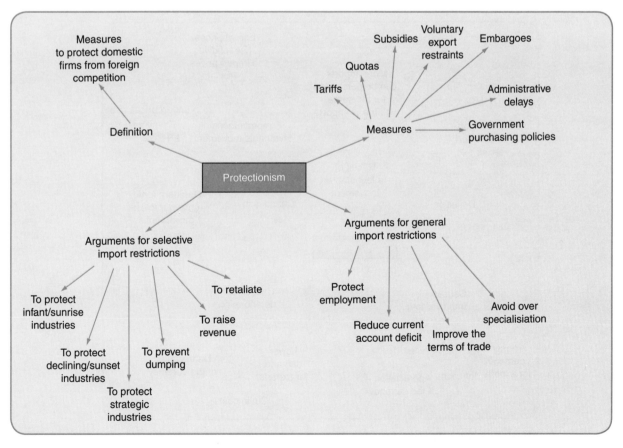

Mind map 8.3   Protectionism

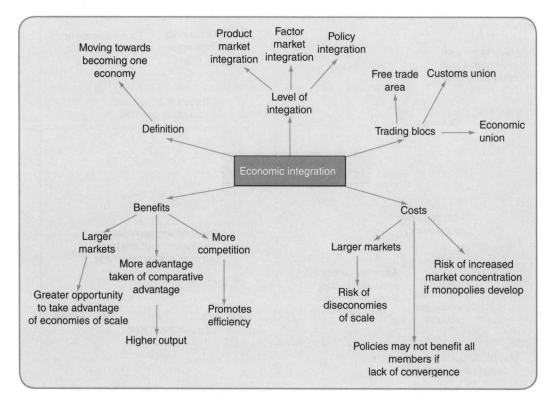

Mind map 8.4 Economic integration

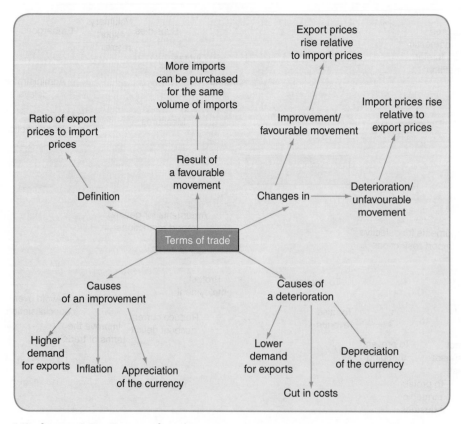

Mind map 8.5 Terms of trade

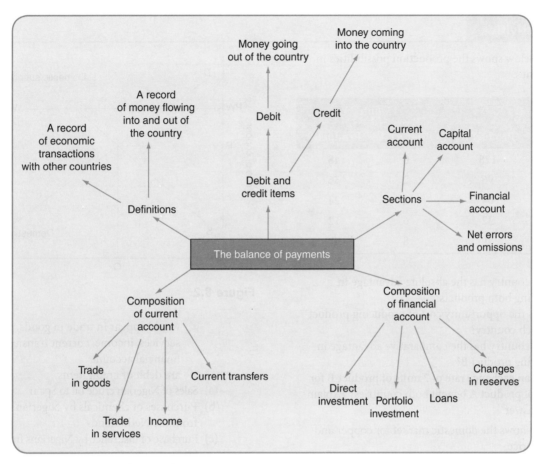

Mind map 8.6   Balance of payments

Short Questions

1. How does international trade differ from domestic trade?
2. Identify three reasons why a supermarket may buy food from abroad.
3. What is the difference between bilateral and multilateral trade?
4. Why may a country have an absolute advantage in producing cars but not a comparative advantage?
5. Explain two ways which can be used to calculate in which product a country has a comparative advantage.
6. How do factor endowments influence comparative advantage?
7. Why might free trade increase consumer surplus?
8. What is the difference between a tariff and a quota?
9. Explain two reasons why a government might want to introduce an embargo.
10. Producers in a country employ child and slave labour. Would it be justifiable for the government of another country to impose trade restrictions on the offending country?
11. What is meant by globalisation?
12. Do trading blocs promote free trade?
13. How does geography influence the composition of trade blocs?
14. If the index of average export prices is 120 and the index of average import prices is 160, what is the terms of trade?
15. Why do the terms of trade tend to move against agricultural producers?
16. What are merchandise exports?
17. What is meant by an invisible trade deficit?
18. What would have to be true for a country to have a balance of trade deficit and a current account surplus?
19. What is meant by portfolio investment?
20. How is the financial account linked to income in the current account?

Revision Activities

1. The table below shows the production possibilities in two countries.

Country X		Country Y	
Product A	Product B	Product A	Product B
0	128	0	48
16	96	2	36
32	64	4	24
48	32	6	12
64	0	8	0

(a) Which country has the absolute advantage in producing both products?
(b) What is the opportunity cost of producing product A in each country?
(c) Which country has the comparative advantage in producing product B?
(d) Would an exchange rate of 7 units of product B for 1 unit of product A benefit both countries? Explain your answer.

2. Figure 8.2 shows the domestic market for copper and the world price.

An economy initially does not engage in international trade. If it then becomes an open economy, explain the effect on the following.
(a) Domestic consumers
(b) Domestic producers
(c) Foreign producers.

3. The following are items which appear in the Nigerian balance of payments. Decide, in each case, whether they:

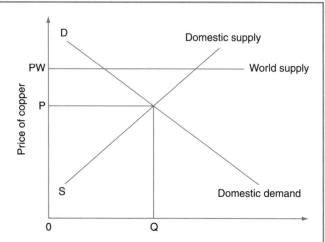

Figure 8.2

✓ would appear in trade in goods, trade in services, income, current transfers or the financial account
✓ are debit or credit items.

(a) Sales of Nigerian crude oil to Spain
(b) Purchases of chemicals by Nigerian firms from the Netherlands
(c) Purchase of insurance by Nigerians from Italian firms
(d) Travel by Nigerian business people on South African airlines
(e) Sales of shares in Nigerian firms to citizens living in Kenya
(f) A UK firm based in Nigeria sending its profits back to the UK
(g) Ghanaians working in Nigeria sending money home to Ghana.

Multiple Choice Questions

1. The table below shows the ability of two countries to produce TVs and chairs in a given time period.

	Production of TVs per worker	Production of chairs per worker
Country X	10	20
Country Y	20	80

What can be concluded from the information?
A Country X has an absolute advantage in TVs and Country Y has a comparative advantage in chairs
B Country X has an absolute advantage in TVs and Country Y has an absolute advantage in chairs

C Country X has a comparative advantage in TVs and Country Y has an absolute advantage in chairs
D Country X has a comparative advantage in chairs and Country Y has an absolute advantage in TVs

2. Figure 8.3 in the following page shows the production possibility curve (RR) for Country X and the production possibility curve (SS) for Country Y.

What does the theory of international trade suggest?
A Country Y should specialise in rice
B Country X has the absolute advantage in cotton
C Country X has the comparative advantage in rice
D There would be no gains from trade

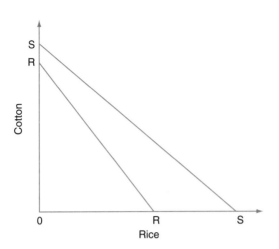

Figure 8.3

3. The table below shows the quantity of cars and units of wheat that two countries R and S can produce with a given quantity of resources.

	Country R	Country S
Cars	10	2
Wheat	800	100

What can be concluded from this information?
A The countries would not gain from international trade
B Country S would import cars and wheat from Country R
C Country R would import wheat from Country S
D Country S would specialise in wheat production and Country R would specialise in car production

4. Country X has a comparative advantage in producing paper and Country Y has a comparative advantage in producing watches. The two countries, however, decide not to specialise and trade. What could explain this decision?
A The exchange rate lies within the countries' opportunity cost ratios
B There is perfect mobility of factors of production between the countries
C Trade is based on absolute rather than comparative advantage
D Transport costs are low relative to the opportunity cost differences between the countries

5. In Country R, a given quantity of imports will produce either 200 units of X or 100 units of Y. In Country S, the same inputs would produce 100 units of X and 25 units of Y. What would the terms of trade have to be for both countries to gain from international trade?

A 1 unit of Y for 2 units of X
B 1 unit of Y for 3 units of X
C 1 unit of Y for 4 units of X
D 1 unit of Y for 5 units of X

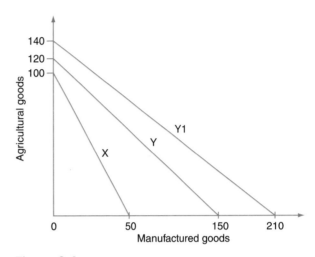

Figure 8.4

6. Figure 8.4 shows the production possibility curves of two countries X and Y.

    As a result of advances in technology in country Y, its production possibility curve shifts from Y to Y1. According to the law of comparative advantage, what should country X do following the change?
A Cease trading with country Y
B Continue to export agricultural goods to country Y
C Import both agricultural goods and manufactured goods from country Y
D Switch to exporting manufactured goods to country Y

7. A country enjoys a comparative advantage over her trading partner in the production of good X. Which diagram illustrates its pre-trade (PQ) and post-trade (PR) consumption possibilities of goods X and Y.

A

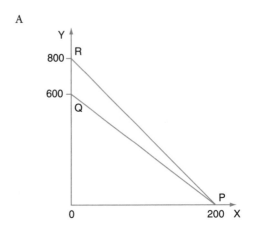

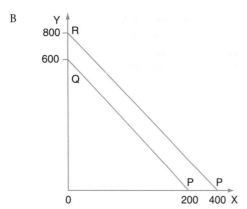

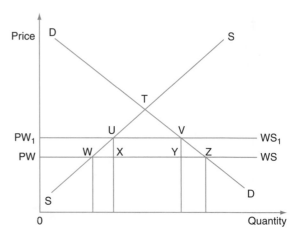

Figure 8.6

10. In Figure 8.6, DD represents domestic demand, DS domestic supply and WS the world supply. The product is imported at a price of PW. If a tariff is imposed which shifts the world supply curve to WS1, what will be the government's tariff revenue?

   A   TUV
   B   UWY
   C   UVXY
   D   YVZ

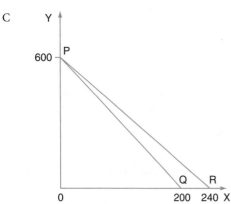

11. Which of the following would reduce protectionism?
   A   The abolition of subsidies given to domestic producers
   B   A reduction in the number of quota free imports
   C   A rise in import duties
   D   A voluntary agreement between the government and foreign governments limiting the number of car imports into the country

12. A government removes a tariff on imports of corn. What is the likely effect on the domestic output and price of corn?

	Domestic output	Price of domestically grown corn
A	Fall	Fall
B	Fall	Rise
C	Rise	Rise
D	Rise	Fall

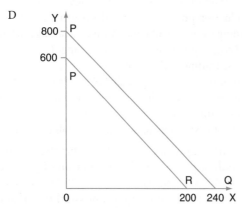

13. What is an ad valorem tariff?
   A   A limit on the quantity of imports
   B   A limit on the value of imports
   C   A percentage tax on imports
   D   A specific tax on imports

8. What would give China a comparative advantage over South Korea in the production of cars?
   A   China can produce a greater volume of cars than South Korea
   B   China's production capacity for cars is greater than South Korea
   C   The environmental costs of car production are lower in China than in South Korea
   D   The opportunity cost of car production is lower in China than in South Korea

14. Which feature is found in a customs union but not a free trade area?
   A   A common currency
   B   A common external tariff
   C   Uniform direct tax rates
   D   Uniform indirect tax rates

9. Which of the following would be a valid reason for a government to impose a tariff on imports?
   A   To prevent dumping
   B   To retaliate against other countries
   C   To increase global output
   D   To achieve the greatest gains from specialisation

15. What would cause a favourable movement in the terms of trade of a country?
   A   Export volume falling more slowly than import volume

B   Import prices falling more than export prices

C   An improvement in the trade in goods balance

D   An improvement in the current account balance

16. The table below shows details about the visible exports and imports of a country.

	Exports		Imports	
	Price per unit	Number of units	Price per unit	Number of units
Year 1	$400	10,000	$500	14,000
Year 2	$560	10,000	$600	14,000

What happened to the balance of trade and the terms of trade from year 1 to year 2?

	Balance of trade	Terms of trade
A	Deteriorated	Deteriorated
B	Deteriorated	Improved
C	Improved	Improved
D	Improved	Deteriorated

17. In which part of the balance of payments does foreign direct investment appear?

A   Capital account

B   Financial account

C   Income balance

D   The trade in goods balance

18. Which of the following would be classified as a credit on the current account of the balance of payments of Pakistan?

A   Money spent in Pakistan by Egyptian tourists

B   Payments made to a French airline by Pakistani business people travelling from Pakistan to France

C   Purchase of vaccinations from Cuba by Pakistani hospitals

D   Reinvestment in Pakistan of profits by Pakistani owners of clothes companies in Sri Lanka

19. The table below shows the details of a country's balance of payments.

	$bn
Trade in goods balance	– 20
Trade in services balance	18
Net income	– 2
Net current transfers	6
Direct investment overseas by domestic firms	– 50
Direct investment into the country	22

What is the current account balance?

A   A deficit of $28bn

B   A deficit of $26bn

C   A surplus of $2bn

D   A surplus of $30bn

20. Which of the following is an invisible export of New Zealand?

A   Expenditure by UK visitors in a New Zealand cinema

B   Expenditure by New Zealand citizens on banking services provided by Italian banks

C   Revenue earned by New Zealand farmers selling lamb to the UK

D   Revenue earned by US multinational companies in New Zealand

Data Response Questions

1.   Trade developments between Australia and Thailand

Table 1

Trade in goods between Australia and Thailand in 2002, all valued in Australian Dollars (A$)			
Australian Exports	A$m	Australian imports	A$m
Total	2510	Total	3140
Of which:		Of which:	
Gold	412	Vehicles	541
Aluminium	348	Petroleum	259
Cotton	186	Heating/cooling equipment	240
Dairy products	137	Seafood	221
Crude petroleum	115	Computers	113

Table 2

Some of the immediate changes in Thailand's tariffs on Australian goods		
Product	Old tariff	New tariff
Beef	51%	40%
Wheat	12–20%	0%
Wine	55%	40%
Large cars	80%	0%
Small cars	80%	30%

In October 2003, Australia signed a trade agreement with Thailand. Under this, more than half of Thailand's 5,000 import tariffs on Australian goods were abolished immediately and others were reduced. These tariffs covered

Australian exports worth more than A$700 million in total. Thailand would abolish the remaining tariffs by 2020. An independent survey predicted that by 2023 these actions would raise the GDP of Australia by A$12,000 million and that of Thailand by A$46,000 million.

(a) (i) Calculate the balance of trade in goods between Australia and Thailand in 2002. [2]

    (ii) What differences were there in the types of goods traded between the two countries? [2]

    (iii) Explain what might have caused these differences. [4]

(b) (i) Name two protective measures, other than tariffs, that restrict free trade. [2]

    (ii) Explain, with the aid of a demand and supply diagram, how the domestic producers of a good are affected by the removal of a tariff on imports of that good. [4]

(c) Discuss whether Australia and Thailand should have abolished all tariffs immediately. [6]

Cambridge 9708 Paper 2 Q1 May/June 2006

2. The problems facing Argentina's beef market

In 2010, Argentina's exports of beef fell to their lowest level for a decade. Beef production was hit by three factors. One was government restrictions on exports. In March 2006, exports of beef were banned for 180 days and in 2010 some restrictions remained. These were designed to protect the domestic price but they created considerable uncertainty among beef importers.

There was a catastrophic drought in 2008 and 2009 which led farmers to slaughter more of their stock, leaving a shortage for the following year. In addition, more than 8,000 Argentine cattle farmers switched from producing beef to producing soyabeans or stopped farming altogether. In 2010, the country failed to meet the amount of beef it is permitted to export to the European Union under the Hilton Quota.

(a) Using a diagram, explain why a ban on exports of beef may protect its domestic price. [5]

(b) What would encourage farmers to switch from beef production to soyabean production? [2]

(c) (i) Why would the European Union have imposed a quota on Argentine beef? [2]

    (ii) What does the data suggest about the effectiveness of the European Union's protectionist measure in 2010? [3]

(d) Discuss whether Argentina is likely to export more beef in the future. [8]

3. The importance of copper production in Chile

Chile is a major producer and exporter of copper. For Chile, copper is a vital export and makes a major contribution to its trade. The Central Bank of Chile forecast a visible trade surplus of US$17 billion for 2006, two-thirds higher than in 2005. Changes in world copper prices in 2006, a year of global growth, were an important influence on Chile's trade performance. However, one problem that China faced in 2006 was a strike for higher wages and better conditions at Escondida, the world's largest copper mine, where 8% of world copper was mined.

Figure 8.7 shows the contribution of copper to Chile's export revenue and changes in world copper prices, 2000–2006.

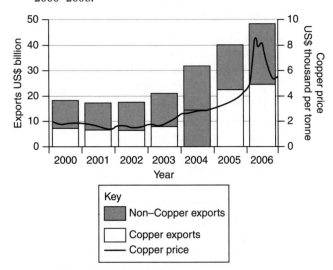

Figure 8.7

(a) (i) Compare the price of copper in the middle of 2003 and the middle of 2006. [2]

    (ii) Explain one change in demand and one change in supply that might have caused this movement in the price of copper. [4]

(b) (i) Calculate Chile's approximate visible trade balance in 2005. [2]

    (ii) What information in addition to that in Figure 8.7 would be required to calculate Chile's current account balance in 2006? [4]

(c) (i) How did the importance of copper as part of Chile's exports change between 2002 and 2005? [2]

    (ii) Discuss whether it is desirable for a country to specialise in the production and export of a single good. [6]

Cambridge 9708 Paper 21 Q1 May/June 2009

Essay Questions

1. (a) Explain why it is usually more difficult to trade internationally than domestically. **[8]**

   (b) Discuss, with examples, how the global distribution of factors of production determines what a country imports and exports. **[12]**

   Cambridge 9708 Paper 21 Q3 Oct/Nov 2009

2. (a) Distinguish between absolute advantage and comparative advantage. **[8]**

   (b) Discuss whether a government should always pursue a policy of free trade. **[12]**

3. (a) Explain what may cause a deterioration in the terms of trade. **[8]**

   (b) Discuss whether the formation of regional trading blocs promotes competition. **[12]**

Theory and Measurement in the Macroeconomy

9

9.1 Employment statistics

Size and components of the labour force

A country's labour force can also be referred to as its workforce or working population. The size of a country's labour force is influenced by a range of factors including the size of its population, the age structure of the population, working age (influenced, in turn, by the school leaving age and the retirement age) and the labour force participation rate.

The labour force participation rate is the proportion of working age people who are economically active. People are economically active, and so are in the labour force, if they are either in employment or are unemployed and actively seeking work.

The economically inactive are people of working age who are not in the labour force. This means they are neither employed nor seeking employment. The major groups who are economically inactive are people who have retired early, those in full-time education, homemakers and those too sick or disabled to work.

The labour force can be divided in a number of ways. These include those working in the primary, secondary and tertiary sectors; skilled and unskilled workers; full-time and part-time workers.

Labour productivity is output per worker hour. It may be increased by education, training, experience and providing workers with more and better quality capital equipment.

Unemployment rate; patterns and trends in (un)employment

People are classified as unemployed if they are without a job and are actively seeking employment.

Two ways of measuring unemployment are the claimant count and the Labour Force Survey (LFS). The claimant count records people as unemployed if they are in receipt of unemployment benefits. The LFS measure is based on the International Labour Organisation's definition of unemployment. This records people as unemployed if they are without a job and have looked for work in the last month or have found a job which they are waiting to start in the next two weeks. This information is found from a random sample of the population.

The unemployment rate is the proportion of the labour force who are without work but who are actively seeking employment. The unemployment rate may fluctuate over time. As unemployment rises, the gap between a country's actual and potential output increases.

The young, the old and the unskilled tend to experience more unemployment than the average worker. The longer people are unemployed, the more difficulty they usually experience in finding a job. This is because they may not keep up with developments in technology and working methods, may lose the work habit and may become less attractive to employers.

As economies develop, employment tends to shift from the primary to the secondary sector and the tertiary sector. Throughout the world, more women are entering the labour force.

Difficulties involved in measuring unemployment

It can be difficult to ensure that any measure of unemployment does not miss out anyone who is unemployed and looking for work and does not include those who are not actively seeking employment or are working in the informal (shadow) economy but are illegally claiming benefits.

The claimant count does not include some people who are unemployed but who are not entitled to unemployment benefits. For example, workers who lose their jobs but have a partner who is employed can claim unemployment benefit for only a limited time. The claimant count also does not include those who are unemployed and actively seeking work but who are too proud to claim unemployment benefit. Some people may have savings they can draw on to support them for a period of time whilst they are seeking a new job. In contrast, the claimant count does include some people who claim the benefits illegally, either because they are not really looking for work or because they already have a job. The claimant count, however, is cheap to collect as the information is recorded every time the unemployment benefit is paid out. It also collects the information relatively quickly.

The LFS captures more of those who are unemployed and, as it is widely used throughout the world, it is good for making international comparisons. It is, however, a relatively expensive measure as it takes time to collect and it can be subject to sampling errors. As well as the risk that the sample may not be representative of the population as a whole, there may be problems in interpreting the information gathered.

9.2 General price level: price indices

An increase in a country's general price level would mean that, on an average, prices are rising. Some prices may be falling but overall the cost of living is rising with people having to pay more in total for the products they buy.

One way governments assess what is happening to the general price level, is to calculate a consumer price index. This is a measure of the average change in the prices of a representative basket of goods and services purchased by households.

To calculate a consumer price index, a base year is selected. This should be a relatively standard year in which there were no unusual events. The base year is given an index figure of 100 and other years are compared to it.

A survey is undertaken of household spending to find out people's spending patterns. The information gathered is used to decide which items to include in the representative basket and what weights to attach to them.

Weights reflect the proportion spent on the products. The more people spend on an item, the more significant is a change in its price. Information on price changes of the products is collected from a range of retail outlets and suppliers. The price change is multiplied by its weight and then all are totalled to give the index figure. For example, if the consumer price index was 100 in 2012 and 108 in 2013, the inflation rate would be 8%.

Most economies measure a consumer price index. The UK government also measures the Retail Prices Index (RPI). This is a form of consumer price index but differs from most countries' consumer price indices in the sense that it includes mortgage payments and local taxation.

9.3 Money and real data

Money data is data measured in current prices, also referred to as nominal value. This means that the data has not been adjusted for inflation. If a person's wages are increased by 12%, this is a nominal increase.

Real data is data measured in constant prices. This means it has been adjusted for inflation. In the above example, if the rate of inflation is 5% whilst the person's money wages have been increased by 12%, their real wages have increased by 12% – 5% = 7%.

9.4 Shape and determinants of aggregate demand

Aggregate demand (AD) is the total demand for a country's output at a given price level. Aggregate demand is composed of consumer expenditure, investment, government spending and net exports (AD + C + I + G + X – M).

The aggregate demand curve slopes down from left to right. This is for three main reasons. One is that as the price level falls, the country's products become more internationally competitive. This is sometimes called the international trade effect. A lower price level also means that the savings people have can buy more – the wealth effect. In addition, a decline in the price level is usually accompanied by a fall in the rate of interest which is likely to increase consumer expenditure and investment – the interest rate effect.

The aggregate demand curve will change its position if any of the components of AD changes for a reason other than a change in the price level. An increase in one or more of the components will cause a rightward shift whereas a decrease will cause a leftward shift (as shown in Figure 9.1)

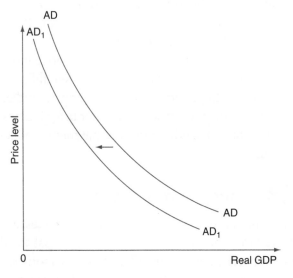

Figure 9.1

Consumer expenditure is spending by households on consumer goods and services. It is the largest component of the aggregate demand of most countries. Among the determinants of consumer expenditure are disposable income, confidence, the rate of interest, wealth, population size and the distribution of income.

Investment is spending on capital goods. It is the most unstable component of AD – it may rise significantly one year and fall by a noticeable amount the next year. The determinants of investment include disposable income, confidence, the rate of interest, advances in technology, corporation tax and government investment subsidies.

Government spending is spending by the government on goods and services. It is influenced by a range of factors. One is the level of economic activity. A government may spend more during a recession in order to stimulate economic activity. Other influences include the amount of tax revenue that can be raised, the dependency ratio and whether the country is at war or peace.

Net exports (export revenue minus import expenditure) are determined by income levels at home and abroad, the exchange rate and import restrictions.

9.5 Shape and determinants of aggregate supply

Aggregate supply (AS) is the total supply that domestic producers are willing and able to sell at a given price level.

The short run aggregate supply (SRAS) curve is drawn on the assumption that the prices of the factors of production (inputs or resources) remain unchanged. It slopes up from left to right because as output increases, average costs rise. This is because whilst the price of factors of production remains unchanged as more is produced, less efficient resources may be used and current workers may be paid overtime rates for additional output. In addition, when the price level rises, output prices usually rise relative to resource prices which increase producers' profits.

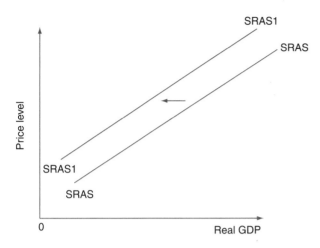

Figure 9.2

The SRAS curve may shift to the left (as shown in Figure 9.2) if there is a rise in resource prices, a decrease in labour productivity, an increase in corporate taxes and if there is a natural disaster.

The long run aggregate supply (LRAS) curve shows the relationship between aggregate supply and the price level when there has been time for the prices of output and resources to adjust fully to changes in the economy (as shown in Figure 9.3). Monetarists/new classical economists argue that the LRAS curve is vertical. They think that, in the long run, the economy will operate at full capacity.

Keynesians argue that the economy can operate at any level of capacity in the long run. When there is considerable spare capacity in the economy, the LRAS curve may be perfectly elastic. This is because it will be possible for firms to employ more resources without raising average costs. As real GDP approaches the full employment level, aggregate supply becomes more inelastic. This is because firms start to experience shortages of resources which increase their price. When the economy reaches full capacity, aggregate supply becomes perfectly inelastic.

LRAS can increase as a result of an increase in the quantity or quality of resources. Specific reasons include immigration of workers, net investment, improved education and training, and advances in technology.

9.6 Interaction of aggregate demand and aggregate supply

An increase in AD occurring when there is considerable spare capacity may have no effect on the price level but will increase the country's output and employment. This is illustrated in Figure 9.4.

An increase in AD which takes place when the economy is beginning to experience shortages of resources will result in a rise in output, the price level and employment. This is shown in Figure 9.5.

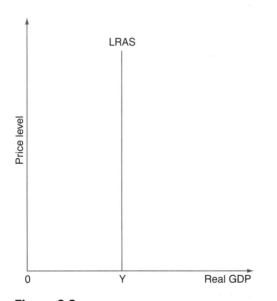

Figure 9.3

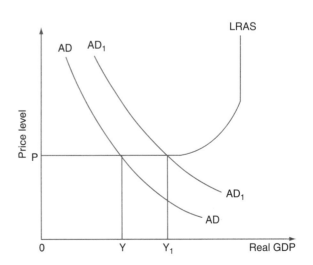

Figure 9.4

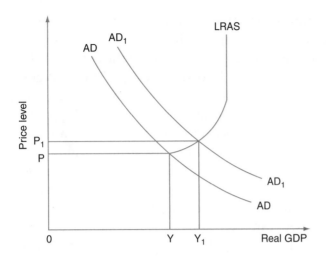

Figure 9.5

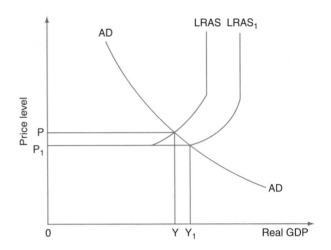

Figure 9.7

When an economy is operating at full capacity, an increase in AD will be purely inflationary. It will increase the price level but have no effect on output and employment as illustrated in Figure 9.6.

An increase in AS when an economy is initially producing at or close to full capacity would increase output and would put downward pressure on the price level. This is illustrated in Figure 9.7.

If an economy has considerable spare capacity, an increase in AS will increase productive capacity but will have no impact on output or the price level. This situation is shown in Figure 9.8.

If the rise in productive capacity is caused by an increase in the labour force but the extra potential workers are not utilised, unemployment will rise.

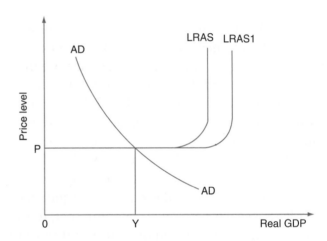

Figure 9.8

Governments aim for sustained economic growth which requires AD and AS to increase in line with each other (as shown in Figure 9.9).

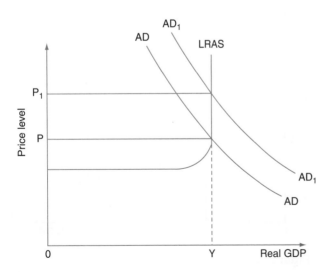

Figure 9.6

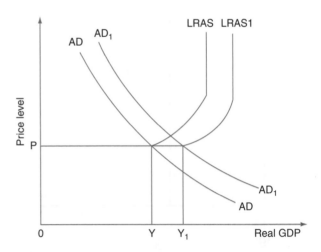

Figure 9.9

9.7 Mind maps

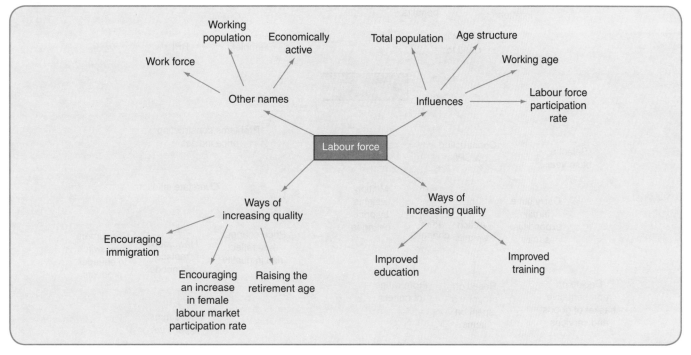

Mind map 9.1   Labour force

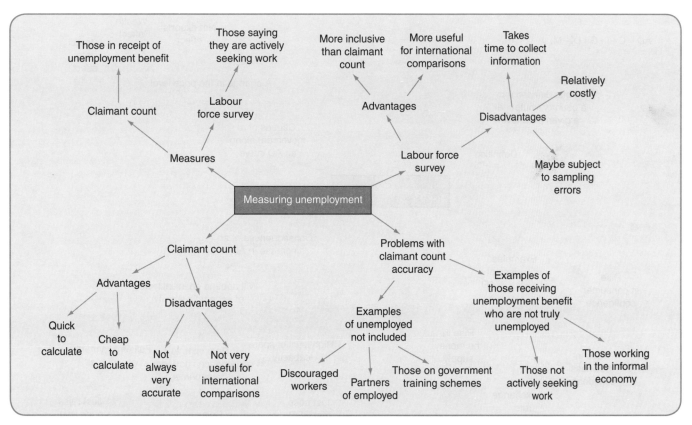

Mind map 9.2   Measuring unemployment

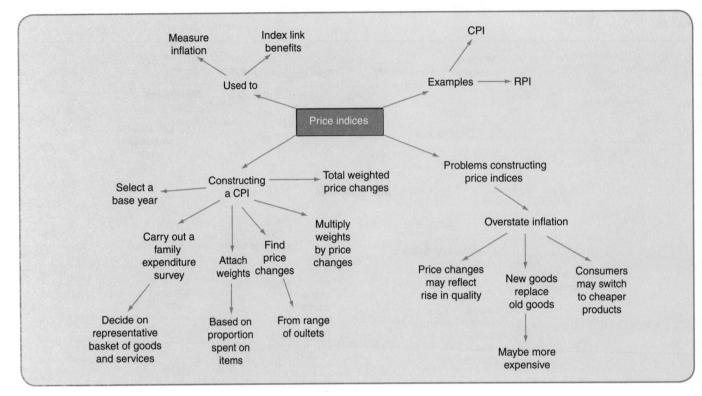

Mind map 9.3   Price indices

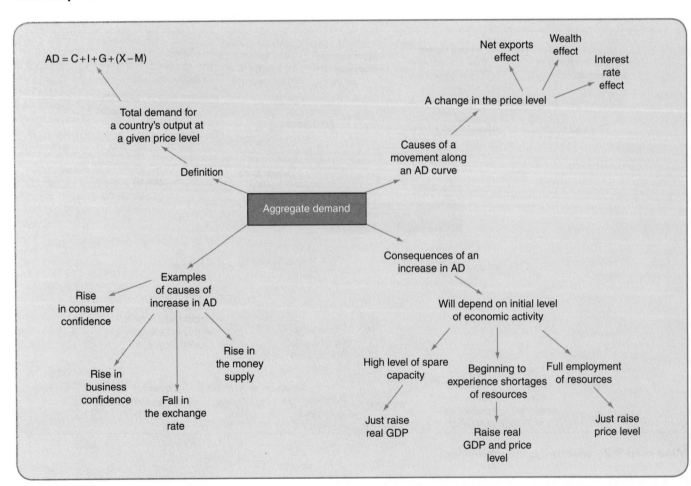

Mind map 9.4   Aggregate demand

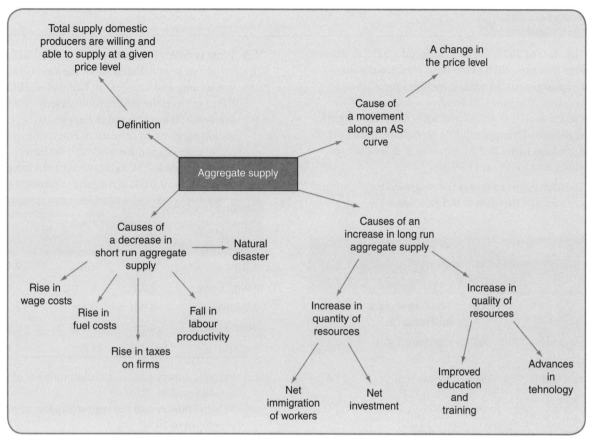

Mind map 9.5   Aggregate supply

Short Questions

1. Explain two reasons why the size of a country's labour force could decrease.

2. What is meant by discouraged workers?

3. Why may a country experience a decrease in production but an increase in labour productivity?

4. What could explain a country experiencing both an increase in employment and unemployment?

5. Identify five reasons why someone may stop being unemployed.

6. Why would unemployment cause unemployment?

7. What is meant by the 'dependency ratio'?

8. What happens to the weight attached to food over time in most countries?

9. What is the key factor which determines whether a country would benefit from a period of inflation?

10. What is meant by 'a fall in the real price of laptops'?

11. Identify the three domestic components of aggregate demand.

12. What effect would a fall in income tax be expected to have on aggregate demand?

13. Why may a recession in one country result in a fall in aggregate demand in another country?

14. What is meant by dissaving?

15. Explain two reasons why a fall in profit levels may reduce investment.

16. What effect would a widespread flood have on a country's aggregate supply?

17. Why does a change in investment have a particularly significant effect on an economy?

18. Compare how an increase in resources may be illustrated on a long run aggregate supply diagram and a production possibility curve diagram.

19. When does macroeconomic equilibrium occur?

20. What combination of shifts in AD and in AS would result in a fall in the price level and a fall in real GDP.

Revision Activities

1. At the start of 2011, India introduced a new consumer index. This uses 2010 as the base year. Unlike the old consumer price index, it includes price trends in services. The new CPI has five major categories. These are food, beverages and tobacco (50% weight), miscellaneous items including electricity, transport and medical care (26.3%), housing (9.5%) and clothing and footwear (4.7%).

   The table below compares the weights in the consumer price indices of Bangladesh and Pakistan.

Bangladesh		Pakistan	
Category	Weight (%)	Category	Weight (%)
Food	58.9	Food, beverages and tobacco	40.3
Clothing and footwear	6.8	Clothing and footwear	6.1
Rent, fuel and lighting	16.9	House rent	23.4
Furniture, finishing and other	2.7	Fuel and lighting	7.3
Medical care and health expenses	2.8	Household furniture and equipment	3.3
Transport and communications	4.2	Transport and communication	7.3
Recreation, entertainment, education and cultural services	4.1	Recreation, entertainment and education	5.0
Miscellaneous goods and services	3.6	Cleaning, laundry and personal appearance	6.0
		Medicine	2.1

(a) What is meant by a base year?

(b) What can be concluded about the amount spent on clothing and footwear in Bangladesh, India and Pakistan from the information given?

(c) What would be expected to happen to the weighting given to recreation, entertainment and cultural services in Bangladesh over time?

(d) Would Bangladesh or Pakistan have been more affected by a 10% rise in the price of furniture in 2010?

2.

Selected labour market statistics 2010			
Country	Size of population	Labour force	Unemployment %
China	1.3bn	780m	9.2
Hong Kong	7.2m	3.7m	3.8
Maldives	0.4m	0.1m	14.5
New Zealand	4.1m	2.3m	6.5
Nigeria	137.2m	48.3m	4.9

(a) Which country had the smallest number of people unemployed in 2010?

(b) Which country had the largest number of people employed in 2010?

(c) Which country had the highest proportion of its population in the labour force in 2010?

(d) Which country had the smallest proportion of its population in the labour force in 2010?

3. Decide whether the following changes in Brazil's economy would have increased the country's aggregate demand, increased the country's aggregate supply, increased both the country's aggregate demand and aggregate supply, decreased the country's aggregate demand, decreased the country's aggregate supply, or decreased both the country's aggregate demand and aggregate supply.

(a) A rise in government spending on education

(b) An increase in net exports

(c) An increase in the money supply

(d) Net immigration

(e) A rise in unemployment.

Multiple Choice Questions

1. Which of the following groups is in the labour force?
   A Homemakers
   B Students in full-time education
   C The retired
   D The unemployed

2. A country has a population of working age of 50 million. Its labour force participation rate is 55% and its unemployment rate is 8%. What are the number of people in the labour force and the number of people in employment?

	No. of people in the labour force (millions)	No. of people in employment (millions)
A	25.3	23.5
B	27.5	25.3
C	31.5	27.5
D	50.0	46.0

3. Why may a rise in employment decrease labour productivity?

	Average skill levels of workers	Average amount of capital equipment used by workers
A	Decrease	Decrease
B	Decrease	Increase
C	Increase	Increase
D	Increase	Decrease

4. Labour productivity in a country's sugar industry grows more rapidly than the country's output of sugar. What must this mean?
    A The amount of labour employed in the sugar industry must have declined
    B The amount of labour employed in the sugar industry must have increased
    C The output of sugar as a percentage of the country's total output must have declined
    D The output of sugar as a percentage of the country's output must have increased

5. Which change would increase the size of a country's labour force?
    A A decrease in net emigration
    B A decrease in unemployment
    C An increase in the school leaving age
    D An increase in the retirement age

6. The table shows the unemployment rate of a country over a period of four years. What can be concluded from this information?

Year	Unemployment rate (%)
2011	6.2
2012	7.9
2013	8.1
2014	10.4

    A The gap between actual and potential output must have increased
    B The number of people unemployed must have increased
    C The output of the country must have declined
    D The size of the labour force must have declined

7. A country has a population of 90 million, 36 million of whom are in employment and 9 million of whom are unemployed. What is the country's unemployment rate?
    A 10%
    B 16.67%
    C 20%
    D 25%

8. Why might the number of people unemployed as measured by the claimant count be lower than the number as measured by the Labour Force Survey?
    A Some people actively seeking employment may not qualify to receive unemployment related benefits
    B Some people actively seeking employment may not be prepared to accept the first job on offer
    C The claimant count does not include people who are cyclically unemployed
    D The claimant count does not include people who are structurally unemployed

9. The annual rate of inflation in a country declines from 5% to 3%. What can be concluded from this information?
    A The cost of living has increased
    B The prices in the retail price index have fallen
    C The value of money has increased
    D The weights in the consumer price index have fallen

10. What would a fall in a country's consumer price index indicate?
    A A decrease in the standard of living
    B A decline in international competitiveness
    C An increase in the cost of living
    D An increase in the value of money

11. A country's consumers spends 10% of their total expenditure on food, 20% on clothing, 30% on housing and 40% on other products. During the year, the price of food rises by 10%, the price of clothing falls by 10%, the price of housing rises by 5% and the price of other products increases by 12%. What is the country's inflation rate?
    A 4.25%
    B 5.3%
    C 9.25%
    D 25.0%

12. The table below shows year on year percentage change in a country's retail price index.

Year/month	Inflation %
2010 June	10.1
December	9.6
2011 June	8.2
December	12.6
2012 June	6.4
December	2.2

What can be concluded from this information?
A The cost of living fell in 2012
B The cost of living was lower in June 2012 than in June 2011
C The price level was at its lowest in June 2010
D The price level was at its highest in December 2011

13. Why are changes made regularly to the weights used in a consumer price index?
A Rates of taxation alter
B Spending patterns change
C The quality of products increases
D There are seasonal fluctuations in prices

14. What is meant by real wages?
A Money wages adjusted for inflation
B Money wages net of tax
C Wages earned by workers in the secondary sector
D Wages in the country relative to wages in another country

15. A country's inflation rate is 9% and the real rate of interest is 3%. What is the nominal (money) rate of interest?
A 3%
B 6%
C 12%
D 27%

16. Which of the following changes would shift the long run aggregate supply curve to the right?
A An increase in the price level
B An increase in wage rates
C An increase in expenditure on education and training
D An increase in consumer expenditure

17. An increase in aggregate demand occurs on the perfectly elastic part of the long run aggregate supply curve. What will be the outcome?
A A decrease in consumption
B A decrease in costs of production
C An increase in the price level
D An increase in real GDP

18. What could have caused the shift in the aggregate demand curve shown in Figure 9.10?
A A decrease in income tax
B A decrease in the money supply
C An increase in imports
D An increase in the rate of interest

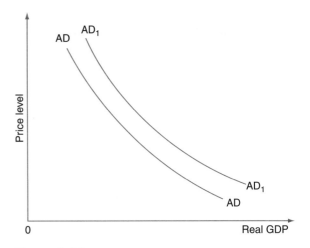

Figure 9.10

19. What effect would net immigration of people of working age have on an economy's aggregate demand and aggregate supply?

	Aggregate demand	Aggregate supply
A	Decrease	Decrease
B	Decrease	Increase
C	Increase	Increase
D	Increase	Decrease

20. Figure 9.11 shows an economy is initially operating at point X. What impact would an increase in net exports have on the country's output and its price level?

	Price level	National output
A	Decrease	Decrease
B	Decrease	Increase
C	Increase	Increase
D	Increase	Decrease

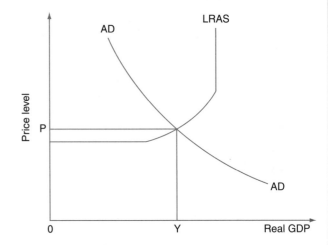

Figure 9.11

Data Response Questions

1. Measuring inflation

    Many countries use consumer price indices to measure inflation. These are weighted price indices which reflect both changes in prices and changes in household spending patterns. There are a number of reasons why household spending patterns alter over time. These include changes in tastes and relative prices. As these factors vary from country to country, weights also vary. Table 1 compares some of the weights in the consumer price indices of Australia, New Zealand and the UK.

    Table 1 Selected weight in the consumer price indices of Australia, New Zealand and the UK 2011

Table 1

	Australia	New Zealand	UK
Food and non-alcoholic beverages	16.8	18.8	11.8
Alcoholic beverages and tobacco	7.1	6.5	4.2
Clothing and footwear	4.0	4.1	6.2
Housing and household utilities	22.3	24.0	13.0
Furnishings, household equipment and services	9.1	4.2	6.0
Health	5.3	5.5	2.4
Transport	11.6	15.0	15.9
Communication	3.0	3.5	2.6
Education	3.2	2.2	1.8
Miscellaneous	17.6	16.2	36.1

The selection of products is revised regularly to ensure that they are representative of the pattern of expenditure by households. Each time they are revised, some new items are added to the 'basket' of consumer products and some are removed. In 2011, tablet computers and e-books, for instance, were added to New Zealand's CPI basket whilst among the products which were removed were dictionaries and envelopes.

It is interesting to note that some economists claim that consumer price indices do not always provide an accurate measure of inflation and so may result in inappropriate monetary policy measures.

(a) What is meant by changes in relative prices? [2]

(b) Apart from changes in tastes and relative prices, explain two reasons why household spending patterns may alter over time. [4]

(c) (i) Can it be concluded from Table 1 that people in Australia spend more on alcoholic beverages and tobacco than people in the UK? Explain your answer. [2]

   (ii) What effect would there be on New Zealand's and the UK's inflation rates if there was a 10% rise in the price of housing and household utilities in each country? [2]

(d) Explain how consumer price indices may influence monetary policy. [4]

(e) Discuss whether consumer price indices provide an accurate measure of inflation. [6]

2. Inflation in Paraguay

    Paraguay is a small South American country with few of the advantages that bigger, better known economies such as Venezuela and Brazil enjoy.

    In January 2008, Paraguay updated its Consumer Price Index (CPI) with new weights based on the latest household budget survey. Table 2 shows the changes in the weights.

Table 2

Weights in the Paraguayan CPI, 1992 and 2008		
	1992	2008
Food	35.1	32.1
Clothing	8.6	4.9
Alcohol and tobacco	2.1	1.2
Furniture	8.0	7.7
Transport	7.4	14.8
Communication	0.7	3.4
Housing	10.4	8.9
Health	4.8	4.1
Entertainment	5.1	6.1
Education	2.7	4.0
Restaurants and hotels	4.8	5.5
Miscellaneous goods and services	10.3	7.3

Performance of Paraguay and Venezuela in controlling inflation as shown in Figure 9.12.

(a) (i) Why are weights used in constructing a CPI? [2]

   (ii) Explain possible reasons why the weights for alcohol and tobacco and transport changed between 1992 and 2008. [4]

(b) (i) What is meant by inflation? [2]

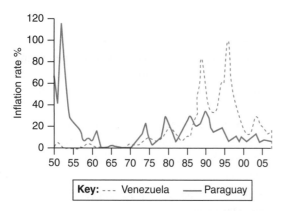

Figure 9.12 Inflation in Paraguay and Venezuela 1950–2005

(ii) Using Figure 9.13 compare Paraguay's inflation between 150 and 2005 with that of Venezuela. [2]

(iii) To what extent did Paraguay achieve a low and stable rate of inflation between 1950 and 2005? [4]

(c) Discuss whether all countries should set annual inflation targets of around 3%. [6]

Cambridge 9708 Paper 22 Q1 May/June 2010

3. India's strengths and weaknesses

One of the strengths of the Indian economy is its young population. The average age of the population is only 25 and 60% of the population is aged below 30. As young Indians have become richer and more educated, they are buying more items including cars, electrical goods and housing. They are also contributing to an increasing labour force.

India is, however, suffering from poor infrastructure. Its roads, railways and telephone system are of poorer quality than its economic rival, China. In an attempt to rectify this, the Indian government is increasing its spending on infrastructure.

Despite the poor infrastructure, a number of Indian industries are becoming world leaders. The country has become one of the biggest outsourcing centres in the world, with IT firms benefitting from a good supply of

high skilled, relatively cheap software experts. There is a rapid growth in business processing with the number of call centres still increasing and more and more pharmaceutical research and the making of spare parts for car manufacturers being located in the country.

Table 3 compares total factor productivity growth in a number of countries. Total factor productivity growth is the percentage increase in output not accounted for by changes in the volume of capital and labour.

Table 3

The annual growth in total factor productivity in a group of selected countries 1990–2010	
Country	Annual productivity growth (%)
China	4
India	2.8
Indonesia	1.7
Japan	1.2
UK	1.1
USA	1.0
Brazil	0.4
Russia	0.2

(a) (i) Define aggregate demand. [2]
  (ii) Using the information provided, explain two reasons why aggregate demand has increased in India in recent years. [4]
  (iii) Using the information provided, explain two reasons why aggregate supply has increased in recent years. [4]

(b) (i) Compare India's growth in total factor productivity with that of the other countries in Table 3. [3]
  (ii) If the capital stock and labour force both increase by 3% and output rises by 5%, what is the total factor productivity growth? [1]

(c) Discuss whether an increase in a country's labour force always benefits an economy. [6]

Essay Questions

1. (a) Explain what determines the size of a country's labour force. [8]

   (b) Discuss whether a widespread shortage of labour might be a major cause of inflation. [12]

   Cambridge 9708 Paper 2 Q3 May/June 2008

2. (a) Explain what factors determine consumer expenditure. [8]

   (b) Discuss how an increase in consumer expenditure may affect output and the price level [12]

3. (a) Explain, with examples, why labour productivity might vary between countries. [8]

   (b) Unemployment can be measured by the claimant count or the labour force survey. Discuss the relative reliability of these two measures. [12]

   Cambridge 9708 Paper 2 Q4 Oct/Nov 2007

Macroeconomic Problems 10

10.1 Inflation

Inflation is a sustained rise in the general price level.

Degrees of inflation

✓ A creeping inflation rate is a low and steady rate of inflation.
✓ Hyperinflation is often taken to mean an inflation rate above 50%.
✓ Accelerating inflation means that the general price level is increasing at a more rapid rate.
✓ Unanticipated inflation arises when the inflation rate is not what people expected.
✓ A fall in the inflation rate means that the price level has risen more slowly than in the previous year.

Causes of inflation

✓ Cost-push inflation is caused by increases in the costs of production. For instance, increases in the cost of oil or increases in wages, not matched by increases in the productivity, may raise costs which firms pass on to consumers in the form of higher

prices. The initial rise in prices can then set off a sustained increase in prices.
✓ Demand-pull inflation occurs when aggregate demand grows at a more rapid rate than aggregate supply. It may, for instance, be the result of a consumer boom.
✓ Inflation may arise if the money supply grows faster than increases in the country's output.

Consequences of inflation

✓ The effects of inflation are influenced by its rate, whether the rate is stable or fluctuating, whether the inflation was correctly anticipated or not, its cause and how the rate compares with other countries.
✓ The costs of inflation include a random redistribution of income, menu costs, shoe leather costs, inflationary noise, fiscal drag, a loss of international competitiveness, administrative costs and discouragement of investment.
✓ The possible benefits of a low and stable inflation rate caused by demand-pull factors are a stimulus to output, a reduction in

debt in real terms and an ability of firms to continue in production and maintain employment in difficult times by cutting real wages.

10.2 Balance of payments problems

Meaning of balance of payments equilibrium and disequilibrium

✓ The balance of payments, in theory, should always balance as credit items (money coming into the country) by debit items (money going out of the country). In practice, because of the large number of transactions involved and the delay in reporting them, a balancing item known as net errors and omissions is added to ensure that the balance of payments balances.

✓ A large debit item on net errors and omissions would indicate that more money has left the country than has currently been accounted for.

✓ Balance of payments equilibrium and balance of payments disequilibrium usually refer to equilibrium or disequilibrium on the accounts within the overall balance.

✓ The current account will be in deficit if earnings from the exports of goods and services, receipts of income and current transfers are less than expenditure on imports of goods and services and income and current transfers sent abroad.

✓ If net transfers in the financial account are in surplus, it means that direct, portfolio and other investment into the country (transactions in liabilities) is greater than direct, portfolio and other investment going out of the country (transactions in assets).

Causes of balance of payments disequilibrium

✓ A trade in goods deficit may be structural and/or cyclical. A structural deficit arises due to a lack of international competitiveness. A cyclical deficit occurs as a result of incomes abroad falling and/or incomes at home rising.

✓ A lack of international competitiveness may arise due to an overvalued exchange rate, a relatively high inflation rate, relatively low productivity and a lack of innovation.

✓ A rise in interest rates abroad and expectation that foreign share prices and foreign government bond prices will increase may result in a net outflow of portfolio investment.

✓ A fall in corporate taxes abroad, a rise in economic prospects in foreign countries are among the reasons why there may be a net outflow of direct investment.

✓ Debit items in the financial account will result in an inflow of income in the form of profit, interest and dividends into the current account in the longer term.

Consequences of balance of payments disequilibrium on domestic and external economy

✓ A current account surplus makes a positive contribution to a country's aggregate demand but involves an opportunity cost in the form of forgone consumption of imports and may put upward pressure on the price level by adding to the money supply whilst involving a net outflow of products.

✓ A current account deficit makes a negative contribution to a country's aggregate demand. It means a country is 'living beyond its means' – consuming more products than it is producing.

✓ How significant a current account disequilibrium is depends on its size, duration and cause.

✓ A current account surplus will, ceteris paribus, put upward pressure on the price of a country's currency while a current account deficit will put downward pressure on the price of the currency.

✓ A current account deficit can be financed by a net inflow direct or portfolio investment, borrowing or drawing on the country's reserves of foreign currency.

10.3 Fluctuations in foreign exchange rates

Definitions and measurement of exchange rates

✓ The nominal exchange rate is the market price of one currency in terms of another or group of currencies.

✓ The real exchange rate is a currency's value in terms of its real purchasing power. It is the nominal exchange rate adjusted by prices at home and abroad. It is, in effect, a measure of international price competitiveness. A rise in the real exchange rate means a country's products have become more expensive relative to foreign products either because the nominal exchange rate has risen and/or because its inflation rate is higher than its trading partners.

✓ If a comparison is made of different countries' GDP figures at the current exchange rate, it may give a misleading indicator of the purchasing power of the currency at home. To overcome this problem, GDP can be converted into a common currency at a purchasing power parity rate. This is a rate at which two currencies would be able to purchase the same quantity of products in the two economies.

✓ The trade weighted exchange rate is also known as the effective exchange rate. It is a weighted average exchange rate which reflects the relative importance of different currencies in terms of their shares in the country's international trade.

Determination of exchange rates – floating, fixed, managed float

✓ A floating exchange rate is one determined by the market forces of demand and supply. A rise in the price of a currency brought about by a rise in demand for the currency and/or a fall in the supply is known as appreciation. A fall in the price of a currency caused by a fall in demand for the currency and/or a rise in its supply is referred to as depreciation.

✓ A fixed exchange rate is one that is set as a particular level and maintained at that level by a government or central bank acting on behalf of the government.

✓ A change from a fixed rate to a higher fixed rate is called a revaluation. In contrast, a devaluation is a reduction in a fixed rate.

✓ A managed float is an exchange rate system where the price of the currency is largely determined by market forces but one in which the government will intervene to avoid large fluctuation in price.

✓ A central bank, acting on behalf of the government, may maintain a fixed exchange rate, or seek to influence a managed float directly by buying and selling the currency or indirectly by raising or lowering the interest rate.

✓ To offset downward pressure a fixed exchange rate or a managed float, a central bank may buy the currency, using reserves of foreign currency, or raise the interest rate.

Factors underlying fluctuations in exchange rates

✓ A country's currency is demanded by traders wishing to buy the country's exports, tourists wishing to visit the country, financial investors wishing to place money in the country's banks and to buy shares in the country's firms and the government's bonds, foreign firms wishing to set up in the country or purchase domestic firms and speculators who anticipate a rise in the value of the currency.

✓ A country's currency is supplied by those wishing to purchase foreign currency in order to buy imports, go on foreign holidays, place money in foreign banks, purchase foreign shares and government bonds, to set up firms or buy firms in foreign countries and to speculate that the price of the currency will fall relative to foreign currencies.

✓ The major causes of fluctuations in exchange rates are changes in international competitiveness, changes in income at home and abroad, changes in the economic performance of the economy, changes in the rate of interest at home and abroad and speculation.

✓ A floating exchange rate will appreciate if net exports increase, hot money flows are attracted into the country by a rise in the rate of interest, foreign direct investment increases or there is speculation that the price of the currency will rise in the future.

Effects of changing exchange rates on the economy

✓ A fall in the price of a currency will reduce export prices, in terms of foreign currency, and will raise import prices, in terms of the domestic currency.

✓ A rise in the price of a currency will raise export prices, in terms of foreign currency, and will reduce import prices, in terms of the domestic currency.

✓ The Marshall-Lerner condition states that the value of the price elasticity of demand for exports plus the price elasticity of demand for imports must be greater than one for a devaluation/depreciation to result in an improvement in the trade balance.

✓ The J-curve shows a trade deficit initially increasing following a devaluation/

depreciation and then reducing. This is because, at first, demand for exports and imports is inelastic as there is no time for buyers to notice and respond to the change in prices. As buyers adjust to the price changes, demand may become elastic.

✓ The reverse J-curve shows an appreciation/, revaluation, at first increasing a trade surplus as demand is inelastic and then reducing a trade surplus as demand becomes more elastic over time.

✓ If demand for exports is inelastic, producers may seek to benefit from a fall in the price of the currency by maintaining their price in terms of foreign currency. This will raise revenue when converted back into the domestic currency.

✓ A fall in the price of a currency which results in an increase in net exports will raise aggregate demand and, if there is spare capacity in the economy, will increase the country's output and employment.

✓ A rise in the price of a currency will tend to reduce inflationary pressure as it will lower import prices and may reduce aggregate demand.

10.4 Mind maps

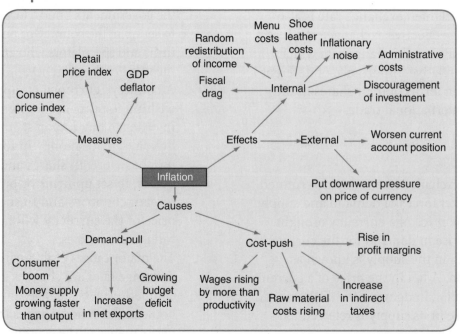

Mind map 10.1 Inflation

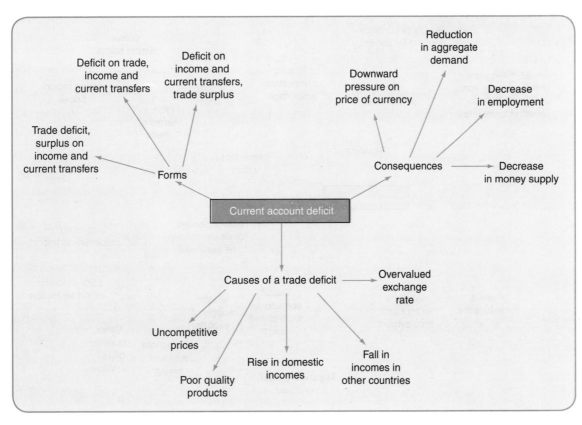

Mind map 10.2   Current account deficit

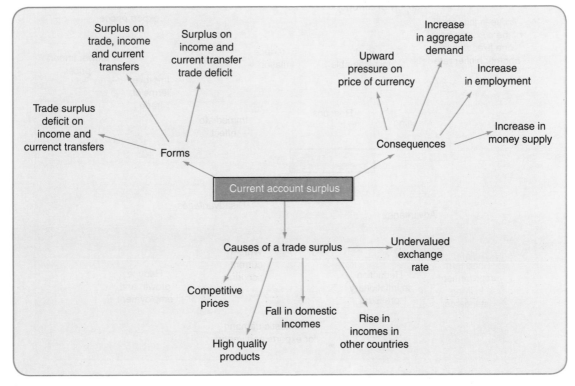

Mind map 10.3   Current account surplus

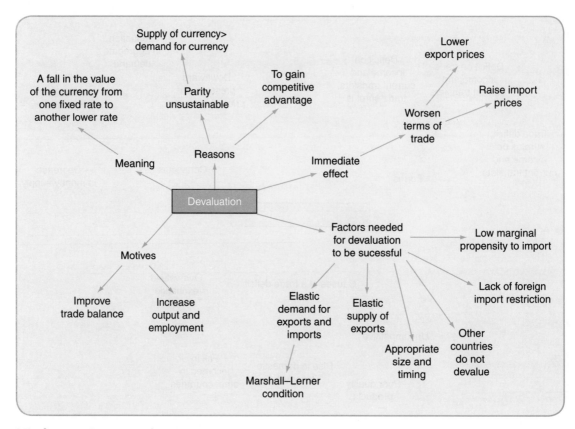

Mind map 10.4   Devaluation

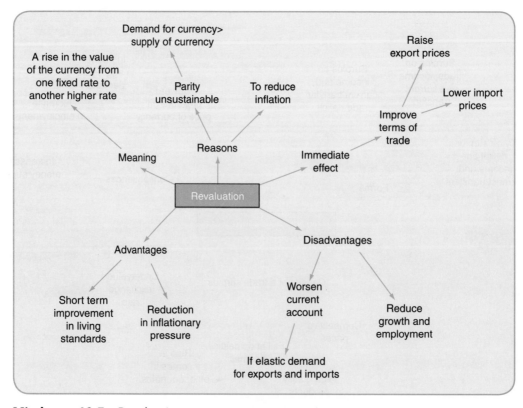

Mind map 10.5   Revaluation

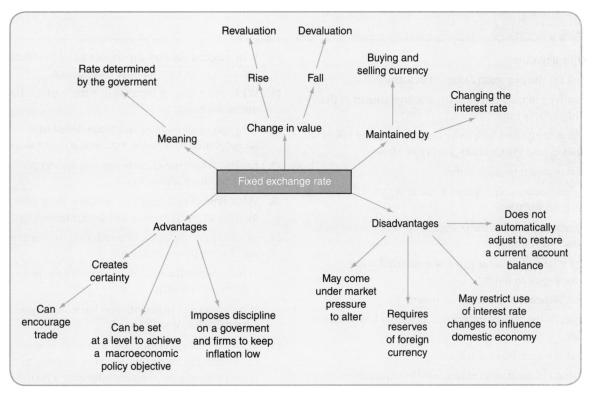

Mind map 10.6   Fixed exchange rate

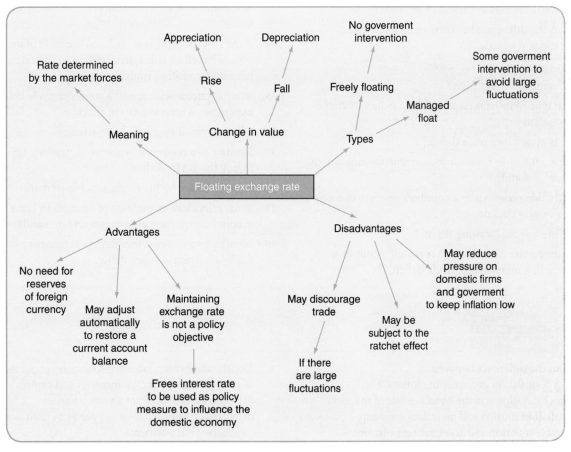

Mind map 10.7   Floating exchange rate

Short Questions

1. Define deflation.
2. What are the two main causes of deflation?
3. What type of inflation may an excessive growth of the money supply cause?
4. What assumptions have to be made to turn the Fisher equation into the Quantity Theory of Money?
5. What is meant by wage drift?
6. In what circumstance would a rise in wages not result in cost-push inflation?
7. Explain why a fall in the price of a currency may result in cost-push inflation.
8. Why might a fall in the price of a currency cause demand-pull inflation?
9. What is meant by a consumer boom?
10. Explain what type of inflation a consumer boom may cause.
11. What is meant by a fiscal boost?
12. In what circumstances may a fiscal boost result in inflation?
13. Explain why the government may gain from inflation.
14. What must happen as a result of inflation?
15. What is the difference between anticipated and unanticipated inflation?
16. When are menu costs significant?
17. What is meant by agflation?
18. Which type of inflation is most likely to be associated with stagflation?
19. What is meant by a trade deficit?
20. What are the links between the current account and the financial account?
21. Identify two reasons why a country's reserves of foreign currency may decline.
22. What are Special Drawing Rights?
23. Why may a rise in a country's real GDP result in an increase in a merchandise trade deficit?
24. Why might a country experience a rise in inflation rate but a fall in its merchandise trade deficit?
25. What effect may an increase in a trade deficit have on unemployment?
26. Why might an increase in a trade deficit be accompanied by a rise in a current account surplus?
27. Explain what type of exchange rate system provides the most certainty for traders.
28. What type of exchange rate system is determined by a mixture of market forces and state intervention?
29. What type of exchange rate system is the concept of purchasing power linked to?
30. What effect will a revaluation have on the price of exports and imports?
31. What effect will hyperinflation have on the internal and external value of a currency?
32. Why may the government of a country operating a fixed exchange rate be under more pressure to reduce inflation than a government operating a floating exchange rate?
33. Explain two factors that influence the price elasticity of demand (PED) of the products produced by a country's firms?
34. If the PED of exports is –0.5 and the PED of imports is –0.2, what effect will a depreciation in the exchange rate have on a country's trade deficit?
35. In what circumstances will a country's trade balance experience a reverse J-curve effect?
36. What is meant by a run on a currency?
37. Identify two reasons why someone may sell US dollars to buy the UAE's dirham.
38. What is meant by the foreign exchange market?
39. What effect will an increase in a country's current account deficit have on its aggregate demand?
40. Identify two causes, other than government policy, of a fall in a current account deficit.

Revision Activities

1. Explain the difference between:
   (a) hyperinflation and creeping inflation
   (b) core inflation and the headline rate of inflation
   (c) inflation illusion and an inflationary gap
   (d) stable inflation and accelerating inflation.

2. Decide whether the following statements about the balance of payments are correct or incorrect and explain why.
   (a) The current account always balances
   (b) Transactions in assets appear as a credit item in the balance of payments

(c) Debit items are balanced by credit items

(d) A government will always seek to avoid a current account disequilibrium

(e) The merchandise balance is the difference between the volume of exports of goods and the volume of imports of goods

(f) Devaluation need not reduce a trade deficit.

3. A country's exchange rate against the US dollar is initially, 10 pesos = $1. The country's firms sell 20 million products at an average price of 20 pesos. The country imports 25 million products at an average price of $2.

The value of the pesos then falls to 15 pesos = US$1. The country's exports rise to 30 million selling at 20 pesos and its imports fall to 15 million bought at an average price of $2.

(a) Calculate the initial trade balance

(b) Calculate the trade balance after the depreciation of the pesos

(c) What does the change in the trade balance indicate about the combined price elasticities of demand for the country's exports and imports?

Multiple Choice Questions

1. The table shows the amount consumers of a country spend on five items and the percentage change in the price of the products in the same year.

Product	Percentage of consumers' expenditure	Percentage price change
Food	20	15
Electricity	5	10
Transport	10	20
Entertainment	5	−10
Clothing	10	30

What was the overall rise in the weighted cost of living index?

A 0%   B 15%   C 16%   D 17%

2. Which one is a possible cause of cost-push inflation?
A An increase in bank lending
B An increase in government spending
C A reduction in the price of the currency
D A reduction in income tax

3. What is meant by fiscal drag?
A Inflation imposing costs on firms as a result of then having to use workers' time in changing prices in catalogues
B Inflation increasing nominal incomes and pushing people into higher tax brackets
C Inflation putting upward pressure on government spending so as to retain its value
D Inflation resulting in a redistribution of income from the government to taxpayers

4. According to the Quantity Theory of Money, what will be the result of a decrease in the stock of money?
A A decrease in the price level
B A decrease in the velocity of circulation
C An increase in bank lending
D An increase in real GDP

5. Between January 2009 and January 2010, the rate of inflation in Pakistan fell from 25% to 10.5%. What can be concluded from this information?
A The internal purchasing power of the Pakistani rupee increased
B The cost of living in Pakistan increased
C Pakistan's consumer price index fell
D Costs of production declined in Pakistan

6. Which of the following is an advantage of inflation for country's firms?
A A reduction in international competitiveness if the country's inflation rate is higher than rival countries' inflation rate
B A reduction in the real cost of employing labour if wages rise by less than inflation
C An increase in uncertainty if the inflation rate is higher than the expected inflation rate
D An increase in the tax burden on firms if the corporation tax rate is not adjusted in line with inflation

7. What is stagflation?
A A fall in the price level
B A combination of inflation, unemployment and low economic growth
C A consistent rate of inflation
D An increase in the price level excluding food and energy prices

8. Which change will affect the trade in services balance of the Maldives?
A The payment of interest on a loan by a firm in the Maldives to an Indian bank
B The purchase of fish caught by fishermen in the Maldives by a Sri Lankan firm
C The sale of insurance by a bank based in Hong Kong to a hotel chain in the Maldives
D The setting up of a factory in the Maldives by an Australian firm

9. The table below shows the balance of payments account for a country.

Balance of payments	$bn
Exports of goods and services	150
Imports of goods and service	130
Net income	−20
Net current transfers	30
Net direct investment	−10
Net portfolio investment	−20
Net other investment	−20
Drawing on reserves	10

Which combination describes the country's balance of payments position?

	Current account balance	Financial account balance
A	Deficit	Deficit
B	Deficit	Surplus
C	Surplus	Surplus
D	Surplus	Deficit

10. Which change is likely to increase a country's current account deficit on the balance of payments?
    A A decrease in consumer spending
    B A decrease in the rate of inflation
    C An increase in the exchange rate
    D An increase in labour productivity

11. Which group would benefit from a depreciation of the Pakistani rupee against the Chinese yuan?
    A Chinese car producers who sell to Pakistani firms
    B Chinese speculators who have holdings of Pakistani rupee
    C Pakistani airlines which carry Chinese passengers
    D Pakistani students studying in China

12. A government operates a managed floating exchange rate system. Which policy might it adopt to raise the value of its exchange rate?
    A Instruct its central bank to raise the rate of interest
    B Order its central bank to sell its currency on the foreign exchange market
    C Raise corporation tax
    D Remove subsidies to exporters

13. The trade weighted exchange rate for the New Zealand dollar is the rate of exchange between the New Zealand dollar and:
    A The currencies of countries it has trading agreements with
    B The currencies of countries operating floating exchange rates

C The currencies of its main trading partners
D The currencies of its neighbouring countries

14. What advantage does a fixed exchange rate have over a floating exchange rate?
    A A greater certainty about the price exporters will receive and importers will pay
    B A greater chance that the exchange rate will be at the equilibrium level
    C An ability to use the rate of interest to influence domestic demand
    D An absence of a need to keep reserves of foreign currency

15. What could cause an increase in the supply of Argentine pesos on the foreign exchange market?
    A An increase in the value of Argentine imports
    B An increase in foreign direct investment in Argentina
    C A decrease in portfolio investment in Argentina
    D A decrease in the value of Argentine exports

16. What is the most likely cause of a depreciation in the value of Bangladesh's currency, the taka?
    A A decrease in foreign direct investment in Bangladesh
    B A decrease in Bangladesh's inflation rate
    C An increase in Bangladesh's interest rate
    D An increase in the number of tourists visiting Bangladesh

17. Which circumstance would increase the likelihood of a fall in the price of a country's currency improving its trade balance?
    A Demand for its exports is income elastic and incomes in its trading partners are falling
    B Demand for its imports is income elastic and incomes at home are rising
    C The combined price elasticities of demand for its exports and imports is less than one
    D The price elasticity of supply of its exports is greater than one

18. Figure 10.1 shows the foreign exchange market for the Egyptian pound.

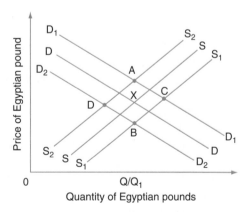

Figure 10.1

The initial position is X. Egyptians then take more holidays abroad and Egyptian firms sell fewer exports. What is the new equilibrium position?

  A A   B B   C C   D D

19. A country's inflation rate rises and its current account deficit declines. What could explain this combination of events?
  A A consumer boom in the country
  B A devaluation of the currency
  C A recession in the country's trading partners
  D A rise in the quality of foreign products

20. What determines the value of a freely floating exchange rate?
  A Central bank manipulation of the rate of interest
  B Central bank purchase and sale of the currency
  C Government manipulation of tax rates
  D Market forces of demand for and supply of the currency

Data Response Questions

1. The appreciation of the Zambian Kwacha

Zambia's currency, the Kwacha, experienced a significant appreciation in the year up to March 2006. This is shown in Figure 10.2. Factors that influenced the exchange rate at this time were an improvement in Zambia's export performance, a reduction in the foreign debt owed by Zambia, an increase in foreign aid received by Zambia and an inflow of foreign investment. The exchange rate is vitally important for Zambia because of its exports of copper, tobacco, maize and cotton are priced in US$ but its costs are paid in Zambian Kwacha.

(a) Identify from Figure 10.2 the greatest monthly appreciation of the Kwacha.
  (i) In which month did this take place? [1]
  (ii) By how much did it appreciate? [1]

(b) Explain what Figure 10.2 suggests about the type of exchange rate system used by Zambia. [3]

(c) Explain how the change in the value of the Kwacha between September 2005 and January 2006 might have been influenced by:
  (i) the improved export performance and
  (ii) the reduction in foreign debt. [6]

(d) How would the appreciation of the Kwacha affect Zambia's terms of trade? [3]

(e) Discuss whether an appreciation of its exchange rate always benefits a country. [6]

Cambridge 9708 Paper 2 Q1 May/June 2008

2. Chinese yuan and UK pound sterling undervalued

At the start of 2010, many economists argued that both the Chinese yuan and the UK pound sterling were undervalued. The Chinese government was intervening in the foreign exchange market to prevent the yuan from rising in value against the US dollar.

The pound sterling had fallen against the US dollar and the euro by approximately 30% since 2007 (see Table 1). The depreciation, however, did not significantly contribute to inflation. This was, in large part, because of the increase in spare capacity in the economy.

Table 1

Changes in the value of the pound sterling 2007–2010		
Year	Exchange rate Currency units per US$	Current account balance
		$bn	% of GDP
2007	0.50	−93.0	−3.2
2008	0.51	−102.4	−4.1
2009	0.61	−44.6	−1.6
2010	0.67	−28.2	−1.1

(a) (i) Identify two ways the Chinese government could intervene in the foreign exchange market to prevent the yuan rising in value against the US dollar. [2]
  (ii) Explain what is meant by a currency being 'undervalued'. [4]

(b) (i) Does Table 1 support the statement that the

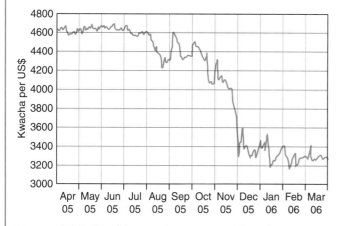

Figure 10.2 Zambian exchange rate (Kwacha per US $), April 2005 to March 2006

value of the pound fell against the US dollar by 30%? [2]

(ii) Explain the relationship between the exchange rate and the current account balance shown in Table 1. [4]

(c) Discuss whether a depreciation in the exchange rate will increase the rate of inflation. [8]

3. US current account and US$ exchange rate

Figure 10.3 shows the US current account balance as a % of GDP from 1980 to 2002, a period in which US GDP rose continuously. Figure 10.4 shows the US$ exchange rate against the euro for the same period.

The figures before 1999 are an imaginary exchange rate based on the values of the main European currencies, as the euro was officially introduced only after 1999.

There is a two-way link between a country's trade performance and its exchange rate. A change in the trade performance can affect the exchange rate while a change in the exchange rate can affect the trade performance.

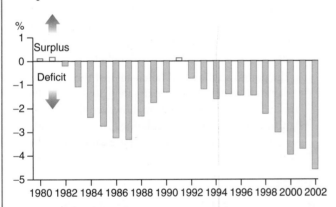

Figure 10.3 US current account balance as % of GDP 1980 to 2002

Figure 10.4 US$ exchange rate against the euro 1980 to 2002

(a) (i) Summarise the performance of the US current account balance between 1980 and 2002. [2]

(ii) Explain how the US might have been able to finance the current account position that it faced between 1992 and 2002. [3]

(b) Suppose a country has a surplus on its current account. Explain how this might affect its exchange rate. [3]

(c) (i) Outline how a depreciation of a country's exchange rate is likely to affect its current account balance. [3]

(ii) Use Figure 10.3 and Figure 10.4 to analyse whether this expected effect of exchange rate depreciation occurred in the case of the US between 1980 and 2002. [3]

(d) Discuss whether a government should try to fix its country's exchange rate. [6]

Cambridge 9708 Paper 2 Q1 Oct/Nov 2007

Essay Questions

1. (a) Explain the difference between cost-push inflation and demand-pull inflation. [8]

(b) Discuss whether a country experiencing inflation will always have a balance of payments problem. [12]

Cambridge 9708 Paper 2 Q4 Oct/Nov 2006

2. (a) Explain the causes of a fall in a country's exchange rate. [8]

(b) Discuss whether a fall in a country's exchange rate will improve its macroeconomic performance. [12]

3. (a) Explain why a government may seek to avoid a current account deficit. [8]

(b) Discuss whether a floating exchange rate will mean that a government can neglect its balance of payments position and concentrate its policies on domestic problems. [12]

Macroeconomic Policies

11

Revision Objectives

After you have studied this chapter, you should be able to:

☞ analyse policies designed to correct balance of payments disequilibrium and policies designed to influence the exchange rate

☞ discuss possible conflicts between policy objectives on inflation, balance of payments and the exchange rate.

11.1 Policies to correct balance of payments disequilibrium and influence the exchange rate

Policies designed to correct a current account deficit may be expenditure dampening or expenditure switching.

Expenditure dampening policies, which can also be called expenditure reducing policies, are designed to reduce a current account deficit by reducing consumers' expenditure. The policies do not discriminate between imports and domestically produced products. Lower consumer expenditure would result in spending on imports to fall and the reduction in spending on domestically produced products may encourage domestic firms to make greater efforts to export their products.

Expenditure dampening policies include raising income tax, cutting government spending and increasing the rate of interest.

Expenditure switching policies are designed to reduce a current account deficit by encouraging domestic consumers to switch their spending from imports to domestically produced products and foreign consumers to switch their spending from foreign made products to the country's exports. For example, expenditure switching policies introduced by Bangladesh would be intended for people, both domestic consumers and foreigners, to buy more

Bangladeshi produced products and fewer products made by other countries.

Expenditure switching policies include trade restrictions, government subsidies, devaluation and trade fairs.

To reduce a current account surplus a government could reduce income tax, lower the rate of interest and increase government spending. It could also raise the value of the currency and remove import restrictions.

A government, through its central bank, can influence the exchange rate by buying and selling the currency and by changing the rate of interest.

To increase the value of the currency, the central bank can buy the currency using its reserves of foreign currency. It can also increase the rate of interest. A higher interest rate would be expected to attract inflows of hot money as foreigners would want to place money into the country's financial institutions.

A government may want to encourage a fall in the value of the currency in order to increase export revenue and reduce import expenditure so as to improve its current account position. Such a policy approach, however, may involve a trade off. A lower exchange rate may increase the inflation rate by increasing aggregate demand and raising the price of imported products.

A rise in the inflation rate may increase a current account deficit which, in turn, may lower the exchange rate.

Expenditure dampening policies may improve the current account position but may have an adverse effect on employment and economic growth as they will reduce consumer expenditure. This may be offset in the longer term by an increase in net exports.

11.2 Mind maps

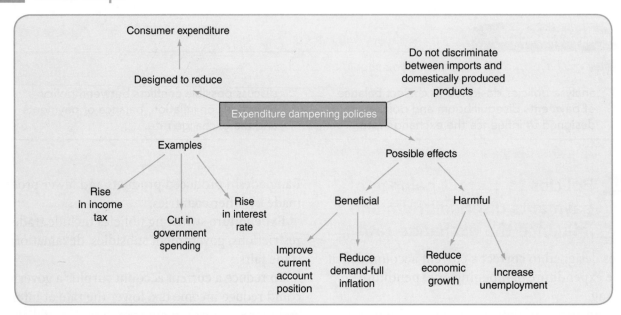

Mind map 11.1   Expenditure dampening policies

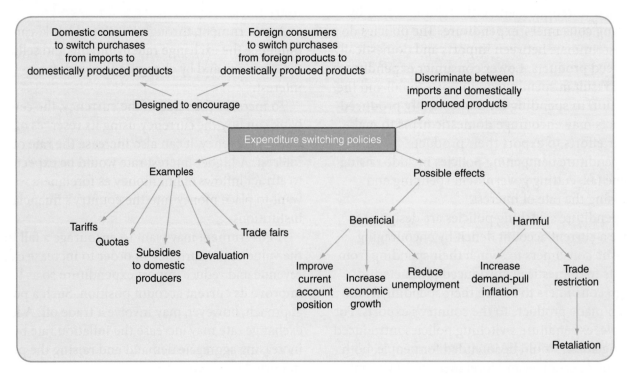

Mind map 11.2   Expenditure switching policies

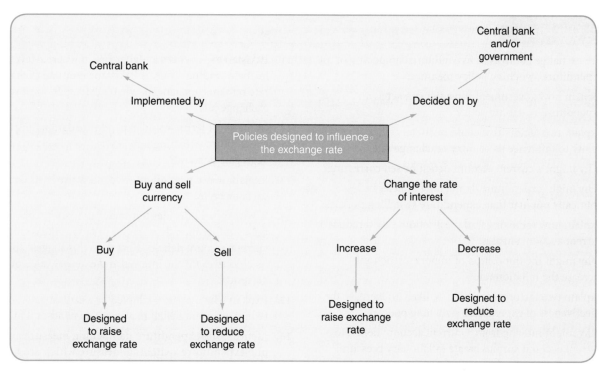

Mind map 11.3   Policies designed to influence the exchange rate

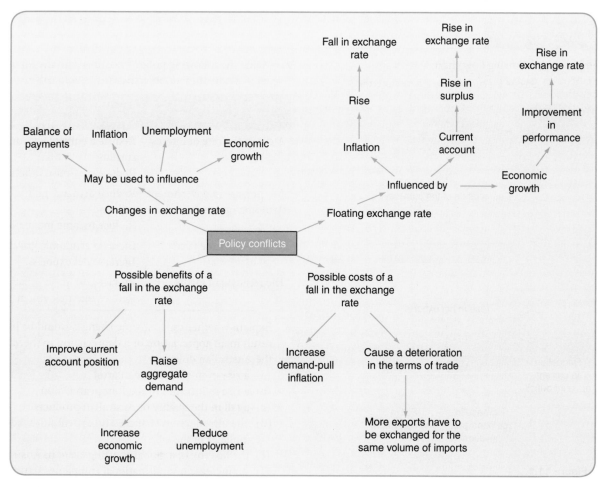

Mind map 11.4   Policy conflicts

Short Questions

1. Are exchange controls an expenditure dampening or an expenditure switching policy measure?
2. Explain how government subsidies can act as an expenditure switching policy.
3. Explain two factors that could restrict a central bank's ability to influence its country's exchange rate.
4. Why might a current account deficit be self correcting?
5. Why might expenditure dampening policies be less politically popular than expenditure switching policies?
6. Explain how removing trade restrictions could reduce a current account surplus.
7. Why might the imposition of import restrictions increase the inflation rate?
8. Explain two factors that would be likely to increase the effectiveness of expenditure switching policies.
9. Why might moving from a current account deficit to a current account surplus create inflationary pressure?
10. Explain why the effect of a rise in the rate of interest on import expenditure is uncertain.
11. Pakistan experiences a higher rise in its price level than its main trading rivals. If Pakistan's nominal exchange rate remains unchanged, what will happen to its real exchange rate?
12. In the short term, would expenditure dampening or expenditure switching policies be more likely to increase aggregate demand?
13. Explain why a firm might welcome a rise in its country's exchange rate.
14. A government decides to use deflationary fiscal policy rather than devaluation to reduce the country's current account deficit. What does this suggest about its objectives for the internal and external value of the currency?
15. Explain why measures adopted by a central bank to reduce a current deficit may cause demand-pull inflation.
16. Identify one expenditure dampening measure and one expenditure switching measure which are not available to a country which is a member of an economic union.

Revision Activities

1. Complete the following flow chart:

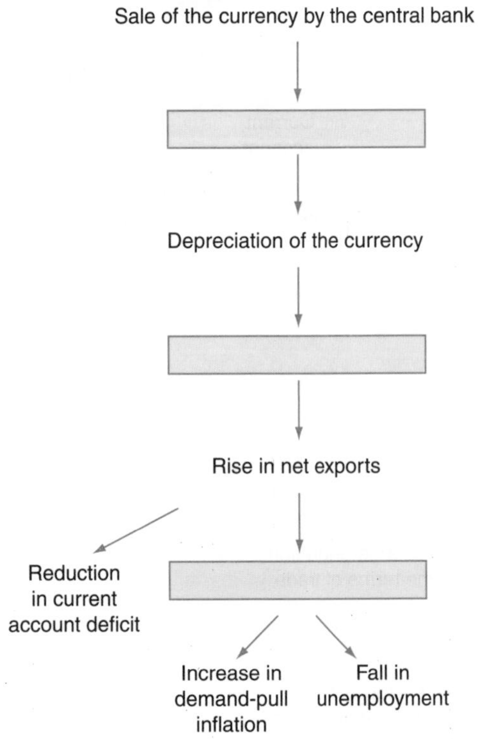

Figure 11.1

2. Match the following policy measures with the following pairs of government objectives they could solve.

Policy measure	Pairs of government macroeconomic objectives
A cut in the rate of interest	Reduce a current account surplus Reduce cost-push inflation
An increase in government spending on education	Reduce demand-pull inflation Reduce income inequality
A rise in income tax	Increase economic growth Increase net exports
The removal of tariffs	Reduce cost-push inflation Reduce unemployment

3. Explain whether the following changes would be likely to result in an appreciation or a depreciation in the value of the Australian dollar:
(a) a rise in the US rate of interest
(b) a rise in incomes in the European Union
(c) a fall in the quality of Australian products
(d) the introduction of successful expenditure reducing policies in the USA
(e) the hosting of a global sporting event in Australia
(f) an increase in multinational companies setting up branches in Australia.

4. Classify the following into internal and external macroeconomic policy objectives.
   (a) Balance of payments equilibrium
   (b) Full employment
   (c) Price stability
   (d) Stable exchange rate
   (e) Steady and sustainable economic growth.

5. Decide whether the following are likely to decrease or increase the ability of a government to reduce a current account deficit by using expenditure switching methods.
   (a) An increase in global incomes
   (b) An increase in trade restrictions abroad
   (c) A high domestic income elasticity of demand for imports
   (d) The adoption of expenditure dampening policies by foreign governments
   (e) The existence of spare capacity in the economy.

Multiple Choice Questions

1. What effect is an increase in interest rates likely to have?
   A A decrease in unemployment
   B A decrease in savings
   C An increase in the exchange rate
   D An increase in investment

2. Which policy measure is most likely to be effective in reducing a balance of trade deficit?
   A Devaluation of the currency
   B An increase in government spending on state benefits
   C A reduction in the rate of interest
   D A reduction in income tax

3. A country's current account position moves from a surplus to a deficit. What will be the result?
   A A decrease in the money supply
   B A decrease in unemployment
   C An increase in the exchange rate
   D An increase in real GDP

4. A government decides to introduce an expenditure switching measure to reduce a balance of trade deficit. Which of the following is an expenditure switching measure?
   A A decrease in state benefits
   B A government subsidy to domestic producers
   C An increase in income tax
   D An increase in the rate of interest

5. What change must be caused by a devaluation?
   A A fall in employment
   B A fall in the budget surplus
   C A fall in the rate of interest
   D A fall in the terms of trade

6. Which policy measure may reduce a current account deficit without causing inflation?
   A A devaluation of the currency
   B A reduction in government spending
   C An increase in import tariffs
   D An increase in indirect taxes

7. A government uses an expenditure dampening measure to correct a balance of payments current account deficit. In the short term, what effect would such a measure have on consumer expenditure and net exports?

	Consumer expenditure	Net exports
A	Decrease	Decrease
B	Decrease	Increase
C	Increase	Increase
D	Increase	Decrease

8. Which combination of economic problems would benefit from an upward revaluation of the currency?
   A A current account deficit and inflation
   B A current account deficit and unemployment
   C A current account surplus and inflation
   D A current account surplus and unemployment

9. A government wants to reduce both the country's inflation rate and its current account deficit. What could be an appropriate government measure to achieve these objectives?
   A Devalue the currency
   B Decrease tariffs on imports
   C Increase the exchange rate
   D Increase income tax

10. Which changes in government spending and the rate of interest is most likely to prevent any inflationary effects resulting from a current account surplus?

	Government spending	Rate of interest
A	Decrease	Decrease
B	Decrease	Increase
C	Increase	Increase
D	Increase	Decrease

Data Response Questions

1. Economic developments in Mexico and Argentina

    There are both similarities and differences in the ways that Mexico and Argentina have tried to develop. In 1994, Mexico joined the North American Free Trade Area (NAFTA) with the US and Canada. A year later Argentina joined Mercosur, a South American customs union, with Brazil, Paraguay and Uruguay. Mercosur is the world's third largest trading bloc after the European Union and NAFTA. Both countries have experienced severe economic crises, Mexico in 1995 and Argentina in 2002. Their experiences of foreign exchange rate movements and inflation have been very different. Figures 11.2 and 11.3 show details of these from 1997 to 1999.

    (a) State one way in which a customs union and a free trade area are the same and one way in which they differ. [2]

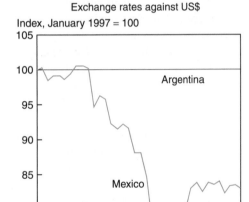

Figure 11.2

    (b) (i) Compare the behaviour of the Mexican and Argentine exchange rates against the US$ in the period shown in Figure 11.2. [2]
         (ii) Explain one possible reason for the trend in the Mexican exchange rate during 1999. [2]
         (iii) Explain how the fixed level of the Argentine exchange rate would have been achieved. [3]
    (c) (i) Compare Mexico's and Argentina's experience of inflation in the period shown in Figure 11.3. [2]
         (ii) Explain how Mexico's inflation rate may have influenced the behaviour of its exchange rate. [3]
    (d) Discuss whether the devaluation of a country's exchange rate will always improve its balance of trade position. [6]

Cambridge 9708 Paper 2 Q1 Oct/Nov 2005

Figure 11.3

2. The Current Account of Swaziland's Balance of Payments

    Swaziland is a small, landlocked economy in Southern Africa. The Swazi currency is the lilangeni (plural emalangeni) and the currency is pegged to the South African rand at a fixed rate of one to one. The country has faced changing international conditions in recent years, as shown in its current account statistics.

Swaziland's current account components, selected years, millions, emalangeni			
	2003	2005	2007
Balance on goods	957.9	– 1641.3	– 1910.1
Balance on services	– 1090.2	– 765.0	– 367.6
Net income	– 317.4	1133.8	449.3
Net current transfers	1136.4	619.8	1366.6

    The Central Bank of Swaziland's report on the 2007 export performance identified the following.

    ✓ Exports grew by 8.4%, with a positive performance by some manufacturing companies
    ✓ Successful exports included sugar, sugar-based products, soft drink concentrates, wood pulp and timber products, textiles and garments, citrus and canned fruits and meat products
    ✓ Global demand and rising export prices led to increased export revenue.

    Export performance was helped by the depreciation of the domestic currency against the US$ and the currencies of other trading partners outside of the Southern African Customs Union (SACU) and Common Monetary Area (CMA), of which Swaziland is a member. In addition, exports of meat and meat products to the European Union (EU) resumed in 2007 after the EU lifted its ban on Swaziland's beef exports. This ban was originally imposed because of Swaziland's failure to comply with the required quality standards.

(a) In Swaziland's current account between 2003 and 2007,

    (i) Which component showed a continuous improvement, and

    (ii) Which component showed a continuous worsening? [2]

(b) How did the current account balance change from 2003 and 2007? [3]

(c) (i) What is comparative advantage? [2]

    (ii) In the light of the Central Bank of Swaziland's report what might be concluded about the nature of Swaziland's comparative advantage and the factors on which it is based? [4]

(d) Explain the conditions necessary for the depreciation of a country's currency to increase its exports revenue. [3]

(e) Discuss the case for and against the use of tariffs by Swaziland to retaliate when the EU banned imports of Swaziland's beef. [6]

Cambridge 9708 Paper 21 Q1 Oct/Nov 2010

3. Two Asian Giants

China has the world's largest population and is experiencing rapid economic growth from a low base. It has recently joined the World Trade Organisation (WTO) with an obligation to open its economy to freer international trade.

In 2002, an economic spokesman for neighbouring Japan stated that this did not mean that China was a threat to Japan's position as 'the factory of the world'. He claimed that China was a labour-intensive, low-cost producer while Japan was ahead in technology-intensive products. He quoted television production, where

Japan specialised in high quality models, while China produced standard models. Engineering and electronics showed similar positions. He argued that Chinese export competitiveness was based on cheap labour which reflected poor labour productivity. He added that half of Chinese exports were produced by subsidiaries of foreign firms, many of them Japanese. However, he warned that the presence of these subsidiaries might lead to an increased transfer of technology to China.

Shows some economic aspects of the two countries in 2000		
	Japan	China
Population (millions)	126	1,260
Gross Domestic Product (GDP) (US$bn)	4,753	1,099
Consumer price inflation	– 0.7	0.4
Average annual % inflation rate in previous 10 years	0.8	7.1
Current account balance (US$bn)	116.9	10.2
Current account balance as % of GDP	2.5	0.9
Exports of goods (US$bn)	459	249
Imports of goods (US$bn)	343	216

(a) Suggest two steps that a country might take to open its economy to freer trade. [2]

(b) (i) Explain why the comparative advantage of Japan and China differs. [4]

    (ii) Compare the importance of international trade to the two countries. [4]

(c) Comment on the inflation figures for the two countries. [4]

(d) Discuss whether China offers an economic threat or opportunity to Japan. [6]

Cambridge 9708 Paper 2 Q1 Oct/Nov 2004

Essay Questions

1. (a) Explain what is meant by a current account deficit. [8]

    (b) Discuss the effectiveness and desirability of imposing tariffs to correct a current account deficit. [12]

Cambridge 9708 Paper 2 Q4 May/June 2008

2. (a) Explain why price elasticity of demand is important in deciding whether currency devaluation will reduce a trade deficit. [8]

    (b) Discuss whether expenditure dampening policies will reduce a trade deficit. [12]

3. (a) Explain the effects of an appreciation of its currency on an economy's inflation rate. [8]

    (b) Discuss whether a rise in an economy's inflation rate will cause the value of its currency to fall. [12]

Part 3   Supplement

Basic Economic Ideas

12

Revision Objectives

After you have studied this chapter, you should be able to:

☞ describe an efficient resource allocation

☞ explain productive and allocative efficiency.

12.1 Concept of economic efficiency

The optimum allocation is the best use of resources. It occurs when the highest quantity of wants are being met with the scarce resources available. To achieve this situation it is necessary for there to be economic efficiency.

Economic efficiency occurs when both productive efficiency and allocative efficiency are achieved.

Productive efficiency is concerned with how resources are used. To attain productive efficiency, both technical efficiency and cost efficiency have to be achieved. Productive efficiency occurs when products are made with the least possible quantity of resources (technical efficiency) and the lowest cost methods (cost efficiency).

In micro terms, productive efficiency is often measured in terms of long run average costs. A firm may be described as productively efficient if it is producing at the lowest point on the long run average cost curve. In this case, a firm would have the lowest unit cost possible and would not be wasting resources.

In macro terms, productive efficiency is achieved when an economy is making full use of resources. When an economy is producing on its production possibility curve, it cannot produce more of one type of product without producing less of another type of product.

Allocative efficiency is concerned with what resources are used to produce. It is achieved when resources are allocated to produce the right products in the right quantities. This means that what is produced reflects consumers' demand. More precisely, allocative efficiency is achieved when price equals marginal cost. Price reflects the value that consumers place on the product and marginal cost is effectively the cost of the last unit.

Pareto efficiency occurs when it is not possible to make someone better off without making someone else worse off. It is also sometimes referred to as Pareto optimality or social efficiency. It does not take into account the distribution of income.

A Pareto improvement occurs when a reallocation of resources makes at least one person better off while making no-one worse off.

Competition may promote efficiency by providing an incentive (in the form of profit) to the producers who make what consumers want at the lowest average cost and a punishment (in the form of going out of business) for those which do not keep the cost low and do not respond to changes in consumer demand.

Competition, however, will not ensure that the socially optimum output is achieved if there are externalities. To be socially efficient, a market should produce where marginal social cost equals marginal social benefit.

12.2 Mind maps

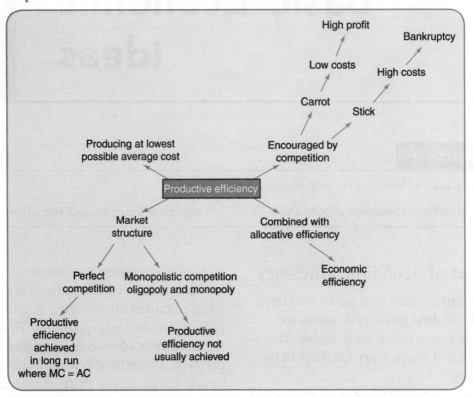

Mind map 12.1   Productive efficiency

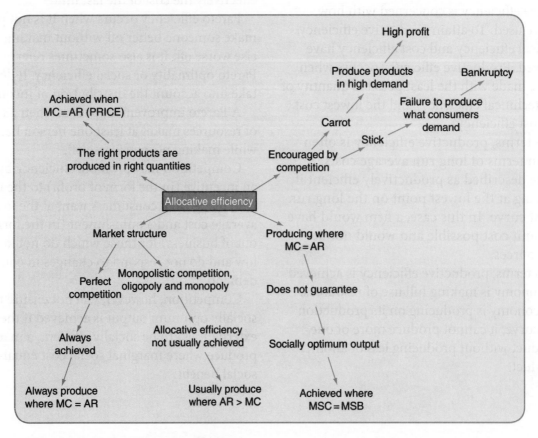

Mind map 12.2   Allocative efficiency

Short Questions

1. What is meant by economic inefficiency?
2. How is productive inefficiency illustrated on a production possibility curve diagram?
3. Explain why a market would be inefficient if output is where price is below marginal cost.
4. Will achieving economic efficiency solve the economic problem?
5. What does the existence of long run shortages in a market indicate about the efficiency of a market?
6. Explain why international trade may promote economic efficiency.
7. What does the existence of unemployment indicate about the efficiency of an economy?
8. Explain what type of efficiency is not achieved in the case of merit goods.
9. Why might the imposition of a tax on a demerit good reduce efficiency?
10. Why is it difficult to achieve allocative efficiency in the case of public goods?

Revision Activities

1. Decide whether the following would affect allocative or productive efficiency and whether they would increase or reduce efficiency.
   (a) A reduction in unemployment
   (b) A shift of resources from producing demerit goods to producing merit goods
   (c) A reduction in surpluses
   (d) A reduction in labour productivity
   (e) A switch from producing less popular to more popular products
   (f) A reduction in organisational slack.

2. Explain whether the following statements about allocative and productive efficiency are true or false.
   (a) A shift to the right of a production possibility curve indicates an increase in productive efficiency.
   (b) If average revenue is above marginal cost, resources are being used inefficiently.
   (c) When allocative efficiency is achieved there is no welfare loss.
   (d) For economic efficiency to be achieved marginal social cost has to be at a minimum for all products.
   (e) Information failure can result in allocative inefficiency.

Multiple Choice Questions

1. Allocative efficiency is achieved where:
   A AC = AR
   B AC = MC
   C MC = AR
   D MC = MR

2. What effect will a more efficient allocation of resources have on an economy operating at full employment?
   A Output will remain unchanged
   B Output will rise
   C Inflation will occur
   D There will be a more even distribution of income

3. When is a economy productively efficient?
   A When the economic problem has been solved
   B When the economy operates on its production possibility curve
   C When there is macroeconomic equilibrium
   D When the economy can produce more of one product without producing less of another

4. Figure 12.1 shows a variety of output positions.
   What is the most efficient output position?
   A   B   C   D

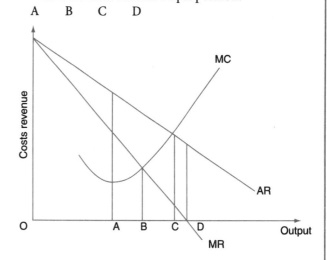

Figure 12.1

5. A firm generates external costs. What would be the effect of a government placing a tax equal to the external costs?
   A The firm's output will increase
   B The firm's demand curve will shift to the left
   C Resource allocation will be unaffected
   D Resource allocation will be improved

6. Figure 12.2 shows a change in the output of an economy from production point X to production point Y.

   What effect will this movement have on allocative and productive efficiency?

	Allocative efficiency	Productive efficiency
A	Remain unchanged	Remain unchanged
B	Remain unchanged	Uncertain
C	Uncertain	Remain unchanged
D	Uncertain	Uncertain

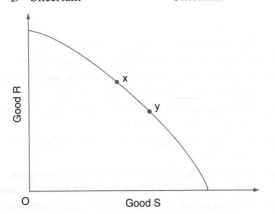

   Figure 12.2

7. Why may international trade promote efficiency?
   A It allows economies to concentrate on products in which they have a comparative advantage
   B It allows economies to operate with spare capacity
   C It eliminates differences in tastes
   D It reduces the mobility of factors of production

8. An economy is employing its resources fully and has allocated them in the most efficient way. In these circumstances how might output be increased?
   A By improving the state of technology
   B By increasing consumption of public goods
   C By setting a maximum price on capital goods
   D By shifting resources from producing capital goods to producing consumer goods

9. Why is an economy inefficient if it is producing inside its production possibility curve?
   A Income is unevenly distributed
   B Prices are higher than necessary
   C It is possible to make more of one product without reducing production of another
   D It is possible to increase output without employing more resources

10. Figure 12.3 shows that the allocation of resources is initially at X. Which reallocation of resources would lead to a Pareto improvement?

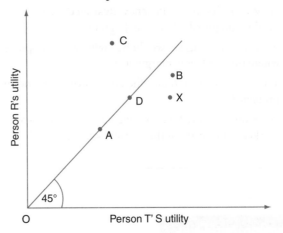

    Figure 12.3

    A A   B B   C C   D D

11. In which circumstance might the imposition a tax on a product result in an efficiency gain?
    A The price of the product is initially equal to marginal cost
    B The price of the product will rise to where average revenue exceeds average cost
    C There is already a tax on a complementary good
    D There is already a tax on a substitute good

12. The diagram below shows a market in equilibrium at a price of P. A government then sets a minimum price of PX and maintains this price by purchasing any surplus supply.

    What is the net loss to society of this policy?
    A Area PPXUV
    B Area NPXU
    C Area YTUZ
    D Area YTUZ

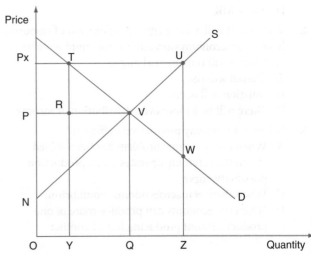

    Figure 12.4

Data Response Questions

1. A number of Latin American countries are seeking to increase the efficiency of their firms. There are thought to be a number of reasons why manufacturing and services are relatively inefficient in the region.

One is poor infrastructure. Overcrowded seaports and airports and badly maintained roads and railways increase transport costs. Indeed, it costs more to transport some products from Latin America to the United States than it costs to send the same products from China to the United States. Higher transport costs can contribute to inflation and unemployment – see Table 1.

Inflation and unemployment rates in selected Latin American countries in 2010		
Country	Inflation %	Unemployment %
Argentina	10.9	7.9
Brazil	5.2	6.9
Chile	1.8	8.3
Colombia	2.5	12.7
Venezuela	28.8	8.2

A lack of competition in a number of industries in Latin America reduces the pressure on firms to keep their costs low and to respond quickly and fully to changes in consumer demand.

Another major cause of inefficiency in the region is thought to be a lack of bank loans available to firms wanting to expand and innovate.

(a) What is meant by firms being efficient? [3]

(b) Explain what evidence there is in the information provided that some Latin American firms are allocatively inefficient. [3]

(c) Analyse two policy measures Latin American governments could introduce to increase efficiency. [6]

(d) (i) Using Table 1, comment on whether high inflation rates are associated with high unemployment rates. [2]

(ii) Discuss whether inefficiency is indicated by:
- A high unemployment rate
- A high inflation rate. [6]

Essay Questions

1. There has been much discussion recently about the effect of climate change and the efficient use of economic resources. Discuss whether the efficient allocation of resources can be achieved only if governments are involved in the process. [25]

Cambridge 9708 Paper 41 Q2 Oct/Nov 2010

2. In 2009, there were huge fires in Australia which destroyed much property and countryside. The government promised to allocate a large amount of money and resources to help with the restoration of the area.

(a) With the help of diagrams explain what is meant by the efficiency in the use of resources. [12]

(b) Discuss the economic implications of the government's approach to the situation. [13]

Cambridge 9708 Paper 41 Q2 Oct/Nov 2010

The Price System and the Theory of the Firm

Revision Objectives

After you have studied this chapter, you should be able to:

- ☞ explain the law of diminishing marginal utility
- ☞ outline the relationship between the law of diminishing marginal utility and an individual demand schedule and curve
- ☞ use figures to illustrate the equi-marginal principle
- ☞ assess the limitations of marginal utility theory
- ☞ draw budget lines to explain the income and substitution effects of a price change
- ☞ distinguish between fixed and variable factors of production
- ☞ calculate total, average and marginal product
- ☞ explain the law of diminishing returns
- ☞ define demand for labour and describe the factors affecting it
- ☞ explain how marginal revenue product theory is related to an individual's demand for labour
- ☞ define supply of labour and describe the factors affecting the supply of labour
- ☞ explain wage determination under free market forces
- ☞ analyse the role of trade unions and government in wage determination
- ☞ discuss the causes of wage differentials
- ☞ explain the nature and influences on economic rent
- ☞ distinguish between the short run and long run production functions

- ☞ explain returns to scale
- ☞ distinguish between an economist's and an accountant's definition of costs
- ☞ calculate marginal cost and average cost
- ☞ draw cost curves
- ☞ distinguish between fixed costs and variable costs
- ☞ explain the shape of the short run and long run average cost curves
- ☞ analyse the relationship between economies of scale and decreasing costs
- ☞ describe internal and external economies of scale
- ☞ explain the survival of firms
- ☞ analyse how and why firms grow
- ☞ explain the relationship between PED and revenue for a downward sloping demand curve
- ☞ distinguish between normal and abnormal profit
- ☞ analyse the objectives of firms
- ☞ assess the different market structures of perfect competition, monopoly, monopolistic competition and oligopoly
- ☞ explain contestable markets
- ☞ discuss the conduct and performance of firms.

13.1 Law of diminishing marginal utility and its relationship to the derivation of an individual demand schedule and curve

Utility is the satisfaction a person gains from consuming a product. Marginal utility is the change in total utility a person experiences as a result of consuming the last unit. Consumers tend to experience less satisfaction for each additional unit of a product consumed – the law of diminishing marginal utility.

Total utility rises when marginal utility is positive, reaches its peak when marginal utility is zero and falls when marginal utility is negative. These relationships are shown in Figure 13.1.

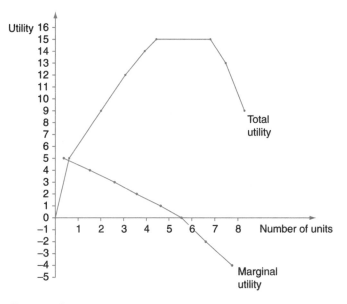

Figure 13.1

An individual demand schedule and curve is based on the marginal utility curve. If, for instance, a person gains a marginal utility of $9 from the fifth unit and $7 from the sixth unit, this will represent the price they would be prepared to pay for the different quantities. This information can be used to draw up a demand schedule and draw a demand curve. For instance, in this case a fall in price from 9 to 7 would result in an extension in demand from 5 to 6.

Equi-marginal product

The equi-marginal principle states that consumers will gain the highest total utility from their income by consuming the combination of products which ensures that the utility from the last $ spent on each product is equal. The formula for this is:

$$\frac{\text{Marginal utility of product A}}{\text{Price of product A}} = \frac{\text{MU of product B}}{\text{P of product B}}$$
$$= \frac{\text{MU of product C}}{\text{P of product C}}$$

Limitations of marginal utility theory

In practice, people do not always behave in the way marginal utility theory suggests. There are a number of reasons for this. One is that not all products can be divided into small units. While

a person can decide whether to buy eight or nine apples, a number of the products they buy are indivisible such as a car. People also often buy on habit and impulse rather than carefully weighing up the marginal utilities of everything they purchase. In addition, for some products marginal utility may increase, the more a person has. For instance, a collector of first edition books may gain more satisfaction from the twentieth book than from the nineteenth book.

Although there are limitations to marginal utility theory, it does help to explain the inverse relationship between price and demand that occurs in the case of most products.

13.2 Budget lines

A budget line shows the various combinations of two products that a consumer can buy with a given income and given prices.

If there is a rise in income, the budget line will shift out parallel to the right. A fall in the price of one product will cause the budget line to pivot outwards. Figure 13.2 illustrates the effect of a fall in the price of apples, with income and the price of oranges remaining unchanged.

Income and substitution effects of a price change

The income and substitution effects explain why the quantity demanded of a product changes

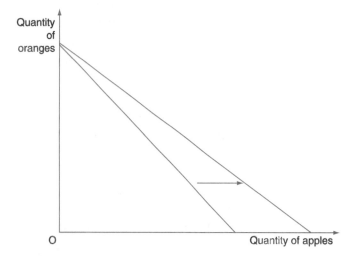

Figure 13.2

when price changes. A higher price will cause the demand for most products to fall for two reasons. One is that people's purchasing power will decrease. They will be able to buy less of all products, including this one – the income effect. The other reason is that people will switch from buying the product to buying a substitute – the substitution effect.

Some products have exceptional demand with the income and substitution effects working in opposite directions. One example is a Giffen good. This is a low quality product which has a direct relationship between price and demand. In this case, a rise in price will cause a rise in the quantity demanded. A Giffen good has the usual substitution effect with the higher price encouraging consumers to switch to a substitute. The substitution effect, however, is more than offset by a negative income effect. With lower purchasing power, people will not be able to afford higher quality products and so will have to buy more of the Giffen good.

Veblen goods have both increasing marginal utility and income and substitution effects which work in opposite directions. A Veblen good is an expensive product which people buy to show how rich they are. A rise in the price of a Veblen good reduces people's purchasing power but people switch from substitutes to buying the product as it now becomes a more impressive purchase.

13.3 Short-run production function

The short run is defined as the period of time when a firm has at least one fixed factor of production. In contrast, the long run is the period of time when a firm can adjust all its factors of production and so all inputs are variable.

A fixed factor of production is one which is not changing in quantity. In most cases the fixed factor of production is capital. In the short run, most firms' factory or office size will be unchanged.

A variable factor of production is one which changes in quantity. For instance, a firm may use more raw materials.

The short run production function shows the relationship between the factors of production a firm uses, at least one of which is in fixed supply,

and the output it produces. Total product is the total output of a good or service produced by a firm.

A firm may be producing a higher output as a result of employing more labour. Average product is total output divided by the number of workers or other input employed. Marginal product is the change in output which results from employing one more worker or other input.

Law of diminishing returns

The law of diminishing returns (also known as the law of variable proportions) states that as more of a variable factor of production is employed with a fixed factor of production, a point will be reached where marginal product will decline. This will cause total output to rise by a diminishing amount.

At first as more workers are employed, marginal product usually increases (increasing returns) but afterwards it declines. Diminishing returns set in because less efficient combinations of factors of production are used. The proportion of the labour employed is too high relative to the capital or land employed.

13.4 Demand for labour

Demand for labour is the number of labour hours firms are willing and able to buy. Demand for labour is a derived demand. Labour is not wanted for its own sake but for what it can produce. If, for instance, the demand for air travel increases, demand for pilots will increase as well.

The aggregate (total) demand for labour is influenced by the level of economic activity. Demand for labour would fall during a recession. An individual firm's demand for labour is affected by a number of factors including demand for the product produced, labour productivity, the wage rate, the price and productivity of other factors of production that can be used as a substitute or a complement to labour.

Derivation of an individual firm's demand for a factor using marginal revenue product theory

The marginal revenue product (MRP) is the change in a firm's revenue which results from employing one more worker. It is found by multiplying the marginal product by marginal revenue.

Marginal revenue product theory states that demand for labour depends on its marginal revenue product. According to the theory, the quantity of labour employed is determined where MRP equals the marginal cost of labour.

13.5 Supply of labour

The supply of labour is the quantity of labour workers are willing and able to sell.

A key influence on the number of hours an individual is prepared to work is the wage rate. The backward sloping labour supply diagram suggests that as the wage rate starts to rise, a person will work longer hours. When a certain wage rate is reached, the person may choose to take more leisure. This change in how a worker responds to a rise in the wage rate is illustrated in Figure 13.3.

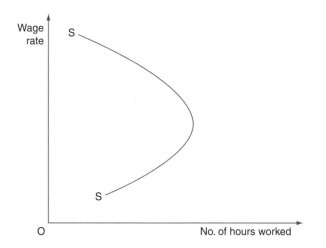

Figure 13.3

A change in the wage rate has both an income and a substitution effect. The income effect of a wage rise is to reduce the number of hours worked, as a person will buy more leisure time. The substitution effect is to increase the number of hours worked as a person substitutes the more financially rewarding work for leisure. The backward sloping supply curve suggests that at low wages, the substitution effect outweighs the income effect whereas at higher wages, the income effect outweighs the substitution effect.

Net advantages and the long run supply of labour

In the long run, when people have time to change jobs, the supply of labour is influenced by the net advantages of jobs and the qualifications and skills required in doing the job. There are a number of pecuniary influences on the supply of labour. As well as the wage rate, a job may also provide overtime pay and bonuses.

Among the non-pecuniary influences are job security, promotion chances, working hours, holidays, fringe benefits (such as a company car), the training on offer, the distance required to travel to work, the convenience of the working hours and the status conferred by the job.

If the qualifications and skills required to do a job are high, there will be a limited supply of labour.

13.6 Wage determination under free market forces

In a free market, the wage rate is determined by the demand for and supply of labour. A decrease in the demand for labour, for example, would reduce the equilibrium wage rate and result in a contraction in demand for labour as shown in Figure 13.4.

The role of trade unions and government in wage determination

A trade union is an organisation which represents workers. It seeks to promote the interests of its members by, for example, bargaining for wage raises, improved working conditions and job security.

A trade union may raise the wage rate of its members through collective bargaining, by supporting measures to restrict the supply of labour and measures to increase the demand for labour. If a trade union pushes the wage rate above the

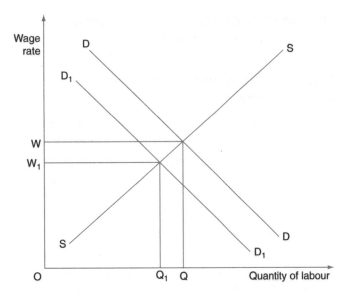

Figure 13.4

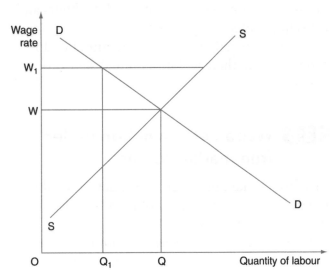

Figure 13.5

equilibrium level, employment will fall as illustrated in Figure 13.5.

Governments influence wages in a variety of ways. They employ workers directly, they pass legislation on trade unions, some run arbitration services and some operate a national minimum wage.

A national minimum wage may increase unemployment if it results in the supply of labour being greater than the demand for labour. There is, however, the possibility that it may be accompanied by an increase in employment if the higher wage increases demand for products, if there

are monopsonist employers in the market and if the rise in pay increases workers' motivation and productivity.

Wage differentiation and economic rent

There are a number of reasons why some workers are paid more than others. One is market forces. Workers whose labour is in high demand and whose supply is limited will be paid more than those whose labour is less highly demanded and whose supply is greater. This is the main reason why skilled workers are paid more than unskilled workers. Other reasons why one group of workers may be paid more than another group is that they have stronger trade union power, that government policy favours their group and they do not experience negative discrimination.

Economic rent arises when a factor of production is paid more than necessary to keep it in its current occupation. Transfer earnings are what the factor can earn in its next best paid occupation. So, for example, if an accountant is paid $40,000 a year and could earn $28,000 a year as a teacher, his transfer earnings would be $28,000 and his economic rent would be $12,000.

Figure 13.6 shows how wages may be divided between economic rent and transfer earnings. The more inelastic the supply of labour is, the greater the proportion of wages will be economic rent.

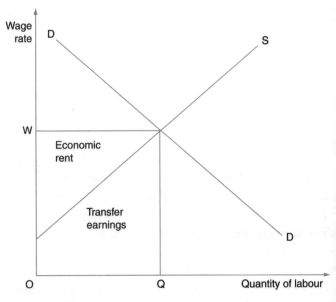

Figure 13.6

13.7 Long run production function

The long run production function shows the relationship between the factors of production a firm uses, all of which are variable, and the output it produces.

In the long run, when all the factors of production can be changed, the most efficient combination of inputs are as following.

$$\frac{\text{Marginal product of Factor A}}{\text{Price of Factor A}} = \frac{\text{MP of Factor B}}{\text{P of Factor B}}$$
$$= \frac{\text{MP of Factor C}}{\text{P of Factor C}}$$

Returns to scale are the changes in output which occurs when a firm changes its scale of operation by altering all its factors of production. Increasing returns to scale are experienced when output increases at a greater rate than inputs. Constant returns to scale are when output and input increases at the same rate. Decreasing returns to scale occur when output increases at a slower rate than inputs.

13.8 Economist's versus accountant's definition of costs

Economists include normal profit in costs of production whereas accountants do not.

Marginal cost and average cost

Marginal cost is the change in total cost when output is changed by one unit. Average cost, which is also called average total cost or unit cost, is total cost divided by output.

Short run cost function – fixed costs versus variable costs

In the short run, total cost is made up of fixed costs and variable costs. Total cost (TC) equals total fixed cost (TFC) plus total variable cost (TVC).

Fixed costs are costs which do not alter when output changes. These costs are incurred even when output is zero and are the costs associated with fixed factors of production. They are also referred to as overheads and indirect costs. Examples include building insurance and rent. Variable costs are costs which are directly related to output. If output increases, variable costs rise. They are also called direct costs or prime costs and include raw material costs.

Average fixed cost (AFC) is total fixed cost divided by output. As total fixed cost does not change as output rises, average fixed cost falls as output rises. Overheads are spread over a larger output. Average variable cost (AVC) is total variable cost divided by output. As output increases, AVC usually falls at first as increasing returns are experienced and then rises when diminishing returns are encountered.

Marginal fixed cost is zero as fixed costs do not change. As a result, marginal cost is equal to marginal variable cost.

Explanation of shape of short run average cost

The short run average cost (SRAC) curve falls at first because both AFC and AVC decline. After it reaches its minimum point, the SRAC curve rises. The increase in AVC outweighs the fall in AFC. Figure 13.7 illustrates AC, AVC and AFC.

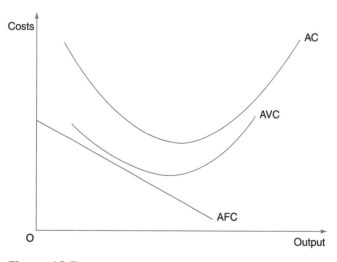

Figure 13.7

The marginal cost curve cuts the average cost curve at its lowest point. When the average cost curve is falling, marginal cost is below average cost whereas when average cost is rising, marginal cost is above average cost as illustrated in Figure 13.8.

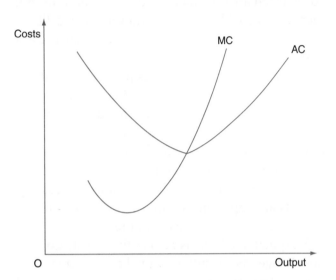

Figure 13.8

13.9 Long run cost function

In the long run, all costs are variable as there is time to alter all the factors of production employed.

In the long run, a firm can change the size of its factory, farm or offices. Every scale of production can be represented by a SRAC curve. On the assumption that a firm will always seek to minimise the cost of producing any given output, the long run average cost curve is found by linking the lowest point on a series of SRAC curves. This is shown in Figure 13.9. The long run average curve (LRAC) is sometimes called the envelope curve of all SRAC curves.

Relationship between economies of scale and decreasing costs

A U-shaped LRAC curve shows that a firm first experiences economies of scale with average costs falling. It then reaches its optimum output which is the lowest average cost. Past this point, the firm experiences diseconomies of scale. Figure 13.10 shows the U-shaped LRAC curve.

It is also possible that a LRAC curve may be downward sloping over a large range of output if economies are very significant. It may also be L-shaped if average costs fall at first and then constant returns to scale are experienced over a large range of output. The minimum efficient scale (MES) is the lowest level of output where all economies of scale have been fully exploited.

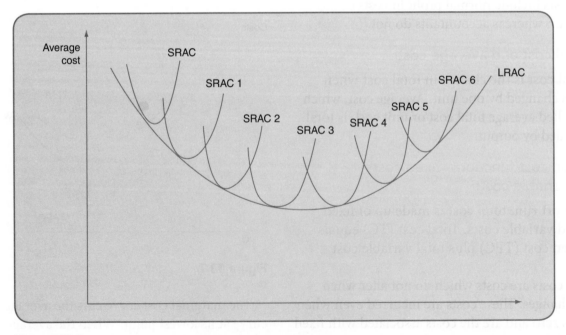

Figure 13.9

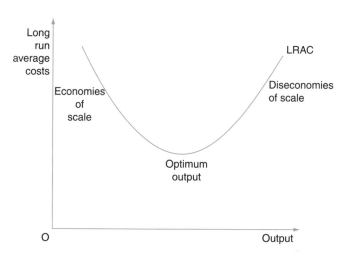

Figure 13.10

Internal and external economies of scale

Internal economies of scale are the benefits, in the form of lower long run average costs, a firm experiences as a result of increasing its output.

There is a range of different types of internal economies of scale including buying economies (discounts for bulk buying), financial (cheaper to and easier to borrow), managerial (employing specialists), marketing (advertising and transporting products more cheaply), research and development (ability to set up a R & D department), risk bearing (diversifying to reduce the risks of changes in market conditions), technical (making use of large, cost efficient equipment) and staff facilities (such as canteens and health care treatment).

External economies of scale are the benefits, in the form of lower long run average costs, available to all the firms in the industry as the result of the industry growing larger. Examples of external economies of scale include the availability of skilled labour, the development of ancillary industries, specialist markets, specialist courses at colleges and universities, disintegration (allowing firms to specialise) and shared research facilities.

International diseconomies of scale are the disadvantages, in the form of higher long run average costs, which result from a firm growing too large. They include management problems of control, co-ordination and communication along with poor industrial relations.

External diseconomies of scale occur when the industry grows too large and as a result firms in the industry experience higher long run average costs. Examples of external diseconomies include competition over factors of production driving up their price and traffic congestion if firms are located close together increasing transport costs.

Whilst internal economies and diseconomies of scale include movements along the LRAC curve, external economies of scale and external diseconomies of scale involve shifts in the LRAC curve. External economies of scale cause the LRAC curve to move downwards and external diseconomies of scale involve an upward shift of the LRAC curve.

13.10 Survival of small firms

Despite the advantages of large firms, most firms in most countries are small. There are a number of reasons why small firms survive including the scope for economies of scale may be limited in some industries, consumers may like the personal service small firms can provide, entrepreneurs may not want to expand, demand for the product may be relatively low, co-operation between small firms, large firms contracting out work to small firms, high transport costs restricting the size of the market, advances in technology reducing the cost advantages of large firms, government assistance for small firms and new firms regularly being set up.

Growth of firms

Firms may seek to increase their size in order to take greater advantage of economies of scale, to gain a larger market share, to prevent other firms taking them over and to achieve greater security by diversifying.

Firms can grow internally and externally. Internal growth involves a firm increasing its output by increasing the resources it employs. This is sometimes referred to as organic growth. External growth involves increasing the size of the firm by taking over or merging with another firm.

There are three main types of mergers. One is a horizontal merger – also called horizontal integration. This involves firms merging which are at the same stage of production and producing the same product. Vertical integration (a vertical merger) takes place when two firms at different stages of production in the same industry combine.

Conglomerate integration (a conglomerate merger) involves the combination of two firms which produce different products.

A horizontal merger will result in the new firm having a larger market share. It may also be able to take greater advantage of economies of scale. There is a risk, however, that diseconomies of scale may be experienced.

Vertical integration forwards enables a firm to take greater control over the marketing of its products and vertical integration backwards helps to ensure supplies. There is no guarantee, however, that the sizes of the two firms will match.

A conglomerate merger enables a firm to diversify but it can be difficult to manage the output of different products. A firm may seek to grow by opening branches in foreign countries. A multinational company has its headquarters in one country but producers in more than one country (see Chapter 16).

13.11 Relationship between elasticity, marginal, average and total revenue for a downward sloping demand curve

Marginal revenue is the change in total revenue resulting from the sale of one more unit. Average revenue (equivalent to price) is total revenue divided by the quantity sold. Total revenue is the total amount earned from selling a product and is price multiplied by the quantity sold.

When a firm is a price maker, it has to lower price in order to sell more of the product. A fall in price will cause total revenue to rise when marginal revenue is positive and price elasticity of demand is greater than one. Total revenue will reach its peak when marginal revenue is zero and PED is unitary. Total revenue falls when marginal revenue is negative and PED is less than one. Figure 13.11 shows these relationships.

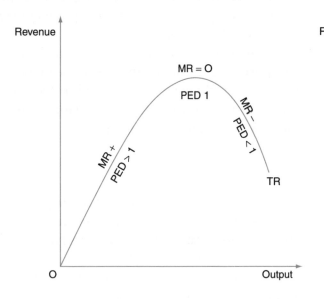

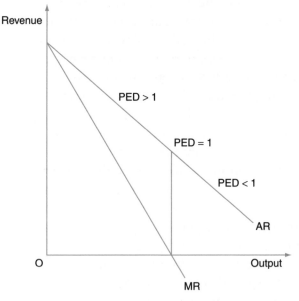

Figure 13.11

13.12 Concepts of firm and industry

A firm is a business organisation. It may be, for instance, a sole trader or a public limited company. A firm may own one or a number of plants. A plant is the production unit, that is a factory, office, farm,

hotel, shop. An industry consists of all the firms that produce the same type of product.

An industry may be narrowly defined or defined more broadly. For instance, there are more firms in the media industry than in the newspaper industry and there are more firms in the global film industry than in the Nigerian film industry.

13.13 Traditional objective of a firm

The traditional objective of a firm is profit maximisation, i.e, making as much profit as possible. Profit is maximised where marginal revenue equals marginal cost.

In practice, it can be difficult for firms to calculate the profit maximisation level of output because it requires precise measurement and market conditions are always changing. Firms may instead add a profit margin to their long run average cost.

Normal and abnormal profit

Normal profit is the minimum level of profit needed to keep the firm producing the product in the long run. It is sometimes referred to as the opportunity cost of supplying capital to the industry. It is included in the costs of production and so is earned where average revenue equals average cost. The level of normal profit varies from industry to industry because there are different levels of risks and stress involved.

Abnormal profit, also called supernormal profit, is any profit earned in excess of normal profit and is earned where average revenue is greater than average cost.

An awareness of other objectives of firms

There is a range of other objectives a firm may pursue including sales maximisation, survival, sales revenue maximisation, driving out rivals, keeping out rivals and profit satisficing.

Sales maximisation involves a firm producing as much as possible up to the point where it is still making normal profit (average revenue equals average cost). There are a number of advantages in aiming for growth. In the short term, aiming for growth may involve some sacrifice of profit but in the longer term, a larger firm may gain higher profit. This is because the firm is able to take greater advantage of economies of scale and gain greater market power. Managers are particularly keen to aim for growth. This is because managers' salaries and status are often more closely connected with the size of the firm rather than its profitability. An increase in size also makes it more difficult for a firm to be taken over, increasing the chance of survival and managers' chances of retaining their jobs. In addition, higher output may enable a firm to take greater advantage of economies of scale.

Survival may become a firm's main objective due to adverse market conditions. A rise in costs of production or a fall in demand may mean that a firm experiences a loss. In such a situation, a firm may strive to stay in the market if it can cover its variable costs and believes that market conditions will improve in the future.

Sales revenue maximisation occurs when a firm produces where marginal revenue is zero. The level of output which ensures the highest total revenue may be higher than maximum profit revenue. Managers' salaries may be linked to sales figures as well as the size of the firm and high total revenue may make it easier for firms to sell shares and to borrow.

Driving out rivals may be a short term objective of a firm. If a firm has market power, it may sacrifice some profit for a while by engaging in predatory pricing. This involves the firm setting a price which it considers will be low enough to drive out a rival firm which has higher costs of production or more reserves of retained profits. If it is successful, it is likely to raise price once its rivals have left the market.

Keeping out rival firms by engaging in limit pricing may become a firm's main objective if it feels threatened by the possible entry of rivals into the market. If it sets price at a relatively low level, potential entrants which are likely to have high costs, at least initially, may be discouraged from joining the market.

Profit satisficing involves aiming for a satisfactory rather than a maximum profit level. Pursuing such an objective may allow a firm to pursue other objectives to satisfy not only the firm's shareholders but also other stakeholders in the firm including managers and workers. Profit satisficing also recognises that it may be difficult and possibly risky to aim for maximum profit.

In the long term, the other objectives mentioned may not conflict with profit maximisation. Increasing in size, maximising sales revenue and reducing the number of its rivals may enable a firm to charge higher prices and lower its costs. Profit satisficing may increase the motivation of managers and workers and so raise productivity and reduce costs of production.

13.14 Different market structures

Market structure describes the main features of a market particularly in terms of the level of competition in the market. The four main market structures which economists assess are perfect competition, monopolistic competition, oligopoly and monopoly.

Moving from perfect competition to monopoly, the level of competition is usually thought to decrease. Economists examine the characteristics, behaviour and performance of market structures.

In perfect competition, there are many buyers and sellers, the product is homogeneous, there is free entry into and exit from the market and buyers and sellers have perfect information.

Monopolistic competition has features of both monopoly and perfect competition. In conditions of monopolistic competition, there are many buyers and sellers, the product is differentiated, there are no or very low barriers to entry into and exit from the market and buyers and sellers do have perfect information.

In conditions of oligopoly, the market is dominated by a few large firms, the product is differentiated or identical, there are barriers to entry into and exit from the market and there is no perfect information.

In a pure monopoly, there is only one seller, the product is unique, there are very high barriers to entry into and exit from the market and firms outside the market will lack perfect information. For purposes of government regulation, a monopoly is sometimes defined as a firm which has a market share of 25% or more and a dominant monopoly as a firm which has a market share of 40% or more.

Contestable markets

The idea of contestable markets focuses on potential rather than actual competition. A contestable market may contain any number of firms but to be a perfectly contestable market it must have no barriers to entry and exit. In the long run, only normal profits will be earned. A monopoly in a contestable market may be forced to act efficiently due to the threat of competition.

Hit and run competition is a feature of a contestable market with abnormal profits attracting firms to enter the market. They may then leave when the profit level falls back to normal in search of higher profits in another industry.

13.15 Conduct of firms: pricing policy and non-price policy

Perfectly competitive firms are price takers. The price is set by the intersection of the market demand and supply curves. An individual firm's output is too small to influence price and its average revenue equals marginal revenue. It cannot charge more than the market price because no-one would buy its product and there is no reason to charge less as it can sell any quantity at the going market price.

Perfectly competitive firms do not advertise as there is perfect knowledge and the products are homogeneous.

Monopolistically competitive firms, oligopolists and monopolists are price makers. Their output influences price and their average revenue exceeds marginal revenue. Advertising is on a small, local scale in the case of monopolistically competitive firms but may be on a large, national scale in the case of oligopolies and monopolists. Such large scale advertising can be used to create and reinforce brand loyalty, attract new customers and act as barriers to entry.

As well as advertising, other forms of non-price competition are a prominent feature of many oligopolistic markets. Non-price competition includes brand names, packaging, free gifts, free delivery and competitions.

Oligopolists may engage in collusion to reduce uncertainty and to drive up price to increase abnormal profit. Collusion is often short term because it is usually illegal, firms have a temptation to cheat by lowering their price to gain extra market share, it is difficult to set a price which will benefit all the firms equally and because not all the major firms may be willing to join the cartel.

Price leadership is a feature of a number of oligopolistic markets. A price leader is a firm which is the first one to change price and which is followed by its competitors. A price leader may be the largest firm, the most profitable firm or the firm which is considered to have the best record in setting the price.

Mutual interdependence is a feature of oligopoly. In deciding its market strategy, a firm takes into account what rival firms are doing and how they might react, for instance, to a change in its price, output or spending on advertising. Some economists make use of game theory in studying the behaviour of oligopolists. Game theory is the study of how people behave in strategic situations.

13.16 Performance of firms: in terms of output, profits and efficiency

Monopolists and oligopolists may restrict output in order to push up price but it may also seek to increase output in order to take greater advantage of economies of scale.

Monopolists and oligopolists can earn abnormal profit in the short run and long run. This is because barriers to entry and exit will prevent new firms entering the industry to compete away the abnormal profit. Monopolistically competitive and perfectly competitive firms can only earn abnormal profit in the short run. In the long run, they will earn normal profit as, in the absence of barriers to entry and exit, will mean that new firms will come into the industry and drive down price and profit.

Perfectly competitive firms always produce where MC = AR (P) and, if these fully reflect all costs and benefits, will be allocatively efficient. In the long run, perfectly competitive firms are also productively efficient, producing at the lowest point on the average cost curve.

Monoplistically competitive firms produce where AR (P) exceeds marginal cost and so is allocatively inefficient. They also produce with excess capacity and are not productively efficient. Oligopolists also fail to be productively efficient and are not allocatively efficient.

Private sector monopolies are both productively and allocatively efficient but may produce at lower average cost than more competitive firms if economies of scale are significant. State owned enterprises may seek to be allocatively efficient.

Non-price competition is a feature of both monopolistic competition and oligopoly. It is particularly significant in oligopolistic markets where advertising and other forms of non-price competition can be on a large scale.

Perfectly competitive firms and monopolistically competitive firms act independently but oligopolists may collude. A pure monopolist would not have another firm to collude with but a legal or dominant monopolist might.

13.17 Mind maps

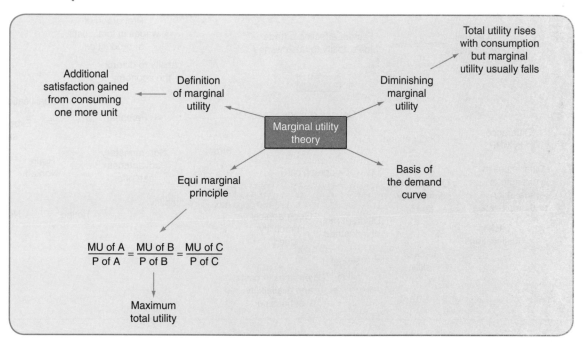

Mind map 13.1   Marginal utility theory

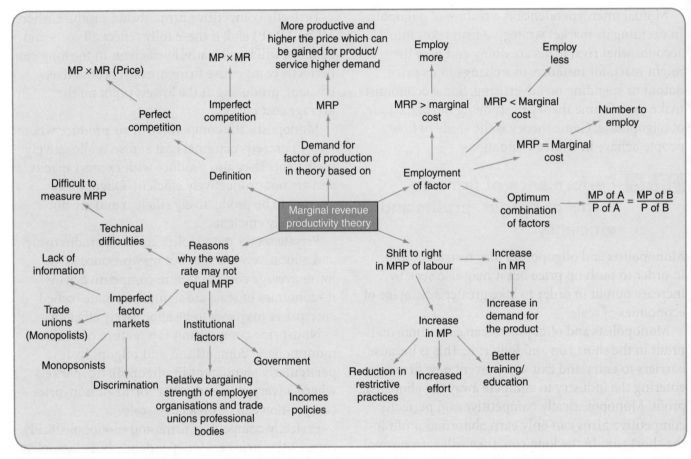

Mind map 13.2   MRP theory

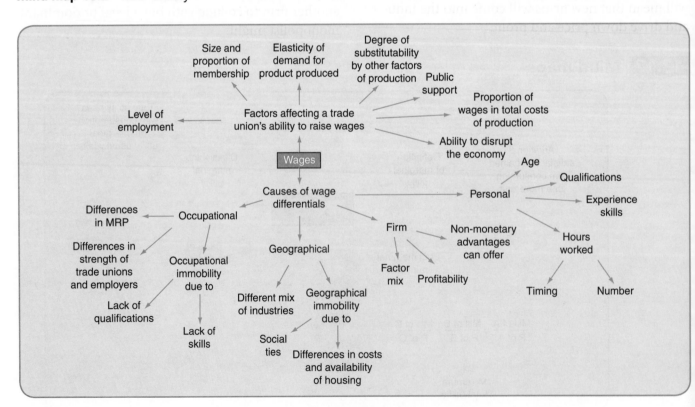

Mind map 13.3   Wages 1

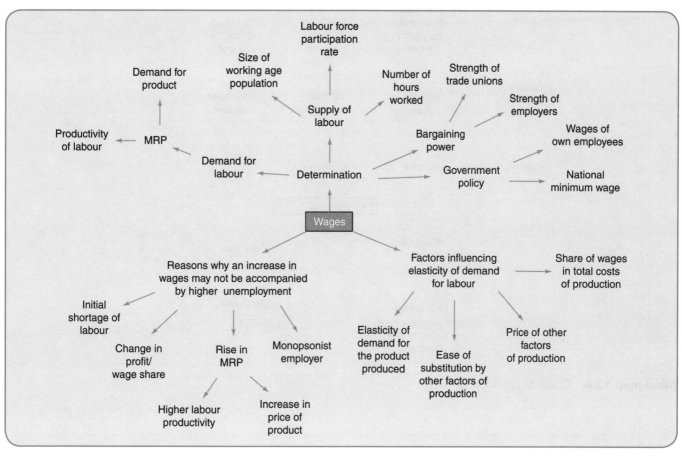

Mind map 13.4 Wages 2

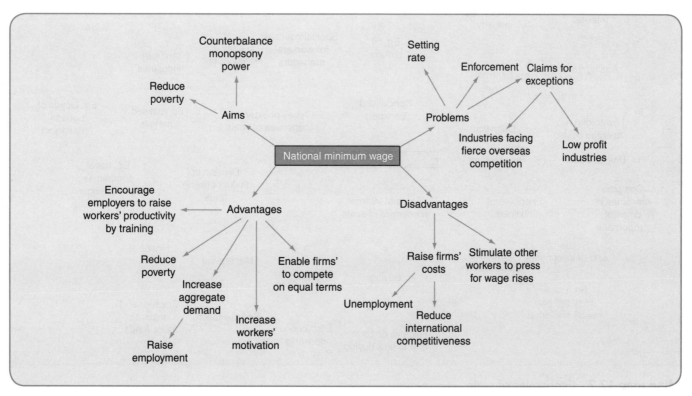

Mind map 13.5 National minimum wage

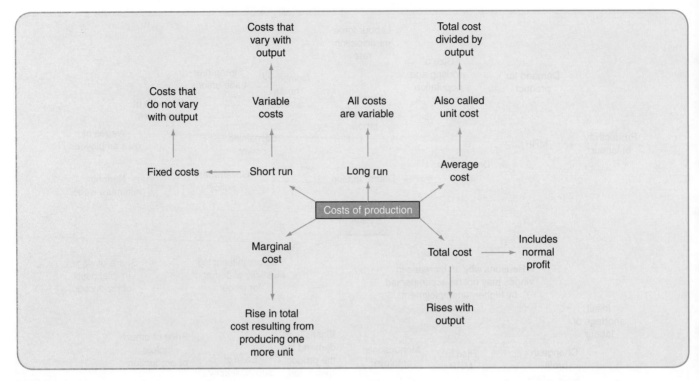

Mind map 13.6 Costs of production

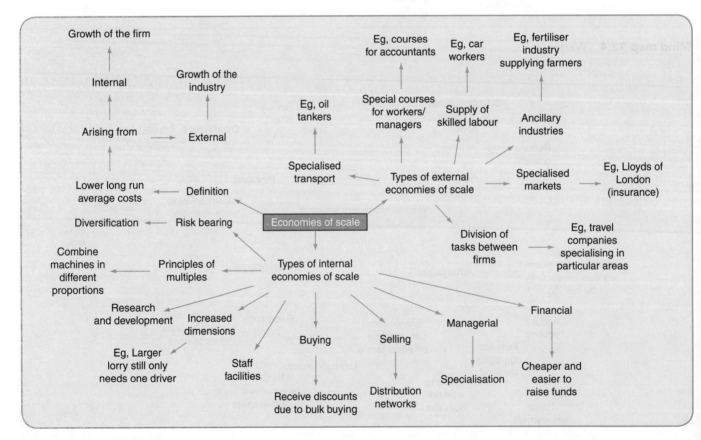

Mind map 13.7 Economies of scale

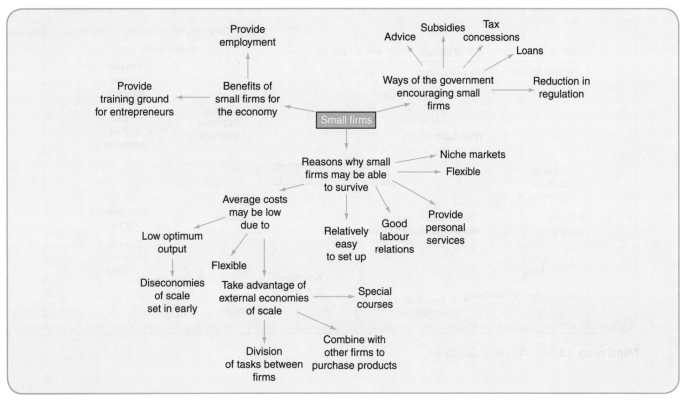

Mind map 13.8   Small firms

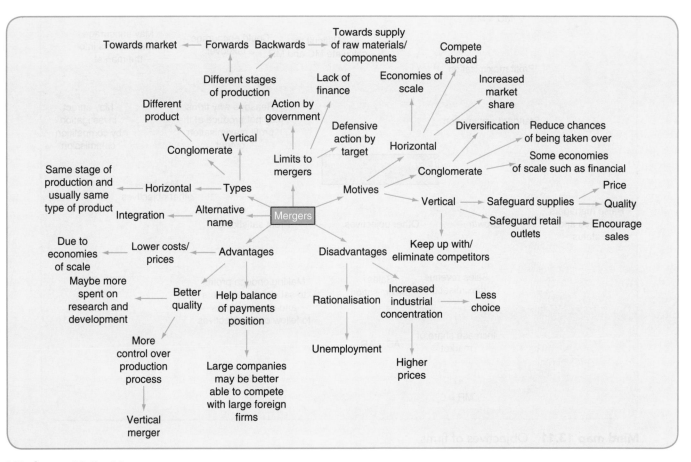

Mind map 13.9   Mergers

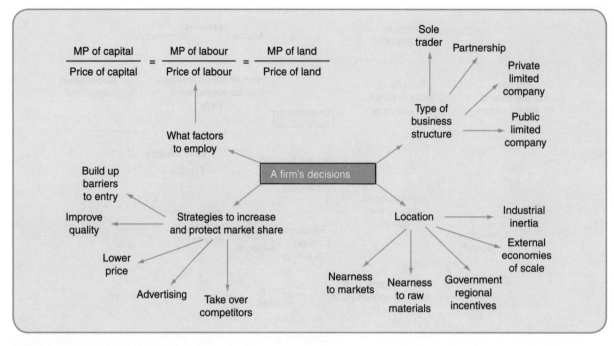

Mind map 13.10 A firm's decisions

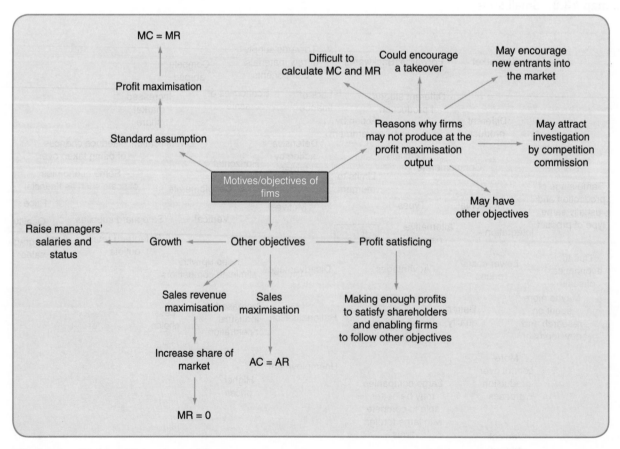

Mind map 13.11 Objectives of firms

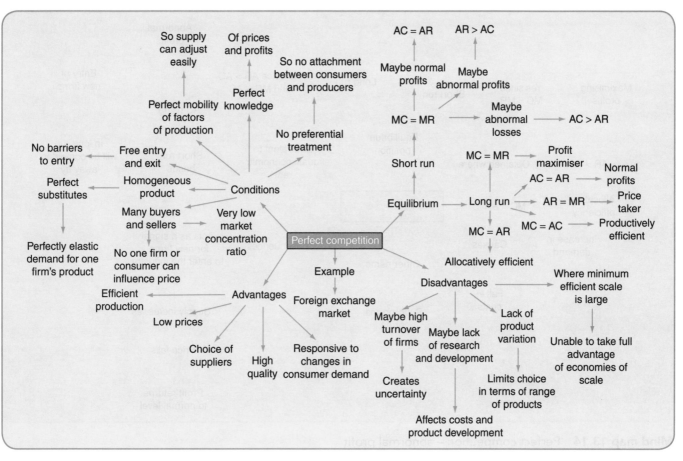

Mind map 13.12 Perfect competition

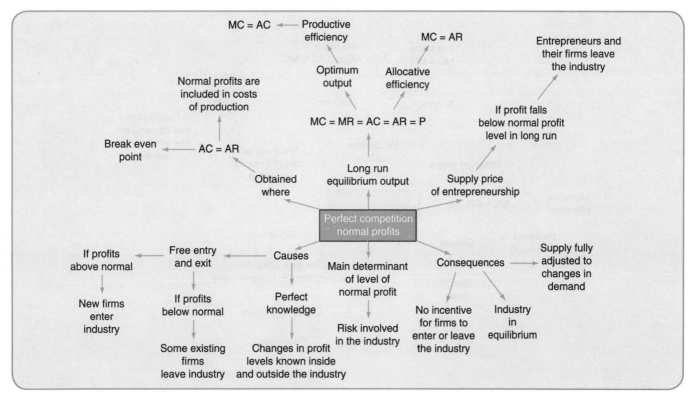

Mind map 13.13 Perfect competition – normal profit

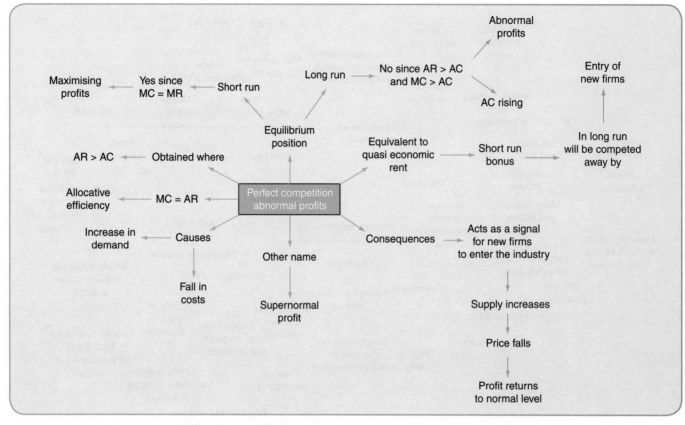

Mind map 13.14   Perfect competition – abnormal profit

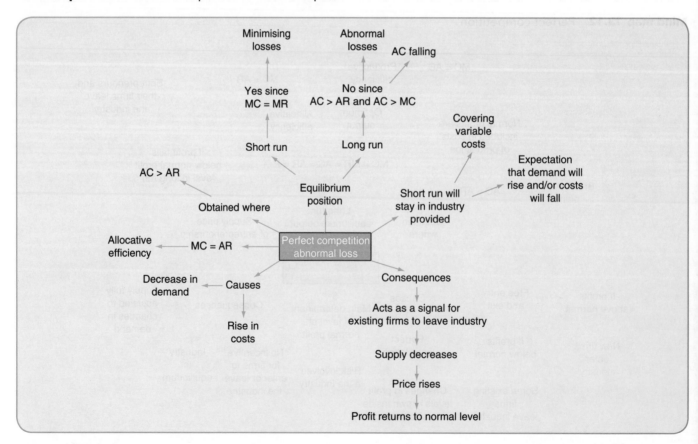

Mind map 13.15   Perfect competition – abnormal loss

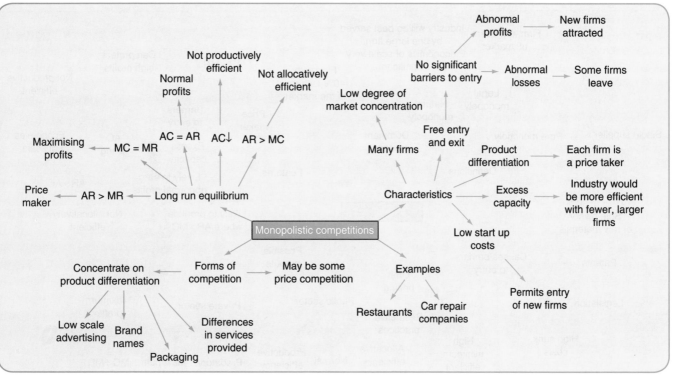

Mind map 13.16   Monopolistic competition

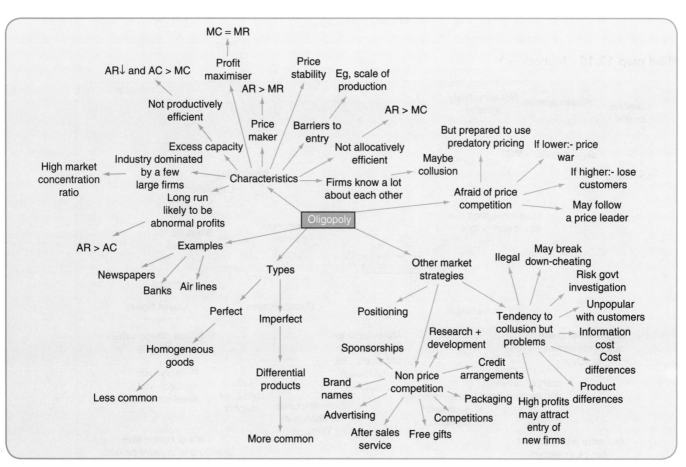

Mind map 13.17   Oligopoly

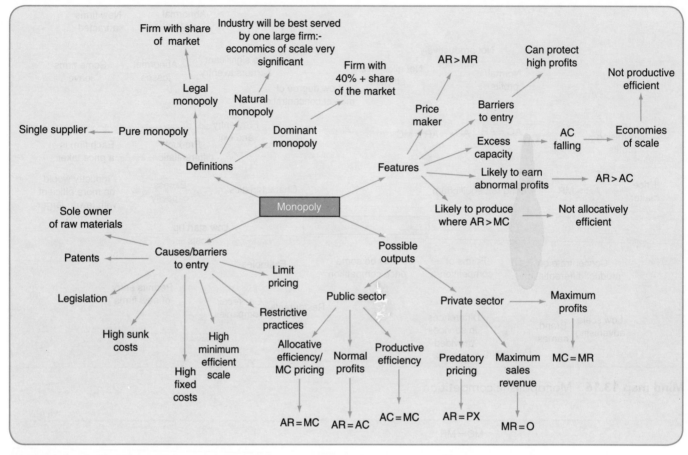

Mind map 13.18   Monopoly 1

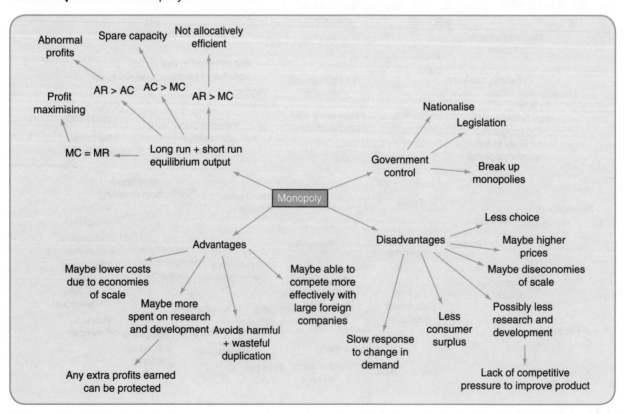

Mind map 13.19   Monopoly 2

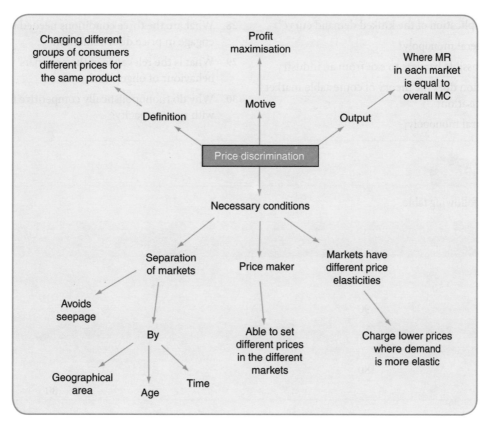

Mind map 13.20   Contestable markets

Short Questions

1. According to the law of diminishing marginal utility returns, what two changes could cause consumers to change their pattern of spending?

2. Why would a person be unlikely to consume a quantity of product where total utility is falling?

3. How does a budget line reflect relative price changes?

4. What are the two main causes of an increase in marginal revenue product?

5. Why have diminishing returns been less of a problem in agriculture than was originally forecast by economists including Malthus?

6. Why is demand for labour a derived demand?

7. Explain two reasons why demand for nurses has increased in most countries in recent years.

8. Identify three reasons why people may switch from a high paid occupation to a lower paid occupation.

9. In connection with trade unions, what is meant by a closed shop?

10. Why may rising educational achievements increase transfer earnings but reduce economic rent?

11. What is the difference between fixed and variable costs?

12. Identify four economies of scale open to a firm that produces a wide range of products.

13. Why are external economies of scale sometimes called economies of concentration?

14. A firm wishes to grow. What may stop it from expanding?

15. How does the investment finance available to a sole trader differ from that available for a large public limited company?

16. How does the relationship between marginal revenue and average revenue of perfectly competitive firm differ from that of a firm operating under conditions of monopolistic competition?

17. Under what conditions might a firm stay in an industry despite making a loss?

18. What is the difference between predatory pricing and limit pricing?

19. Why might a firm not seek to maximise profit?

20. What is second best theory?

21. Is perfect competition perfect?

22. What is meant by a monopsony and oligopsony?

23. What is the implication of the kinked demand curve?
24. What is a bilateral monopoly?
25. Explain two possible barriers to exit from an industry.
26. What implication does the theory of contestable market have for privatisation?
27. What is a natural monopoly?

28. What are the three conditions needed for a firm to engage in price discrimination?
29. What is the relevance of the prisoners' dilemma for the behaviour of oligoplists?
30. Why do monopolistically competitive firms produce with spare capacity?

Revision Activities

1. Complete the following table.

Output	Total cost	Fixed cost	Variable cost	Average cost	Average variable cost	Average fixed cost	Marginal cost	Marginal variable cost
0	100							
1		50						
2			90					
3	201							
4								19
5			180					
6							80	

2. Decide which of the following statements are true and which are false and briefly explain why.
    (a) The income effect of a price change relates to the ability to buy a product and the substitute effect relates to the willingness to buy it.
    (b) Average fixed cost and average variable cost always fall as output increases.
    (c) Demand for skilled labour is more inelastic than demand for unskilled labour.
    (d) If the supply of labour is perfectly elastic, all of the workers' earnings will be economic rent.
    (e) Trade unions always seek to increase members' wage rates.
    (f) Small firms may continue to survive if they provide for a niche market.
    (g) If a firm is a price maker, total revenue will be maximised when marginal revenue is zero and price elasticity of demand is unitary.
    (h) A perfectly competitive firm can sell any quantity at the market price whereas a monopolist can only sell more by lowering price.
    (i) A perfectively competitive market is a contestable market and so a contestable market is a perfectly competitive market.

3. Complete the following table.

Comparison of market structures				
Feature	Perfect competition	Monopolistic competition	Oligopoly	Pure monopoly
No. of firms in the market	Very many		Dominated by a few large firms	One
Market concentration ratio		Low	High	
Barriers to entry and exit		None or low		
Type of product produced				Unique
Influence on price			Price maker	
Ability to earn supernormal profit in the long run	No			

Multiple Choice Questions

1. The marginal utility of a product to a consumer is zero. What does this indicate?
   A Total utility is also zero
   B Total utility is maximised
   C The product is free
   D The consumer is in equilibrium with the product

2. A person allocates her expenditure between three products, X, Y and Z. The table below shows the prices of the products and the marginal utilities the person gains.

Product	X	Y	Z
Price ($)	10	1	2
Marginal utility (units)	50	8	10

   How should the person's spending alter in order to maximise her utility?

	X	Y	Z
A	More	Less	More
B	More	Less	Less
C	Less	More	More
D	Less	More	Less

3. Basic food products such as rice and bread are cheaper than diamonds. This is despite basic food products being more essential to people's lives than diamonds. What could explain this paradox of value?
   A Basic food products are a necessity whereas diamonds are a want
   B Diamonds are in greater supply than basic food products
   C There is a difference in market structures
   D There is a difference between total and marginal utility

4. In Figure 13.12 the consumer's initial budget line is MN

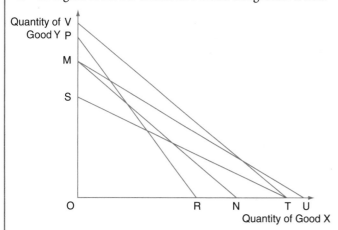

Figure 13.12

   The consumer's money income decreases, there is no change in the price of good X but good Y becomes cheaper. What could be the consumer's new budget line?
   A PR
   B ST
   C UT
   D MV

5. The table shows the short run marginal cost of producing product X.

Units of X	Marginal cost ($)
1	40
2	30
3	30
4	60
5	80

   Total fixed cost is $20. At what level of output is short run average total cost lowest?
   A 2 units
   B 3 units
   C 4 units
   D 5 units

6. When a firm produces five units of a product, its average fixed cost is $20 and its average variable cost is $90. When it increases its output to six units, its average fixed cost falls to $18 and its average variable cost falls to $80. What is the marginal cost of producing the sixth unit?
   A −$12
   B $8
   C $30
   D $38

7. A firm's total fixed cost is $8,000. Its average total cost is $10 and its average variable cost is $8. What is the firm's output?
   A 800 units
   B 1,000 units
   C 4,000 units
   D 8,000 units

8. When do diminishing returns occur?
   A Adding extra units of a variable factor of production to a fixed factor causes a fall in marginal output
   B Changing all the factors of production employed results in a less than proportionate increase in output
   C Increasing demand starts to slow down
   D Increasing output results in costs rising more rapidly than revenue

9. What does the law of diminishing returns imply?
    A All factors of production used can be varied in the short run
    B At least one factor of production used is fixed in the short run
    C Total fixed cost increases at an increasing rate
    D Total variable cost increases at a decreasing rate

10. The table below shows what happens to the output of apples as a farm takes on more workers.

No. of workers	Output (tonnes of apples per week)
1	100
2	250
3	450
4	600
5	700

Between which levels of output do diminishing returns set in?
    A 100 and 250 tonnes
    B 250 and 450 units
    C 450 and 600 units
    D 600 and 700 units

11. What would cause the demand curve for labour to shift to the right?
    A A decrease in demand for the final product
    B A decrease in labour mobility
    C An increase in labour productivity
    D An increase in wages

12. The table below shows the output of handbags of a firm in relation to the number of workers it employs.

No. of workers	Total output per day	Output per worker
1	20	20
2	50	25
3	90	30
4	120	30
5	140	28

What is the marginal product of the third worker employed?
    A 5 units
    B 10 units
    C 30 units
    D 40 units

13. A firm employs two variable factors of production, Y and Z. Factor Y costs $5 per unit and factor Z $8 per unit. The marginal product of factor Y is 35 units and the marginal product of factor Z is 40 units. What should the firm do to minimise its costs of production?

A Employ more of factor Y and less of factor Z
B Employ more of factor Z and less of factor Y
C Employ less of both factor Y and factor Z
D Employ more of both factor Y and factor Z

14. Figure 13.13 shows a supply curve for labour.

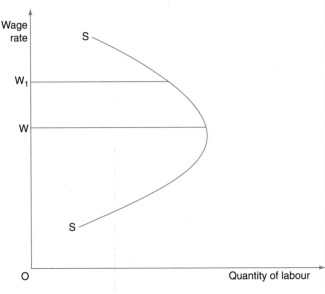

Figure 13.13

What happens when the wage rate rises from W to W_1?
A Capital is substituted for labour
B Hours of leisure become more highly valued
C The occupational mobility of labour decreases
D The substitution effect becomes stronger relative to the income effect

15. Figure 13.14 shows a labour market in which a trade union has successfully negotiated a wage rise from W to W_1.

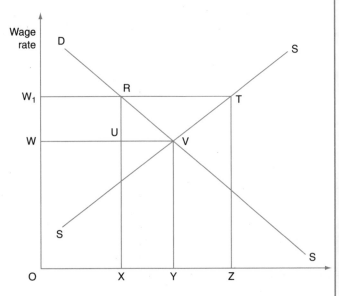

Figure 13.14

What does the diagram indicate will be the result of the rise in the wage rate?

A   Economic rent enjoyed by workers will increase by RUV
B   Employment will fall from Z to Y
C   The supply of labour will change from SS to W₁TS
D   The total wage bill will change from OWVY to OW₁TZ

16. Figure 13.15 shows a firm operating in a perfectly competitive market for its product and a monopolistic labour market. What will be the wage rate paid to workers and the quantity of labour employed?

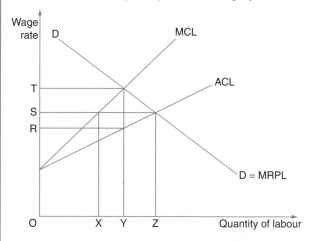

Figure 13.15

	Wage rate	Quantity of labour
A	R	Y
B	S	Z
C	T	Y
D	S	X

17. Which of the following may have monopsony power in a national labour market?

A   A national trade union
B   The army
C   The sole seller of a particular brand of orange juice
D   Workers who possess skills in high demand

18. Figure 13.16 shows the effect of an increase in demand for accountants on the market for accountants.

What is the increase in economic rent earned by accountants?

A   NPST
B   QRST
C   NPSVUT
D   TSUV

19. In which circumstance would the introduction of a national minimum wage be accompanied by a rise in employment?

A   It increases demand for labour
B   It is at a level below the equilibrium level
C   Demand for labour is elastic
D   Supply of labour is inelastic

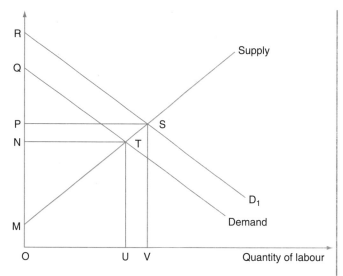

Figure 13.16

20. What is the relationship between average cost and marginal cost of production if average cost is falling?

A   Marginal cost must be below average cost
B   Marginal cost must be above average cost
C   Marginal must be equal to average cost
D   Marginal cost may be above or below average cost

21. The owner of a small private beach decides to charge tourists for its use. There is a fixed cost but no variable cost in allowing people to use the beach. What should the owner do to maximise her profits?

A   Attract as many tourists as possible
B   Charge the highest price possible
C   Minimise total cost
D   Maximise total revenue

22. In an industry wage rates increased from $10 an hour to $13 an hour, output per worker increased by 15% and the number of hours worked decreased from 40 hours to 38 hours and raw material costs increased by 5%. What happened to labour costs and total output?

	Labour costs	Total output
A	Decreased	Decreased
B	Decreased	Increased
C	Increased	Increased
D	Increased	Decreased

23. A firm's average revenue is less than average cost. What condition is necessary for the firm to continue in business in the short run?

A   Average revenue is greater than average fixed cost
B   Average revenue is greater than average variable cost
C   Average revenue is greater than marginal cost
D   Average revenue is greater than marginal revenue

24. What is the short run supply curve of a perfectly competitive firm based on?

A   Its marginal cost curve above its average variable cost curve
B   Its marginal cost curve above its average total cost curve

C   Its average variable cost curve

D   Its average total cost curve

25. What is a characteristic of monopolistic competition?

A   Average revenue exceeds average cost in the long run

B   Firms can make supernormal profit in the long run

C   High barriers to entry and exit

D   Products are differentiated

26. Figure 13.17 shows a firm producing under conditions of monopolistic competition in the short run. Which area shows the firm's total cost at the profit maximising output?

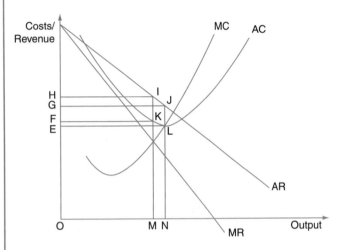

Figure 13.17

A   OFKM

B   FHIK

C   OECN

D   EGJL

27. If a profitable monopolist changes its objective from profit maximisation to sales maximisation, what change will it make to price and output?

	Price	Output
A	Decrease	Decrease
B	Decrease	Increase
C	Increase	Increase
D	Increase	Decrease

28. Firms in an industry collude and charge a common price of P. Each firm agrees to restrict its output to a production quota determined by the industry cartel. The firm is given a quota of Q as shown in Figure 13.18.

What is the maximum short run increase in supernormal profit the firm could receive by cheating on the agreement?

A   JKLM

B   JKLM − HMNG

C   JKLM + HMNG

D   HMNG

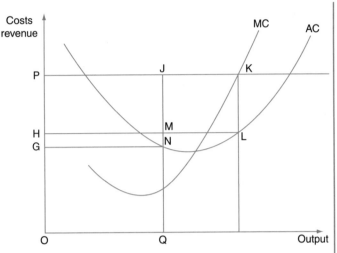

Figure 13.18

29. Figure 13.19 shows the output of a monopolist. If it produces at the profit maximising output, what is the gap between its output and the allocatively efficient output?

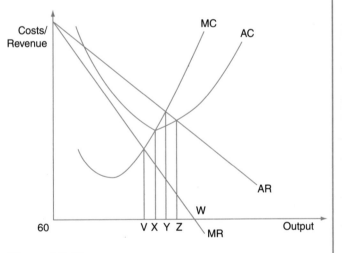

Figure 13.19

A   VW

B   VY

C   XY

D   XZ

30. What is a defining characteristic of a contestable market?

A   It is possible to transfer the fixed capital in the industry to alternative uses

B   The firms in the industry spend large amounts on advertising

C   There is only one firm in the industry

D   Whilst there are no barriers to entry into the industry, there are barriers to exit from the industry

Data Response Questions

1. The North Sea oil industry

The UK drills for oil in the North Sea. In 1999, production was at 4.5 million barrels a day. However, as the oil in the easily accessible oil fields was used up, production fell steadily from 1999 until it reached 3.3 million barrels a day in 2005.

Nevertheless, in 2008, many in the oil industry did not consider it to be in decline. Indeed, some of the industry's problems were associated with an expanding industry, not a declining one. The managers complained, for example, about the shortage of skilled labour and the high rents of the oil rigs. Investment increased by 30% in 2006 and it was hoped that production would rise in 2007.

The UK government was keen to keep production high as the industry supported 250,000 jobs and had a large impact on the trade balance of the country. One way to keep production high is to explore new oil fields. Another way is to use every drop of oil from the existing oil fields. Sometimes a large company leaves an oil field which still has substantial amounts of oil because although it is technically possible to extract the oil, it is not profitable to do so. However, a smaller company is often able to drill profitably for the oil when it is unprofitable for a large company.

In view of this, the UK government encouraged smaller and more enterprising firms. In 2005, 152 licences to drill were given to 99 companies. These new licences gave companies exclusive rights to develop the more inaccessible oil fields for six years rather than four years as before. Furthermore, the government also changed the rules on access to pipelines – which are often owned by large companies – so that smaller companies could get their oil to the market more easily.

The Chancellor of the Exchequer (Minister of Finance) also made changes. In 2002, he raised the tax on oil company profits to 40% from the usual company rate of 30%, and raised it again in 2005 to 50%. The justification given was that the oil companies made huge profits.

The industry is more dependent than ever on large expensive technological advances and innovation. Most of the government's reforms recognised this.

(a) How might the North Sea oil industry have an impact on the national income of the UK? **[4]**

(b) From the information given, consider whether the North Sea oil industry appears to be declining or expanding. **[6]**

(c) What evidence is there that might explain why the North Sea oil industry is still prepared to increase investment? **[4]**

(d) Discuss whether the increase in the number of

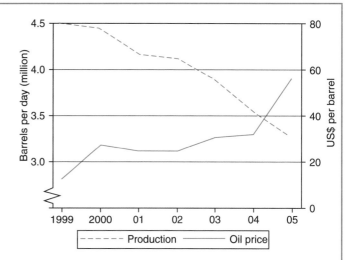

Figure 13.20

small firms drilling for oil contradicts the usual explanation of the existence of small firms. **[6]**

Cambridge 9708 Paper 4 Q1 May/June 2008

2. The Success of Supermarkets

In some countries supermarkets dominate food shopping. In the UK, 75% of the food bought for home use comes from supermarkets. A third of that comes from one supermarket, Tesco. Tesco makes billions of pounds profit, one third of which goes to the government in taxes. It employs 110,000 people in the UK and many more in developing countries.

In the past, UK shoppers queued to buy expensive food from many small shops with limited choice and restricted opening times. Now, in supermarkets, they have the benefit of a wide choice, reasonable prices, international dishes, organic produce, fair trade items, clear labels of the contents of the products and because of intense competition between the supermarkets, some open 24 hours.

However, the media complains that supermarkets are not competitive but monopolies. They say their profits are too high, they have caused small shops to close and forced suppliers in developing countries to accept low prices and to pay low wages.

It must be remembered that supermarkets grew because they gave the customers what they wanted and aimed at certain types of shoppers. One supermarket, which started as a small shop, insisted on selling only high quality products while another offered customers low prices.

Supermarkets also adapted to changing market trends. One began to supply products with its own brand name which were sold more cheaply because there were no advertising costs. A further brilliant idea in expanding their business was the introduction of store loyalty card. Shoppers with a loyalty card are given discounts which encourages them to continue to shop in the same supermarket.

The most successful supermarkets expanded their businesses by buying large sites to build huge stores (they are criticised for such building, especially if it destroys parts of the countryside or environment). They expanded into non-food products to fill these stores, making it possible for consumers to buy many household items from clothes to kitchen utensils to electrical goods in the same shop. Supermarkets have also recently introduced on-line shopping and home deliveries.

Many small shops have closed. There are bound to be casualties in retailing. How can the blame for that be the fault of the supermarket? Their size should not be a concern. It is, after all, the consumer who decides where to shop and what to buy.

(a) How has the type of market structure in food retailing in the UK changed? [3]

(b) To what extent does the article support the view that the consumer is sovereign in food retailing? [4]

(c) Explain what the various objectives of a firm might be. [5]

(d) Do you agree with the conclusion of the article that the size of a firm should not be a source of concern? [8]

Cambridge 9708 Paper 41 Q1 Oct/Nov 2010

3. Cola wars

Coca-Cola and Pepsi-Cola dominate the carbonated soft drink (fizzy drink) market in the United States of America. In 2011, Coca-Cola had a 42% share and Pepsi-Cola had 30% of the market. That same year Coca-Cola held the top two market share spots for individual drinks. The classic Coca-Cola drink stayed in top place and Diet Coke overtook Pepsi-Cola to claim second place.

Both companies have seen the total sales of fizzy drinks decline as US consumers have been switching to healthier drinks including bottled water, juices, sports and energy drinks. In the light of this change in the market, Coca-Cola and Pepsi-Cola have been following rather different strategies. Coca-Cola has and continues to spend more on advertising than its rival. It is also specialising in soft drinks and is increasing its market share of fruit drinks.

Pepsi-Cola tends to change its objectives and market strategy more frequently. In 2010, for instance, it decided not to advertise its drinks at the Super Bowl sporting event and instead launched an online competition for the nomination of good causes that Pepsi-Cola might finance. This approach did result in more money being given to charities but did not generate many more sales. As a result, in 2011 Pepsi-Cola went back to advertising at the Super Bowl. In 2012, it increased its spending on marketing and advertising but still not the levels of Coca-Cola.

Whilst Coca-Cola has some diversification in terms of soft drinks, Pepsi-Cola has diversification not only in terms of soft drinks but also in terms of snack foods. In 2012, it launched a new drink, Pepsi Next, which has 60% less sugar content than the classic Pepsi drink. This is in keeping with its long term objective to transform the company into a producer of healthier drinks and snacks.

(a) Does the information indicate that Coca-Cola had a monopoly of the fizzy drinks market in 2011? Explain your answer. [2]

(b) In what circumstances may a firm benefit from engaging in a price war? [4]

(c) What evidence is there in the information that Pepsi-Cola is becoming more allocatively efficient? [4]

(d) Comment on whether increasing spending on advertising will increase a firm's profits. [4]

(e) Discuss whether Pepsi-Cola's approach to diversification is more likely to be successful than Coca-Cola's approach. [6]

Essay Questions

1. Changes in consumer expenditure alter firms' output. A decrease in consumer expenditure may reduce output and costs of production.

   (a) Explain why, according to utility theory, consumers would change their spending patterns. **[12]**

   (b) Discuss whether reducing the scale of production will reduce costs of production. **[13]**

2. Discuss what might cause inequalities in wage rates in an economy. **[25]**

   Cambridge 9708 Paper 4 Q3 May/June 2009

3. (a) Compare monopoly and monopolistic competition. **[12]**

   (b) Discuss whether a monopoly always disadvantages consumers. **[13]**

4. In 2007, the cost of a single ticket on London trains bought at the time of travel was £4. The same ticket bought in advance was £2.50 if used up to 19.00 hours and £2 after 19 hrs. Children could travel free at any time, and those over 60 could travel free after 09.00 hrs.

   (a) Explain what is meant by price discrimination and analyse what evidence there is of price discrimination in the above statement. **[12]**

   (b) Discuss how the output and pricing policy adopted by a firm might differ depending on the market structure in which it operates. **[13]**

   Cambridge 9708 Paper 41 Q3 May/June 2010

Government Intervention in the Price System

14

Revision Objectives

After you have studied this chapter, you should be able to:

☞ analyse the sources of market failure
☞ define deadweight loss
☞ explain how market imperfection can give rise to market failure
☞ describe the objectives of government microeconomic policy

☞ evaluate the effectiveness of government policies to correct market failure and policies towards income and wealth redistribution
☞ assess privatisation.

14.1 Sources of market failure

The causes of market failure include the existence of externalities, merit and demerit goods, public goods, information failure, factor immobility and the abuse of market power.

14.2 Meaning of deadweight loss

In general terms, deadweight loss is the loss in economic welfare caused by an increase in price above the efficient level and a fall in output below the efficient level. Deadweight loss can arise because of taxes, tariffs and monopoly.

The deadweight loss of a monopoly is the loss in consumer surplus which arises because a monopolist restricts output and pushes up price. Consumers would benefit from output increasing to the allocatively efficient level.

Figure 14.1 shows a monopolist producing the profit maximising output of Q. This is below the allocatively efficient level of QX and creates a deadweight loss of Z.

Market imperfections – existence of monopolistic elements

Monopolists use their market power to convert some consumer surplus into producer surplus. Figure 14.2 shows that an industry moving from perfect

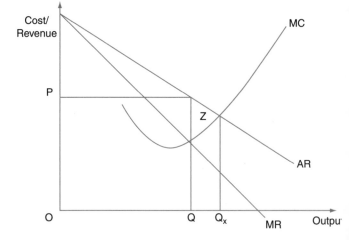

Figure 14.1

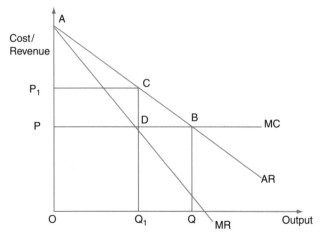

Figure 14.2

competition to monopoly would result in a rise in price from P to P1 and a fall in output from Q to Q1.

There is a fall in consumer surplus of PP1CB. Of this PP1CD is converted from consumer surplus to producer surplus. DCB just disappears – it is a deadweight loss.

14.3 Objectives of government microeconomic policy: efficiency, equity

Government microeconomic policy seeks to increase economic efficiency and promote equity. The government may intervene if it believes there is market failure and if it believes its intervention will improve allocative efficiency and productive efficiency.

Market forces may not ensure productive efficiency if there is a lack of competition or if there is information failure. They may also not result in allocative efficiency for a number of reasons. Too many resources are allocated in the case of demerit goods. Too few resources are allocated in the case of merit goods and there is a risk that no resources will be used to produce public goods. A monopolist may restrict output below the allocatively efficient level in order to drive up price and increase its profit.

Market forces can also result in significant differences in incomes. Those who have accumulated wealth and those with skills in high demand can earn high incomes whilst the sick, elderly and unemployed may find it difficult to obtain any income.

Most governments consider that everyone should have access to a minimum level of income and essential services.

14.4 Policies to correct market failure: regulation

Regulation involves legally enforced requirements and standards implemented by governments and international organisations. Examples include a ban on smoking in public places, the requirement for car drivers and passengers to wear seat belts and the need for restaurants to work under hygienic conditions.

Regulation has the benefits of being backed up by law, may have an immediate effect and may be simple to understand. There is a problem of what is the right level of regulation. For instance, what level of air and noise pollution should airplanes be permitted to emit.

For regulation to be effective, governments need accurate information. Regulation is more likely to be successful if most people agree with it. If not, it can be expensive to enforce. It may take time to draw and pass legislation. Some forms of regulation are only effective if adopted globally or at least by a large number of countries; example, limits on air pollution emitted by firms producing steel.

It does not raise revenue except through fines on those who do not obey the regulation. It may result in informal markets developing with banned products being sold illegally. It is a blunt instrument which works against the market.

Policies towards income and wealth distribution

Among the policies a government may use to make income and wealth more evenly distributed, are progressive taxes, payment of state benefits and the direct provision of goods and services.

A progressive tax is one which takes a higher proportion of the income or wealth of the rich than of the poor. This is because as income or wealth rises, the marginal tax rate increases. An example in most countries of a progressive tax is income tax.

A proportional tax takes the same proportion of income or wealth from the rich and poor. In this case, the marginal tax rate remains constant.

A regressive tax places a greater burden on the poor who pay a higher proportion of their income or wealth. The marginal tax rate falls as income or wealth rises. In most countries, sales tax (VAT) is regressive.

State benefits may be universal, i.e., paid to everyone in a certain group, such as child benefit paid to all parents. They may also be means tested, i.e., paid only to people who have below a certain income.

A government may provide education and health care free to people in a bid to ensure that everyone has access to these crucial services. Such an approach may enable the poor to increase their skills and health and so their earning potential.

Effectiveness of government policies

The use of government policies may increase efficiency by offsetting market failure. There is, however, a risk of government failure. This occurs when government intervention reduces economic efficiency.

Government failure may arise due to a lack of information, inaccurate information, households and firms reacting to policy measures in unexpected ways, policy measures having an adverse effect on incentives, decisions being influenced by the desire to be politically popular and possibly corruption.

14.5 Privatisation

Privatisation involves the transfer of assets and activities from the public sector to the private sector. It may take the form of denationalising an industry, the sale of government shares in private sector firms and contracting out services.

If a government is concerned that there is a risk that privatising an industry may not increase efficiency or may cause inequity, it may decide to regulate the industry. This may involve placing restrictions on price rises. For instance, a government may pass a law which limits price rises to a figure equivalent to a rise in the price level minus one or two percentage points. It may also seek to stop the firm or firms from using their market power to prevent the entry of new firms.

Arguments in favour of privatisation include increasing efficiency by reducing bureaucracy, increasing the profit motive and possibly increasing competition. Other arguments are to raise government revenue in the short term, to widen share ownership and to reduce government expenditure in the case of loss making state owned enterprises.

Arguments against privatisation include long term loss of government revenue in the case of profitable state owned enterprises, the possibility of abuse of market power if the industry becomes a private sector monopoly, private sector firms not taking externalities into account and a general loss of government control over the economy.

14.6 Mind maps

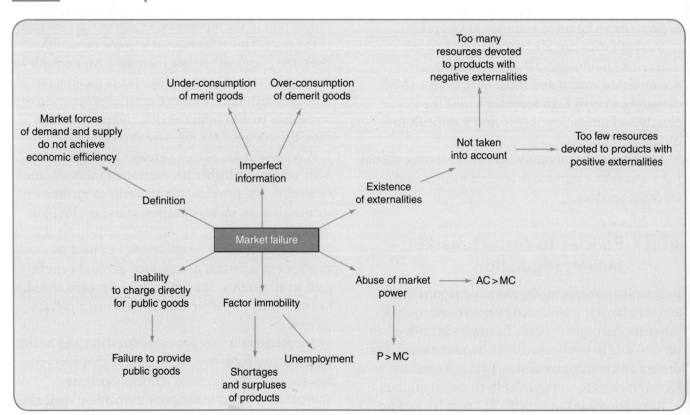

Mind map 14.1 Market failure

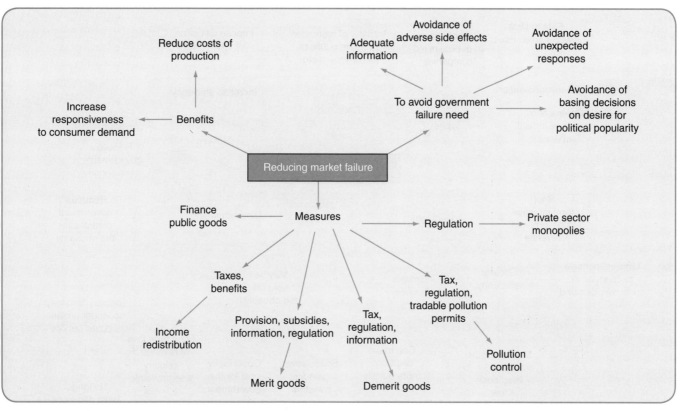

Mind map 14.2   Reducing market failure

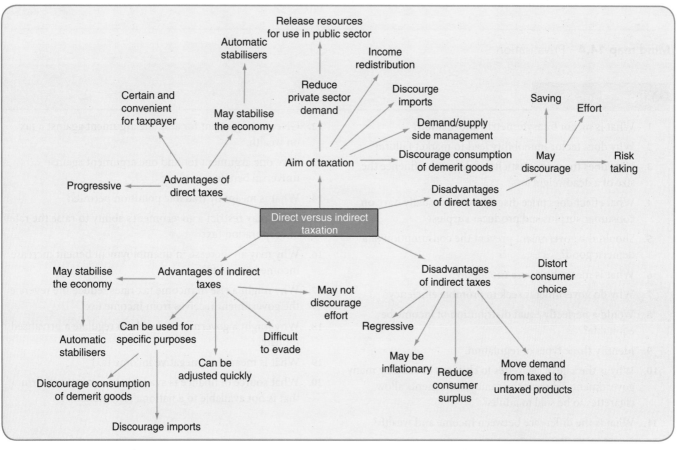

Mind map 14.3   Direct versus indirect taxation

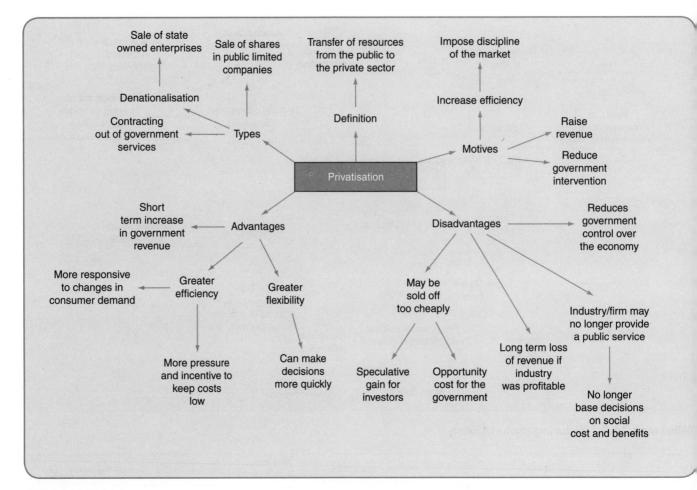

Mind map 14.4 Privatisation

1. What is meant by asymmetric information?
2. Why does factor immobility lead to market failure?
3. How does the price elasticity of demand influence the size of a deadweight loss?
4. What effect does price discrimination usually have on consumer surplus and producer surplus?
5. Should the government prevent the consumption of a demerit good?
6. What is meant by a pure monopoly?
7. Why do governments seek to promote efficiency?
8. Would a perfectly equal distribution of income be equitable?
9. Identify three types of regulation.
10. Why is the sale of cigarettes to children banned by many governments whereas the same governments allow cigarettes to be sold to adults?
11. What is the difference between income and wealth?
12. Give one argument for and one argument against a tax on wealth.
13. Give one argument for and one argument against universal benefits.
14. What is meant by tradable pollution permits?
15. What may restrict a government's ability to raise the rate of corporation tax?
16. Why may an increase in unemployment benefit increase income inequality?
17. Why might a rise in income tax rates reduce the revenue the government receives from income tax?
18. Why might a government decide to regulate a privatised industry?
19. What is meant by a negative income tax?
20. What source of finance is available to a privatised firm that is not available to a nationalised industry?

Revision Activities

1. Using Figure 14.3, identify the following.

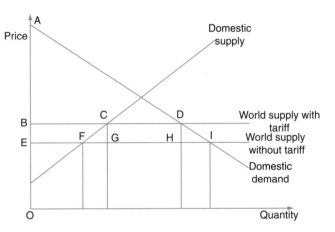

Figure 14.3

(a) The area of consumer surplus before the tariff is imposed
(b) The area of consumer surplus after the tariff is imposed
(c) The area of consumer surplus lost due the tariff
(d) The tax revenue gained from the tariff
(e) The increase in domestic producer surplus resulting from the tariff
(f) The deadweight loss resulting from the tariff.

2. The Peruvian government, like a number of Latin American governments, is facing a conflict between energy security and environmental protection. It is seeking to increase the country's output of electricity to meet the country's increasing demand for fuel and to avoid disruptive blackouts. To generate more electricity, the government is considering building dams and flooding part of its south eastern jungle. Opponents of the proposed scheme claim that it would displace 10,000 people belonging to the Ashanikas, an Amazonian tribe, and destroy important wildlife habitats. They suggest that the government should instead promote biomass and wind power and encourage energy saving measures, in part by increasing the tax on electricity.

(a) What is meant by 'energy security'?
(b) Identify a first party who might benefit and a third party who might suffer from the government's proposed scheme.
(c) Using a demand and supply diagram, explain how the imposition of an indirect tax gives rise to a deadweight loss.

3. Explain in each case whether government intervention is likely to be designed to:
      (i) increase efficiency
     (ii) increase equity
    (iii) increase efficiency and equity.
(a) the provision of unemployment benefits
(b) the imposition of a tax on cigarettes
(c) the privatisation of a profit making electricity industry
(d) the provision of state education free to consumers
(e) a cut in the top rate of tax
(f) the granting of a subsidy to providers of public transport.

Multiple Choice Questions

1. In which circumstance would a government have to intervene to correct market failure?
    A Firms making unpopular products going out of business
    B A cartel breaking down when a member reduces prices
    C Private sector firms basing their output decisions on social benefits and social costs
    D Street lighting not being provided by the private sector

2. Which of the following is a possible source of market failure?
    A An absence of external benefits
    B Lack of provision of public goods
    C Overprovision of merit goods
    D Private costs equalling social costs

3. Figure 14.4 shows an industry producing under conditions of constant marginal cost. Under conditions of perfect competition, the industry produces an output of Y. It then becomes a monopoly producing an output of X. Which area represents the deadweight loss which occurs?

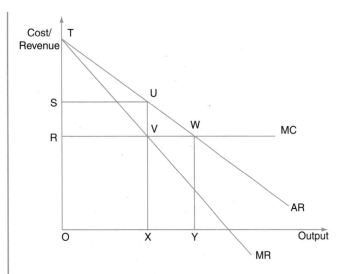

Figure 14.4

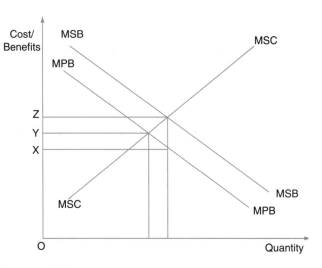

Figure 14.5

A STU
B RSUW
C UVW
D XVWY

4. Which consequence of the formation of a monopoly is not a cause of government concern?
    A A decrease in employment
    B A decrease in the rate of innovation
    C An increase in the exploitation of economies of scale
    D An increase in prices

5. Why may government intervention be necessary to achieve the allocatively efficient level of training?
    A Average costs of firms that do not provide training are higher than those that do
    B External benefits are gained by firms that do not provide training
    C Private benefits are gained by firms that provide training
    D Social costs of training equal private costs of training

6. Which situation would result in the socially optimum level of pollution being achieved?
    A The government covers the full cost of eliminating pollution
    B The private sector covers the full cost of eliminating pollution
    C The marginal social cost and marginal social benefit of pollution reduction are equal
    D The marginal social benefit of pollution reduction exceeds the marginal social cost of pollution reduction

7. Figure 14.5 shows the market for a product. What should be the subsidy per unit given by a government to achieve the socially optimum output?

A OX
B YZ
C XY
D XZ

8. A government decides to ban rather than regulate monopolies on a case by case basis. What may justify this action?
    A Monopolies always charge a higher price than competitive firms
    B Monopolies always produce at a higher cost per unit than competitive firms
    C The benefits of investigating monopolies may accrue more to consumers than producers
    D The costs of investigating might exceed the benefits

9. When is a tax defined as regressive?
    A All income earners pay the same proportion of their income in tax
    B A low income earner pays less tax than a high income earner
    C The average rate of tax falls as income rises
    D The marginal rate of tax is constant as income rises

10. Which type of tax is the most regressive?
    A A flat rate value added tax
    B A flat rate corporation tax
    C A poll tax
    D A proportional income tax

11. The table below shows the demand and supply schedules before the imposition of a tax.

    If a tax of $3 is imposed on the product, what proportion of the tax will be borne by producers?
    A The whole proportion
    B Two-thirds
    C Half
    D One-third

Price ($)	Quantity demanded	Quantity supplied
10	20	1280
9	60	1000
8	150	850
7	260	600
6	400	400
5	600	150
4	900	50

12. A government decides to reduce income tax and raise indirect taxes. It raises the same amount of tax revenue. What is the likely effect on the distribution of income and work incentives?

	Distribution of income	Work incentives
A	Less equal	Decrease
B	Less equal	Increase
C	More equal	Increase
D	More equal	Decrease

13. Figure 14.6 shows the relationship between income and the amount paid in tax. Which type of tax is illustrated?

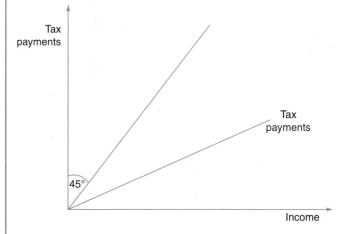

Figure 14.6

A Flat rate
B Progressive
C Proportional
D Regressive

14. What effect may a taxation policy which redistributes income from the rich to the poor have on an economy?
A It will increase aggregate demand if the marginal propensity to consume of households is lower at high levels of income
B It will increase a current account deficit if the marginal propensity to import is higher at high levels of income

C It will increase real gross domestic product if progressive taxes act as a disincentive to work and effort
D It will increase savings if the marginal propensity to save is higher at high levels of income

15. What is meant by regulatory capture?
A It is when a firm gains control of the market and uses its market power to push up price
B It is when a regulatory agency protects the interests of producers
C It is when a regulatory agency restricts the prices that producers can charge
D It is when shareholders influence a firm's objectives more than managers do

16. Figure 14.7 shows the cost and revenue curves of an industry. Why would government intervention be necessary to achieve allocative efficiency?

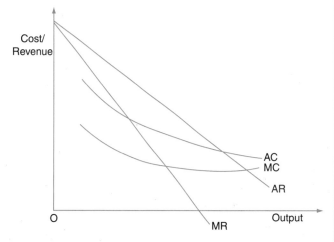

Figure 14.7

A A profit maximising private sector firm would produce a higher output
B A profit maximising private sector firm would produce where average cost equals average revenue
C The producer would make a loss at the allocatively efficient output
D The producer would make supernormal profit at the allocatively efficient output

17. Why might government intervention to correct market failure, lead to greater inefficiency?
A The government bases its decisions on private benefits rather than social benefits
B The government bases its decisions on social rather than private costs
C There is no adverse effect on incentives as a result of redistributing income from the rich to the poor
D There is no time delay between recognising a problem and implementing a policy measure

18. In what circumstance is privatisation likely to increase economic efficiency?
    A Firms in the industry collude to set price
    B Firms in the privatised industry produce where average revenue exceeds marginal cost
    C The privatised industry becomes more responsive to changes in consumer demand
    D The privatised industry expands and encounters diseconomies of scale

19. A government imposes a maximum price, set below the equilibrium price, on the rent of private accommodation to help the poor. Why might this activity actually disadvantage some of the poor?
    A It would increase producer surplus
    B It would increase the price elasticity of demand
    C It would reduce the quantity of rented private accommodation
    D It would reduce the price elasticity of supply

20. A product is under-consumed. In which circumstance would the introduction of a government subsidy to consumers fail to move the quantity consumed to the allocatively efficient level?
    A Demand for the product is perfectly price elastic
    B Demand for the product is perfectly price inelastic
    C The subsidy is shared between the consumer and the producer
    D The subsidy per unit is less than the marginal external benefit

Data Response Questions

1. Counting the cost

    India has transformed itself from a primary agricultural economy into a major industrial one in less than 60 years. Some argue that industrialisation results in increased wealth and a better standard of living. Certain areas of India are better suited to this industrialisation than others. The state of Odisha is one.

    Odisha is a state of contrasts – 48% of its people live below the poverty line; it is the most heavily indebted Indian state; its literacy level is below the national average; the level of infectious disease is high and malnutrition is alarming; it is subject to natural disasters such as floods. But, it has unrivalled natural resources including one of Asia's largest deposits of coal, large areas of forest and extensive mineral reserves. These attract big industrial companies. A new steel plant has been established which will produce six million tonnes of steel a year.

    In Odisha, a balance has to be achieved between potential profits and benefits from industrialisation and its cost to the people and the environment. The steel company is recruiting from local engineering colleges and is able to offer employment and new opportunities. People have increased incomes. Farm workers earn less than half of what the factory pays its workers. One worker said 'I used to work on the land and was at the mercy of the weather. Now I do not have to pray for rain. The company also employs my sons and we are much better off.'

    However, the factory is not labour-intensive and is unlikely to employ the huge numbers of people seeking work. Also, mining and infrastructure development destroyed some of the forest. Industrial production results in soil erosion and pollutes the air with the dark smoke from factory chimneys which causes acid rain. The rivers have also become contaminated with toxic waste which is posing a threat to wildlife, such as elephants, tigers and deer. As well as to local people. The industrialisation meant that some people lost their homes and had to be resettled elsewhere. The new houses had safe water provision and drainage, unlike some of the original homes, but people complained about the poor way the houses were built in 2007, 13 were killed during a protest about the lack of compensation for the loss of their homes.

    (a) What is meant by industrialisation? [2]
    (b) Why might some argue that the economic costs of the exploitation of the natural resources in Odisha are too high? [5]
    (c) Comment on whether the development of the steel plant is likely to have been a benefit for the workers in Odisha. [5]
    (d) It is said that the standard of living in Odisha continues to be very low. Discuss whether the evidence you have been given is sufficient to support this view. [8]

    Cambridge 9708 Paper 41 Q1 Oct/Nov 2009

2. The Deepwater Horizon disaster

    In April 2010, an explosion on BP's Deepwater Horizon rig off the Gulf of Mexico, 50 miles from Louisiana, killed 11 people and started a fire on the platform. After two days it sank causing the biggest oil spill in America's history. More than 200 million gallons of oil gushed into the sea. There were a number of failed attempts to stop the leak but it was not until 19 September 2010, that it was finally plugged and the temporary ban on oil drilling in the area was lifted.

Hundreds of workers who tried to control the spill reported mystery illnesses and doctors in the area reported an increase in respiratory infections, headaches, nausea and eye irritations. The Gulf coast's economy is reliant on energy, seafood and tourism. The shrimp and the tourism industries were particularly badly affected. It was also estimated that more than 6,000 turtles, 82,000 birds and 26,000 dolphins and whales were killed.

BP has been criticised for trying to cut its costs by using contractors whose safety standards it has not sufficiently checked out. It is likely to be more careful in the future as the disaster had a noticeable effect on its profits. It spent over $13.6 billion in responding to the disaster and sealing the rupture. It has also set aside $20 billion dollars in a compensation fund.

(a) Explain two negative externalities created by the Deepwater Horizon disaster.   [4]
(b) Does the existence of negative externalities necessarily mean that too many resources are devoted to oil production?   [5]
(c) What effect is the disaster likely to have on BP's profits?   [3]
(d) Discuss whether a government should permit the drilling of oil off its coast by foreign oil companies.   [8]

3. The taxi market

In most cities in the world, there is usually a large number of taxi firms with each firm having a relatively small share of the market. The service provided by taxi firms is usually similar. There are not many barriers to entry into and exit from the market and it is possible to start a taxi firm with just one or two vehicles. Those barriers which do exist usually arise from regulation. The regulation of the taxi market can take three main forms. These are limits on the number of firms that can operate in an area, controls on the fares that can be charged and on the imposition of quality standards.

The taxi market in a number of cities and countries has been deregulated or at least partially deregulated. In April 2012 the Greek government passed legislation reducing the price of the licence needed to operate a taxi firm from C80,000 to C3,000 and increased the number of licences issued.

Deregulation may provide a number of benefits. It may make a taxi market more contestable and may increase economic efficiency. Taxi fares may fall, waiting times may be reduced and there may be innovations, for example, the introduction of taxis that cater for people in wheelchairs.

There are, however, arguments for keeping some regulation. If controls on fares are removed, passengers may find it difficult to search out the most competitive prices. Quality controls can benefit consumers. For example, a government can set standards for the road worthiness of taxis and its agencies can check that these standards are being met and that drivers possess a clean driving licence. Eliminating restrictions on the geographical coverage of taxi firms may also cause problems. There may be an over-supply of taxis and a tendency for taxi firms to concentrate their vehicles in city centres and key tourist spots. Lower profits which may be caused by greater competition may encourage taxi firms to cut out low value trips and reduce quality.

(a) What is meant by deregulation?   [2]
(b) To what extent does the information suggest that taxi firms operate under conditions of monopolistic competition?   [5]
(c) Comment on whether a more contestable market will increase competition.   [5]
(d) Discuss whether deregulation will increase economic efficiency in the taxi market.   [8]

Essay Questions

1. (a) Explain what is meant by an efficient market equilibrium.   [12]
    (b) Discuss how the market mechanism might fail in the allocation of resources.   [13]
    Cambridge 9708 Paper 41 Q2 May/June 2010

2. (a) Explain why the price mechanism may generate negative externalities.   [12]
    (b) Discuss what policy measures a government could use to reduce pollution.   [13]

3. Discuss whether economic efficiency can be improved if governments are involved in the regulation and provision of goods and services when there is market failure.   [25]
    Cambridge 9708 Paper 41 Q2 May/June 2011

Theory and Measurement in the Macroeconomy

Revision Objectives

After you have studied this chapter, you should be able to:

☞ explain the use of national income statistics as measures of economic growth and living standards

☞ describe the GDP deflator

☞ compare economic growth rates and living standards over time and between countries

☞ assess other indicators of living standards and economic development

☞ distinguish between broad and narrow money supply

☞ assess the government's budget position and deficit financing

☞ explain the circular flow of income

☞ distinguish between the Keynesian and monetarist schools of thought

☞ explain the meaning of aggregate expenditure, the components of aggregate expenditure (AE) and their determinants

☞ assess income determination using the AE-income approach and the withdraw/injections approach

☞ explain the multiplier and the accelerator

☞ describe the sources of the money supply in an open economy

☞ assess the Quantity Theory of Money

☞ compare the Liquidity Preference Theory and the Loanable Funds Theory of interest rate determination.

15.1 National income statistics

Use of national income statistics as measures of economic growth and living standards

National income (NI) statistics cover measures of the country's output in a year.

Gross Domestic Product (GDP) is the total output produced in a country in a given time period. It can be measured in three ways. These are the output, income and expenditure methods. As these names suggest, the output method measures output directly, the income method measures all the factor payments earned in producing output and the expenditure method totals up all spending on domestically produced products.

Gross National Product (GNP) is the total output produced by the country's population wherever they produced it. Net property income from abroad is added to GDP to calculate GNP. Net property income from abroad covers profit, interest, dividends and rent earned on the ownership of foreign assets minus payments on assets in the country owned by foreigners.

Gross National Income (GNI) is GDP plus primary incomes received from the rest of the world minus primary incomes payable to non-residents. Primary income includes compensation of employees (employment income from cross-border and seasonal workers), taxes less subsidies on production and property, and entrepreneurial income. GNI is very similar to GNP – it just makes two further adjustments in relation to GDP.

Net National Product (NNP) is GNP minus depreciation. So whilst GNP includes all investment, NNP only includes net investment that is additions to the capital stock.

Net Domestic Product (NDP) is GDP minus depreciation. It can also be calculated as NNP minus net property income from abroad.

National income figures are initially measured at market prices. This means they are measured in terms of the prices charged in shops and other outlets. The figures are then converted into factor cost, i.e., in terms of the factor incomes earned in producing the products. To convert NI figures from market prices to factor cost, indirect taxes are subtracted and subsidies are added.

National income figures are also measured at both current prices and constant prices. GDP at current prices (also referred to as money GDP) is measured in terms of the prices operating in the year in question. GDP at constant prices (also called real GDP) is GDP adjusted for inflation.

GDP deflator

The GDP deflator is the price index used to adjust money GDP and real GDP. GDP deflator includes price changes of both consumer and capital goods and takes into account the prices of exports but not the prices of imports.

Comparisons of economic growth rates and living standards over time and between countries

Actual (short run) economic growth is measured by changes in real GDP. Living standards have traditionally been compared using real GDP per head (also referred to as real GDP per capita). More recently, GNI per head has become a popular measure.

Real GDP per head and real GNI per head are readily available indicators of living standards but they have a number of limitations. They are narrow measures as they only take into account one factor that influences living standards – income. They do not consider, for instance, pollution, the types of products pollution, leisure hours and political freedom. GDP/GNI per head may also be high but if income is very unevenly distributed, only a small proportion of the population may benefit. GDP/GNI per head may also be low but living standards might be relatively high if there is a large informal economy with significant unrecorded economic activity.

Other indicators of living standards and economic development

The Human Development Index (HDI), published by the United Nations, measures more indicators of living standards than real GDP/GNI per head. It includes three major components. These are education (measured since 2010 as mean years of education received by adults aged 25 and expected years of schooling for children of school going age), life expectancy at birth and GNI per head.

The Index of Sustainable Economic Welfare (ISEW) is a wider composite measure. It takes into account income inequality and adds items that increase economic welfare, such as the value of unpaid housework, and deducts items which reduce economic welfare including pollution, crime and traffic accidents.

15.2 Money supply

A country's money supply is the total amount of money in an economy. Broad money includes items which act both as a medium of exchange and store of value. Narrow money, as its name suggests, is narrower in scope and focuses on money as a medium of exchange.

Government accounts – government budget, deficit financing

The government's budget shows the relationship between its revenue, largely tax revenue, and its expenditure. A budget surplus arises when government revenue exceeds government expenditure; a balanced budget when government revenue equals government expenditure; and a budget deficit when government expenditure exceeds government revenue.

If a government spends more than it raises in tax revenue, it can finance the deficit in four main ways. It can borrow from the central bank (sometimes referred to as resorting to the printing press), from commercial banks, from the non-bank sector and from abroad.

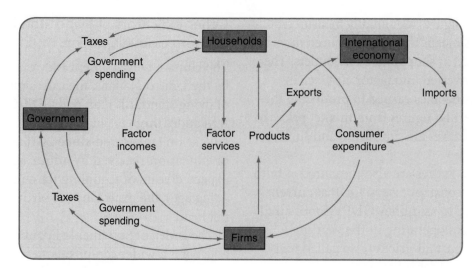

Figure 15.1

Borrowing from the non-bank sector by selling government securities to households and firms does not increase the money supply. It makes use of existing money. The other three methods do, however, increase the money supply.

15.3 The circular flow of income between households, firms, government and international economy

The circular flow of income describes how expenditure and income moves around an economy and the relationship between firms, the government and the international economy as shown in Figure 15.1.

Injections are additions to the circular flow. The three injections are investment, government spending and exports. Leakages are withdrawals from the circular flow. The three leakages are saving, taxes and imports. Figure 15.2 illustrates injections and leakages.

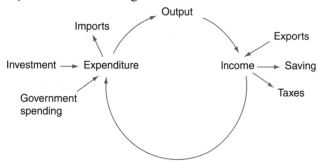

Figure 15.2

15.4 Main schools of thought on how the macroeconomy functions: Keynesian and Monetarist

Keynesian economists think there is a high risk of market failure and that there is no guarantee that economies will operate at full capacity. In contrast, monetarists think markets usually work efficiently and that economies will either be operating at full employment or moving towards full capacity.

Monetarists think that inflation is caused by an excessive growth of the money supply. Keynesians argue that inflation can cause an increase in the money supply, with firms and households borrowing more money to keep pace with higher costs and prices. Keynesians think inflation may be caused by cost-push or demand-pull inflation.

Monetarists think that government borrowing can result in crowding out with higher government spending leading to a reduction in firms' investment. If the government borrows more, private sector firms may find it more difficult and expensive to obtain funds. Keynesians, however, claim that government borrowing may result in crowding in, encouraging more investment. If higher government spending causes income to rise, more saving will provide the finance for investment and higher consumer expenditure will encourage more investment.

15.5 Aggregate expenditure function

Aggregate expenditure (AE) is the total planned expenditure at different levels of income. It is composed of consumer expenditure (consumption), investment, government spending and net exports.

Income determination using AE – income approach and withdrawal/injection approach

Income is determined where aggregate expenditure equals output. Figure 15.3 shows this would be at an output of Y. If aggregate expenditure rises, output will increase until the full employment level of output (YFe) is reached.

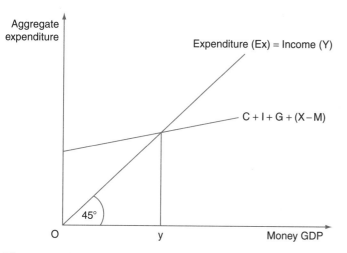

Figure 15.3

The equilibrium level of income is also where planned injections equal planned leakages (also called withdrawals), in other words, where I + G + X = S + T + M. An increase in injections would cause income to raise until injections again equal withdrawals as shown in Figure 15.4.

Inflationary and deflationary gaps

An inflationary gap occurs when aggregate expenditure exceeds the full employment level of income as shown in Figure 15.5. At YFe, there is an inflationary gap of ab.

A deflationary gap is experienced when aggregate expenditure is below the full employment level of income. Figure 15.6 shows a deflationary gap of cd.

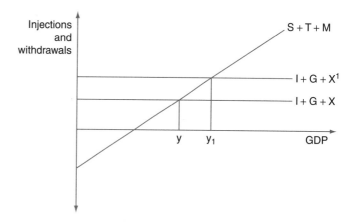

Figure 15.4

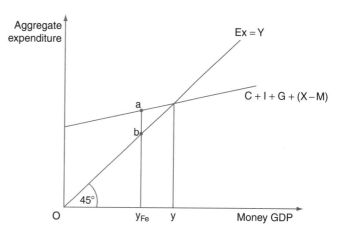

Figure 15.5

The multiplier

The multiplier is the relationship between a change in an injection and the final change in GDP. A multiplier of 3, for instance, means that an increase in investment of $10 million would result in an increase in GDP of $30 million.

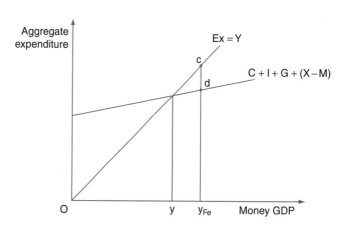

Figure 15.6

The multiplier can be calculated after a rise in autonomous expenditure has worked through the economy by using the following formula.

$$k = \frac{\text{change in GDP}}{\text{change in injection}}$$

It can also be estimated before a change in autonomous spending by using the following formula.

$$k = \frac{1}{\text{mps} + \text{mrt} + \text{mpm}}$$

Here mps is the marginal propensity to save, mrt is the marginal rate of taxation and mpm is the marginal propensity to import. This is the full version of the multiplier. In a closed economy with no government sector, the multiplier is 1/mps and in a closed economy with a government sector it is 1/mps + mrt.

Figure 15.7 shows that the multiplier is Y1-Y/r-s.

Autonomous and induced investment – the accelerator

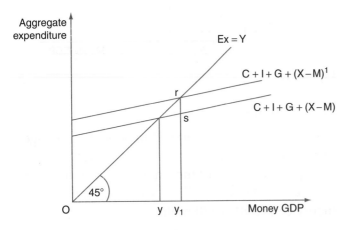

Figure 15.7

Autonomous investment is investment which occurs independently of changes in income. Advances in technology, for instance, might result in more investment at the same level of income.

Induced investment is investment undertaken because of an increase in income. With more income, firms would expect consumer expenditure to rise and so would be likely to undertake more investment.

The accelerator theory suggests that net investment depends on the rate of change in GDP and that a change in GDP will cause a greater percentage change in net investment.

The money supply may be increased by commercial banks lending more, a government financing a budget deficit by borrowing from the central bank and commercial banks, and more money entering than leaving the country.

When someone deposits money in a bank, it enables the bank to use it as the basis for loans of a greater value. Banks can lend more and so create money, because only a small proportion of deposits are cashed. Most deposits are transferred within the banking sector. The credit multiplier shows the value of new deposits that can be created as a result of a change in a bank's liquid assets. It is calculated by the the following formula.

$$\frac{100}{\text{liquidity ratio}}$$

If a government borrows from the banking sector or abroad, it will add to the money supply. If, however, it borrows from the non-bank, domestic sector it will be making use of existing money.

The total currency flow is the overall balance on the current, capital and financial accounts excluding the reserves. If there is a total currency flow surplus, the money supply will increase.

Relationship between money supply, price level and output as explained by the Quantity Theory of Money

The Fisher equation of exchange is MV = PT (sometimes expressed as MV = PY). M is the money supply, P is the general price level, V is the velocity of circulation and T or Y is the value of transactions/output.

Monetarists developed the Fisher equation into the Quantity Theory by assuming a change in the money supply will not affect V and T. With V and T constant, a change in the money supply will cause a proportional change in the price level.

Keynesians reject the Quantity Theory, arguing that V and T can be influenced by changes in the money supply and so the equation cannot be used to predict how a change in the money supply will affect the price level.

15.7 The demand for money

Interest rate determination; Liquidity Preference Theory and Loanable Funds Theory

The Loanable Funds Theory suggests that the rate of interest is determined by the demand for loanable funds (from households, firms and the government wanting to borrow) and the supply of loanable funds (savings). If savings increase, the availability of loanable funds rises. With greater availability of funds to borrow the rate of interest will fall. A fall in investment would also reduce the rate of interest. In this case, it would be because demand for funds to borrow would decline.

In contrast, according to the Liquidity Preference Theory, the rate of interest is determined by the demand for and supply of money. There is thought to be three main reasons why households and firms demand money (choose to hold their wealth in a money form). These are the transactions motive (money held to buy products), precautionary motive (money held in case of unexpected expenses and opportunities) and the speculative motive (money held to take advantage of changes in the price of government bonds and the rate of interest). The Liquidity Preference Theory claims that a fall in the rate of interest can be caused by two possible changes. One is an increase in the money supply and the other is a fall in the demand for money.

15.8 Mind maps

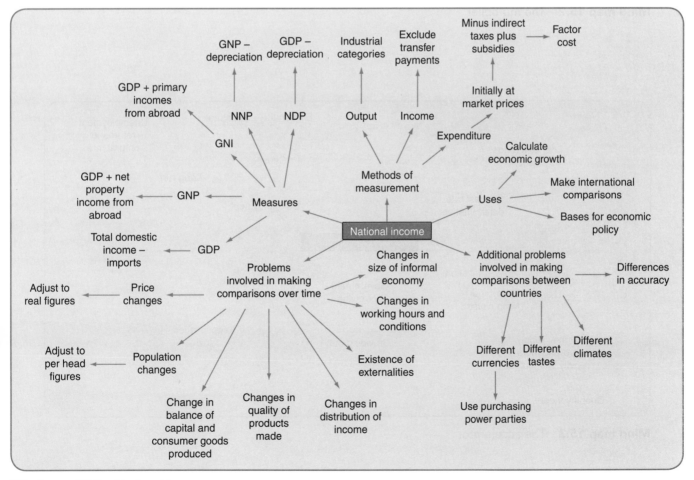

Mind map 15.1   National income

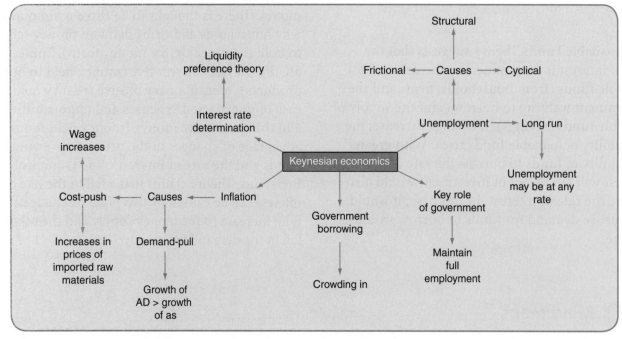

Mind map 15.2 The multiplier

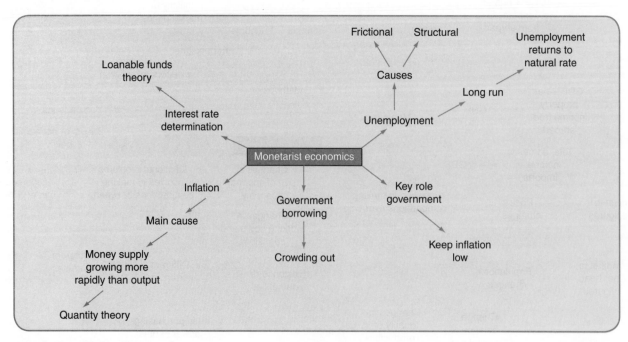

Mind map 15.3 The accelerator

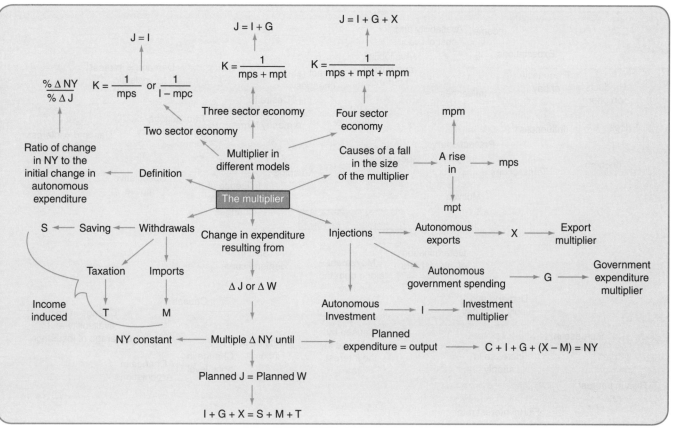

Mind map 15.4   Liquidity preference

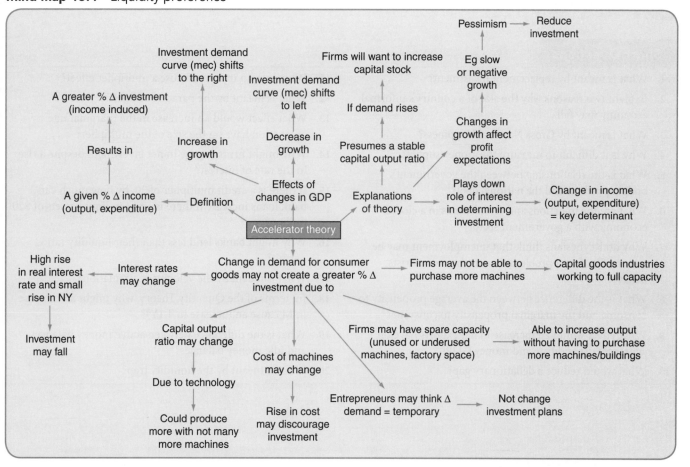

Mind map 15.5   Keynesian economics

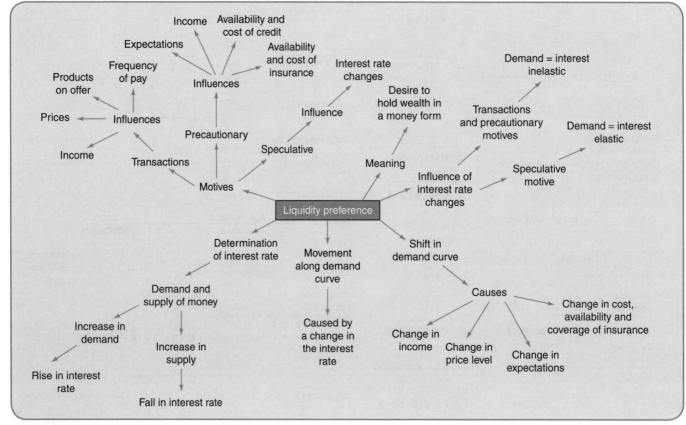

Mind map 15.6   Monetarist economics

1. What is meant by replacement investment?

2. Explain two reasons why the size of a country's informal economy may fall.

3. What is meant by Gross National Happiness?

4. Why is it difficult to measure the money supply?

5. What is the relationship between the government's budget position and the national debt?

6. What are the injections and withdrawals in a closed economy with a government sector?

7. Why do Keynesians think that unemployment may be a more significant problem than as the monetarists' believe?

8. What is the difference between the average propensity to consume and the marginal propensity to consume?

9. What effect would an increase in net exports have on aggregate expenditure and money GDP?

10. What would reduce a deflationary gap?

11. Why does an injection cause a multiplier effect?

12. What is meant by the paradox of thrift?

13. What effect would an increase in the marginal rate of taxation have on the size of the multiplier?

14. Why might firms engage in net investment despite a rise in the rate of interest?

15. If there is a credit multiplier of 20, by how much can bank loans increase if there is a rise in liquid assets of $20 million?

16. Why might banks lend less than their liquidity ratios permit?

17. What is meant by the velocity of circulation?

18. In terms of the Quantity Theory, why might an increase in M cause an increase in T (Y)?

19. What is the difference between active money balances and idle money balances?

20. What is meant by the liquidity trap?

Revision Activities

1. (a) The marginal propensity to consume is 0.75 of disposable income. Initially there is no government sector and the country is a closed economy.
        If investment is $50bn, what is the value of national income?

    (b) A government sector does develop with a marginal tax rate of 20% and government spending is $70bn. Calculate:
        (i) the new multiplier figure
        (ii) the new level of national income

    (iii) the government's budget position.

    (c) The country then engages in international trade. 1/8th or 12.5% of consumer expenditure is spent on imports and export revenue is $40bn. Calculate:
        (i) the new multiplier effect
        (ii) the new national income
        (iii) the trade balance
        (iv) the budget position.

2. Complete the following table.

	Keynesians	Monetarists
View on market failure	Significant	
View on government failure		Significant
View on Quantity Theory		Support
Cause of inflation		Excessive growth of the money supply
Main causes of unemployment		Frictional and structural
Effects of government borrowing	Crowding in	
Shape of LRAS curve	Horizontal, then upward sloping and then vertical	
Macroeconomic policy	Favour demand management	
Government intervention		To be kept to a minimum. Main responsibilities = remove market imperfections and keep inflation low

3. Decide whether the following statements relating to money and banking are true or false.
    (a) Current (sight) accounts are included in both narrow and broad measures of the money supply.
    (b) Banking is based on trust.
    (c) In banking, there is a conflict between profitability and liquidity.
    (d) The Fisher equation is a truism.
    (e) A budget deficit will always increase the money supply.
    (f) A credit crunch involves a surplus of bank loans which stimulates economic activity.

Multiple Choice Questions

1. An economy's gross domestic product is $90 billion and its net domestic product is $75 billion. What can be concluded from this information?
    A Consumer expenditure was $15 billion
    B Depreciation was $15 billion
    C Net exports were $15 billion
    D Net property income from abroad was $15 billion

2. An economy's GDP increased from $20bn in 2002 to $39bn in 2012. Over the same period, population increased from 20 million to 22 million and the price level increased from 100 to 125. Assuming other influences remained unchanged, what happened to living standards over this period?

    A Decreased as real GDP fell
    B Decreased as real GDP per head fell
    C Increased as real GDP rose
    D Increased as real GDP per head rose

3. Why may an increase in national income per head measured at constant prices underestimate the change in a country's standard of living?
    A The average number of hours worked may be reduced
    B The rate of inflation may have increased
    C The size of the informal economy may have fallen
    D The size of the population may have increased

4. The table below shows the level of consumer expenditure at different levels of disposable income.

Disposable income ($ billion)	Consumer expenditure ($ billion)
100	120
200	200
300	270
400	320
500	350

What happens to the average and marginal propensities to consume as disposable income increases?

	Average propensity to consume	Marginal propensity to consume
A	Decrease	Decrease
B	Decrease	Increase
C	Increase	Increase
D	Increase	Decrease

5. What is the monetary base?
   A Notes and coins in circulation
   B Notes and coins in circulation plus the cash reserves of the banking system
   C Notes and coins in circulation, the cash reserves of the banking system and sight accounts
   D Notes and coins in circulation, the cash reserves of the banking system and sight and time deposits

6. Which of the following changes could result in a reduction in the money supply?
   A A decrease in commercial bank's liquidity ratios
   B A decrease in the rate of interest
   C An increase in the budget surplus
   D An increase in the current account surplus

7. Which way of financing a government's spending is likely to result in the greatest increase in the money supply?
   A Increasing income tax rates
   B Decreasing income tax rates
   C The sale of government bonds to the non-bank sector
   D The sale of treasury bills to the banking sector

8. Which of the following ideas is an important concept of the monetarist school of thought?
   A An increase in the money supply will increase productive capacity
   B If the money supply increases by a greater percentage than real GDP, prices will rise
   C The government cannot take any action which will reduce unemployment
   D The government can reduce unemployment by increasing aggregate demand

9. Which combination of policy measures would be considered to be a Keynesian approach to reducing unemployment?

A An increased budget deficit, increased unemployment benefits and tax cuts
B A reduction in unemployment benefits, tax cuts and deregulation
C A reduction in the growth of the money supply, privatisation and a rise in the exchange rate
D An increase in information about job vacancies, a reduction in total government spending and a rise in the rate of interest

10. According to the concept of crowding out, what effect would an increase in public sector net borrowing have on the interest rate and private sector investment?

	Interest rate	Private sector investment
A	Decrease	Decrease
B	Decrease	Increase
C	Increase	Increase
D	Increase	Decrease

11. Figure 15.8 shows an economy in which the full employment level of national income is Yfe.

Which distance represents the deflationary gap?
   A RS
   B TR
   C TS
   D XYFe

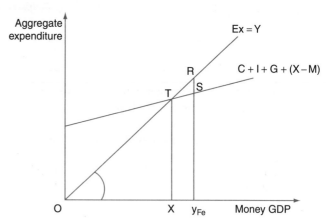

Figure 15.8

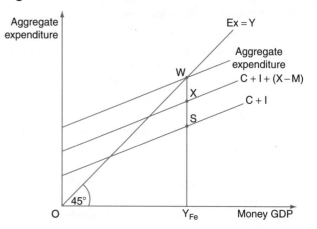

Figure 15.9

12. Figure 15.9 shows aggregate expenditure in an open economy. What does the distance WX represent?
    A An inflationary gap
    B Government spending
    C Savings
    D Taxes

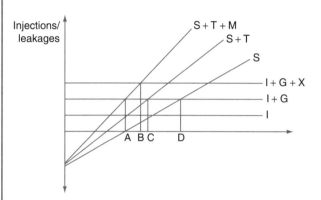

Figure 15.10

13. What is the equilibrium level of income (as shown in Figure 15.10) in a closed economy with a government sector?
    A A   B B   C C   D D

14. Which of the following is an injection into the circular flow of income in an economy?
    A Saving by citizens of the country
    B Spending by domestic firms on capital goods
    C Spending by citizens of the country on domestically produced goods
    D Tax revenue received by the government

15. If mps is 0.1, mrt is 0.05 and mpm is 0.05, what is the size of the multiplier?
    A 2
    B 4
    C 5
    D 10

16. Government spending in an economy is initially $15 billion and national income is $90 billion. Out of every increase of $100 in national income, $10 is saved, $10 is taken in taxes and $20 is spent on imports. To raise national income to the full employment level of $140 billion, by how much will the government have to increase its spending?
    A $5 billion
    B $20 billion
    C $35 billion
    D $50 billion

17. What does the accelerator theory suggest?
    A Income is a function of the growth of investment
    B Investment increases at a faster rate than saving
    C Investment is a function of the growth of income
    D Saving increases at a faster rate than investment

18. According to the Loanable Funds Theory, what will cause the rate of interest to rise?
    A A decrease in the money supply
    B A decrease in the level of savings
    C An increase in liquidity preference
    D An increase in inflation

19. Why will someone hold money for speculative reasons?
    A They expect the price of government bonds to fall
    B They expect their income will fall
    C They expect the rate of interest will fall
    D They want protection against unforeseen expenditure

20. Figure 15.11 shows the market for money is initially in equilibrium at X. Commercial banks then give more loans and the transactions demand for money increases.

    What is the new equilibrium position?
    A A   B B   C C   D D

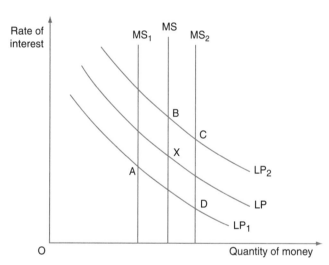

Figure 15.11

21. In a closed economy, government spending is $40m, consumer expenditure is 0.75Y and investment is $(70 – 2r) m. The full employment level of income is $400m. What rate of interest is required to obtain this level of income?
    A 2%
    B 5%
    C 10%
    D 15%

22. In an economy, 10% of extra income is paid in taxes, 10% is saved, 5% is spent on imports and the rest is spent on domestically produced products. If exports increase by $200m, what is the rise in consumer expenditure?
    A $150m
    B $160m
    C $600m
    D $640m

23. Figure 15.12 shows planned investment and planned savings in an economy at different levels of income.

What is the level of actual investment at an income level of Y?

A MNB   NPC   PYD   NY

24. According to Keynesian theory, there are a number of factors that influence demand for money including interest rates, the price level and real income. Which combination of changes in these influencing factors would always increase the demand for money?

	Interest rates	The price level	Real income
A	Decrease	Constant	Decrease
B	Decrease	Increase	Constant
C	Increase	Decrease	Decrease
D	Increase	Constant	Increase

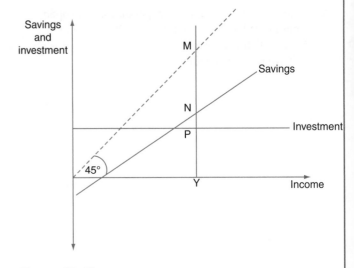

Figure 15.12

Data Response Questions

1. Interest rates, inflation and growth

Between 1st July and 1st October 2007, the GDP of the US rose at an annual equivalent rate of 4.0%. This was faster than the forecast rate of 3.1%. The rise was caused by an increase in consumer spending and by rising exports.

By November 2007, however, there were increased signs of a housing market slump, a rise in oil prices and a fall in the value of the US dollar. These changes presented the Federal Reserve (the US central bank) with a problem about interest rates.

America's housing market is slumping...

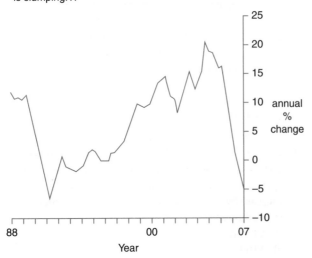

Figure 15.13   House price index

... so the Federal Reserve acted to cut borrowing costs

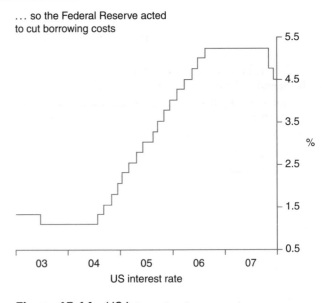

Figure 15.14   US interest rate

The Federal Reserve had already cut interest rates in October 2007 and it reduced the interest rate again in November in order to help defend the US economy against the worsening housing market.

Further interest rate cuts were thought unlikely, as there was anxiety over the rising price of oil, which by November 2007 had reached a record level. The Federal Reserve said, 'recent increases in energy and commodity prices may result in further inflation.'

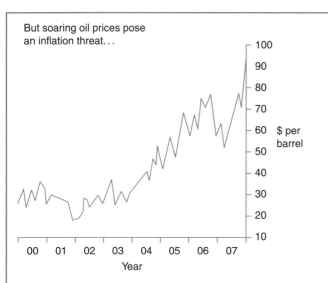

But soaring oil prices pose an inflation threat...

Figure 15.15   US Crude oil price

(a) Name two components of aggregate demand not mentioned in the first paragraph of the extract.   **[2]**
(b) Calculate the percentage increase in the GDP of the US between 1 July and 1 October 2007.   **[2]**
(c) Why does Figure 15.15 refer to a 'falling dollar' when the trend of the line is upward?   **[2]**
(d) Discuss the likely effectiveness of a reduction in interest rates as a solution to a housing market slump.   **[6]**
(e) To what extent does the data support the view that the US economy was facing 'conflicting policy objectives'?   **[8]**

Cambridge 9708 Paper 41 Q1 May/June 2010

...and so does the falling dollar

Figure 15.16   Dollars to the EURO

2. The challenges facing Brazil

In 2008, the Brazilian government thought that the economy might go into recession. The government injected extra spending into the economy to reduce its deflationary gap and to avoid a fall in real GDP. It was largely successful but it built up government debt.

In 2010, Brazil had a growth rate of 7.5%, its highest rate since 1986 and moved Brazil to the position of seventh largest economy in the world. The government was concerned about the economy overheating. Unemployment was at its lowest rate since records began, inflation was at 6% which was above the central bank's target of 4.5% and the currency was strong.

At the start of 2011, the government was concerned about the high level of aggregate demand and was worried that its large budget deficit would lead to a high interest rate. Table 1 shows the budget position and the interest rate for a number of countries. The government was planning to cut its spending and introduce some new taxes. Some economists also suggested imposing stricter rules on banks' liquidity ratios in order to reduce consumer demand.

(a) How is economic growth measured?   **[2]**
(b) Using an aggregate expenditure diagram, explain how injecting government spending into the economy might reduce a deflationary gap?   **[5]**
(c) (i) Why might a large budget deficit lead to a high interest rate?   **[3]**
    (ii) Explain whether the information in Table 1 confirms this relationship.   **[4]**
(d) Discuss whether 'stricter rules on banks' liquidity ratios' would reduce consumer demand.   **[6]**

Table 1

The budget position and interest rate of selected countries, 2010		
Country	Budget balance as % of GDP	Interest rate %
Brazil	−2.3	11.25
China	−2.2	4.77
India	−5.1	7.14
Russia	−4.6	7.75
UK	−10.1	0.80
USA	−8.9	0.26

3. Decline in Pakistan's rate of economic growth

Between 1999 and 2008 Pakistan's average economic growth rate was 7.8%. In 2009, the country's economic growth rate fell because of the global recession. Floods in 2010 and 2011 also had an adverse effect on the country's output. In 2010, for instance, more than $2.6bn worth of capital stock was destroyed.

In 2011, the country's economic growth rate was below its neighbours' economic growth rates. Some economists blamed this slowdown not only on the adverse effects of the floods but also on the country's lack of infrastructure and underdeveloped markets. Investment in Pakistan is lower as a proportion of its GDP than in its neighbouring countries. Table 2 compares the components of GDP in Bangladesh, China, India and Pakistan.

As well as the composition of the GDPs varying, the size of the four countries, multipliers also differ.

The main industries in Pakistan are textiles and food processing. The services sector is expanding although most of it is relatively small scale and some of it operates in the informal sector.

Pakistan has a number of economic advantages. In 2011 it had a relatively small negative output gap. It has a young population with an average age of 21.7, compared to 24.2 in Bangladesh and 25.1 in India. It also has a range of natural resources, including copper and gas and a potential for labour productivity to increase significantly.

(a) Explain the effect that a fall in capital stock will have on aggregate supply. [3]

(b) (i) Can it be concluded from the data in Table 1 that the output of Pakistan's economy was $163bn in 2011? Explain your answer. [3]

(ii) What was the level of domestic demand in India in 2011? [2]

(c) Analyse why the size of the multiplier varies between countries. [4]

(d) What is meant by a negative output gap? [2]

(e) Discuss whether consumer expenditure is likely to increase in Pakistan. [6]

Table 2

Country	The components of GDP in selected countries 2011				
	Consumer expenditure ($bn)	Investment ($bn)	Government ($bn)	Exports ($bn)	Imports ($bn)
Bangladesh	69	22	5	17	24
China	1695	2393	648	1346	1097
India	771	496	165	275	330
Pakistan	130	31	13	21	32

Essay Questions

1. (a) For what purposes do people demand money? [10]

(b) Discuss the effect of an increase in the supply of money on interest rates and national income. [15]

Cambridge 9708 Paper 4 Q7 May/June 2008

2. (a) Explain how the impact of the Keynesian multiplier process will change if a free-market closed economy becomes a mixed economy with foreign trade. [12]

(b) Analyse how a change to the equilibrium level of income resulting from the multiplier process might lead to unemployment or inflation. [13]

Cambridge 9708 Paper 41 Q6 May/June 2011

3. In 2010, the GDP per head of the Netherlands was four times greater than that of Chile. Discuss whether this means that people in the Netherlands enjoyed living standards four times as great as that of people in Chile. [25]

Macroeconomic Problems 16

Revision Objectives

After you have studied this chapter, you should be able to:

- ☞ distinguish between economic growth and development
- ☞ assess indicators of comparative development and underdevelopment
- ☞ describe the characteristics of developing economies
- ☞ distinguish between actual and potential economic growth

- ☞ define full employment and the natural rate of unemployment
- ☞ analyse the causes of unemployment
- ☞ discuss the consequences of unemployment
- ☞ analyse the links between macroeconomic problems.

16.1 Economic growth and development

Definition of economic growth and development

Economic growth is, in the short run, an increase in real GDP and, in the long run, an increase in productive capacity.

Economic development is an improvement in economic welfare. It is wider than economic growth and involves, for instance, a reduction in poverty, increased life expectancy and a greater range of economic and social choices.

Economic growth is often but not always associated with economic development. Higher output can lead to more employment which can move people out of absolute poverty. Increases in the quantity and quality of housing may increase life expectancy and some of the increase in tax revenue can be spent on education and health care which can also enable people to live longer and enjoy a higher quality of life. However, a country may experience economic growth but if it is accompanied by more pollution, longer working hours and worse working conditions, it may not experience economic development. Similarly, it is possible that economic development may occur without economic growth.

For instance, whilst income may not increase, a more even distribution of income may promote economic development. Greater political freedom will give people more choices. Reducing any tensions with other countries may give people a greater sense of security.

Indicators of comparative development and underdevelopment in the world economy

The United Nations classifies countries into those with very high development, high development, medium development and less development, according to the Human Development Index. It also classifies them according to high, middle and low incomes.

Characteristics of developing economies

It is important to remember that developing economies are not all the same and any one economy is unlikely to share all of the characteristics above.

Developing economies may have low income per head, high population growth, high dependency ratios, uneven distribution of income, a high proportion of output and employment accounted for by the primary sector, a reliance on a narrow range of mainly primary exports, high rates of rural to urban migration, net emigration, low levels of literacy, low life expectancy and low productivity.

Actual versus potential growth in national output

Actual (short run) economic growth involves an increase in real GDP. This can be illustrated by a movement of the production point from within a production possibility curve towards the curve or the rightward shift of the aggregate demand curve towards the full employment level.

Potential (long run) economic growth involves an increase in productive capacity. This can be illustrated by a shift to the right of a production possibility curve or of a long run aggregate supply curve.

Factors contributing to economic growth

If there is spare capacity in the economy, actual economic growth can be achieved by an increase in aggregate demand. A rise in consumer expenditure, investment, government spending or net exports may encourage producers to increase their output.

Potential economic growth enables real GDP to increase over time. It can be caused by an increase in the quantity and/or quality of resources. The quantity of resources may increase due to net investment, immigration of workers. The quality of resources may increase due to advances in technology, improved education and training.

An economic (trade or business) cycle arises when economic growth fluctuations around trend (potential) economic growth.

Governments aim for steady economic growth. This enables firms and households to plan ahead and avoids inflation (during a boom) and unemployment (during a recession).

Governments also aim for sustainable economic growth. This is economic growth which does not endanger future generations' ability to grow by depleting resources and creating pollution.

Costs and benefits of growth; including using and conserving resources

Economic growth can raise living standards, reduce unemployment, make the country more powerful in global institutions and in global negotiations, increase confidence and increase tax revenue enabling governments to spend more on education, health care and reducing poverty.

There is a possibility that living standards may fall in the short run in order to achieve economic growth. If a country is operating at full capacity, the output of some consumer goods may have to be sacrificed in order to produce more capital goods. In the long run, of course, more capital goods will enable more consumer goods to be produced.

Economic growth may be accompanied by increased pollution, stress, longer working hours, depletion of natural resources and some structural unemployment.

Economies often face the question of whether to use or conserve non-renewable resources such as oil. Using resources now will contribute to economic growth and so generate income and raise living standards. It will also contribute to the country's export earnings and so its trade balance. Exploiting resources now may be a wise decision if it is thought that demand for the resources may fall in the future, for instance, due to the development of synthetic substitutes. Conserving resources, however, may enable future generations to enjoy income from them and may mean that the country will not become dependent on other countries for resources.

16.2 Unemployment

Full employment and natural rate of unemployment

Full employment is the highest possible use of a factor of production. The term is often used in relation to labour. Full employment of labour does

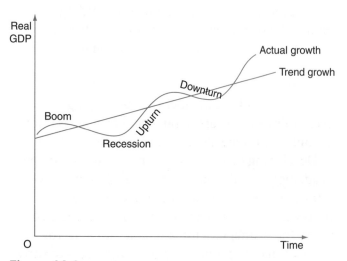

Figure 16.1

not mean zero unemployment as there will always be some people changing jobs. Some economists suggest that full employment occurs when there is approximately 3% unemployment although the rate is likely to vary from country to country.

The natural rate of unemployment (also called the non-accelerating inflation rate of unemployment (Nairu)) is the level of unemployment which exists when the labour market is in equilibrium. In this situation, the total demand for labour equals the total supply of labour and there is no upward pressure on wages and the price level.

Causes of unemployment

The three main types of unemployment are frictional, structural and cyclical unemployment.

Frictional unemployment exists even when there is full employment. It is short term unemployment which occurs when workers are in between jobs. There are many forms of frictional unemployment. These include voluntary, search, casual and seasonal unemployment.

Voluntary unemployment happens when unemployed people choose not to take up the job vacancies on offer. Search unemployment occurs when workers who have lost one job do not take the first job offered to them. Instead they search around for a better job. Casual unemployment occurs when workers have periods of employment followed by periods of unemployment; for example, film actors and festival organisers. Seasonal unemployment is when people work during certain periods of the year and then are unemployed during off peak time. For instance, tour guides will find work during peak holiday times but may be out of work for the rest of the year.

Structural unemployment lasts longer than frictional unemployment and can be on a significantly larger scale. It exists due to the occupational and geographical immobility of labour. The structure of economies is always changing with some industries and occupations declining whilst other industries and occupations are developing and expanding. If workers cannot move easily from declining to growing sectors they will be unemployed.

Structural unemployment can take the forms of international, regional and technological unemployment. International unemployment arises when workers lose their jobs because industries decline due to competition from foreign industries. Regional unemployment refers to a situation where industries and occupations decline in particular areas of the country. Technological unemployment occurs when industries and occupations disappear due to advances in technology.

Cyclical unemployment (also called demand deficit unemployment) can last for years and may be on a very large scale. It occurs due to a lack of aggregate demand and so affects most industries and occupations.

Consequences of unemployment

The effects of unemployment depend on the cause of unemployment, the duration of the unemployment, the scale of unemployment, the groups affected and the support given to the unemployed.

Unemployment can reduce inflationary pressure, can enable firms to expand and give the unemployed time to reflect on what they really want and to research job opportunities.

The costs of unemployment, however, are generally thought to far outweigh any benefits. The economy will lose potential output and so living standards will be lower than possible. The government will lose potential tax revenue and will have to spend more on unemployment benefits. The unemployed are likely to suffer a fall in income and may experience health problems. The longer people are unemployed, the more difficult they usually find it to obtain another job. This is because their skills may become out of date and firms tend to become more reluctant to employ them.

16.3 Inter-connectedness of problems

Relationship between internal and external value of money

A rise in the inflation rate will reduce the internal value of money. As a result of reducing the price competitiveness of the country's products, it is also likely to result in a fall in demand for exports and a rise in demand for imports. These changes, in turn, will reduce demand for the currency and increase its supply and so cause a fall in the exchange rate (the external value of the currency).

A rise in the external value of money will raise its internal value. An appreciation of the currency will reduce the price of imports, meaning that people will be able to buy more of them. The lower price of imported raw materials will also reduce the costs of producing some domestic products.

Relationship between balance of payments and inflation

A current account surplus may contribute to inflationary pressure by raising aggregate demand and the money supply.

A higher inflation rate than rival countries may cause an economy to experience deterioration in its current account position. This is because its products will become less price competitive.

Relationship between inflation and unemployment

The traditional Phillips curve suggests that it is possible for a government to reduce unemployment by increasing aggregate demand but at the expense of a higher inflation rate. Figure 16.2 shows an economy initially operating with a 9% unemployment rate and a 4% inflation rate. Expansionary fiscal or monetary policy may reduce unemployment to 3% but at the cost of an increase in the inflation rate to 11%.

The expectations augmented Phillips curve suggests that there is no long run trade off between

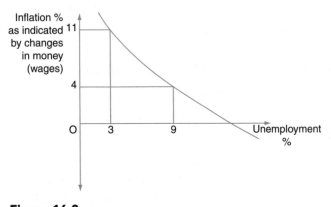

Figure 16.2

unemployment and inflation. It implies that unemployment may be reduced in the short run as a result of an increase in aggregate demand but in the long run, unemployment will return to the natural rate. Figure 16.3 shows that the natural rate of unemployment is 6%. An attempt to reduce unemployment to 4% by increasing aggregate demand may succeed in the short run at the cost of an inflation rate of 3%.

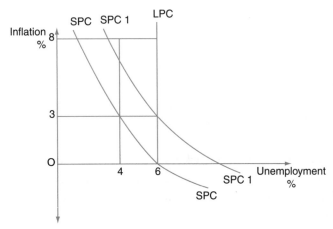

Figure 16.3

In the longer run, however, the economy will return to the natural rate of unemployment as workers and firms adjust to the new inflation rate. Some workers realising their real wages have not risen may resign. Others will press for wage rises which will raise firms' costs causing them to lower output and employment. The economy is now on a higher short run Phillips curve with expectations of inflation built into the system. A new attempt to reduce unemployment to 4% would push up the inflation rate to 8%.

Some economists argue that with increased global competition and advances in technology it is possible to reduce unemployment close to full employment without causing inflation.

A reduction in inflation reduces unemployment as it tends to increase international competitiveness of the country's products. This may increase net exports and so raise aggregate demand and reduce cyclical unemployment.

16.4 Mind maps

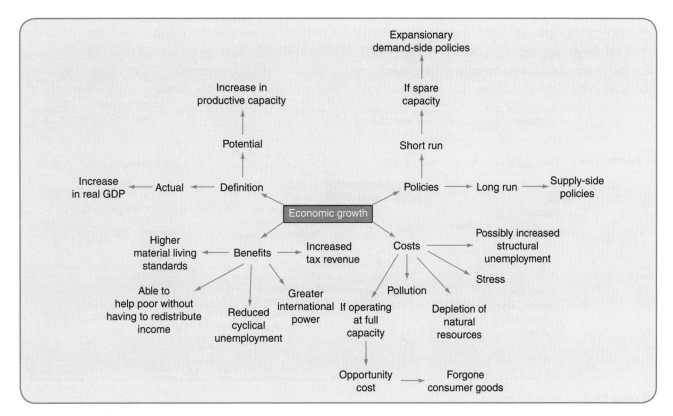

Mind map 16.1   Economic growth

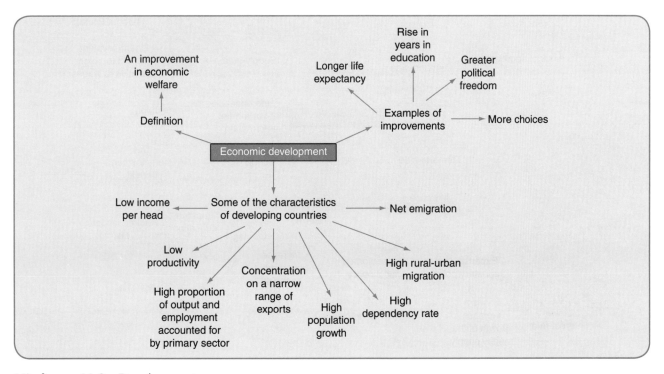

Mind map 16.2   Development

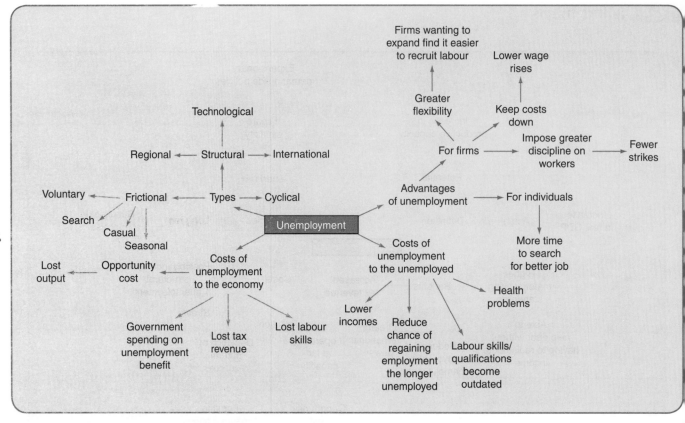

Mind map 16.3 Unemployment 1

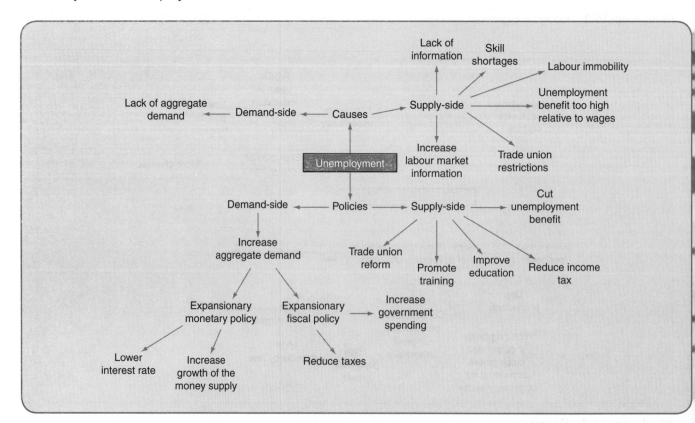

Mind map 16.4 Unemployment 2

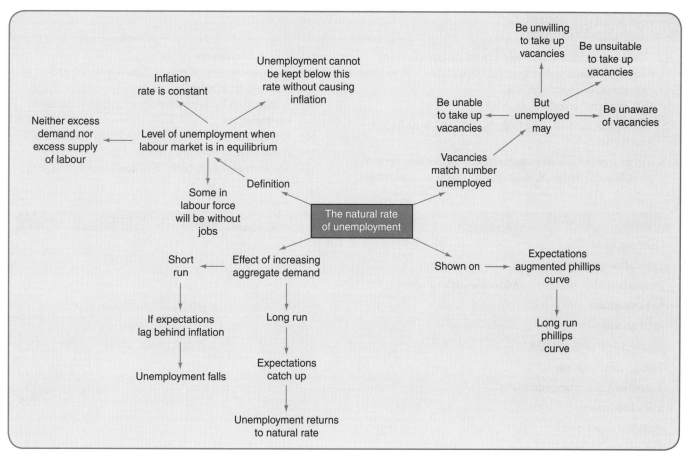

Mind map 16.5 Natural rate of unemployment

Short Questions

1. Identify the two main reasons why a country may have a higher GNI than another country but a lower value on the HDI.

2. Why do countries with a low income per head tend to have a high birth rate?

3. Explain an advantage a rich country could gain from providing foreign aid to a poor country.

4. How is the Malthusian theory of population related to the law of diminishing returns?

5. What is meant by a natural increase in population?

6. How is the concept of optimum population linked to economic development?

7. Distinguish between a supply constraint and a demand constraint in connection with economic growth.

8. Why is investment so significant in connection with economic growth?

9. Why might zero net investment be associated with economic growth?

10. Explain two reasons why external debt might hinder economic growth.

11. Identify two disadvantages of cutting down rainforests to increase real GDP.

12. What is meant by voluntary unemployment?

13. Explain two factors that could reduce frictional unemployment.

14. Why might unemployment increase despite a rise in real GDP?

15. Why might advances in technology create rather than destroy jobs?

16. What is meant by discouraged workers?

17. Why may the internal value of money decline but its external value rise?

18. What does the slope of the traditional Phillips curve suggest?

19. What does the long run Phillips curve show?

20. What would a traditional Phillips curve show?

Revision Activities

1. Using a production possibility curve diagram and an aggregate demand and aggregate supply diagram, explain:
   (a) actual economic growth
   (b) potential economic growth.

2. Complete the following table by writing low or high in each box.

3. In each of the following cases, decide whether the changes would cause frictional, structural or cyclical unemployment.

(a) A global recession
(b) The introduction of new technology which reduces the need for labour
(c) A reduction in the availability of information about job vacancies
(d) A reduction in business and consumer confidence
(e) A decrease in demand for fish
(f) The increased popularity of biofuels over oil as a fuel.

Characteristic	Developed economy	Developing economy
GDP per head		
Population growth		
Proportion of population employed in agriculture		
Urbanisation		
HDI ranking		
Foreign debt as % of GDP		
Energy consumption		
Enrolment in tertiary education		
Life expectancy		
Population per doctor		
Labour productivity		
Savings ratio		

Multiple Choice Questions

1. What is a common characteristic of a developing country?
   A A large population
   B A large primary sector
   C A low infant mortality rate
   D A low rate of rural – urban migration

2. What is included in the Human Development Index?
   A The level of pollution
   B The distribution of income
   C Time spent in education
   D Time spent on leisure activities

3. A multinational company based in India owns a subsidiary company in Germany which in 2013 made a profit of $11m. $6m of that profit was set back to the parent company in India and the remaining profit went towards the financing of a $13m investment project in Germany. The remaining finance for the investment project was borrowed from banks, $3m coming from German banks and the remaining money coming from India's banks. What effect would these transactions have had on India's current account, financial account and net currency flow?

	Current account	Financial account	Net currency flow
A	+ $11m	−$13m	+ $6m
B	+ $5m	−$7m	−$2m
C	+ $6m	−$5m	+ $1m
D	+ $8m	−$3m	−$6m

4. Selected data on four countries

	GNI per head ($)	Life expectancy (years)	Time spent in education (years)
Country A	1,800	45	6
Country B	2,000	62	10
Country C	2,200	82	14
Country D	2,400	70	10

On the information provided, which country had the highest standard of living?

A A   B B   C C   D D

5. Why may developing countries seek to rely less on primary products?

A The high price elasticity of primary products

B The high income elasticity of demand for primary products

C The low cross elasticity of demand between primary products and their artificial substitutes

D The low price elasticity of supply of primary products

6. What does an optimum population allow an economy to achieve?

A An even distribution of income

B Equality between the birth rate and the death rate

C Full employment of resources

D The highest possible level of real income per head

7. What is meant by the optimum population?

A The largest population of any country in the world

B The largest population that can be supported with existing resources

C The population which gives the highest average output with existing resources

D The population which will exist in the absence of immigration

8. A country's real GDP per head rose but its HDI fell. What could explain this combination of changes?

A Life expectancy decreased

B Pollution increased

C Population increased

D Working hours increased

9. What could cause an economy to experience a rise in real GDP despite a decrease in its productive potential?

A A decrease in consumer expenditure

B A decrease in investment

C An increase in taxation

D An increase in the budget deficit

10. Which change would directly cause long run economic growth?

A A shift to the left of the aggregate demand curve

B A shift to the right of the aggregate demand curve

C A shift to the left of the long run aggregate supply curve

D A shift to the right of the long run aggregate supply curve

11. The growth in demand for electronic books results in a number of bookshops going out of business. How is the type of unemployment of bookshops classified?

A Casual

B Cyclical

C Search

D Structural

12. What is meant by the natural rate of unemployment?

A The level of unemployment which exists when the demand for and supply of labour are equal

B The level of unemployment which exists when there is zero economic growth

C Unemployment caused by a change in the economic structure in an economy

D Unemployment caused by a lack of aggregate demand

13. What would be most likely to reduce an economy's natural rate of unemployment?

A An increase in the gap between paid employment and unemployment benefit

B An increase in the gap between potential and actual output

C An increase in net exports

D An increase in the money supply

14. What is the opportunity cost to the economy of unemployed workers?

A Imports of products purchased

B State benefits paid to the unemployed

C The extra leisure time some of the unemployed may gain

D The output which the unemployed would have produced if they had been in work

15. A country experiences an increase in cyclical unemployment. The government does not change tax rates or rates of unemployment benefit. What will be the effect of the higher unemployment on government spending and tax revenue?

	Government spending	Tax revenue
A	Decrease	Decrease
B	Decrease	Increase
C	Increase	Increase
D	Increase	Decrease

16. Which change would cause inflation in a fully employed economy?

A An increase in government spending on pensions

B An increase in labour productivity

C An increase in saving

D An increase in imports

17. What will be the effect of a country's potential output increasing more rapidly than its actual output?

A An increase in the inflation rate

B An increase in the size of the output gap

C A decrease in the unemployment rate

D A decrease in real GDP

18. Which of the following changes would increase the circular flow of income?

A A decrease in government spending

B A decrease in investment

C An increase in exports

D An increase in savings

19. An economy is experiencing demand-pull inflation, a current account deficit, a falling exchange rate and unemployment. The government decides to increase its spending. What does this indicate is its main, short term, macroeconomic aim?

    A   To achieve price stability

    B   To improve the current account position

    C   To increase the value of the currency

    D   To reduce unemployment

20. What effect would a high and accelerating rate of inflation be expected to have on the internal and external value of money?

	Internal value of money	External value of money
A	Decrease	Decrease
B	Decrease	Increase
C	Increase	Increase
D	Increase	Decrease

Data Response Questions

1. **Tourism and local workers**

   Here are two accounts of the tourist industry in Africa.

   Article 1

   The Gambia, on the west coast of Africa, is ranked 160th out of 173 on the United nations Human Development Index. Over half the population lives on less than one US dollar a day and survives on subsistence agriculture and cannot compete with subsidised American farmers.

   However, things are changing as The Gambia expands its tourist sector, which is now a significant part of the national economy, accounting for 7.8% of GDP. It employs 5,000 directly and creates work for 6,000 others. Large European tour operators play a vital role in marketing, assuring quality, and providing flights and accommodation. Tourism has the major advantage that developed countries cannot place tariff barriers on tourist exports. Tourists spend on average US$40 to US$50 a day on meals, souvenirs, crafts and tours. One third is spent in the informal sector, where income is not recorded, providing a livelihood for taxi drivers, craft workers and local guides.

   However, there are disadvantages. Tourism is highly seasonal, and tour operators negotiate low prices which keep profit margins low – so low, in fact, that many hotels have closed because they could not cover costs.

   Article 2

   A union official in Tanzania said that the tourism sector is expanding in Tanzania but the return to the country's economy is low and the benefit to the workers is about 0.5% of the total industry's income. In hotels, the lowest wages are around US$50 a month, from which tax and rent have to be deducted. Someone who has been working 5 years or more receives no extra pay or promotion. Contracts are short-term lasting for a year.

   Hotel operators oppose workers joining trade unions. Most tour operators come from outside Tanzania. Only 10% of each US dollar earned by the tourist industry remains in the country, and most of that goes to the management not the workers. Top managers are usually foreign workers paid two or three times as much as a Tanzanian manager.

   (a)   What evidence is there that The Gambia is a developing country? [4]

   (b)   What does Article 1 mean when it says that tourism created work for 6,000 others? [3]

   (c)   Article 2 mentions the low pay of hotel workers. Why might many hotel workers receive low pay? [5]

   (d)   Does the evidence provided enable you to conclude that tourism merely exploits resources and is of little benefit? [8]

   Cambridge 9708 Paper 4 Q1 Oct/Nov 2007

2. **Poverty in the USA**

   In 2000, the US was experiencing positive economic growth. Output increased most rapidly in the tertiary sector which employs 80% of the labour force. The country did, however, have an unemployment rate of 10.1% and was experiencing a relatively high poverty rate. The proportion of people living in poverty in 2008 was 13.2%. In 2010, it was reported that 46.5m out of the population of 310m were living in poverty. This figure was based on a poverty line of $22,025 for a family of four. That same year the average income in the country was $52,029.

   In a country which was ranked 4th in the Human Development Index in 2010, the number of homeless people was increasing in many cities like Detroit and Las Vegas. The number of children in these who qualify for free school meals increased to nearly 50%. Economists expressed concern that children were growing up in poverty as a significant proportion of these children are likely to be poor when they are adults.

   The problem of such a high proportion of the population living in poverty was surprising in a country which is classified as a developed country. The country had an average life expectancy of 79.6 years in 2010 but the poor have a shorter life expectancy.

(a) What happened to the percentage of people living in poverty in the US between 2008 and 2010?   [2]

(b) Explain the link between unemployment and poverty.   [3]

(c) What evidence is there in the extract that USA is a developed country?   [5]

(d) Explain why the children of the poor tend to be poor when they are adults.   [4]

(e) Discuss whether achieving the macroeconomic objectives of full employment and economic growth would reduce poverty.   [6]

3. The following extract is adapted from an article that appeared in 2003.

Diversification in Botswana

The Botswana government wishes to use the wealth that the country obtains from diamonds to diversify its economy. Botswana is the world's largest producer of diamonds, which accounted for 77% of the country's exports earnings and 45% of GDP in 2002.

The diamond industry is controlled by one company, Debswana, which is a partnership of De Beers (a private sector company) and the Botswana government. Each owns 50% of Debswana but De Beers keeps 75% of the profits.

Botswana has various incentives for investors. The corporate tax is one of the lowest in Africa at 15% and profits can easily be sent back to the home country because there are no exchange controls. Wage rates are relatively low and the workforce is the most educated in Africa. Botswana is also perfectly located to become a financial hub for the 200 million people in the 14 African countries in the Southern African Development Community.

However, there are problems for investors. Usable water is in short supply and transport costs are high. Labour is also in short supply because the population is only 1.7 million and life expectancy is only 39.

Despite its problems, Botswana remains an example of prosperity in conflict-ridden Africa. US President George W Bush said 'Botswana has demonstrated sure, sound economic administration and a commitment to free market principles.'

(a) Identify two advantages of Botswana for foreign investors.   [2]

(b) What evidence is there in the article that Botswana is a developing country?   [4]

(c) What type of market structure exists in the Botswana diamond industry? Use the article to explain your answer.   [3]

(d) President Bush said 'Botswana has demonstrated a commitment to free market principles.' What did he mean, and does the article support this view?   [5]

(e) Discuss the advantages to a developing economy if it becomes more diversified.   [6]

Cambridge 9708 Paper 4 Q1 May/June 2005

Essay Questions

1. (a) Explain the differences between economic development and economic growth.   [12]

(b) Discuss whether multinational companies promote economic growth in developing countries.   [13]

2. (a) Explain how a government might seek to reduce the natural rate of unemployment.   [12]

(b) Discuss whether a decrease in unemployment will cause inflation.   [13]

3. For some years governments of developed countries have been promoting Fair Trade, which means paying a fair price for primary products bought from African developing countries. Now the governments of developed countries, anxious to conserve resources, are complaining that the transport of products around the world increases pollution and should be limited. They support instead the purchase of goods produced at home. These are often more expensive to produce. African farmers may be left with products that their local people do not eat.

(a) Explain what might determine whether a country is classified as developed or developing.   [12]

(b) Discuss whether the old and the new approaches to trade of the developed countries would help achieve the conservation of resources.   [13]

Cambridge 9708 Paper 41 Q5 May/June 2010

Macroeconomic Policies

17

Revision Objectives

After you have studied this chapter, you should be able to:

☞ describe the objectives of macroeconomic policy
☞ assess policies towards developing economies

☞ discuss fiscal, monetary, exchange rate and supply-side policies
☞ evaluate policy options to deal with problems arising from policy conflicts.

17.1 Objectives of macroeconomic policy: stabilisation, growth

Governments seek to achieve low and stable inflation, low unemployment, a balance of payments equilibrium, sustained and sustainable economic growth and avoidance of exchange rate fluctuations.

Low and stable inflation is sometimes referred to as price stability. The lowest level of unemployment possible is known as full employment.

Sustained economic growth occurs when increases in aggregate demand are matched by increases in aggregate supply. Sustainable economic growth is growth which does not damage future generations' ability to enjoy increases in output.

The avoidance of fluctuations in economic activity promotes greater certainty which in turn stimulates investment.

17.2 Policies towards developing economies; policies of trade and aid

International trade can bring a variety of benefits to developing countries, including opportunities to specialise, take advantage of economies of scale, increased aggregate demand, increased economic growth and increased employment.

Developing countries experience a number of disadvantages in international trade. They rely on the exports of primary products which have a lower income elasticity of demand than manufactured goods. The Prebisch-Singer hypothesis argues that the terms of trade move against developing countries resulting in them having to export more products to purchase the same quantity of imports. In addition, developed countries impose a number of trade restrictions on developing countries whilst putting pressure on the developing countries to remove trade restrictions against them.

The International Monetary Fund (IMF) aims to encourage the growth of world trade, promotes economic stability and helps countries with balance of payments difficulties.

The World Bank provides financial assistance for investment projects in developing countries. To obtain loans from the World Bank and the IMF, developing countries sometimes have to agree to structural adjustment programmes which require them to increase the role of market forces in their economies, to remove trade restrictions and to increase macroeconomic stability.

The World Trade Organisation seeks to promote free trade. It encourages member countries to remove trade restrictions and it presides over trade disputes between members.

Governments may have a number of motives for giving aid. These include altruistic, political and commercial reasons. Aid may be tied or untied, bilateral or multilateral. Untied, multilateral aid tends to be the most beneficial. Aid may compensate for a lack of saving and may provide the finance for investment in capital goods and human capital and so contribute to economic growth and development.

There is a risk that countries may become dependent on aid. Funds may flow from developing to developed countries, if the servicing of the debt arising from aid is greater than the new aid received and if the aid is used for unproductive projects.

17.3 Types of policy

Fiscal policy

Fiscal policy covers government decisions on taxation and government spending. It aims to influence aggregate demand. Reflationary/expansionary fiscal policy seeks to increase the aggregate demand whereas deflationary/contractionary fiscal policy seeks to reduce the aggregate demand.

Discretionary fiscal policy covers government decisions to change government spending and taxes rates and/or tax coverage.

Automatic stabilisers are changes in government spending and tax revenue that occur as a result of changes in real GDP and which offset economic fluctuations.

To reduce cyclical unemployment and to increase economic growth when there is a negative output gap, a government may implement reflationary fiscal policy by increasing government spending and/or reducing taxes.

To reduce demand-pull inflation and to reduce a current account deficit, a government may implement deflationary fiscal policy by reducing government spending and/or increasing taxes.

A government imposes taxes for a variety of reasons. As well as seeking to influence economic activity, taxes are levied to raise revenue, to discourage the consumption of demerit goods, to reduce income and wealth inequalities and to dissuade people from buying imports.

Direct taxes are taxes on income and wealth. They are paid directly to the tax authorities. Indirect taxes, sometimes known as outlay taxes, are taxes on spending. The incidence of indirect taxes is often shared between producers and consumers.

Taxes are progressive, proportional or regressive. A progressive tax is one which takes a higher percentage of the income of the rich; a proportional tax takes the same percentage as income rises and a regressive tax takes a larger percentage of the income of the poor. In the case of a progressive tax, the marginal rate of tax is higher than the average rate of tax. The marginal and average tax rates are the same in the case of a proportional tax. In the case of a regressive tax, the marginal rate of tax is lower than the average rate.

The canons of taxation are the principles of taxation. Adam Smith identified four qualities of a good tax: equity, certainty, convenience and economy. Since Adam Smith's time, two additional qualities have been added. These are efficiency and flexibility.

Government spending can be divided into current expenditure, capital expenditure and transfer payments. Current expenditure is spending on the day to day running of, for instance, state owned hospitals and schools. Capital expenditure is government spending on the building of schools and hospitals. Transfer payments are state benefits.

There are various upward pressures on government spending. These include an ageing population, advances in medical technology, military conflicts and increases in the number of students going to university.

The poverty trap is a situation where people find it difficult to move out of poverty because when their income rises they lose state benefits and begin to pay income tax.

Monetary policy

Monetary policy covers government and/or central bank decisions on money supply and the rate of interest.

Monetary policy aims to influence aggregate demand. Reflationary/expansionary monetary policy seeks to increase aggregate demand whilst deflationary/contractionary monetary policy seeks to reduce aggregate demand.

To reduce cyclical unemployment and to increase economic growth when there is a negative output gap, the money supply may be increased or the rate of interest reduced.

To reduce demand-pull inflation and to reduce a current account deficit, the growth of the money supply may be reduced or the rate of interest increased.

Among the measures a central bank can take to influence the money supply are open market operations. These include the sale and purchase of government bonds. If the central bank sells government bonds, it will lower their price and so raise the rate of interest and will reduce banks' liquid assets and so their ability to lend. If, on the other hand, the central bank buys government bonds, it will raise their price which will lower the rate of interest. It will also increase banks' liquid assets and so increase their ability to lend.

Exchange rate policy

The manipulation of exchange rates is sometimes included in monetary policy and is sometimes treated as a separate policy.

A fall in the exchange rate may improve the current account position if the combined PED for exports and imports is greater than one. A lower exchange rate may reduce unemployment and increase short run economic growth.

To reduce demand-pull inflation, a government or central bank may seek to raise the exchange rate. A government or central bank may try to avoid significant fluctuations in the exchange rate in order to provide trade and foreign direct investment.

Supply side policy

Supply side policy seeks to increase aggregate supply by increasing the quantity and quality of resources. Among supply side measures are education, training, cuts in income tax, cuts in unemployment benefit, cuts in corporation tax, privatisation, deregulation and a reduction in trade union power.

Supply side policy has the potential to improve all the government's macroeconomic objectives in the long run.

17.4 Evaluating policy options

Some fiscal policy measures work automatically to offset economic fluctuations, fiscal policy measures can have a multiplier effect and some can affect both aggregate demand and aggregate supply.

Fiscal policy measures can experience a time lag, may have unexpected results, the multiplier may be

miscalculated, some forms of government spending are inflexible, some fiscal policy measures may have an adverse effect on incentives and some may have adverse side effects.

The rate of interest can be changed relatively quickly, can influence three of the components of aggregate demand and may affect aggregate supply by altering investment.

A change in the rate of interest can take between eighteen months and two years to work through the economy and affect aggregate demand, commercial banks may not pass on interest rate changes to their customers, firms and households may not react in the way expected, there may be adverse side effects and if interest rates are already low, a further fall will not have much of an impact.

It can be difficult to decide what to include in measures of the money supply and it can be difficult to control the growth of the money supply.

A change in the exchange rate can offset changes in the domestic price level and influence aggregate demand. Attempts to influence the exchange rate can be offset by movements in other countries' exchange rates and by speculation.

Supply side policy measures can be selective and can increase the efficiency of markets. Increasing aggregate supply whilst there is a lack of aggregate demand will not raise economic performance. Some supply side measures take a long time to have an effect, are expensive and some may not work.

Policies that increase aggregate demand may increase short run economic growth and reduce unemployment but may increase demand-pull inflation. Promoting economic growth may increase a current account deficit.

Increasing incentives for firms and workers may result in a more uneven distribution of income. Increasing economic growth may reduce absolute poverty but increase relative poverty. To achieve a range of macroeconomic policy objectives, a government may need to implement a range of policy measures.

If an economy is experiencing both a high inflation rate and a high unemployment rate, a government may decide both to raise the rate of interest and to spend more on training.

17.5 Mind maps

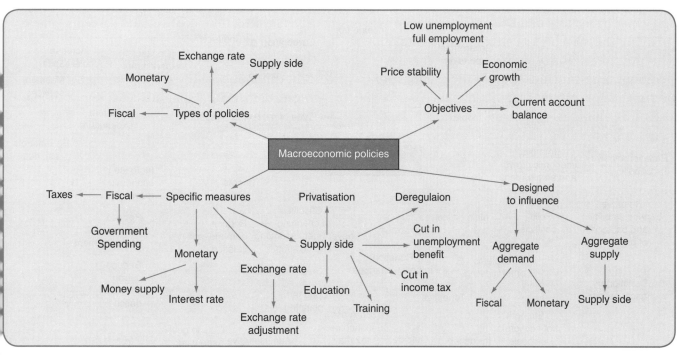

Mind map 17.1 Macroeconomic policies

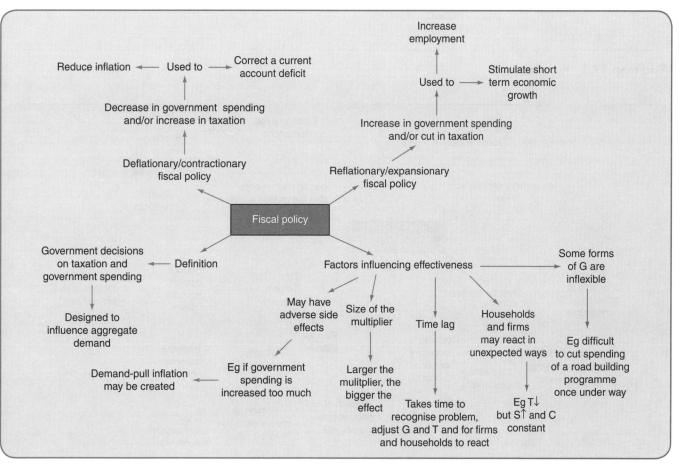

Mind map 17.2 Fiscal policy 1

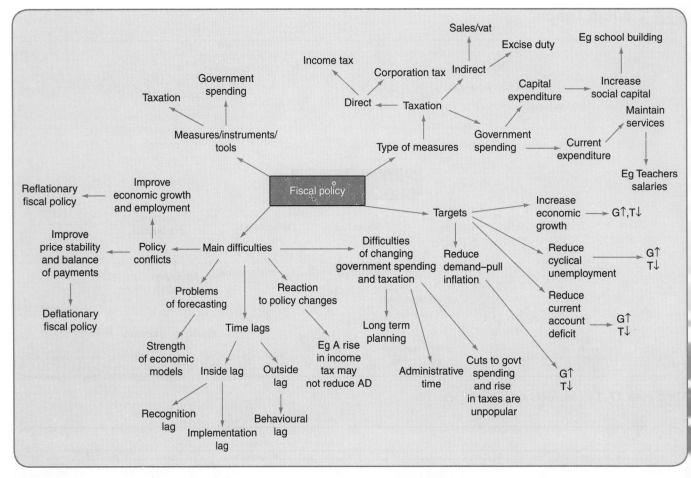

Mind map 17.3   Fiscal policy 2

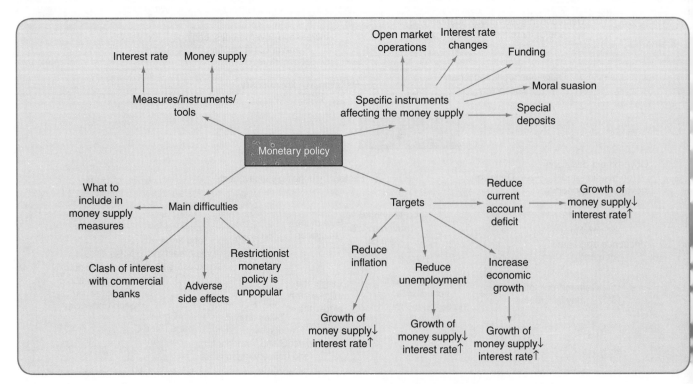

Mind map 17.4   Monetary policy 1

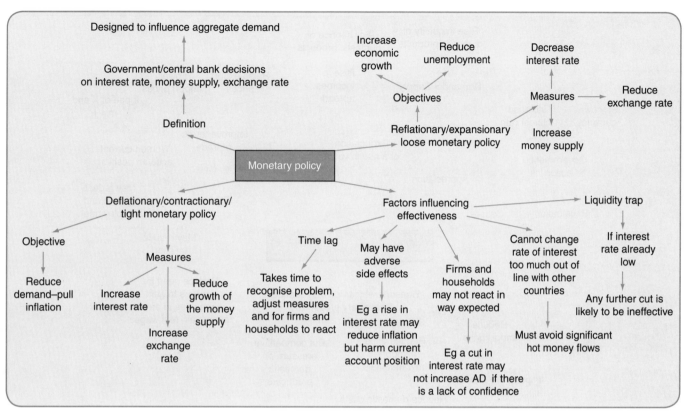

Mind map 17.5   Monetary policy 2

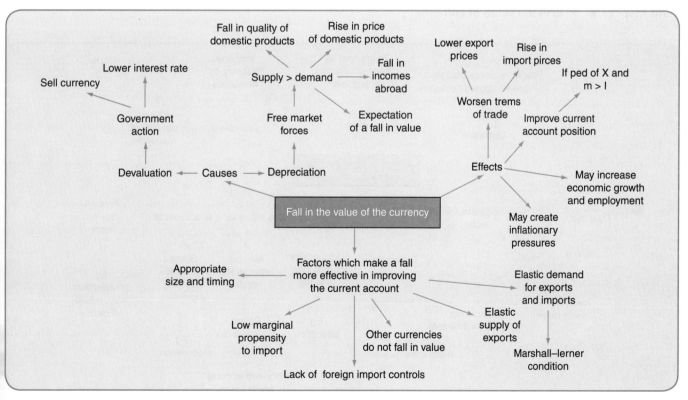

Mind map 17.6   Fall in the value of currency

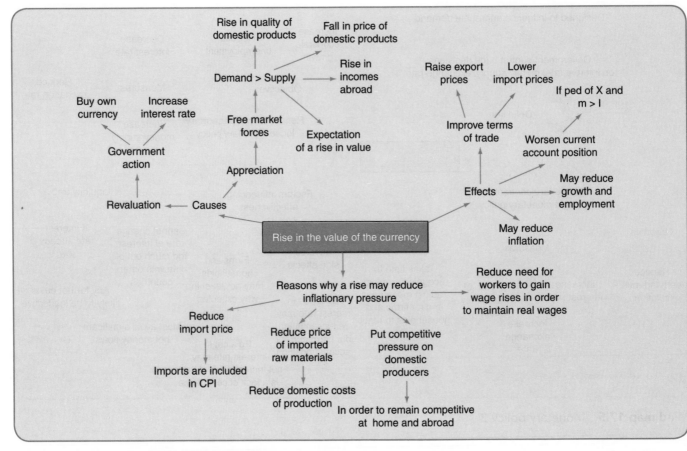

Mind map 17.7 Rise in the value of currency

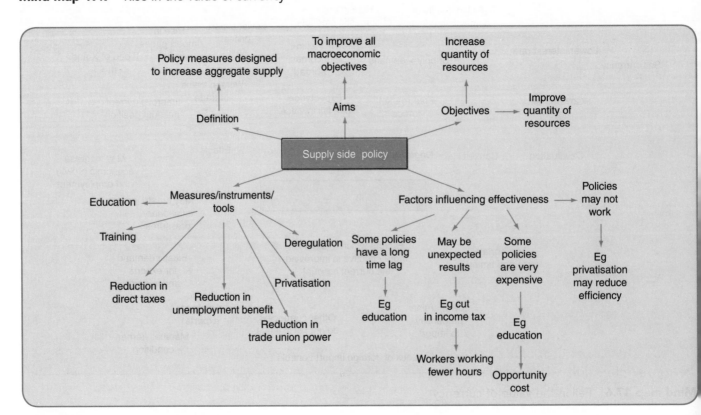

Mind map 17.8 Supply side policy

Short Questions

1. Why do governments seek to keep unemployment low?
2. Why may unemployment increase even if there is a positive economic growth?
3. What is meant by price stability?
4. What is the relationship between a positive output gap and inflation?
5. What is the difference between a recession and a depression?
6. What are the stages of the economic cycle?
7. Explain one reason why a current account deficit now may lead to a current account surplus in the future.
8. Why is it more useful to compare countries' current account balances as a percentage of GDP than in absolute terms?
9. Identify three financial flows from developed to developing economies.
10. Explain one way a government of a developed country could open up its markets to firms in developing countries.
11. What is meant by the Prebisch-Singer hypothesis?
12. Identify two reasons why a multinational company from a developed country may set up in a developing country.
13. Explain one disadvantage of a government defaulting on its overseas loans.
14. Explain two reasons why workers' remittances may increase.
15. What is meant by the 'canons of taxation'?
16. What is a 'poll tax'?
17. What is meant by a 'flat tax'?
18. What does the Laffer curve show?
19. Why might a decrease in commercial banks' liquidity ratios not result in an increase in bank lending?
20. Identify one reason why a cut in a central bank's interest rate may not increase bank lending.
21. What is meant by 'quantitative easing'?
22. Explain one way in which fiscal and monetary policy are connected.
23. What is meant by globalisation?
24. Should tariffs be imposed on products produced by industries that use child labour?
25. Why may relaxing immigration controls be regarded as a supply side policy measure?
26. Explain whether cutting income tax is a fiscal policy measure or a supply side policy measure.
27. Why may globalisation put downward pressure on governments to cut their spending on education?
28. Why may globalisation put pressure on governments to increase their spending on education?
29. What could cause a conventional Phillips curve to shift to the left?
30. What does the expectations augmented Phillips curve suggest about the relationship between unemployment and inflation?
31. What is meant by 'the vicious circle of poverty'?
32. Identify three reasons why income may become less evenly distributed in an economy.
33. Identify two reasons why a government may want its country's exchange rate to fall in value.
34. Explain two reasons why a government may move from operating a fixed to a floating exchange rate.
35. Why might an increase in unemployment reduce a current account deficit in the short run?
36. Is it possible for a government to reduce unemployment to zero?
37. What is meant by an economy's trend growth rate?
38. Why may a government seek to reduce the economy's economic growth rate?
39. What is meant by a counter cyclical policy measure?
40. Explain one benefit of inflation targeting.

Revision Activities

1. A government spends $20bn on goods and services and provides benefits which are equal to 10% of real GDP. The rate of taxation is 30%.
   (a) At which level of real GDP would the government have a balanced budget?
   (b) What would be the budget position if real GDP is $250bn?
   (c) Explain two reasons why government spending on goods and services may increase as real GDP rises.

2. (a) In 2007, Spain's budget position changed from a surplus to a deficit. By 2009, the deficit had increased to 10% of GDP. In the same year, GDP was reduced by 4.2% and its unemployment had increased to 18%. The Spanish government had started a large public works project and had experienced a large fall in tax revenue. Mr Zapatero, the Spanish Prime Minister, said, 'I have always thought of fiscal policy as an instrument that would

respond to changes in the state of the economy.' He promised to cut the budget deficit to 3% of GDP by raising taxes.

    (i) Using the information provided, explain why Spain's budget moved from surplus to deficit.

    (ii) Comment on Mr Zapatero's statement.

(b) In 2009, the Indian central government owned 246 enterprises. The enterprises employed nearly 1.6m people and accounted for 8.3% of the country's GDP. Between 1999 and 2004, it sold a number of state owned enterprises to the private sector. In 2009, it was again considering privatisation largely because it was facing a fiscal deficit of about 7%. Some economists claimed that a better reason to privatise is to invite others to run the businesses more efficiently.

    (i) What is meant by a 'fiscal deficit'?

    (ii) Explain two arguments in favour of privatisation.

(c) A study published in 2010 found that in only six of the thirty members of the OECD, a club of mainly rich countries, had income inequality increased over the decade. The six were Italy, Poland, Portugal, Turkey, the UK and the US. Income inequality increased less in the UK than the OECD average. Wealth in the UK, however, is distributed far more unequally than income with the richest tenth in the country holding assets worth one hundred times those of the poorest. Wealth inequality also increased between 2000 and 2010.

    (i) Identify, from the information given, one reason why income inequality increased in the UK.

    (ii) Explain one government policy measure to reduce income inequality.

(d) In January 2010, the Venezuelan government devalued the country's currency, the bolivar, by nearly 50%. The government was hoping to stimulate export-led growth. Opponents, however, expressed concern that the devaluation would increase the country's already high inflation rate. Some claimed it could raise the inflation rate from 27% to as high as 40%.

    (i) Explain why devaluation may promote 'export-led growth'.

    (ii) Why might devaluation increase a country's inflation rate?

3. A tax system allows people to earn $5,000 tax free. It then imposes a tax rate of 20% on the first $10,000 of taxable income and 40% on taxable income above $10,000.

    (a) Calculate the marginal and average tax rates on an income of $18,000.

    (b) Is the tax system proportional, progressive or regressive?

    (c) If a flat tax of 15% was introduced on all income, how would the amount of tax paid by a person earning $18,000 change by?

Multiple Choice Questions

1. Without any change in government policy, what will be the effect of a consumer boom on direct and indirect tax revenue?

	Direct tax revenue	Indirect tax revenue
A	Decrease	Decrease
B	Decrease	Increase
C	Increase	Increase
D	Increase	Decrease

2. What is meant by a regressive tax?

  A One which reduces economic activity

  B One where the marginal tax rate is higher than the average tax rate

  C One which involves high income earners paying a lower proportion of their income in tax than low income earners

  D One which involves high income earners paying less in tax than low income earners

3. Which government policy measure would be most likely to reduce a current account deficit and unemployment?

  A Devaluation

  B Decrease in income tax

  C Increase in the rate of interest

  D Increase in corporation tax

4. Which of the following is not an automatic stabiliser?

  A Income tax

  B Sales tax

  C State retirement pension

  D Unemployment benefit

5. What is an example of a supply side policy measure?

  A Deregulation

  B Devaluation

  C An increase in corporation tax

  D An increase in the rate of interest

6. What does the traditional Phillips curve suggest?
   A A direct relationship between unemployment and raw material costs
   B A direct relationship between unemployment and inflation
   C An inverse relationship between unemployment and economic growth
   D An inverse relationship between unemployment and money wages

7. A government reduces unemployment benefit. What type of unemployment is it most likely to be seeking to reduce?
   A Cyclical
   B Frictional
   C Structural
   D Technological

8. Why is a rise in government spending matched by an equal rise in tax revenue likely to increase aggregate demand?
   A A proportion of the income taken in tax would have been saved
   B A significant proportion of the government spending may be devoted to imports
   C Government spending usually accounts for a larger proportion of aggregate demand than consumer spending
   D Taxes are a withdrawal from the circular flow whereas government spending is an injection

9. Figure 17.1 shows a long run Phillips curve (LPC) and three short run Phillips curves (SPC). The economy is initially at the natural rate of unemployment of 8% with an inflation rate of 5%.

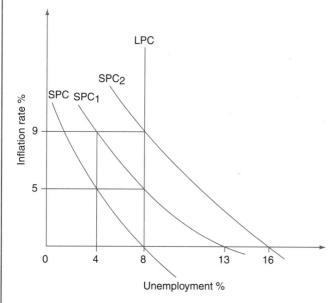

Figure 17.1

If inflationary expectations stay at 5% and the government wants to eliminate inflation, what rate of unemployment will it have to accept in the short run?
   A 4%   B 8%   C 13%   D 16%

10. Which type of unemployment comes from a lack of aggregate demand?
    A Cyclical
    B Frictional
    C Structural
    D Voluntary

11. Which measure is a deflationary monetary policy measure?
    A A deliberate reduction by the central bank in the rate of interest
    B A deliberate reduction by the government of its budget deficit
    C A sale of government securities by the central bank
    D A sale of state owned enterprises to the private sector

12. A government replaces a progressive income tax system with a regressive income tax system. What is the most likely outcome of this action?
    A A more equal distribution of income
    B A greater incentive to work
    C An increase in the rate of taxation on higher income groups
    D A reduction in saving

13. Which combination of problems is most likely to result in a government increasing taxation and cutting its spending?
    A Demand-pull inflation and a current account deficit
    B A current account deficit and unemployment
    C Unemployment and a current account surplus
    D A current account surplus and demand-pull inflation

14. Which policy measure may increase the natural rate of unemployment?
    A A decrease in income tax
    B A decrease in the rate of interest
    C An increase in unemployment benefit
    D An increase in labour market flexibility

15. Which policy measure may reduce demand-pull inflation but increase cost-push inflation?
    A A rise in the rate of interest
    B A rise in government spending on training
    C A decrease in corporation tax
    D A decrease in import tariffs

16. Which method of training of financing government spending would increase the money supply?
    A Borrowing from the central bank
    B Increasing income tax

C   Selling government bonds to the non-bank general public

D   Selling state owned enterprises to the private sector

17. What determines whether a change in income tax rates is a fiscal or supply side measure?

A   Whether the change is intended to be short term or long term

B   Whether the change is introduced by the government or the central bank

C   Whether the intention is designed to influence aggregate demand or aggregate supply

D   Whether the intention is to reduce or increase aggregate supply

18. What would reduce the effectiveness of expansionary fiscal policy?

A   An absence of a time lag

B   An absence of a recession in trading partners

C   A high marginal propensity to save

D   A low marginal propensity to import

19. Why might a decrease in income tax reduce a budget deficit?

A   It may stimulate a rise in real GDP

B   It may increase tax evasion

C   It may cause fiscal drag

D   It may increase the tax base

20. What is meant by the incidence of taxation?

A   The extent to which a tax is progressive

B   The extent to which a tax reduces incentives

C   Who bears the tax

D   Who benefits from the tax

Data Response Questions

1. Poverty and Economic Growth

At a time when there has been rapid economic growth in some parts of the world, there are many countries in Africa where severe poverty and infectious diseases are still widespread.

In dealing with poverty in developing countries much is written about the need for aid from the rich developed countries. The high profile given to aid hides the fact that financial, human and other resources are continuously taken from developing countries by these wealthy nations which are seeking their own economic growth.

Economic growth in the last 50 years has occurred at the same time as a widening gap both between rich and poor people in a country and between rich and poor countries. Economic growth and development should be reconsidered to include a more even progress in areas such as basic living conditions, access to health and education.

Despite economic growth, almost half the world's population still lives in poverty on less than US$2 a day. This reveals a weakness in the argument that economic growth is the solution to poverty.

(a) Explain briefly what is meant by economic growth. [3]

(b) Explain the possible link between developed and developing countries in achieving economic growth. [5]

(c) Analyse whether Figure. 17.2 supports the statement that 'almost half of the world's population still lives on less than US$2 a day. [5]

(d) The extract concludes with the statement that 'there is a weakness in the argument that economic growth is a solution to poverty.' Discuss whether this is a correct conclusion from the evidence given. [7]

Cambridge 9708 Paper 4 Q1 Oct/Nov 2008

2. Private Sector Money and Development

The amount of money that workers employed in foreign countries send home is worth US$200 billion a year and, therefore, the potential benefits of this to developing countries are huge. In some countries, the amount of money sent home by those working abroad is very significant when compared with official development aid. For example, in Bangladesh and Kenya, recent figures were as following.

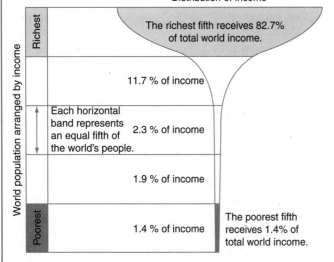

Distribution of income

The richest fifth receives 82.7% of total world income.

11.7 % of income

Each horizontal band represents an equal fifth of the world's people.

2.3 % of income

1.9 % of income

1.4 % of income

The poorest fifth receives 1.4% of total world income.

World population arranged by income

Richest

Poorest

Figure 17.2

	Kenya 2004	Bangladesh 2005
Money sent home by workers	US$464m	US$2.2bn
Money given in aid	US$625m	US$1.4bn

Even small amounts of money sent by individuals can have an impact on development. The primary education of many children is paid for by money sent home by relatives working abroad. Some governments actively encourage workers to send money home. India has offered non-resident Indians special investment opportunities, perhaps influenced by the example of China where investment from Chinese living abroad has been a big factor in the development of the economy. The same story is true for Ghana and Nigeria, which are developing 'remittance partnerships' with the UK to make it easier and cheaper for people from those countries working in the UK to send money home.

Further, the money could be worth even more if it is directed not just within the family but into community projects. In Mexico, local governments are matching any private money from abroad invested in community development with their own funds.

However, money is not the simple solution to poverty. The director of an economic research centre rejects the idea that income is the most significant contributor to well-being. Instead, in the research centre, attention is being turned to measures that seem to have little to do with economists' traditional indicators about production and consumption. When asked to rank what was important to them, poorer people said religion, relationships and inner peace were more important than income.

(a) Explain whether the money sent to Kenyans working in the UK would be included in Kenya's GDP. [2]

(b) Analyse why many people from developing countries might go to work in developed countries. [5]

(c) How far does the article support the view that economic development is largely the result of private actions rather than government policy? [5]

(d) Discuss the suggestion that 'traditional indicators about production and consumption' are of little use as measures of welfare. [8]

Cambridge 9708 Paper 4 Q1 May/June 2009

3. The rise of the BRICs

The term 'BRICs' to refer to Brazil, Russia, India and China was first used by Jim O'Neele, an investment analyst working for the investment bank, Goldman Sachs. The BRICs are forecast to be the dominant economies by 2050 and are already the four largest economies outside the OECD (Organisation for Economic Co-operation and Development).

The reasons why the BRICs are expected to overtake the USA and other developed economies in a few decades is their relatively rapid economic growth rates. The BRICs avoided going into recession towards the end of the first decade of the twenty first century but all experienced a cyclical budget deficit.

When Jim O'Neill coined the term BRICs he was not thinking of them as a trading bloc but they are beginning to think of themselves as a group and they held their official summit meeting in June 2009.

As the BRICs have continued to grow the value of their exports and imports has also increased. Brazil and India have recently experienced small current account deficits whilst China and Russia have continued to experience large current account surpluses. All the countries impose a range of trade restrictions. Exports account for more than a third of the gross domestic product (GDP) of China and Russia but less than a fifth of the GDP of Brazil and India. The table shows a range of economic data on the BRICs.

Selected economic data 2010			
	Economic growth Rate (%)	Inflation rate (%)	Unemployment rate (%)	Income per head (US$)
Brazil	5.5	5.2	7.4	6,860
Russia	3.5	7.0	8.6	9,080
India	7.7	12.0	10.7	1,050
China	9.7	3.5	9.6	2,430

(a) Explain what is meant by a cyclical budget deficit. [4]

(b) Analyse how economic growth may affect the value of a country's imports. [4]

(c) How far does the extract suggest the BRICs are open economies? [6]

(d) Discuss the extent to which the extract supports the view that the BRICs have a similar macroeconomic performance. [6]

Essay Questions

1. Analyse why the aims of government policy might conflict with each other and discuss which of these aims ought to be given priority. **[25]**

   Cambridge 9708 Paper 4 Q7 Oct/Nov 2007

2. (a) It is feared that if the government increases taxes the level of national income will fall. Explain whether this is necessarily true. **[10]**

   (b) Discuss whether a fall in the level of national income is a good indicator that there has also been a decline in the standard of living in the country. **[15]**

   Cambridge 9708 Paper 4 Q6 May/June 2008

3. In 2006, it was reported that a country's unemployment rate had remained steady and that its central bank, through its interest rate policy, had prevented an increase in inflation despite a sharp rise in oil prices.

   (a) Explain what might cause unemployment. **[12]**

   (b) Discuss how interest rate policy might prevent a rise in inflation. **[13]**

   Cambridge 9708 Paper 4 Q5 May/June 2009

Answers

Chapter 1

Activity 1.1

	Demand-pull inflation	Cost-push inflation
Definition	An excess of aggregate demand	A rise in the costs of production
Illustrated by	A shift to the right of the AD curve	A shift to the left of the AS curve
Examples of causes	A consumer boom Increase in net exports Rise in government spending	Rise in wages Rise in price of imported raw materials Rise in indirect taxes
Impact on real GDP	Increase up to the FE level	May reduce
Policy to reduce	Deflationary fiscal policy Deflationary monetary policy	Supply side policy

Activity 1.2

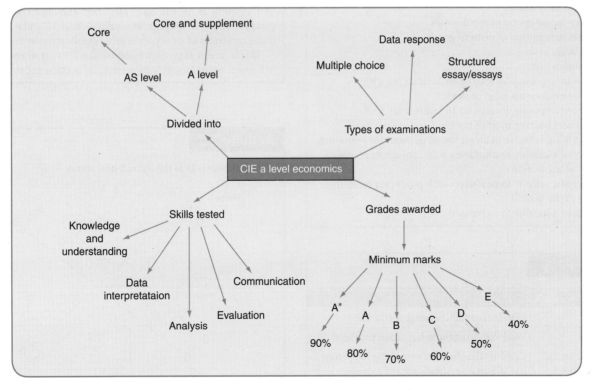

Mind map 1.1

Activity 1.3

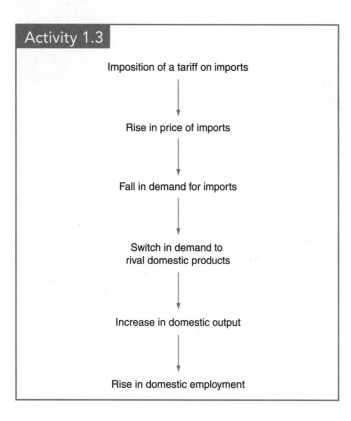

Imposition of a tariff on imports

↓

Rise in price of imports

↓

Fall in demand for imports

↓

Switch in demand to
rival domestic products

↓

Increase in domestic output

↓

Rise in domestic employment

Activity 1.4

Diagrams for Section 3 Core
1. Negative externalities
2. Positive externalities
3. Under-consumption of merit goods
4. Over-consumption of demerit goods
5. Maximum price
6. Minimum price
7. Effect of imposing/increasing an indirect tax with consumers bearing most of the tax
8. Effect of imposing/increasing an indirect tax with producers bearing most of the tax
9. Effect of an increase in direct tax on demand for a product.
10. Effect of a subsidy to producers with consumers receiving most of the benefit
11. Effect of a subsidy to producers with producers receiving most of the benefit
12. Effect of a subsidy to consumers

Activity 1.5

Directive words	Definitions
Analyse	Examine carefully, bringing out the links
Calculate	Work out using the information provided
Comment on the extent to which	Consider likelihood of something happening, size of a change, significance of a change
Compare	Bring out the similarities and differences

Define	Give the precise meaning
Describe	Give an account of the main characteristics
Discuss	Examine advantages and disadvantages, reasons for and against, qualifying factors in a critical way
Explain	Make clear
Evaluate	Assess a theory, policy, causes or consequences
Identify	Select and state
Justify	Show adequate reasons for answer given
Outline	Briefly describe the main features
Summarise	Select the main points

Chapter 2

Activity 2.1

1st question:	D. Supply rises by 25% (100/400 × 100) whilst price rises by 20% ($2/$10 × 100), so PES = 25%/20%.
2nd question:	B. A perfect competitor is a price taker whereas a monopolist is a price maker.

A is incorrect as both a perfect competitor and a monopolist can earn supernormal profit in the short run. It is in the long run that a monopolist can earn supernormal profit whereas, due to free entry and perfect knowledge, a perfect competitor cannot.

C is wrong as AC will equal MC under conditions of perfect competition in the long run. It is unlikely that AC will equal MC under conditions of monopoly in either the short run or the long run.

D is incorrect as profit maximisation will occur where MC = MR under conditions of both perfect competition and monopoly.

Activity 2.2

The answer is D as the figure below shows.

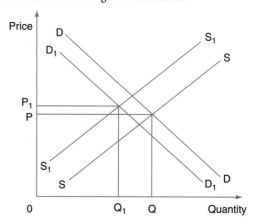

Chapter 3

Activity 3.1

There is some evidence to suggest that India will consume more sugar than the US in 2030. The information provided mentions that some people in the sugar industry are predicting such an outcome. India's population is likely to continue to grow, albeit at a slower rate, and incomes are likely to continue to rise.

The information does, however, state that 'some Indians are becoming more health conscious. If this trend spreads, sugar consumption per head might fall. The information does not give a figure for the current growth in the US nor provides predictions on the relative growth of population and income of the US or the views of others in the sugar industry. The information is not sufficient to conclude that India will still exceed the US in sugar consumption in 2030.

Activity 3.2

Four points from:

- ✓ The unemployment % fell consistently over the period in Egypt
- ✓ The unemployment % fluctuated over the period in South Africa
- ✓ The unemployment % was lower at the end of the period in both countries
- ✓ The unemployment % remained lower in Egypt over the period
- ✓ It is not shown how unemployment was measured in the two countries – different measures might have been used.

Activity 3.3

You should first recognise that the value of the Yen has risen over this period. There is a tendency to think that a falling line must mean that it a value has declined. When you look at the vertical axis, however, you should recognise that what was happening was that the number of Yens that had to be given to buy $1 was falling, meaning that the value of the Yen increased.

You should mention that whilst the trend was for the Yen to rise, it did actually fall in value in 2010.

You could then explain that a rise in the value of a currency would lead to higher export prices and lower import prices. This might reduce a current account surplus if demand for exports and imports is elastic – the Marshall-Lerner condition. If PED < 1, it may increase a current account surplus.

You should add evaluative points. For instance:

- ✓ The current account consists of more than trade in goods and services. An increase in the combined trade in goods and services balance may be offset by an increase in a deficit on the income balance and/or current transfers balance.

- ✓ Figure 3.1 only shows the value of the Yen against one currency. This is a major currency but it is possible that whilst the Yen was rising in value against the US dollar, it might have fallen against the euro and the Chinese yuan (renminbi). It would be more informative to compare the value against a trade weighted basket of currencies.

Chapter 4

Activity 4.1

The accelerator theory is useful in explaining net investment. It concentrates on the main influence on investment which is changes in demand for consumer goods. Firms will invest when the expected exceeds the cost of investment. If demand is increasing, firms will expect to sell more goods and hence receive more revenue.

However, the accelerator theory does not provide a complete explanation of the behaviour of net investment. Demand for consumer goods may rise without a greater percentage rise in demand for capital goods. Indeed, there may be no change in investment. Firms will not invest to expand capacity if they do not believe that the increase in demand for consumer goods will last. Expectations are a significant influence on investment. If firms are pessimistic about the future they may not even replace machines as they wear out.

Firms may also not buy new capital goods if they have spare capacity. They will be able to respond to the rise in demand by making use of previously unused or underused capital equipment and plant.

So spare capacity in consumer goods may result in no or a smaller change in demand for capital goods. In contrast, it may be an absence of spare capacity in the capital goods industry which may prevent firms from being able to purchase more capital equipment. The consumer goods industries may want to buy, example, more machines but the capital goods industries may not have the resources to produce them.

Changes in technology may also mean that an increase in demand for consumer goods may bring about a smaller percentage increase in demand for capital goods. A new machine, embodying advanced technology, will be able to produce more goods than the machine or machines which it replaces.

Other influences on investment may change. For example, there may be changes in the cost of machinery, corporation tax or government subsidies. Investment is also influenced by changes in the rate of interest. An increase in the rate of interest will increase the opportunity cost of investment.

Activity 4.2

Plan

(a) 1. Introduction – define the balance of payments and identify the accounts that comprise it.
2. Current account – trade in goods, trade in services, income and current transfers.

3. Capital account.
4. Financial account.
5. Net errors and omissions.
6. Conclusion – current account receives the most attention; financial account often involves the largest figures.

(b) 1. Introduction – explain what is meant by a deficit and mention that it is not always harmful.

2. Effects of a deficit – fall in AD, outflow of money.
3. Overall effect will depend on size and duration. A small deficit which last a short term may not be a problem. Large, long term deficit may build up international debt and/or reduce reserves.
4. Overall effect influenced by cause – lack of price competitiveness equals bad; low incomes abroad/import of raw materials or capital goods may be self-correcting.
5. May be covered by inflow of FDI on financial account.
6. May reduce demand-pull inflation.
7. Government measures to reduce deficit may reduce AD in short term and AS in long term.
8. Conclusion – whether harmful or not depends on size, duration and cause and how government responds.

Chapter 5
Answers to short questions

1. Two disadvantages of barter are the need to establish a double coincidence of wants and the problem of deciding on the value of products. When exchanging products, a person has to find someone who not only wants the product s/he wishes to exchange but also has something s/he wants in return. Finding such a double coincidence of wants can be very time consuming.

   When directly swapping products, there is also the problem of deciding, for example, how many chairs should be exchanged for a personal computer.

   (Note there is also the problem of giving change, example, if it is thought that a personal computer is worth two and a half chairs.)

2. Liquidity means the ability to turn an asset into cash quickly and without loss. If, for instance, a bank has a lack of liquidity, it may experience problems if a large number of its depositors seek to withdraw money from the bank.

3. Fixed capital is a term that covers capital goods which stay in the business for some time and do not change their form during the process of production. For instance, in a firm that produces toys, the factory, warehouse and machinery would be classified as fixed capital.

   Working capital, in contrast, covers capital goods which are used up in the process of production. For instance, the toy firm would classify as working capital the plastic it uses to make toy soldiers and the card and wood it uses to produce jigsaw.

4. Fixed capital formation is total spending on (investment in) fixed capital goods including buildings, machinery and vehicles used by firms.

5. Most products are economic goods. Resources have to be used to produce economic goods and so economic goods have an opportunity cost. The resources used to provide a course

on economics might have been used to provide a course on accountancy.

   In contrast to economic goods, free goods take no resources to produce them. As a result, they do not have an opportunity cost. There are only a few free goods. Air and sunshine are examples.

6. Health care is an economic good and not a free good. In some cases, health care is provided free at the point of use. For instance, in the UK people receive free medical treatment from the National Health Service. The treatment they receive, however, takes resources to produce it and these resources could have been used for another purpose. The provision of health care, therefore, involves an opportunity cost.

7. The primary sector covers industries which are involved in the extracting or growing of raw materials. Examples of industries in the primary sector include agriculture, fishing, fuel extraction, forestry and mining.

   Secondary sector industries change raw materials into finished and semi-finished goods. The secondary stage of production covers construction and manufacturing. Examples of industries in the secondary sector include building, chemical production, food processing, printing and textiles.

   The tertiary sector is the term used for those industries which provide services. Tertiary industries include banking, health care, insurance, tourism and transport.

8. Teachers work in the tertiary sector, farmers in the primary sector and car producers in the secondary sector.

9. Microeconomics is concerned with economics on a small scale. It covers the study of individual markets, industries, firms and households. In contrast, macroeconomics is economics on a large scale. It is concerned with the economy as a whole and includes, for example, the study of unemployment, inflation and economic growth.

10. The theory of the firm is a micro topic as it is concerned with how individual firms perform.

Answers to revision activities

1.

A comparison of a market economy and a planned economy		
Features	**Market Economy**	**Planned Economy**
Allocative mechanism	The price mechanism	State directives
Key sector	Private	Public
Key decision makers	Consumers	The state
Other names	Free enterprise, laissez faire, private enterprise	Centrallyplanned, collectivist, state owned
Example	Hong Kong	North Korea
Ownership of means of production	Privately owned	State owned
Provision of public goods	No	Yes
The profit motive	Present	Not present

2. (a) A constant opportunity cost – as more capital goods are produced, the same amount of consumer goods would have to be given up and as more consumer goods are made, the opportunity cost in terms of capital goods will not change.

(b) The movement of the PPC from AB to AC would mean that the economy's ability to produce consumer goods has increased whilst its ability to produce capital goods is unchanged.

(c) The PPC will have shifted either because the economy has fewer resources or lower quality resources. For instance, the decline in the quality of education may have decreased the productivity of labour and so have reduced the quantity of products that the economy can produce.

3. (a) Land: locations where films are shot.
Labour: camera operators, film directors.
Capital: cameras, editing equipment.
Enterprise: shareholders in the film company, producers.

(b) Capital intensive industry: the oil industry.
Labour intensive industry: hotel and catering.

(c) A range of factors influence the supply of labour to a particular occupation including: the wage rate, bonuses, working conditions, working hours, holidays, promotion chances, job security, pensions.

(d) Enterprise and opportunity cost are linked as entrepreneurs will take into account the return in the next most profitable industry when considering whether to keep using their skills in a particular industry. If profit falls below the normal profit level, some entrepreneurs will switch their resources to producing other products.

(e) The rent of land is usually higher than that of land in rural areas as land is scarcer there relative to the demand for it. In a city there are usually many competing uses for a relatively small area of land. In contrast, there is usually less competition for the use of land in rural areas.

4. (a) Cheques are not money. They are a means of transferring bank deposits from one person to another.

(b) The function of money which allows products to be bought on credit is a standard of deferred payments.

(c) To act as a medium of exchange money has to be divisible in order that payments of different values can be made and change can be given.

(d) A sight deposit (current account) is more liquid than a time deposit (deposit account).

(e) The general acceptability/durability and durability/general acceptability of money allows it to act as a store of value.

(f) Money acts as a unit of account when the value of products is compared.

Answers to multiple choice activities

1. A   Opportunity cost is the best alternative forgone. By deciding to go to university, the woman has given up the opportunity of earning income by working.

2. C   A normative statement is a statement of opinion. It is debatable whether the government should play a larger role in a mixed economy. A, B and D are all positive statements.

3. B   Complete specialisation means devoting all resources to producing one type of product. B shows that all available resources are employed providing capital goods. A and C show a combination of consumer and capital goods being produced – with A indicating some resources are unemployed. D is an unattainable position.

4. C   The fundamental economic problem arises due to scarcity – infinite wants but finite resources. Due to the economic problem, economies have to decide how to use their resources. A, B and D are issues that may face certain economies at certain times but not all economies at all times.

5. C   A mixed economy has both a private sector and a public sector. The private sector covers firms that are owned by individuals. In this sector, decisions on the allocation of resources are based on the market forces of demand and supply. The public sector covers state run organisations in which the allocation of resources is determined by state planning. A and B are likely to apply to market, mixed and planned economies whereas it is unlikely that D would be found in any type of economic system.

6. D   In a planned economy, it is the government/state that determines what products are produced, how they are produced and for whom.

7. C   As not all wants can be met, choices have to made.

8. D   A store of value enables people to save money to use in the future. A enables people to compare the value of products and assets, B enables them to buy and sell products and C permits them to borrow and lend.

9. B   Land covers natural resources. A beach is an example of a natural resource. In contrast, A, C and D are human made goods used in the production of the holiday and so are capital goods.

10. B   To raise the production of capital goods from 20m to 50m, resources have to be switched away from producing capital goods. The output of consumer goods has to fall from 100m to 60m and so 40m consumer goods have to be forgone in order to make 30m more capital goods.

Answers to data response question

1. (a) Removal of price controls, removal of subsidies and privatisation (sale of state owned enterprises).
(b) Primary
(c)

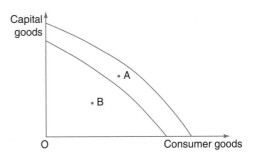

The ability of the Russian economy to produce products is likely to have decreased, shifting the PPC to the left. This is because life expectancy fell which probably reduced the size of the labour force. The rise in unemployment would have moved the production point further inside the curve.

(d) The inhabitants of Poland enjoyed a higher life expectancy in 2007 than those in Russia. The economic growth rate, however, was higher in Russia than in Poland. Both countries experienced a lower rate of economic growth and a higher rate of inflation than China and Slovenia.

Whilst the economic growth rate figure covers six years, the data on life expectancy and unemployment is for only one year. It would have been useful to have additional data including the countries' current account position and inflation rate.

(e) Reasons why the sale may benefit the economy

✓ There may be competitive pressure on firms in the private sector to be allocatively, productively and dynamically efficient.

✓ Private sector companies may provide more choice, higher quality and lower prices for consumers.

✓ There will be saving of government revenue if the firms had been loss making.

Reasons why the sale may not benefit the economy

✓ There may be a reduction in social provision for workers if an adequate state welfare system is not developed.

✓ Private sector firms may not take externalities into account.

✓ Private sector monopolies may develop which charge high prices and produce low quality products and do not provide choice.

✓ There will be a loss of government revenue in the long run if the state owned enterprises had been profitable.

Answers to essay questions

1. (a) A production possibility curve (PPC) shows the amount of two types of products that can be produced with existing resources and technology. In practice, most PPC are bowed outwards. This is because the opportunity cost of producing one type of product increases the more of it that is produced. The best resources are employed first, so as more is made, the more resources that have to be used and so the more of the other product that has to be given up.

The figure below shows the PPC of an economy that can produce agricultural and industrial goods.

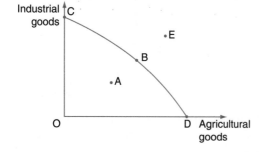

Production point A shows unemployed resources. Points B, C and D are all productively efficient points and point E is unattainable.

A PPC will shift to the right if the quantity or quality of resources used in producing both types of products increases. If the ability to produce only one type of product increases, the shape of the PPC will change. If there is an increase in the productivity of agricultural workers, the amount of agricultural goods that can be produced will increase. The figure below shows that the rise in the productive capacity of producing agricultural goods will pivot out the PPC on the horizontal axis.

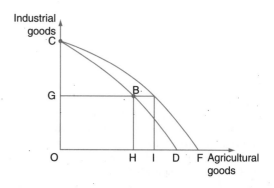

The maximum amount of industrial goods that can be produced remains at C whilst the maximum amount of agricultural products that can be produced increases from D to F. If originally, the economy had selected to produce at point B, it would have made G quantity of industrial goods and H quantity of agricultural goods. After the increase in the productivity of agricultural workers, G quantity of industrial goods could now be combined with a higher quantity of agricultural goods – I amount.

(b) It is very unlikely that the problem of scarcity will ever be solved as whilst the quantity and quality of resources tend to increase over time, so do wants. Different economic systems seek to manage the problem of wants exceeding resources in different ways.

In a market economy the market forces of demand and supply determine how resources are allocated. These forces work through the price mechanism. Those whose services are most in demand will earn the highest wages and will have high purchasing power. If demand for a product decreases, its price will fall which will result in a contraction in supply. As a result, scarce resources will not be wasted producing products that consumers are not willing and able to buy. In contrast, if demand for a product increases, the resulting rise in price and profitability will encourage producers to make more of the product.

In coping with the economic problem, the price mechanism has a number of advantages. It rations out goods and services. Those who can afford the products can buy them. It provides an incentive for firms and workers to be efficient. The more productive workers are, the more they will earn and the more responsive firms are to changes in consumer demand, the higher the profits they will earn. The price mechanism also works automatically and quickly, moving resources away from products declining in popularity towards those increasing in popularity.

In practice, though, there may be market failure with the price mechanism not resulting in productive, allocative and dynamic efficiency. Monopoly power may develop which can distort the allocation of resources. A monopoly firm may create an artificial shortage in order to drive up prices. A pure market system would fail to produce public goods, and would under-produce merit goods as they would be under-consumed and would over-produce demerit goods as they would be over-consumed. Linked to merit goods and demerit goods, consumers and producers in a pure market system would also not take into account positive and negative externalities. The greater the extent of market failure, the less efficiently resources will be used and the greater will be the problem of scarcity.

In addition, in a pure market economy, different groups will be affected by the problem of scarcity to different degrees. Whilst the rich may be able to satisfy most of their wants, the poor may not be able to meet even their basic needs. The price mechanism reflects effective demand, that is not only the willingness to buy products but also the ability to buy them. It may be claimed that there is consumer sovereignty but consumers do not have equal purchasing power. It is the poor which experience scarcity to the greatest extent.

An advantage claimed for a planned economy is a relatively even distribution of income. State planning may also mean that externalities are taken into account, wasteful duplication may be avoided, there may be full employment of resources, public goods will be produced and the consumption of merit goods will be encouraged whilst the consumption of demerit goods will be discouraged.

Relying on state planning and directives to allocate resources, however, is not guaranteed to be more efficient in tackling the problem of scarcity than a market economy. Whilst all resources may be employed, they may be underemployed or not employed in the most productive uses. There may be an overconcentration on capital goods. State planning can be slow and may not accurately assess consumers' wants. The lack of incentives for workers and SOEs may also mean that output is lower than possible and so scarcity is more of a problem than necessary.

Which type of economic system is more effective in dealing with the economic problem depends on the degree of market failure and government failure. As there advantages and disadvantages of market and planned economies, most countries operate mixed economies. Such an approach may result in greater equity and a more beneficial provision of products than in a market economy and may provide more incentives and less bureaucracy than in a planned economy. No type of economic system, however, can eliminate the problem of scarcity.

2. Money can be defined as any item which carries out the functions of money. There are four functions of money. One is to act as a medium of exchange. This means that money enables people to buy and sell products and resources and overcomes the double coincidence of wants. A man can sell a carpet he has made and use the money he receives to buy a rare book. In the absence of money, he would have to find someone who is both willing and able to sell a rare book and who wants a carpet.

Another function is to act as a store of value. People can save money over time. They can also use money to measure and compare the value of products. This function is referred to as a unit of account. The fourth function is to act as a standard of deferred payments. This means that money allows people to establish the price of future claims and payments so that they can agree contracts and can borrow and lend.

To act as money, an item has to possess a number of characteristics. A crucial one is that it has to be generally acceptable and so must be limited in supply. An item also has to be durable, divisible, portable, recognisable and stable in value.

To operate as a medium of exchange, an item clearly has to be generally acceptable. People will only accept an item in payment for a product if they can use it to buy other products. They also have to recognise the item as money and be able to carry around an item to use it to buy products (portability). Money can act as a store of value only if it is durable, that is it will not perish, and stable in value as otherwise savings in money could become worthless. To act as a unit of account, which is sometimes known as a measure of value, money must be divisible and stable in value so that the price and cost of items can be compared. In carrying out the function of a standard of deferred payments, two important characteristics are general acceptability and stability in value. People have to think that money will be acceptable in the future and that it will not lose its value.

If an item loses the necessary characteristics, it will stop acting as money. During a period of hyperinflation, a currency will be losing its value at a very rapid rate. If it ceases to be generally acceptable as means of payment, saving, comparison of value and arranging future claims and payments, it will stop acting as money.

3. (a) The basic economic problem is scarcity. Wants exceed resources meaning there are not enough resources to produce everything that people want. As a result of the problem, all economies have to make choices. The three fundamental choices they have to make are what to produce, how to produce and for whom to produce.

If there was enough land, labour, capital and enterprise to make all the products people desire to have, economies would not have to answer the question of what to produce. In practice, economies have to decide what they are going to use their resources to make. If, for instance, an economy decides to make more capital goods, the opportunity cost would be consumer goods. The following figure shows that making full use of its resources, an economy is producing 30m capital goods and 40m consumer goods. To produce additional 10m capital goods, 12m consumer goods have to be forgone.

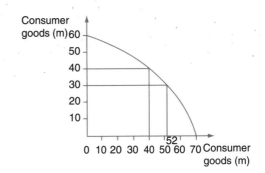

An economy also has to answer the question of how to produce. It has to decide which resources to use and in what combinations. Its decision will be influenced by what resources are available to it. For instance, an economy which has a plentiful supply of labour and a shortage of capital is more likely to make use of labour intensive methods of production rather than capital intensive methods. It is important to use the most efficient methods so that the gap between what is produced and want people what is minimised.

Economies have to answer the question of who should receive the output produced as consumers will be competing for a limited number of goods and services.

Different economic systems answer the three fundamental economic questions in different ways but the existence of the economic problem means that no type of economic system can avoid finding answers to these questions.

(b) A market economy relies on the price mechanism to answer the three fundamental questions which arise because of the economic problem of scarcity.

A market economy is one in which capital is privately owned, the role of the state is minimal and the market forces of demand and supply, operating through the price mechanism, determine the allocation of resources, the methods of production and the distribution of income.

Consumers indicate their choices by the prices they pay and the possibility of earning profit provides the incentive for firms to respond to their choices. The price of resources and their productivity influence the methods of production and goods and services are distributed on the basis of who can pay for them.

The price mechanism is not the only way to tackle the basic economic problem. State planning is used in a planned economy and a combination of state planning and the price mechanism is used in a mixed economy. No mechanism can solve the basic economic problem as wants will always exceed resources.

The price mechanism does have a number of advantages. It is an automatic method of allocating resources and one which enables consumers to decide what is produced. If a product becomes more popular, consumers will compete for it, bidding up its price. A higher price will encourage firms to produce more of it.

The price mechanism also promotes efficiency, innovation, enterprise and hard work. If firms produce what consumers are willing and able to buy at the lowest possible prices, they can earn high profits. If, however, they produce unpopular products or fail to keep their costs down, they may be driven out of business. Entrepreneurs that develop new products and new methods of production and workers who are very skilful and diligent can be well rewarded.

The price mechanism does also, however, have disadvantages. The price mechanism does not take into account externalities. In cases where there are negative externalities such as pollution, there will be overconsumption and overproduction with too many resources being devoted to such products. Merit goods will be under-consumed and demerit goods will be over-consumed. There will be no financial incentive for private sector firms to produce public goods. There may be abuse of market power, with monopolies restricting output and driving up prices. People who have the highest incomes are able to buy the most products and the poor may not be able to purchase many products.

The price mechanism has the potential to tackle the basic economic problem in an efficient manner but it cannot solve it and there is no guarantee that relying on market forces will result in the best outcome. The extent to which the price mechanism is effective depends on the degree of market failure which occurs.

Chapter 6

Answers to short questions

1. Demand curves slope down from left to right, showing that as price falls, more is demanded. This is for two reasons. One is that as a product becomes cheaper, people become more able to buy it – their purchasing power increases. The other reason is that people will become more willing to buy the product as it will become more price competitive and relative to substitutes.

2. Derived demand is when demand for one product depends on demand for another product. For instance, demand for the labour services of doctors depends on demand for health care. Similarly, demand for air travel depends, in part, on demand for foreign holidays.

3. Three factors that influence demand for air travel are the price of air travel, the price of foreign holidays and income. A rise in the price of air travel would be expected to cause a contraction in demand for air travel. Foreign holidays are a complement to air travel. If foreign holidays rise in price it is likely that demand for foreign holidays will contract and so demand for air travel will decrease. Air travel has positive income elasticity of demand and so a rise in income would usually result in an increase in demand for air travel.

4. A contraction in demand for ice cream is caused by an increase in the price of ice cream. Less ice cream is demanded because it is more expensive. In contrast, a decrease in demand for ice cream is caused by a change in any influence on demand for ice cream other than a change in its price. For instance, a decrease in demand for ice cream may be caused by a spell of cold weather.

5. If a product has a perfect substitute, it will have perfectly elastic demand. If the firm producing the product raises its price, it will lose all of its sales.

6. Two factors that could make demand more price inelastic is a reduction in the degree to which other products are substitutes and an increase in the extent to which the product is seen to be a necessity.

7. As price rises, demand becomes more elastic. A higher price makes consumers more price sensitive.

8. An income elasticity of demand which is both positive and greater than one means that as income rises, demand increases by a greater

percentage. Demand for a number of services is both positive and elastic, making them superior products.

9. The cross elasticity of demand between one model of car and petrol would be negative as they are complements. In contrast, the cross elasticity of demand between the model and other models of cars is positive. This is because they are substitutes.

10. Three factors that could cause an increase in the supply of rice are the granting of a government subsidy, a fall in the costs of production and a period of suitable weather. The granting of government subsidy would provide an incentive for farmers to grow more rice. Lower costs of production would make farmers more willing and able to grow rice. In addition, good climatic conditions would be likely to increase the amount of rice grown successfully.

11. A fall in the price of blankets would lead to a contraction in the supply of blankets. Some resources used to produce blankets are likely to be shifted to producing duvets. This would increase the supply of duvets.

12. The more perishable a product is, the more inelastic supply tends to be. This is because it is difficult to store the product and so should price rise, supply cannot be raised by drawing on stocks. Similarly, should price fall, it will be difficult to reduce supply.

13. The burden of an indirect tax in a market where demand is inelastic and supply is elastic will fall mainly on consumers as shown in the figure below. The price rises by most of the tax.

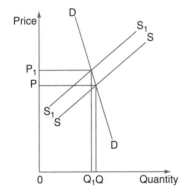

14. Adjustments in demand and supply move a market from disequilibrium to equilibrium. If, for instance, supply exceeds demand there will be a surplus. This surplus will drive price down causing supply to contract and demand to extend until the two are equal again.

15. The laws of demand and supply are predictions about how market forces work. For instance, one 'law' is that an increase in demand will cause a rise in price and an extension in supply.

16. An increase in supply and a decrease in demand will reduce price. The effect it will have on the quantity bought and sold is uncertain. It will depend on the size of the relative changes. If the decrease in demand is greater than the increase in supply, the quantity traded will fall. In contrast, if supply increases by more than demand falls, the quantity traded will rise.

17. Joint demand arises when two products are complements. The products are demanded to be used together such as a laptop and a printer. In contrast, composite demand is when a product is demanded for two or more purposes. For instance, Soya may be grown to be used in both food products and in biofuels.

18. The amount of consumer surplus received from the purchase of a product differs between consumers because people have different purchasing power and value the product differently.

19. The price mechanism rations products by limiting their consumption to those who can afford to pay the prices. If demand for a product increases, initially demand will exceed supply. Price will rise until demand equals supply again.

20. The market price is the price charged in a free market. In the short term it may not equate demand and supply, although in the longer term market price should move to the equilibrium price.

Answers to revision activities

1. (a) An increase in the cost of paper will raise the cost of producing newspapers. This will lead to a decrease in supply, which in turn will raise price and cause demand to contract as shown in the figure below.

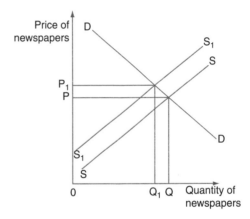

   (b) Newspapers and internet websites are substitutes, a reduction in the quality of internet websites may encourage people to switch to newspapers. This would increase demand for newspapers, raise their price and cause supply to extend.

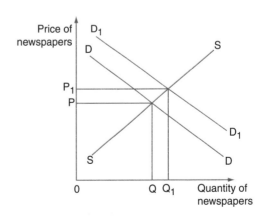

   (c) Free gifts provided by newspapers are designed to increase demand. Higher demand will push up price and cause a rise in quantity supplied.

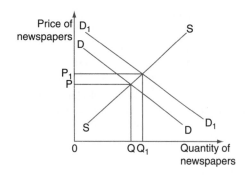

(d) The introduction of more efficient printing presses will reduce costs of production. Lower production costs will increase supply which will lower price and cause demand to extend.

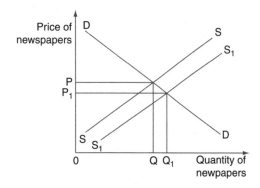

2. (a) PES = 20%/25% - 0.8.

(b) Supply is inelastic.

(c) The car firm would want to make its PES more elastic. It would want to be able to take full advantage of a rise in price and to minimise the risk of making a loss if price falls.

(d) The firm could seek to make its supply more elastic by cutting its production time. This might be achieved by introducing more advanced equipment and training workers.

3. (a) It would be expected that luxury wrapping paper would be more elastic than standard wrapping paper. The information does, however, suggest that the standard wrapping paper has a closer substitute as it has a higher positive XED figure.

(b) To raise revenue the firm should raise the price of the luxury wrapping paper as demand will fall by a smaller percentage than the rise in price. In contrast, it should lower the price of standard wrapping paper, as demand is elastic.

(c) The firm should lower the price of the firm's standard gift tags as this would significantly raise sales of its standard wrapping paper. It should leave the price of the luxury gift tags unchanged.

(d) In the long run, the firm should concentrate on the luxury wrapping paper. This is because the product is more income elastic and would, therefore, benefit to a greater extent from rises in income over time. It also faces less competition in producing this type of wrapping paper.

Answers to multiple choice questions

1. C The only factor that can cause a movement along a demand curve is a change in the price of the product itself.

2. D A shift to the right of the demand curve shows an increase in demand. Popular films will encourage people to buy more cinema tickets. A and C would cause a movement along the demand curve and B a shift to the left of the demand curve.

3. D If total expenditure remains unchanged when price falls by 8%, it must mean that the quantity demanded has risen by 8%. An equal percentage change in price and quantity demanded means that PED is unitary.

4. C 80 units is 40% of 200. To raise sales by 40%, price would have to fall by − 0.8 = 40%/? = - 50%. So price would have to fall by 50% i.e. by $5.

5. A If demand is inelastic, a fall in price will cause a smaller percentage fall in demand and so total revenue will fall. Demand is not perfectly inelastic and so a rise in price will cause demand to contract.

6. B The lower demand is, the more consumers will react to any price change. The more consumers demand a product, the less they will tend to reduce their demand when price rises.

7. D The existence of close substitutes would mean that a rise in price would cause a greater percentage fall in demand and a fall in price would cause a greater percentage rise in demand. A, B and C would be likely to cause a low PED.

8. B Demand becomes more elastic, the higher the price and the lower the quantity demanded. As price rises, fewer people will be willing and able to buy the product.

9. B The price of Product X rises by 50%. Demand for Product Y increases by 20%. So XED = 20%/50% = 0.4.

10. B A negative figure shows two products are complements. − 0.2 is low and so this suggests the two products are distant complements.

11. D This is a straightforward question. Demand for inferior goods falls as income rises.

12. B The shorter the time it takes to produce a product, the greater the supply can be adjusted to market conditions. A would make demand rather than supply more elastic. C and D would make supply more inelastic.

13. C Price rises by 50%. With PES being 0.8, supply would increase by 40%. This means supply rises to 2,800. The firm's revenue is $30 × 2,800 =$84,000.

14. C The figure below shows supply exceeding demand.

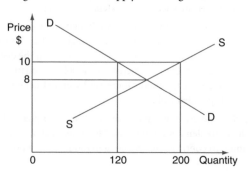

There is a surplus of 80 and price exceeds the equilibrium level by $2.

15. C

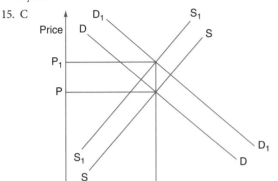

Figure shows that an increase in demand and a decrease in supply could raise price but leave the quantity traded unchanged.

16. B   A rise in incomes would increase demand for laptops, shifting the demand curve to the right. A reduction in the cost of producing laptops would increase supply, shifting the supply curve to the right. The new curves intersect at B.

17. D   Producers' revenue is initially $7 × 70 = $490. When the $3 per unit tax is imposed, producers' revenue falls to $6 × 50 = $300. There is a decline of $190. Consumers spend $450 after the tax is imposed but $150 of this goes to the government in the form of tax revenue.

18. C   A subsidy to producers causes an increase in supply. A fixed sum subsidy would cause a non-parallel shift.

19. C   The consumer surplus is initially PWZ. After the rise in price, the consumer surplus falls to P1WX. This is a decrease of PP1XZ.

20. C   Consumers bear most of an indirect tax when demand is inelastic and supply is elastic. When demand is perfectly inelastic A consumers would bear all of the tax. When demand is elastic C and D, producers bear most of the tax.

Answers to data response questions

1. (a) (i) The formula to calculate income elasticity of demand is:

$$YED = \frac{\% \text{ change in quantity demanded}}{\% \text{ change in income}}$$

   (ii) Table 1 indicates that air travel has both positive income elasticity of demand and income elastic demand. This means that it is a normal good and a special type of normal good, that is a luxury or superior good.

   (b) (i) Demand for business flights is more price inelastic than demand for leisure flights. Business passengers have less choice as to where they are travelling to and when they are travelling than leisure passengers. A rise in fares for a flight from, for instance, London to Colombo in Sri Lanka may not discourage a buyer from a UK clothes shop who is seeking to sign a contract with a Sri Lankan clothes manufacturer. It may, however, result in a holiday maker changing her holiday destination to, for example, the Maldives.

   (ii) Demand for short-distance flights is more price elastic than demand for long-distance flights. This is because there are usually closer substitutes available for short-distance flights. For instance, a person could drive, take a train of fly from New Delhi to Mumbai but, in terms of time, there is not really a close substitute to air travel in the case of travelling from New Delhi to New York.

   (c) The tables suggest that an airline would benefit from cutting the fare of short-distance leisure flights. This is because as demand is elastic, a fall in price would result in a greater percentage change in quantity demanded and so a rise in revenue. In contrast, cutting the fares on long-distance international and short-distance business flights would reduce revenue. This is because demand is inelastic and so a fall in price would cause a smaller change in quantity demanded. In the case of long-distance leisure flights, a fall in price would cause an equal percentage rise in quantity demanded and so leave revenue unchanged. Of course, an airline would be advised to check that these figures are an accurate reflection of current demand conditions before making any change in prices.

   (d) An increased demand for air travel would benefit the airline industry. Revenue would increase and costs would fall if the higher output enables the firms to take advantage of internal economies of scale and benefit from external economies of scale. For instance, airlines may be able to borrow more easily and at a lower rate of interest (financial) and use larger, more efficient aircraft (technical). The expansion of the airline industry would contribute to the trade in services balance of the balance of payments. Employment in the industry is also likely to rise with more pilots and cabin crew being taken on. In addition, expanding airports will create employment in the local area with, for instance, greater demand for the sources of taxi drivers.

   There is, however, a risk that the increased demand for air travel might be at the expense of demand for other forms of transport. If this is the case, there may not be a net positive contribution to economic growth and employment. An increase in air travel may also increase air and noise pollution and may deplete supplies of oil more quickly. Countries which do not have major airlines will experience a larger debit on its trade in services balance.

   Whether an increased demand for air travel will have a net beneficial of harmful effect on an economy will depend on a number of factors. It will be more likely to have a beneficial effect if the increased demand adds to total demand for transport, if new, less polluting airplanes are employed and the economy operates major airlines.

2. (a) Pakistan's share of the global cotton market was 9.3m/101m = 9.21%.

   (b) A fall in the price of cotton may be caused by a rise in the supply of cotton and/or a decrease in demand for cotton.

   (c) Spices are an ingredient in national colouring. A rise in their price would increase the cost of producing natural clothing and so increase its price.

   (d) The information suggests that the supply of spices is inelastic. It mentions that it 'takes some time to increase the quantity supplied'. Indeed, it states that it takes five years to grow nutmeg. With such a long growing period, it is difficult to adjust supply to changes in demand.

(e) The price of spices rose in 2011 because of an increase in demand and an increase in supply. Demand increased due to an increase in the popularity of spices and their wider use. Supply decreases due to bad weather conditions and flooding. The figure below shows the increase in demand and decrease in supply pushed up price.

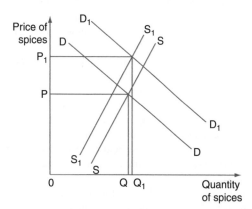

(f) Whether farmers will benefit from a rise in the price of their products will depend on the cause of the rise and the duration of the higher prices. A rise in price which results from an increase in demand is likely to be more beneficial than one which has been caused by a decrease in supply. Higher demand will raise revenue and may reduce cost per unit. A decrease in supply which results from bad weather and flooding would raise the price and possibly revenue of those farmers who are not so badly affected. It may, however, destroy most or not all of the crops of other farmers.

A rise in price may encourage farmers to plant more crops. If price remains high, they may be able to increase their revenue. If, however, price falls in the future they may receive less revenue than they expected and may make a loss.

3. (a) Derived demand means that demand for a product is dependent on demand for another product. The demand for OTR tyres is influenced by demand for trucks. The more trucks demanded, the more tyres are needed. In turn, demand for trucks is dependent on demand for minerals.

(b) The supply of OTR tyres was highly inelastic in 2006. This means that a rise in the price of the tyres would have resulted in a smaller percentage rise in quantity supplied. One reason why supply was relatively unresponsive to price change was that the tyre firms were working at full capacity. The firms did not have spare resources available to make more tyres. It was also not possible to switch resources from making car tyres to making OTR tyres. In addition, stocks were very low and so could not be drawn on to alter supply in any meaningful way and there were no new factories predicted to open before the end of 2007.

(c) In 2006, demand for tyres increased by estimated 50%. This large increase in demand would have caused price to rise

from P to P1 and the quantity supplied to extend from Q to Q1 as shown in the following figure. The rise in price would have been more significant than the rise in quantity supplied because supply was inelastic.

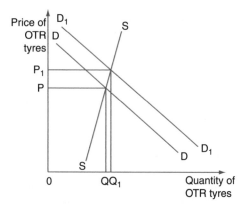

(d) The productivity of mining companies would decline. Productivity is output per factor unit per hour. With the problem of obtaining new tyres, some trucks may have to work with worn out tyres. This may mean that the trucks break down more often, have to respond more frequently and may have to be driven more slowly.

(e) Some might argue that the shortage of OTR tyres might require government intervention. This is because the market is not working efficiently and this is restricting the growth of mineral production. This, in turn, may restrict economic growth and employment in the countries producing the minerals.

On the other hand, the shortage may be corrected in the longer term by market forces. The rise in price will encourage existing firms to expand their capacity and may attract new firms to enter the industry.

There is also no guarantee that a government will have the knowledge and expertise to improve efficiency. Indeed, it may move the market from a shortage to a surplus.

Answers to essay questions

1. (a) Cross elasticity of demand (XED) is a measure of the responsiveness of demand for one product to a change in the price of another product. It is measured by the formula:

$$XED = \frac{\% \text{ change in quantity demanded of product A}}{\% \text{ change in the price of product B}}$$

Cross elasticity of demand indicates not only whether products are substitutes, complements or independent goods but also the extent of any relationship. The sign shows the relationship and the figure the extent of the relationship.

Substitutes have positive XED. For instance, a 10% rise in the price of holidays in Thailand may cause a 2% demand for holidays in Sri Lanka. This would give a positive XED of 0.2. Some people who might have visited Thailand may now switch to Sri Lanka. The following figure shows this positive relationship.

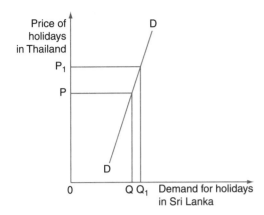

Price of holidays in Thailand — P_1, P, 0, Q Q_1 Demand for holidays in Sri Lanka

The figure of 0.2 indicates that the two holiday destinations are not very close substitutes. The more similar consumers view products to be, the higher the degree of positive XED they are likely to have.

Complements, in contrast, have negative cross elasticity of demand. This means that a change in the price of one product will cause demand for the related product to change in the opposite direction. For instance, the 10% rise in the price of holidays in Thailand may result in a 20% decrease in demand for flights to and from Thailand. This would give an XED figure of – 2.0. This would suggest that the two products are close substitutes. The following figure shows this inverse relationship between the changes in the price of holidays in Thailand and flights to and from Thailand.

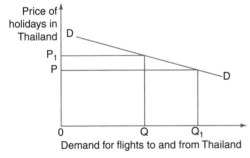

Demand for flights to and from Thailand

If products are unrelated, there will be a XED of 0. The following figure shows that the rise in the price of holidays in Thailand leaves demand for pineapple juice unchanged.

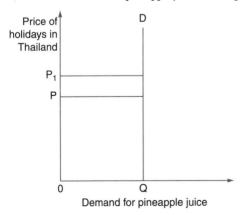

Demand for pineapple juice

(b) Price elasticity of demand (PED) is the responsiveness of demand to a change in price. The formula for calculating PED is:

$$\frac{\% \text{ change in quantity demanded}}{\% \text{ change in price}}$$

The PED for mobile phones has varied within countries over time and continues to vary between models and countries. When mobile phones were first introduced they were seen as something of a luxury and most were quite expensive. This made their demand relatively elastic. A rise in their price caused a greater percentage fall in demand.

Now in many countries, however, people have become reliant on their mobile phones for an increasing range of functions. For many people a mobile phone has become a necessity and this has made the demand for mobile phones in general to be inelastic. In most countries if the price of all models of mobile phones were to rise, there would be a smaller percentage fall in demand.

More expensive models of mobile phones which have substitutes in the form of other mobile phones have elastic demand. A rise in the price of one model may cause a significant number of consumers to switch to a rival brand.

In countries, such as Zimbabwe, where a relatively small proportion of the population own mobile phones, demand is likely to be relatively elastic. As more and more people buy mobile phones and the phones carry out more functions, demand will become more inelastic.

2. (a) Consumer surplus is the difference between the amount consumers are prepared to pay for a product and the amount they actually pay. If a woman was willing and able to pay $20 for a product but is only charged $16, she would receive consumer surplus of $4.

A subsidy to producers will increase supply. Producers will be encouraged to supply more because of the extra payment. The supply curve will shift to the right. The increase in supply will lower price from P to P1 and the quantity traded will rise from Q to Q1 as shown in the figure. The area of consumer surplus will change from PAB to P1AC.

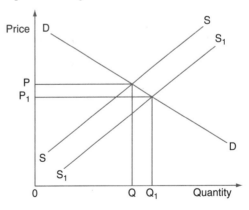

How much consumer surplus will increase will depend on the size of the subsidy and the price elasticity of demand. The larger the subsidy and the more inelastic the demand, the greater will be the increase in consumer surplus.

(b) Whether a fall in the price of a product is accompanied by a reduction in the quantity traded will depend on the cause of the fall. A fall in price caused by a decrease in demand will result in a contraction in supply. The following figure shows that the decline in demand results in a reduction in the quantity traded of 50,000 units.

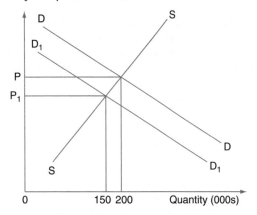

A fall in price caused by an increase in supply would, however, lead to a rise in the quantity traded. The figure below illustrates how a shift to the right of the supply curve causes price to fall to P. And the quantity traded to increase by 80,000 units.

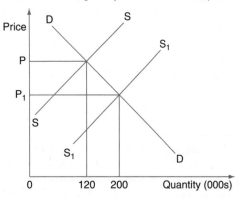

It is also possible that a fall in price and an increase in the quantity traded may be caused by a combination of an increase in supply and an increase in demand. For this to occur, supply would have to increase by more than demand.

The removal of a minimum price set above the equilibrium price would increase the quantity traded. The figure below shows that price is initially set at PX. At this price, there is a surplus of 10,000 units. 25,000 units are supplied but only 15,000 units are purchased. Removing the minimum price would permit the price to fall to the equilibrium price of P. The quantity traded now rises by 6,000 units to 21000 units.

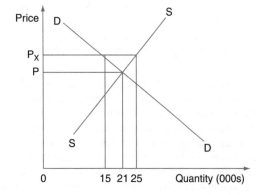

The relationship between a fall in price and the quantity traded will depend on the cause of the fall.

3. (a) Equilibrium price occurs where demand and supply are equal. If the price is above or below the equilibrium, market forces will move it to the equilibrium level. The figure shows a market initially in disequilibrium with supply exceeding demand at a price of P.

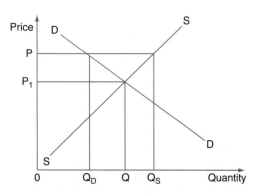

The unsold surplus will drive down price to P1. At this price, there will be no tendency for price to change. At P1 firms will be able to sell all that they are prepared to supply and consumers will be able to buy all that they are willing and able to purchase.

If price is below the equilibrium level, there will be excess demand. The figure shows demand being greater than supply at a price of P. The shortage created will lead some unsatisfied consumers offer to pay higher prices and so the price will be driven up. This will cause demand to contract and supply to extend until demand and supply are again equal. In this case, this is at a price of P1.

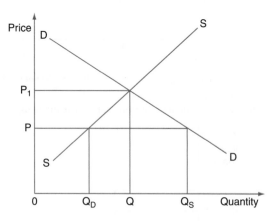

Over time price moves from one equilibrium to another equilibrium level and so on. What causes the equilibrium price to alter is changes in market conditions. If, for instance, demand increases, there will initially be a shortage. The following figure shows that at what was the equilibrium price of P, demand (Qa) will now exceed the quantity supplied (Q). As indicated above, this situation will drive up the price to a new equilibrium of P1.

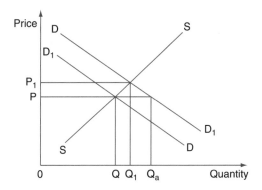

A decrease in supply will also cause the market to move to a new, higher equilibrium price. A lower equilibrium price will occur if either supply increases and/or demand decreases.

(b) A change in income can have a major influence on demand for a firm's products. Some firms produce products which have income elastic demand. This means that a change in income causes a greater percentage change in demand.

To decide how firms will be affected by a change in people's income, it is necessary to examine not only the degree of income elasticity of demand their products have but also the type of income elasticity of demand they possess and whether the country is experiencing economic growth or a recession.

Inferior goods have negative income elasticity of demand. This means that demand falls as income rises. In contrast, normal goods have positive income elasticity of demand, with demand and income moving in the same direction. Normal goods which have an income elasticity of demand of more than one are referred to as superior goods or luxury goods.

During a recession, a firm's revenue would increase if its product is an inferior one. Demand will increase as income falls and this increases the firm's revenue. Most years, however, income rises and during times of economic growth, a firm's revenue earned from the sale of a normal good will increase and may increase significantly if the product is a superior good. For instance, if income elasticity of demand is 3, a 10% rise in income will lead to a 30% increase in demand which will have a noticeable impact on total revenue. As income usually rises, a firm is likely to want to concentrate on products with positive income elasticity of demand.

Over a given income range, some products may have zero income elasticity of demand. This means that as income rises, demand does not change. For instance, a rise in income may have no effect on demand for toothpaste in a developed country.

As income changes over time, what are perceived to be superior, normal and inferior goods may change. For example, as income continues to rise in China, bicycles are beginning to be seen as inferior goods, as more people start to buy cars. Technological advances can also affect how products are perceived and so their income elasticity of demand. In a number of countries, digital radios have positive income elasticity of demand whilst analogue radios now have negative income elasticity of demand.

Chapter 7

Answers to short questions

1. To correct market failure arising from positive externalities, a subsidy should be set at a level equal to marginal external benefit (MEB). If MEB can be calculated accurately and a subsidy is given equal to this, the quantity traded should be based on social benefit.

2. 'The polluter pays principle' is the idea that those who create an external cost should pay for it. In other words, an external cost should be converted into an external benefit.

3. If social costs are equal to private costs, it means there are no external costs.

4. If private benefits form a high proportion of social benefits, students might be expected to pay a high proportion of tuition fees.

5. In carrying out a cost-benefit analysis it is often difficult to measure external costs and external benefits. Where market prices do not exist, estimated prices have to be used. These estimated prices are called shadow prices.

6. Education is a private good because it is both excludable and rival. In some countries education is provided free to children by the state. It is, however, possible to charge for education and there are many private schools and universities throughout the world. Those who are not willing to pay, can be prevented from gaining access to education. Education is rival in the sense that if one person is given a place in a university or in a school, that place is not available for someone else.

7. Electronic road pricing is charging drivers according to where and when they drive. A person will pay more for driving into a city centre at peak times than driving on an uncongested road. Being able to charge drivers in this way indicates that those who do not pay can be excluded from consuming road space.

8. Merit goods are a special category of private goods. In the case of, for instance, education and health care it is possible to exclude non-payers and consumption is rival.

9. A product may be treated as a demerit good in one country but not in another because governments have different views on how harmful some products are and how aware the public are of the harmful effects. For instance, Singapore bans the import and sale of chewing gum (except medicinal gum). Most other governments allow people to chew gum despite the problems caused by people disposing of chewing gum.

10. A government may provide free primary school education because parents may undervalue the private benefits their children will receive. Another reason is that parents will not take into account the external benefits of education when they make decisions about whether to send their children to school. In addition, a government may want to provide free education on the grounds of equity. Education is such an important product, that a government may consider that every child should have access to it, whether their parents could afford to buy it or not.

11. A maximum price is set below the equilibrium price. This will lead to a shortage, with demand exceeding supply. Some consumers, anxious to purchase the product may be prepared to buy from traders selling the product at a price above the legally set level.

12. A buffer stock may fail to stabilise the price of a commodity if not all producers are members of the scheme. In such a situation, if the non-members' output accounts for a noticeable proportion of the total, changes in the supply may lead to fluctuations in price. In addition, some members of the buffer stock may cheat and sell the product at a lower price in order to capture more of the market.

13. A government may impose a tax on a product for two main reasons. One is to raise revenue and the other is to discourage its consumption.

14. The state might stop producing a product if the product changes from being a public good into a private good. Initially, it was difficult to prevent people who did not pay from receiving TV broadcasts. For some time now this has been possible and more private sector firms are transmitting TV programmes throughout the world and charging directly for them. The state might also stop making a product if it considers that the public's desire for the product has declined.

15. The level of government intervention in a country's markets is determined by both the perceived level of market failure and also its view on whether intervention will be successful.

16. A first and second party involved in the production and consumption of air travel are the airline and a passenger. A third party is someone living near the airport who experiences air, noise and visual pollution.

17. A government may not go ahead with an investment project despite a CBA indicating it would provide a positive net present value if a rival project would give a higher net present value or would be more politically popular.

18. External costs being internalised means that they are changed into private costs. For instance, taxing a polluting firm would increase its private costs which may move its output to the allocatively efficient level.

19. Operators of a buffer stock more often buy the commodity than sell it because the limits are usually set too high, relative to the long run equilibrium level. This means that they have to intervene to buy the commodity to stop its price falling.

20. A government does not have to run a country's prisons. It does have to finance their provision as they are a public good but it could pay a private sector firm or firms to run them.

Answers to revision activities

1. (a) The information indicates that ice cream in China has positive income elasticity of demand and is a normal good. As income increased in the past, demand for ice cream increased.
   (b) One external cost arising from the consumption of high fat and high sugar foods to the increased burden placed on health services. Those who do not consume the foods may have to pay higher taxes or higher prices because of the increased burden and may also experience delays in medical treatment.
   (c) Firms selling dental products and diabetes testing devices will benefit from increased sales. This rise in their sales is a spillover effect resulting from the increased consumption of the high fat and high sugar products.

2. (a) Private benefits: 2, 7.
   Private costs: 6, 9.
   External benefits: 5.
   External costs: 1,3, 4, 8.

(b)

Output	Social benefit	Marginal social benefit	Social cost	Marginal social cost
20	100		90	
21	120	20	100	10
22	150	30	120	20
23	200	50	150	30
24	240	40	190	40
25	260	20	220	30

2. The optimum output is 24 units since this where MSB = MSC.

3. (a) A buffer stock
   (b) A smoking ban
   (c) Regulation
   (d) Government provision
   (e) Government subsidy to dentists

Answers to multiple choice questions

1. C   If more people travel by train, others will benefit from their action. These include road users. A is a private cost and B and D are private benefits.

2. B   External costs are usually not taken into account in consumption and production decisions. Their existence means that costs are higher than the market reflects, leading the price being too low and consumption being too high.

3. D   If there is a net social benefit, it means that social benefits exceed social costs. Social benefits are private benefits plus external benefits. In this case, social benefits are $600m + $700m = $1,300m. As private costs are $500m, it means that external costs must be less than $800m.

4. D   A definition question. A public good is both non-rival and non-excludable. A defines a merit good and C is a free good. There is no specific term for a product that has no external benefits or external costs.

5. C   As it is not possible to charge directly for a public good, there is a risk that some people will consume the product without paying for it.

6. B   A private good is both excludable and rival. It may be produced by both private firms and state owned enterprises. It may also generate external benefits as well as private benefits.

7. B   A government may decide to provide a private good such as education so that everyone can consume it. If the government thinks that people overvalue the benefits of consuming a product, they may seek to discourage its consumption. Private sector firms base their production decisions on private costs rather than social costs and are able to charge for a private good.

8. A   The equilibrium price is $3. Setting a maximum price above the price will have no effect on the market.

9. C   The maximum price is set above the equilibrium price and so is ineffective. The price and the quantity will be determined where demand and supply are equal.

10. A   At a maximum price of PT consumers would demand Q2 amount but would only be able to consume Q1 amount. Therefore, the total expenditure is OPTXQ1.

11. **C** A demerit good is over-consumed and so its output is too high. Its price is too low as it does not fully reflect the costs its consumers experience and does not take into the harmful effects on third parties.

12. **A** A government is likely to tax a demerit good in order to discourage its consumption. A public good will not be provided by the market and so a government is likely to either provide it directly or pay a private sector firm to provide.

13. **B** At a price of PF, supply will exceed demand. Consumers will purchase YZ amount in order to maintain the price of PF.

14. **C** A tax on a product is known as indirect tax. A non-parallel shift means that the tax as an actual amount increases the higher the price.

15. **C** The subsidy per unit is represented by the distance between the original and the new supply curve.

16. **C** The tax per unit is $5 and so the tax revenue is $5 × 100.

17. **A** A subsidy granted to the consumers of the product would increase demand and so shift the demand curve to the right. B would shift the supply curve to the right, C the demand curve to the left and D the supply curve to the left.

18. **C** If demand is elastic, the rise in price which will result from the imposition of a tax will cause demand to fall by a greater percentage. The tax should be set at a rate equivalent to marginal external cost.

19. **B** The allocatively efficient level of output would be achieved if people demand Z amount of the product. People will only buy this quantity at a price parallel to X. To reduce price to this point, a subsidy of WX per unit would be needed.

20. **B** To improve efficiency, a government should seek to encourage the consumption of a product which generates external benefits. Placing a tax on such a product would have the opposite effect. A, C and D should all increase efficiency.

Answers to data response questions

1. (a) (i) With limited resources, if the BBC decides to produce more educational programmes, it would have to produce fewer popular television programmes. The diagram shows that to produce twenty more educational programmes, the BBC would have to give up the production of ten popular television programmes. The opportunity cost of one educational programme is half a popular television programme.

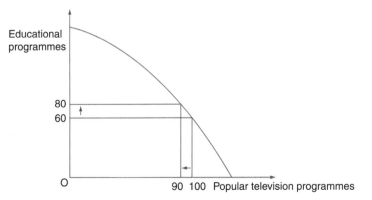

If more money is available to the programme makers from increased licence fees, more resources may be purchased by the BBC. This will enable more of both types of programmes to be made, shifting out the production possibility curve to the right.

(ii) A television licence and a television set are complements and so have negative cross elasticity of demand. If a television licence becomes more expensive, demand for television licences will contract. Demand for television sets will decrease. The demand curve for television sets will shift to the left. The lower demand for television sets will cause their price to fall and supply to contract.

(b) Figure 7.10a shows that some viewers would be prepared to pay a higher licence fee than that charged. They are getting extra value in the form of consumer surplus. The area above the price line and below the demand curve represents consumer surplus as shown in the version of the diagram shown here.

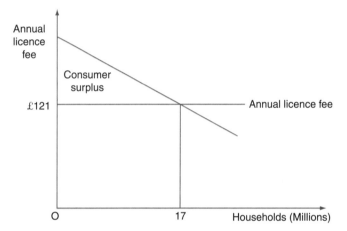

(c) (i) Figure 7.10b suggests that national defence has the characteristics of a public good to a greater extent than police services. It implies that it is completely impossible to exclude free riders from enjoying the benefits of national defence and that supplying the services for one person has little effect on the ability to supply it to someone else. In contrast, police services have a lower degree of non-excludability and a significantly lower degree of non-rivalry.

(ii) Education and health care are not public goods. They are private goods as, in both cases, they are excludable and rival. It is possible to prevent people who are not willing to pay directly for the services from enjoying them. Indeed, in a number of countries, education and health care are sold through the market. Using resources to provide education or health care for one person stops someone else from consuming the services produced by those particular resources. For instance, if one person is operated on, the operating theatre, doctors and nurses cannot be used to treat someone else at this time. In addition, the marginal cost of supplying the services is not zero and they are rejectable.

2. (a) Removing a subsidy will lead to a decrease in supply. The reduction is supply will raise price and cause demand to contract as shown in below.

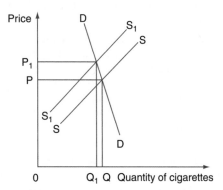

How much price rises and demand contracts will be influenced by how large the subsidy was and the price elasticity of demand. As demand for cigarettes is price inelastic, the main impact is likely to be on price rather than quantity.

(b) The price of soap was expected to rise less in actual amount than the price of toothpaste but by a greater amount in percentage terms – 600% as opposed to 300%.

(c) The information does not suggest that subsidy given to soap and toothpaste was sufficient to achieve the price the government wanted. Although the government had subsidised the products and imposed maximum prices, the products were sold illegally on shadow markets.

(d) It is probably more justifiable to subsidise soap than potatoes. Soap is a necessity. It may also be considered to be a merit good. Some people may not fully appreciate the need to wash on a regular basis. Keeping clean can cut down on the transmission of diseases and reduce health costs. Some people may also not be able to afford soap.

Potatoes are eaten in greater quantities by the poor than the rich. So there may be a case for subsidising them on the grounds of equity. There are, however, substitutes for potatoes which provide greater health benefits.

3. (a) Figure 7.11 shows that the demand for and price of copper move in the same direction. When the demand for copper rose between 2004 and 2008 and between 2009 and 2011, so did the price. In addition, when demand for copper fell between 2008 and 2009, so did the price. This may suggest that changes in demand caused the changes in price. Of course, however, price may also have changed as a result of changes in supply conditions.

(b) (i) An external benefit and an external cost are both effects on third parties that arise from the production and consumption decisions of others. Whereas an external benefit is a good effect on a third party, that he or she does not pay for, an external cost is a harmful effect on a third party for which he or she receives no financial compensation.

(ii) Two external costs created by the operation of the multinational company's copper mine are respiratory diseases and the destruction of crops. People living near the copper mine are experiencing deterioration

in their health, not through their own actions but as a result of the multinational company's mining activities. They are losing crops not because of their own production and consumption decisions but because of the way the multinational company is mining.

(c) The Zambian government could regulate the copper mining industry. It could set limits on the amount of sulphur dioxide that mining firms can emit. Any firm that exceeds the limit could be fined and/or its owners could face a prison sentence.

The Zambian government could also nationalise the mine. In this case, it could run the mine on the basis of social costs and benefits rather than private costs and benefits.

(d) The financing of the building and running of a local school by the multinational copper company is likely to generate both private and external benefits. The company may gain from being able to employ more educated workers. Such workers are likely to be skilled and hence more productive and so enable the company to produce at a lower average cost of production.

The local population will experience an external benefit from the provision of education for their children. The increased knowledge, skills, better job opportunities and possibly higher pay that the children may gain will be the result of the expenditure by the mining company.

Answers to essay questions

1. (a) A normal demand curve slopes down from left to right indicating that as price falls. The quantity demanded rises. All the consumers in a market usually pay the same price for a product. They do not all, however, have the same willingness to buy the product. In the figure below, the price charged is $10 and 50 units are sold.

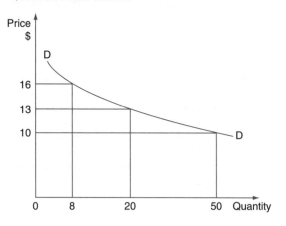

Eight units, however, could have been sold at $16. People prepared to pay this amount would have enjoyed a consumer surplus of $6. The term consumer surplus refers to the difference between what people are prepared to pay for a product and the amount they actually pay. The figure above also shows that 20 units would have been demanded at a price of $13. Those willing and able to pay $13 but actually charged $10, in effect get a bonus of $3. People only prepared to pay $10, however, would experience no consumer surplus.

The area of consumer surplus is that above the price line and below the demand curve as illustrated by the stated area in the following.

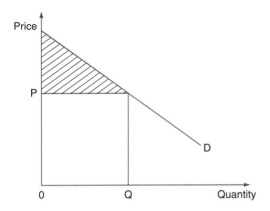

A fall in price will cause consumer surplus to rise. The figure below shows the reduction in price from P to P1 leading to a rise in consumer surplus from PAB to P1AC.

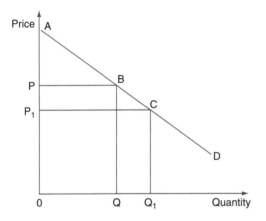

(b) The imposition of an indirect tax means that producers have to pay the government money. This effectively raises firms' costs of production and so decreases supply. The fall in supply raises price, passing some of the tax on to consumers. The more inelastic demand is the greater the proportion of the tax borne by consumers.

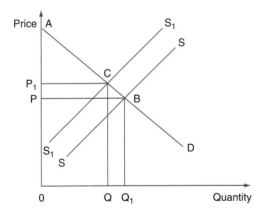

The diagram seems to suggest that consumers will suffer as a result of the introduction of the indirect tax. They pay more, consume less and experience a decrease in consumer surplus from PAB to P1AC.

If, however, the tax is put on a product which is a demerit good, consumers may benefit. The higher price may move the market closer to the allocatively efficient level. Consumers may experience less harmful effects from their own consumption of the product and from other people's consumption. For example, a tax on cigarettes may discourage some people from smoking. This will improve their health, reduce expenditure on smoking related illnesses and increase labour productivity.

Consumers may also benefit from government expenditure financed by the indirect tax on the product. This spending might go on merit goods and public goods. Merit goods provide a greater benefit to consumers than they themselves realise and provide external benefits. Public goods will not be provided by the private sector as it is not possible to exclude free riders from enjoying them.

Some consumers might also benefit from the introduction of an effective maximum price. To be effective a maximum price would have to be set below the equilibrium price. The figure below shows that the effect of the maximum price will be to create shortage.

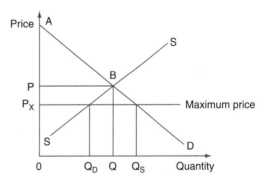

The quantity demanded will exceed the quantity supplied. Some consumers will benefit as they will be able to purchase the product at a lower price and so will be able to enjoy more consumer surplus. Other consumers, however, will not be able to obtain the product. The excess demand may result in a shadow market developing, selling the product above the legally enforced maximum price.

The following figure shows that the consumers who are able to purchase the product at the maximum price of PX gain an additional consumer surplus of Y. Those who are unable to buy the product lose consumer surplus of X. With area Y being greater than area X, there is a net gain to consumers.

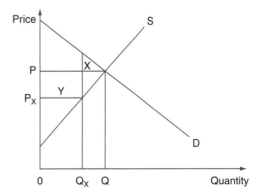

The more elastic demand and supply are, the more chance there is that there will be a net loss in consumer surplus. The figure below shows the loss in consumer surplus (X) is greater than the gain in consumer surplus (Y).

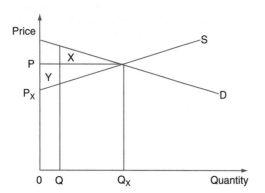

2. (a) Information failure can take a number of forms. Consumers may lack information or they may have inaccurate information. There may also be asymmetric information with producers having more information than consumers.

Information failure can result in market output being above or below the allocatively efficient level. Information failure often occurs when there are negative and positive externalities. A lack of awareness of effects on third parties can result in an inefficient use of resources. If not all costs are taken into account, output will be too high. In contrast, if not all benefits are considered, output will be too low.

Information failure can be particularly significant in the case of both merit and demerit goods. These products generate externalities but are also more beneficial or harmful to consumers than they themselves realise. Consumers lack information or have inaccurate information about merit goods and so they under-consume them. The figure below shows that actual demand is DD whereas demand based on full and accurate information about the merit good is DXDX.

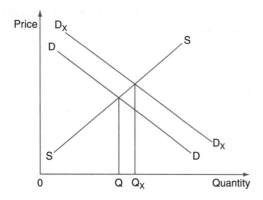

The allocatively efficient output is QX but the market output is below this at Q. The product is under-consumed and so under-produced.

In the case of demerit goods, there is information failure in terms of their harmful effects. As people do not fully appreciate the adverse effects, consumption of such products may have on themselves and on third parties, there is over-consumption and so overproduction. The figure below shows the market demand is DD whereas demand based on full information would be DXDX.

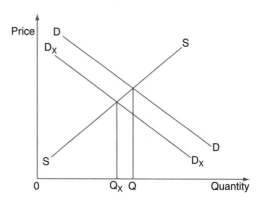

Too many resources are devoted to the product.

(b) To correct market failure in the case of merit and demerit goods a government may provide information, grant a subsidy or impose an indirect taxation or use regulation.

In the case of a merit good, a government may provide information, subsidise its production or consumption or make its consumption compulsory. Providing information about the beneficial effects of a merit good is designed to increase demand. The following figure shows the demand curve shifting to the right, moving the market closer to the allocatively efficient level of QX.

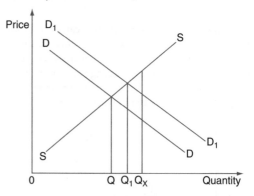

The provision of information works with the market and can help to reduce information failure. To be effective, however, the government must itself have accurate information. It can be debated who is a better able to judge how beneficial a product is – consumers themselves or the government? There is also no guarantee that consumers will be persuaded by the information.

A subsidy given to consumers is also designed to increase demand. Higher demand will again move the market closer to the allocatively efficient level. It may, however, be difficult for a government to calculate the amount of the subsidy. If it is too high, the product ay end up being over-consumed.

A subsidy given to producers would be designed to encourage higher consumption by reducing its price. The following figure shows a subsidy would shift the supply curve to the right, lower price and cause demand to extend.

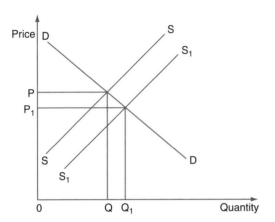

A subsidy to a producer is a market based solution and the resulting lower price can particularly benefit the poor. The more elastic demand is, the larger the impact on the quantity bought and sold is likely to be. A subsidy has to be set at the right level and may involve a relatively high cost. It may also take time to introduce and will be less effective if demand is inelastic.

If it is thought that the product is very important for people to consume, a government may make it compulsory for people to consume it. For example, a government may require people to send their children to school, wear seat belts in cars and install smoke alarms in their homes. Using regulation has the benefits of being backed by law and simple to understand. It does, however, require monitoring and if a significant proportion of the population are not in favour of the law, may require a high cost of enforcement.

To discourage consumption of a demerit good, a government may provide information about the harmful effects of consuming the product. This may discourage demand and move the market to the allocatively efficient level. As with providing information about a merit good, however, the government must have accurate information and consumers may not respond in the way expected.

A government may also use regulation by banning the consumption of the product or not permitting children to consume the product. Sometimes a ban can have a significant impact on people's behaviour. For instance, the smoking ban in public places in Ireland has reduced smoking. There is a risk, however, that a ban may give rise to a shadow market. For example, the ban on the sale of ivory has resulted in an illegal trade in ivory.

In addition to providing information and using regulation, a government may impose a tax on a demerit good. The imposition of an indirect tax will cause the supply curve to shift to the left, price to rise and demand to contract as shown in the figure.

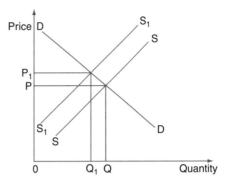

This measure works with the market and raises revenue for the government. It may not, however, be very effective if demand is inelastic, will fall more heavily on the poor and may contribute to inflation.

All policy measures have some advantages and some disadvantages, and governments often use a combination of measures. For example, governments run anti-smoking campaigns using TV advertisements, tax cigarettes and some require cigarette producers to put health warnings on their packets, ban children from smoking and ban smoking in public places.

3. (a) A public good is one which is both non-rival and non-excludable. Non-rival means that one person's consumption of the product does not stop another person consuming the product. Non-excludable means that it is not possible to prevent those not willing to pay for it, from consuming it. For instance, defence is a public good. Defending one more person will not reduce the benefit other people gain from the defence. As well as being non-rival, it is also non-excludable. It is not possible to provide defence for some people whilst excluding non-payers and so it is non-excludable.

The existence of free riders means that private sector firms will lack an incentive to produce public goods. The market will fail to make public goods as it is not possible to charge for them directly. This is why the government has to finance the production of public goods and pay for the output from tax revenue. Advances in technology, however, are converting a number of products from public goods to private goods. For example, electronic road pricing has the potential to turn road space into a private good.

(b) Cost-benefit analysis (CBA) is not perfect but it can help to improve economic decision making. A CBA is usually applied in the case of large scale public investment projects.

A private sector investment appraisal takes into account private costs and private benefits. A private sector firm will go ahead with an investment project if it estimates that the revenue it will earn from the project will exceed the costs it will incur. Such an approach does not take into account the impact the project is likely to have on those not directly involved in producing or consuming the product.

A CBA has the advantage that it seeks to include the effects on the whole of society. As well as considering private costs and private benefits, it also takes into account external costs and external benefits. A CBA will recommend a project if social benefits exceed social costs, giving rise to a positive net present value.

Conducting a CBA, however, is not an easy process. Those likely to be affected by the project have to be identified. The likelihood of costs and benefits occurring has to be established and then a monetary value has to be placed on the costs and benefits. This is particularly difficult in the case of external costs and external benefits. There is a degree of subjectivity in seeking to value such effects. For instance, whilst the value of noise experienced by households living close to a new rail line might be estimated by considering the cost of sound proofing the accommodation; it is much harder to decide how to value the visual pollution the residents may experience. It is especially difficult to value the effect on the environment. Asking people how they will be affected will not necessarily provide reliable evidence as beneficiaries tend to understate the positive effects they receive and those who will be disadvantaged will tend to overstate the harmful effects.

Future costs and benefits are discounted to take into account the greater value derived from revenue received now and the more significant costs experienced now are. It can be difficult to decide on the rate of discount as, for instance, it will involve future interest rates.

A CBA may also be influenced by political factors. A government may be influenced by lobby groups and may be reluctant to accept the findings of a CBA if it thinks it will make it politically unpopular.

In practice it is very unlikely that a CBA will cover all the costs and benefits that will arise from an investment project. It is also very unlikely that it will succeed in measuring accurately those costs and benefits it does identify A CBA, nevertheless, does consider more of the effects than a private sector investment appraisal and is more likely to come closer to a more efficient decision.

Chapter 8

Answers to short questions

1. International trade may differ from domestic trade in a number of ways. It involves the exchange of products between countries, may involve a greater distance and so higher transport costs, may involve dealing with different currencies, may mean working in different languages, may encounter trade restrictions and may involve different tastes, cultures, rules and regulations.

2. A supermarket may buy food from abroad because it may be cheaper, may be of a better quality and may not be available in the domestic market.

3. Bilateral trade is trade between two countries whereas multilateral trade is trade with many nations.

4. A country may have an absolute advantage in producing a product but not a comparative advantage if, whilst it is better at producing it than another country, it is even better at producing another product.

5. It is possible to calculate in which product a country has a comparative advantage by working out the opportunity cost ratios of producing different products and by considering the difference in the quantity it can produce per resource unit compared with other countries. A country will have a comparative advantage if it can produce it at a lower opportunity cost ratio than another country and if it is even better or not so bad at producing the product.

6. Factor endowments influence what a country is good at producing. For instance, if a country has good quality, fertile land with a good climate, it may have a comparative advantage in producing agricultural products.

7. Free trade may increase competition. This may drive down price and so increase consumer surplus.

8. A tariff is a tax on imports whereas a quota is a limit on imports.

9. A government may introduce an embargo on products that it regards to be harmful for people to consume. It may also impose a general embargo on trade with a country that it is at war with or has a dispute with.

10. Imposing trade restrictions on a country that employs child and slave labour may make the offending country poorer and worsen its people's living conditions. It is, however, generally thought that it is legitimate to impose trade restrictions on countries which engage in illegal labour market conditions.

11. Globalisation is the development of the world into one market by breaking down barriers to the free movement of products, capital and labour.

12. Trading blocs promote free trade between member countries but not necessarily with countries which are not members of the trading bloc.

13. Trading blocs consist of countries which are geographically close to each other.

14. The terms of trade is:

$$\frac{\text{The index of average export prices} \times 100}{\text{The index of average import prices}}$$

In this case, the terms of trade are:

$$\frac{120 \times 100}{160} = 75$$

15. The terms of trade tend to move against agricultural products because demand (and the price) of these products tend to rise less than manufactured goods and services when income rises. In addition, agricultural producers usually have less bargaining power than producers of manufactured goods and services.

16. Merchandise exports are exports of goods.

17. An invisible trade deficit means that a country has earned less from the sale of services to other countries than it has spent on imports of services.

18. A balance of trade deficit means that the value of imported goods is greater than the value of exported goods. If a country has a balance of trade deficit and a current account surplus, it must have a greater overall combined surplus on the three other parts of the current account balance. The three other parts are trade in services, income and current transfers.

19. Portfolio investment is spending on shares and debt securities of foreign countries and foreigners' spending on domestic shares and debt securities.

20. Direct and portfolio investment and loans appear in the financial account. These give rise to profit, interest and dividends which appear as income in the current account balance.

Answers to revision activities

1. (a) Country X has the absolute advantage in producing both products as it can make more of both.
   (b) In Country X the opportunity cost of product A is 2B whereas in Country Y it is 6B. In Country X an increase in 16 units of A involves sacrificing 32 units of B, a ratio of 1:2. In Country Y, to produce 2 more units of A involves giving up 12 units of B, a ration of 1:6.
   (c) Country Y has the comparative advantage in producing product B as it can produce it for a lower opportunity cost – one sixth as opposed to a half.
   (d) An exchange rate of 7 units of product B for 1 unit of product A would not benefit both countries as it lies outside the countries' opportunity cost ratios. The exchange rate has to be more than 2 units of product B and less than 6 units of product B to benefit both countries.

2. Figure 8.2 shows that the country produces copper more cheaply than the world price.
   (a) Domestic consumers would be faced with a higher price and lower quality.
   (b) Domestic producers would benefit from selling a higher quantity at a higher price.
   (c) Foreign producers would face more competition and their share of the market would be reduced.

3. (a) Trade in goods, credit item.
   (b) Trade in goods, debit item.
   (c) Trade in services, debit item.
   (d) Trade in services, debit item.
   (e) Financial account, credit item.
   (f) Income, debit item.
   (g) Current transfers, debit item.

Answers to multiple choice questions

1. C   Country Y has an absolute advantage in both products as it can make more of both products. It has a comparative advantage in producing chairs as it can make these at a lower opportunity cost – 1 chair equals a quarter of a TV as opposed to 1 chair equals half a TV. Country X has the comparative advantage in producing TVs – 2 chairs as opposed to 4 chairs.

2. A   Country Y has the absolute advantage in both products. It is even better at producing rice. Country X is not as bad at producing cotton – it has the comparative advantage in this product. Country Y should specialise in rice and Country X should specialise in cotton.

3. C   Country R has the absolute advantage in producing both products. It is even better at producing wheat making 8 times as much wheat but only 5 times as many cars. Country R has the comparative advantage in making cars. Country R will concentrate on making wheat and will export it in exchange for cars from Country S.

4. B   If there is perfect mobility of factors of production between the countries, the resources may move rather than the products. A and D would make trade more likely. In the case of C, trade is based more on comparative rather than absolute advantage.

5. B   To benefit both countries, the terms of trade have to lie between the opportunity cost ratios of the two countries. This

means that it must be more than 2 units of X (200/100) and less than 4 units of X (100/25).

6. B   Country Y has the absolute advantage in producing both products. Initially, its comparative advantage is in producing manufactured goods and X's comparative advantage in producing agricultural goods. In Country Y, the opportunity cost of producing 1 manufactured good is 3/5 of an agricultural good whereas it is 2 agricultural goods in Country X. The opportunity cost of producing agricultural goods is lower in Country X – ½ manufactured goods as opposed to 1 ¼ in Country Y. After the change in technology, the opportunity cost ratios change for Y but X continues to have the lower opportunity cost in producing agricultural products (½ manufactured good as opposed to now 1 ½ in Country Y). Country Y also continues to have the comparative advantage in producing manufactured goods (now 2/3 as opposed to 2 agricultural goods).

7. A   If a country has a comparative advantage over its trading partner in the production of X, it would be able to exchange the product for more of Y through international trade. This would increase the quantity of Y it could consume.

8. D   A country is said to have a comparative advantage in a product if it can produce it at a lower opportunity cost than another country.

9. A   Dumping is regarded as unfair competition as it involves selling a product in a foreign market at less than cost price. B is not regarded as a valid reason and C and D are unlikely to occur.

10. C   The tax per unit is UX and XY quantity of goods are imported so tax revenue is UVXY.

11. A   The abolition of subsidies would reduce the protection given to domestic producers. They would now have to compete on more equal terms with foreign producers. B, C and D would all increase protectionism.

12. C   A tariff on imports of corn will raise the price of imported corn. This will encourage consumers to switch to domestically grown corn. The higher demand for domestically produced corn and the reduction in competitive pressure is likely to increase the price of domestically produced corn.

13. C   A tariff is a tax on imports. Ad valorem means to the value or percentage.

14. B   The main difference between a customs union and a free trade area is that a customs union requires member countries to impose the same tariff on non-members whereas a free trade area does not. A, C and D are possible features of an economic union.

15. B   The terms of trade is a measure concerned with export and import prices. A favourable movement occurs when the number gets larger. If initially the index of average export prices is 100 and the index of average import prices is 100, the terms of trade would be 100. Then if export prices fall to 90 and import prices fall to 60, the terms of trade would increase to 150.

16. C   In year 1, the terms of trade was 400/500 x 100 = 80 and in year 2 the terms of trade changed to 560/600 x 100 = 93. The balance of trade changed from a deficit of $3m ($m - $7m) to a deficit of $2.8m ($5.6m - $8.4m).

17. B   Foreign direct investment appears in the financial account of the balance of payments.

18. A   A credit item involves money coming into the country. Money spent in Pakistan by Egyptian tourists brings money into Pakistan and would appear in the trade in services balance of the current account. B and C would appear in the current account (trade in services and trade in goods) but as credit items. D is a credit item but one which would appear in the financial account.

19. C   The current account is the visible trade balance plus trade in services balance plus net income plus net current transfers.

20. A   Expenditure by UK visitors in a New Zealand cinema would appear in the trade in services (invisible balance) of New Zealand's current account. B is an invisible import and C is a visible export. D would not appear in the current account until any profit that arises from the revenue is sent back to the US. It would then appear as a debit item in the income part of the balance of payments.

Answers to data response questions

1. (a) (i) The balance of trade in goods between Australia and Thailand in 2002 was $2,510m – $3,140m = a deficit of $630m.
    (ii) Australia exported minerals and agricultural products. Thailand imported mainly manufactured products with the exception of petroleum and seafood.
    (iii) The differences in the products traded may have been explained by comparative advantage. A country has a comparative advantage in a product when it can produce it at a lower opportunity cost than its trading partner. Australia has factor endowments which favour the production of minerals and agricultural products. It has a plentiful supply of minerals and has a large quantity of fertile land. Thailand possesses good supplies of capital equipment with which to produce capital goods.
   (b) (i) Two protective measures are quotas and embargoes.
    (ii) The following figure shows the effect of removing a tariff.

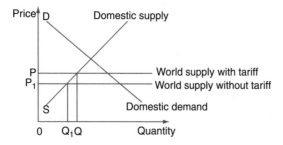

   Price falls from P to P1as there is more competition from imported products. Domestic producers will reduce their supply from Q to Q1 as they will be less willing and able to supply the product at a lower price.
   (c) There are arguments both for and against abolishing all tariffs immediately. Such an approach makes it easier for both countries to exploit fully their comparative advantage. With Australia concentrating on minerals and agricultural products to a greater extent and Thailand specialising more on manufactured goods, output should rise as the data suggests. Consumers should experience higher quality products, lower prices and so more consumer surplus because of the increased competition.

Abolishing all tariffs, however, would not necessarily remove all protection. Other measures of protection, including quotas and exchange control, may still exist and may indeed increase. Reducing tariffs immediately may also not give domestic firms time to adjust to the new competitive challenge. For instance, Thailand's car producers would have had to cope with the tariff on large cars being abolished.

There are also arguments in favour of maintaining some tariffs. These include to protect new industries known as infant industries. They may need protection to become established as they will not initially be able to take advantage of economies of scale. Thailand might have been imposing a tariff on wheat from Australia if it was trying to establish wheat production. Another argument for imposing a tariff is to prevent dumping. If Thailand thought that Australia was dumping beef in the country and the country has beef farmers, it may be justified in keeping its tariff.

Abolishing tariffs has the potential to increase efficiency and welfare. It may, however, be difficult to abolish all tariffs immediately and there may be some justification for keeping some tariffs.

2. (a) A ban on exports of beef is intended to increase the supply of beef on the domestic market. The following figure shows that a shift to the right of the supply curve will lower price and cause demand to extend.

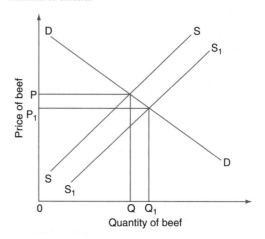

At a time when the overall supply was decreasing because of the drought and a reduction in the quantity of land devoted to beef production, this measure might actually have stopped supply from decreasing and so may have prevented price rising.
   (b) A rise in the relative profitability of soyabean production will encourage farmers to switch from beef production to soyabean production.
   (c) (i) The European Union would have imposed a quota on Argentine beef to protect its own farmers. By restricting the amount of Argentine beef that can be sold in the EU, it is likely that there would be more demand for EU produced beef.
    (ii) The data suggests that in 2010 the European Union's quota was not actually needed. Argentina's exports were below the quota and so the quota did not have any impact on the quantity of imports.
   (d) It is uncertain whether Argentina will export more beef in the future. It may do so if the government removes restrictions on exports and climate conditions remain favourable. It will also

depend on whether demand for beef continues to rise and also the price and quality competitiveness of Argentine beef.

There is a possibility, however, that Argentina may export less beef in the future. Demand for beef may fall if there is a global recession or a rise in vegetarianism. Importing countries may impose more trade restrictions, including tariffs and embargoes. Other countries, such as Uruguay, may capture more of the global market. Comparative advantage can change over time. For instance, Uruguay's farmers may become more productive as a result of training and the quality of cattle stock may be increased by selective breeding. The reputation of Uruguayan beef may also increase over time and the country may be better at negotiating free trade agreements with other countries.

Argentina's comparative advantage may switch to other products. The country may become more productive in other agricultural products such as soyabeans or may switch more of its resources to manufactured goods and services.

3. (a) (i) The price of copper in the middle of 2006 was higher by approximately $6,000 compared to the middle of 2003.

   (ii) An increase in demand may have caused the rise in the price of copper. Demand might have increased due to higher global output. If countries were producing more products they would have bought more copper.

   Alternatively, the price of copper may have been driven up by a decrease in supply. For example, an increase in the cost of producing copper would have shifted the supply curve of copper to the left, pushing up its price.

   (b) (i) The demand for copper is likely to have increased as 2006 was 'a year of global growth'. With higher incomes, demand for products using copper in their production would increase.

   As well as higher demand, a decrease in supply is also likely to have contributed to the higher price. The data mentions there was a strike at the largest mine in 2006. This would have reduced supply.

   (ii) In addition to the information given in Figure 8.8, details would be needed about the value of imports of goods, the trade in services balance, income balance and current transfers balance.

   (c) (i) Copper, as part of Chile's exports, became more important between 2002, and 2005. In 2002 copper exports accounted for just over 40% of total exports but approximately 60% in 2005.

   (ii) Specialising in the production and export of one single product can allow a country to exploit fully its comparative advantage. Concentrating on one product can enable considerable advantage to be taken of economies of scale, lowering unit costs. The country may also gain a reputation for producing the product, with buyers automatically associating the country with the product.

   Relying on producing and exporting one product is, however, a high risk strategy. Demand for a product may decrease due to a change in tastes. Rival countries may also develop a comparative advantage in the product or may impose trade restrictions on the product. There is also a risk that supply problems may be encountered. This is particularly true of agricultural products which can be adversely affected by bad weather and disease.

Concentrating on one product may also result in resource depletion which may endanger future generations' living standards. In addition, production of the product may not create high quality jobs which develop workers' skills and increase their incomes.

Most countries export a range of products to spread their risks.

Answers to essay questions

1. (a) Trading internationally often involves sending goods longer distances than when trading domestically. Longer journeys involve higher transport costs. This, however, is not always the case. For instance, a Chinese firm based in Nanning will be sending its product a shorter distance to Hanoi in Vietnam than to Beijing.

   Engaging in international trade, unlike domestic trade, may include dealing with other currencies. A Pakistani firm, for instance, that buys imports from Germany will probably pay in euros and one that exports to Japan will probably be paid in yen. Changing these currencies into Pakistani rupees (PRs) will involve time and effort. As well as these transaction costs, there is a risk involved in exchanging currencies. For instance, the firm may agree to pay 50,000 euros for some imports when the exchange rate is 1 euro equals 100PRs. If by the time the payment has to be made, the value of the PR has fallen to 1 euro equals 125 PRs, the firm will have to pay 6.25mPRs instead of 5mPRs.

   Trade restrictions may occur when trading internationally whereas they are absent in domestic trade. When a firm sells a product within its own country, it is competing on equal terms with other firms in the country. It may, however, be at a competitive disadvantage when selling to other countries if the governments of those countries impose trade restrictions. A tariff may be imposed or a subsidy given to domestic producers in a bid to make imports less price competitive. A quota, exchange control or voluntary export restraints may make it difficult for foreign firms to sell the quantities they want. Even more significant would be an embargo which would close a market to imports.

   International trade provides importers with a greater range of firms to buy from but also provides exporters with both a larger market but also more of a challenge. They will face more competition and may have to advertise and provide information in a different language. Different standards and laws may apply in different countries. Tastes may also vary from country to country. With globalisation, however, there is an increase in the similarity of tastes with some brands becoming well known throughout the world.

   (b) The global distribution of factors of production determines what a country imports and exports to a certain extent. The four factors of production are land, labour, capital and the entrepreneur. Factors of production are not evenly distributed throughout the world. For instance, Saudi Arabia is rich in oil, Germany in advanced capital equipment and Hong Kong in workers with well developed skills in the financial sector.

   If a country has plentiful supplies of fish and workers skilled in fishing, it may concentrate on fishing. Another country with a plentiful supply of relatively low skilled labour may specialise in labour intensive, heavy industry. The theory of comparative advantage states that countries will benefit from specialising in products they have a lower opportunity cost in making. So opportunity cost, in this case, reflects factor endowment.

Factor endowment, however, does not explain all of international trade. A country may have, for instance, large supplies of natural gas or may have the capital equipment and workers who are skilled in producing cars but may still import the products if the country's demand exceeds its supply.

Trade restrictions can also distort the pattern of international trade. A country may be relatively efficient in producing a product but may not be able to export it if other countries impose tariffs, quotas, embargoes or other protectionist measures. The existence of trading blocs may mean that member countries trade more with other members and less with non-members.

The exchange rate may also lie outside the opportunity cost ratios which will mean that a country may not benefit from trading. In practice, it may be difficult to assess where a comparative advantage lies as so many products and countries are involved and the quantity and quality of factors of production change over time.

Countries may want to produce products which do not reflect factor endowments because they want to protect strategic industries or build up infant industries.

So whilst factor endowments and comparative advantage explain some international trade, they do not explain all of it.

2. (a) Absolute advantage occurs when a country can produce more of a product per resource unit. In the example below, Country A has the absolute advantage in producing cars whilst Country B has the absolute advantage in growing wheat.

Output per resource unit		
	Cars	Wheat
Country A	50	100
Country B	20	400

It is thought, however, that comparative advantage explains a greater proportion of international trade. A country is said to have a comparative advantage in a product when it can produce it at a lower opportunity cost and is even better at producing the product or not so bad at producing it. In the example below, Country X has the absolute advantage in producing both TVs and rice.

Output per resource unit		
	TVs	Rice
Country X	20	500
Country Y	5	250

Country X has the comparative advantage in producing TVs. The country can make four times as many TVs than Country Y whereas it can only make twice as much rice. The country has a lower opportunity cost in producing TVs, 25 rice as opposed to 50 rice in the case of Country Y.

Country Y has the comparative advantage in producing rice. It can make only a quarter as many TVs but half as much rice. It has a lower opportunity cost – 1/50 of a TV as opposed to 1/25 of a TV in the case of Country X.

Comparative advantage theory suggests that international trade is beneficial for two countries even if one of them has the absolute advantage in both products produced as long as two conditions are met. These are that there should be a difference in the relative efficiencies and that the exchange rate should lie between the two countries' opportunity cost ratios. In the example above, the exchange rate would have to be 1 TV exchanges for more than 25 rice and less than 50 rice.

(b) The main reason why a government may wish to pursue a policy of free trade is that unrestricted international trade has the potential to increase the living standards of its citizens.

Free trade can allow a country to exploit fully its comparative advantage. It can concentrate on producing what it is best at producing and import products at a lower opportunity cost. If other countries follow a policy of free trade, there can be an efficient global allocation of resources. Global output can rise and people can consume more products.

Higher level of competition which results from free international trade can drive down prices, raising consumer surplus. It can also raise the quality of what is produced.

There are, however, a number of reasons why it may not always be to a country's advantage to engage in free trade. If other countries are imposing trade restrictions, the country may need to protect its industries. If other countries are subsidising their firms or dumping products in the country, its firms may not be able to compete even if they have a comparative advantage.

It may also be thought that some of the country's firms may be able to grow and develop a comparative advantage. An infant (sunrise) industry may experience a fall in average cost when it expands as it would be able to exploit economies of scale to a greater extent. If it is not protected at the start, it may be competed out of the market. It may, however, be difficult to spot which new industries have the potential to develop a comparative advantage. There is also a risk that even if the industries with potential are selected, the protection may make them become complacent and so they may not go on to lower their average costs significantly.

A government may also want to protect a declining (sunset) industry to prevent a sudden and large increase in unemployment. By allowing an industry to decline gradually, the number of workers can be allowed to fall through what is called natural wastage. Over time workers who leave through retirement and moving to other jobs may not be replaced. There is a risk, however, that those involved with the industry will fight to keep the industry going. It may also be argued that it would be better to keep unemployment down by giving support to infant rather than declining industries.

In addition, a government may want to protect its strategic industries, including agriculture, to ensure their survival. A government may not want to become reliant on imports for essential products in case, for whatever reason, the supplies are cut off.

In theory, free trade has the potential to create significant benefits but in practice there are reasons why a government may not always pursue a policy of free trade. Indeed, no government in the world pursues a policy of completely free trade.

3. (a) The terms of trade is the ratio of average export prices relative to average import prices. A deterioration in the terms of trade means that the number gets smaller and results from export prices falling relative to import prices.

There are a number of possible causes of deterioration. One is devaluation. A government may decide to lower the value of its currency from one fixed rate to a lower one in order to reduce a current account deficit.

Export prices may also fall relative to import prices as a result of rising labour productivity in the country or higher inflation in other countries from which the imports are purchased. As with devaluation, these causes are likely to increase the volume of exports sold and reduce the volume of imports purchased. Whether export revenue rises and import expenditure falls will depend on the price elasticity of demand for exports and imports.

A less beneficial cause of deterioration is a fall in the relative price of exports resulting from a decrease in demand. In this case, fewer exports will be sold at a lower price, causing export revenue to decline. A deterioration caused by higher demand for exports may also have a harmful effect on the current account of the balance of payments, although the effect is more uncertain. This is because some of the imports may be capital goods or raw materials which may later contribute to the country's exports.

(b) The effect that the formation of regional trading blocs will have on competition will depend on the nature of the trading blocs and may differ between those countries inside the blocs and those outside.

A free trade area seeks to achieve free trade between member countries. All the member countries agree to remove trade restrictions on the free movement of products between each other. This should promote competition within the free trade area as now competition will not be distorted by protectionist measures. If the firms in one member country produce a product at a higher cost, the country's government cannot discourage its citizens from purchasing imports by, for instance, imposing tariffs. The member countries are, however, free to set whatever restrictions they want on non-members. This means that the formation of a free trade area may have a neutral effect in terms of competition with non-members.

A customs union involves not only removing trade restrictions on fellow members but also agreeing on imposing the same tariff on the products of non-member countries. Again this trading bloc should increase competition between the members.

An economic union takes integration even further than a customs union. As well as removing trade restrictions on other member countries and imposing a common external tariff, it involves the countries moving towards operating as one economy. It has not only free movement of products but also the free movement of labour and capital. People can work in other member countries, with their qualifications recognised, and firms can set up branches in other member countries.

An economic union involves the use of a common currency. This means that a member country cannot gain a competitive advantage over its fellow member countries by devaluing its currency. A common currency contributes to a level playing field of competition and reduces transaction costs and promotes price transparency. The harmonisation of economic policies, such as minimum wage rates and similar tax rates, also make it difficult for one government to protect its industries from competition from other member countries.

The effect that the formation of regional trading blocs will have on non-members will depend on whether any external tariff is higher or lower than the average tariff member countries were charging before their formation.

It will also be influenced by whether the increased internal competitive pressures raise the efficiency of member countries' firms. If they do, firms in one trading bloc may be able to compete more effectively with firms from other trading blocs. Those countries outside of major trading blocs may, however, have greater difficulty competing.

Chapter 9
Answers to short questions

1. The size of a country's labour force could decrease as a result of net emigration of workers and a rise in the school leaving age.

2. Discouraged workers are people who are willing and able to work but give up the search for employment after failing to find work.

3. A country may experience a decrease in production but an increase in labour productivity as a result of a rise in unemployment. With fewer people in work, total output may fall but if the best workers have been retained, output per worker hour may increase.

4. A country may experience both an increase in employment and unemployment if there is a rise in the labour force. More people entering the labour force can increase both the numbers employed and the numbers of unemployed.

5. Among the reasons why someone may stop being unemployed are s/he may find another job, may emigrate, may retire, may enter full-time employment and may become a homemaker.

6. Unemployment could cause unemployment as experiencing a period of unemployment may result in people losing the work habit, their skills becoming out of date and the longer people are out of work, the less attractive they will appear to employers.

7. The dependency ratio is the number of people who are too old, too young or too sick to work as a percentage of the labour force. In other words, it is those who have to be supported as a percentage of those who will support them.

8. In most countries, the weight attached to food declines over time. This is because as people get richer, they may spend more in total on food but a smaller proportion as they are likely to spend even more on other items.

9. The key factor which determines whether a country would benefit from a period of deflation. It is likely to be beneficial if it has arisen due to an increase in aggregate supply. If, however, it has been caused by a decrease in aggregate demand, it is likely to have an adverse effect on the economy.

10. A fall in the real price of laptops would mean that laptops have risen in price less than the inflation rate.

11. The three domestic components of aggregate demand are consumer expenditure, investment and government spending.

12. A fall in income tax would be expected to increase aggregate demand. Disposable income would rise which would increase consumer expenditure, a component of aggregate demand. Higher consumer expenditure would encourage firms to expand and so increase investment, another component of AD.

13. A recession in one country would reduce spending in that country including spending on imports. A main trading partner may experience a decline in its net exports and so in its aggregate demand.

14. Dissaving is spending more than income. Dissaving can occur by drawing on past savings and/or by borrowing.

15. A fall in profit levels will reduce the willingness and ability of firms to invest. Firms will have less incentive to invest and will have lower amounts to spend on investment.

16. A widespread flood would cause a decrease in aggregate supply as it would damage and destroy some of the country's resources. The top soil would be removed from farmland, factories and other capital goods would be damaged and some workers may lose their lives.

17. A change in investment has a particularly significant effect on an economy as it increases both aggregate demand and aggregate supply.

18. An increase in resources would shift both the aggregate supply curve and the production possibility curve to the right.

19. Macroeconomic equilibrium occurs when aggregate demand equals aggregate supply. At this point there will be no tendency for real GDP or the price level to change.

20. A decrease in aggregate supply and an even greater decrease in aggregate demand could cause a fall in the price level and a fall in real GDP as shown in the figure below.

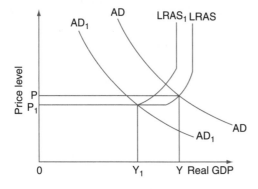

Answers to revision activities

1. (a) A base year is a year whose index number is set at 100. Other years are then referred back to this year. If a consumer price index is, for instance, 108 two years later, it means that the price level has increased 8% over the two years.
    (b) Nothing can be concluded about the actual amount spent as weights reflect the proportion spent. It can, however, be concluded that people in Bangladesh spent the greatest proportion of their spending on clothing and footwear out of these three countries and India the least, although there is not much difference in the percentages.
    (c) The weighting given to recreation, entertainment, education and cultural services in Bangladesh would be expected to increase over time. This is because income would be expected to rise and as people get richer, they spend more on recreation, entertainment, education and cultural services.

    (d) Pakistan would have been more affected by a 10% rise in the price of furniture as people in Pakistan spend a greater proportion of their income on furniture.

2. (a) The Maldives – 15,950.
    (b) China – 708.24m.
    (c) China – 60%.
    (d) Maldives – 25%.

3. (a) An increase in AD and later an increase in AS.
    (b) An increase in AD.
    (c) An increase in AD.
    (d) An increase in AD and an increase in AS.
    (e) A decrease in AD.

Answers to multiple choice questions

1. D   The labour force consists of both the employed and the unemployed i.e., those who are economically active. A, B and C are all economically inactive.

2. B   55% of the country's working age population is in the labour force i.e. 55% of 50m = 27.5m. Of the people who are in the labour force 8% are unemployed and so 92% are employed. 92% of 27.5m is 25.3m.

3. A   Labour productivity is output per worker hour. A rise in employment will decrease labour productivity if the newly employed are less skilled and if they work with less capital.

4. A   If labour productivity in the sugar industry has increased, it means that output per worker hour has increased. If the output of sugar has fallen it must mean that less labour is being employed in the industry. This may be because fewer workers are employed and/ or the average number of hours worked per worker has decreased.

5. D   An increase in the retirement age would mean that people would work for a longer period of their life. People will stay in the labour force for longer. C would reduce the size of the labour force. B does not change the size of the labour force, just its utilisation. A would mean that the size of the labour force is still decreasing but at a slower rate.

6. A   B, C and D may not necessarily have occurred. The number of people unemployed may have decreased if the size of the labour force had fallen. The unemployment percentage may have increased and the output of the country may have risen if labour productivity has increased. The unemployed are part of the labour force and the unemployment percentage may have increased whilst the size of the labour force may have decreased, stayed the same or increased.

    The only thing that must have resulted from the rise in the unemployment percentage is the gap between the output the country is producing and the maximum output that the country is capable of producing.

7. C   The country's labour force is 45m, so the labour force is 9m/45m x 100 = 20%.

8. A   There are a number of groups who may be unemployed but who are not entitled to receive benefits. C and D are incorrect and B is included in both the claimant count and the labour force survey.

9. A   Prices were still rising but rising more slowly. This will mean that the cost of living has increased and the value of money has decreased. Weights in the consumer price index may change but in total they will stay at 100 or, in some cases, 1,000.

10. D   If the price levels falls, each unit of money will purchase more products. The cost of living will fall. More information would be needed to conclude what has happened to the standard of living and international competitiveness, including what has happened to wage rates and other countries' price levels.

11. B

Category	Weight		Price change	Weighted price change
Food	1/10	x	10%	1%
Clothing	1/5	x	−10%	−2%
Housing	3/10	x	5%	1.5%
Other products	2/5	x	12%	4.8%
				5.3%

12. C   The inflation rate was positive throughout the period. This means that the price level increased throughout the period. The cost of living rose throughout the period and the price level was highest in December 2012.

13. B   The weights in a consumer price index reflect spending patterns. These spending patterns change over time and so the weights have to be changed.

14. A   A definition question. 'Real', in economics, means adjusted for inflation.

15. B   The real rate of interest is the nominal rate of interest minus the inflation rate i.e. 9%−3% = 6%.

16. C   An increase in expenditure on education and training could cause labour productivity and so increase the maximum potential output of the economy. A and D could cause a movement along the long run aggregate supply curve. B would result in a shift to the left of the short run aggregate supply curve.

17. D   As the following figure shows, an increase in aggregate demand which occurs when there is considerable spare capacity in the economy will increase country's output but leave the price level unchanged.

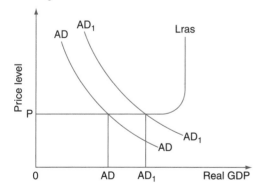

18. A   Figure 9.10 shows an increase in aggregate demand. A decrease in income tax would raise disposable income which would be likely to increase consumer expenditure. B, C and D would all be likely to reduce aggregate demand.

19. C   More people in the country would raise consumer expenditure and may also increase investment and consumer expenditure. More people of working age would also increase the labour supply and raise productive potential.

20. C   An increase in net exports would cause a shift to the right of the aggregate demand curve. As the economy is approaching full capacity, it would raise output and push up the price level.

Answers to data response questions

1. (a) A change in relative prices means that the price of some products has altered at a different rate to others. A relative fall in the price of car travel, for instance, may not necessarily mean that car travel is actually falling in price, but is rising more slowly than bus and train travel.

   (b) Two reasons why household spending patterns may alter over time are changes in income and changes in the products available. A rise in income is likely, for example, to reduce the proportion of household expenditure devoted to food and non-alcoholic beverages whilst increasing the proportion devoted to furnishings, household equipment and services. The introduction of new goods and services can result in spending on them taking up a sufficient share of household expenditure for them to appear in the CPI basket. This was clearly the case with tablet computers in New Zealand in 2011.

   (c) (i) It cannot be concluded from Table 1 that people in Australia spend more on alcoholic beverages and tobacco than people in the UK. The Australians spend a higher proportion of their total expenditure on this category, but if their total expenditure is less, this might be a smaller actual amount.

      (ii) A 10% rise in the price of housing and household utilities would have more of an effect on the UK's inflation rate than on New Zealand's inflation rate. This is because spending on housing and household utilities accounts for 22.3% of total spending in New Zealand but only 13% in the UK. With a weighting of 22.3/100, a 10% rise in the price of housing and household utilities would contribute 22.3/100 x 10% = 2.23% points to the inflation rate of New Zealand. With a weighting of 13/100, it would only contribute 13/100 x 10% = 1.3% point contribution to the UK's inflation rate.

   (d) A central bank examines consumer price indices to consider if its own country's inflation rate is sufficiently low and stable and how it compares with other countries' inflation rates. If a central bank considers that its CPI is increasing too rapidly, especially relative to rival countries, it may decide to raise the rate of interest. In this case, the intention would be to reduce consumer expenditure and so lower aggregate demand and demand-pull inflation. In contrast, if the consumer prices index indicates that the price level is below the government's target inflation rate, a central bank is likely to lower the rate of interest.

   (e) No consumer prices index provides a totally accurate measure of inflation. There may be sampling errors and whilst weights are reviewed on a regular basis, there is nevertheless a time lag. This can give rise to what is called a substitution bias. A CPI has a fixed basket of goods and services for at least one year. This means that it cannot reflect within that time period the likelihood that consumers will switch away from buying products which are becoming relatively more expensive to those which are becoming relatively cheaper. Prices are selected from a range of retail outlets but again these may not fully reflect short-term changes in where consumers buy their products from, including from cheaper outlets.

There is also the problem that the quality of products changes over time. For instance, a television set produced now may be more expensive than one produced a year before but it may have more advanced features, so like is not really being compared with like.

The CPI is based on a basket of goods and services purchased by a 'typical' household. Different types of households and households in different regions are, however, likely to experience differences in the changes to their cost of living. Pensioners, for example, will suffer a higher rise in their cost of living than the inflation rate indicates if they spend more than the average on food and the price of food rises significantly.

In addition, there are variations in how different countries measure their consumer price indices with some differences in components. This can make international comparisons difficult.

2. (a) (i) Weights are used in constructing a CPI because a change in the price of items which people spend a high proportion of their total expenditure on is more significant in influencing the cost of living than items on which they do not spend much on.

(ii) The weight for alcohol and tobacco fell between 1992 and 2008. This would have been because people were spending a smaller proportion of their total expenditure on those items over the period. People may have become more health conscious or higher taxes may have discouraged consumption. In contrast, the weight for transport doubled, indicating that transport accounted for 14.8% of total expenditure in 2008 but only 7.4% in 2008. People may have been travelling more because of rises in income or they may have been spending more due to rise in the price of transportation.

(b) (i) Inflation is a sustained rise in the general price level.

(ii) Paraguay's inflation rate was higher than Venezuela's from 1950 to approximately 1987. After that period Venezuela's inflation rate was higher.

(iii) Paraguay's inflation rate was neither low nor stable at the start of the period. It was lower during the 1960s but then rose in approximately 1974, albeit not to the same high levels as at the start of the period.

(c) There are a number of advantages in setting inflation targets. An inflation target may convince people that the central bank will keep inflation low. Firms may moderate prices, workers may ask for lower wage rates and consumers may not rush large item purchases. An inflation target also makes the central bank accountable – it would have to explain if it failed to achieve the target.

3% is a low rate of inflation and one which most governments would find acceptable. Low inflation has a number of benefits including maintaining international price competitiveness, encouraging investment, avoiding a random redistribution of income, and menu and shoe leather costs.

For some governments, however, 3% may be too ambitious a target in the short term. If a country has been experiencing a very high rate of inflation for some time, an inflation rate of, example, 7% might be more achievable.

There is also a risk that setting an inflation target may make a central bank concentrate on controlling inflation at the expense of reducing economic growth and increasing unemployment. For instance, a central bank's decision to raise the rate of

interest when the inflation rate is 5% may lead to an increase in unemployment which may create greater problems for the economy than the 5% inflation. In addition, to be effective, people would have to possess confidence in the central bank.

3. (a) (i) Aggregate demand (AD) is total demand for a country's products at a given price level. It consists of consumer expenditure, investment, government spending and net exports.

(ii) The information gives two reasons why aggregate demand has increased in India in recent years. These are a rise in consumer expenditure and an increase in government spending. Consumer expenditure has been rising due to increasing incomes. Higher incomes enable people to purchase more goods and services. Consumer expenditure is the largest component of AD. The government has been spending more on infrastructure. Government spending is another component of AD.

(iii) Aggregate supply (AS) has increased due to the higher government spending on infrastructure and an increasing labour force. Government spending on infrastructure increases not only AD but also AS. More and better infrastructure will lower firms' costs of production and make it possible for the country to produce more goods and services.

A larger labour force also increases productive capacity. With more workers, firms can increase their output and so the country's maximum potential output rises.

(Note the rise in total factor productivity also increases AS as more can be produced with the same quantity of resources).

(b) (i) India has the second highest growth in total factor productivity. Its growth rate was 1.2% points below China's, 1.1% points above Indonesia's, more than double Japan's, UK's and USA's. It was significantly higher than Brazil's and Russia's.

(ii) Total factor productivity growth would be 2% (5%−3%).

(c) An increase in a country's labour force will benefit an economy if the extra workers are employed. In this case, real GDP will increase and inflationary pressure may be reduced. The effects will be more beneficial, the more skilled the extra labour is.

The increase may not, however, be beneficial, at least in the short run, if the extra labour is not employed. In this case, the increase will raise the country's productive potential but not its actual output as shown here.

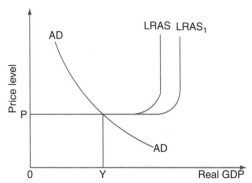

The number unemployed will increase which will raise the burden on state finances.

Answers to essay questions

1. (a) There are a range of factors that determine the size of a country's labour force. Countries with large populations are likely to have large labour forces. It is probable, however, that two countries with the same population will nevertheless have different sized labour forces. This is because there is likely to be a difference in their age structures, school leaving ages, number of people who go to university and attitudes to women working.

   A country with children of the old forming a significant proportion of its population will have fewer people of working age than one with a higher proportion of its population aged between approximately 16 and 65.

   A country with a high school leaving age and a low retirement age will also have a smaller working age population than one with a low school leaving age and a high retirement age. The larger the working age population, the greater the potential labour force will be. It will, however, also be influenced by the labour force participation rate. This is the proportion of people of working age who are either employed or seeking employment. The labour force will be smaller if more people go on to higher education but it will be of a higher quality. People choosing to retire early will reduce the size of the labour force as will a greater proportion of people being too ill or disabled to work.

   Attitudes to women working also influence the size of the labour force. Countries with fewer women working have a greater potential labour force than those which do not.

   (b) A widespread shortage of labour may be a cause of inflation but, at any one time, may not be the major cause.

   If firms are experiencing difficulties in recruiting workers, they are likely to raise wages in a bid to attract the scarce labour. This may push up their costs of production, causing a wage/price spiral. The following figure shows that the rise in the wage rate leads to a decrease in SRAS and pushes up the price level to P1. This is not a one off increase. The higher wage rate will increase aggregate demand. The higher AD will, in turn, increase demand for labour and push up the wage rate again and so on.

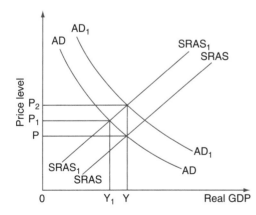

   A widespread labour shortage, however, is not the only cause of inflation. A country may have a plentiful supply of labour but may still experience a rise in the price level if there is another cause of an increase in the costs of production. For instance, there may be higher fuel costs.

As well as cost-push factors, the price level may rise as a result of demand-pull factors. If an economy is approaching full capacity, an increase in aggregate demand will cause demand-pull inflation as shown in the figure below.

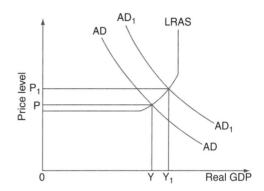

   Higher aggregate demand may be caused by, for instance, an increase in consumer expenditure resulting from a rise in consumer confidence or a rise in net exports following an increase in incomes abroad. It may also arise as a result of an increase in the money supply. If the money supply grows faster than output, the general price level will rise.

   A widespread shortage of workers is one of a number of possible causes of inflation. It may be the major cause but it may also be making a smaller contribution than, for instance, a rise in raw material costs or a consumer boom started before the rise in wages.

2. (a) In many countries, consumer expenditure is the largest component of aggregate demand. There are a number of influences on the amount households spend on consumer goods and services.

   Possibly the most important influence is income. As income rises, people spend more in total. Once income reaches a certain level, however, a further rise may reduce the proportion that people spend. This is because they can now afford to save. For instance, people with a disposable income of $20,000 may spend $18,000 i.e. 90% of their income. If their disposable income should rise to $30,000 they may now spend $24,000. The actual amount they are spending has risen by $4,000 but the proportion has fallen to 80%.

   Another important influence on consumer expenditure is consumer confidence. If people feel optimistic about future economic prospects and anticipate a rise in income, they may spend more now. In contrast, if consumers anticipate a recession, they may decide to spend less and save more to protect themselves against hard times.

   Changes in the rate of interest can also affect consumer expenditure. A rise in the rate of interest may reduce consumer expenditure for three key reasons. One is that it will increase the reward for saving. Another is that it will make it more expensive to borrow. The third is that anyone who has borrowed in the past at a variable interest will have less money to spend.

   Other factors that influence consumer expenditure include population size, wealth and the distribution of income. An increase in the number of people in the population will increase consumer expenditure. The more wealth people have, the more likely they are to spend. They will be aware that wealth can generate income and they will know that they can borrow against their wealth. Consumer expenditure tends to increase if income becomes more evenly distributed. Raising the income of the poor will enable them to spend more whilst reducing the income of the rich will probably not cause them to cut back significantly on their spending.

(b) An increase in consumer expenditure will increase aggregate demand, provided it is not more than an offset by a fall in any of the other components of aggregate demand.

The impact of a higher aggregate demand will depend on the initial state of the economy. If the economy is initially producing with considerable spare capacity, an increase in consumer expenditure may increase the country's output and leave the price level unchanged as shown in the figure here.

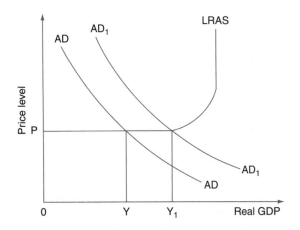

In the case of an economy operating at a point where the economy is beginning to experience shortages of resources, an increase in AD may raise both output and the price level. The following figure shows the increase in aggregate demand from AD to AD1 causing real GDP to increase from Y to Y1 and the price level to increase from P to P1.

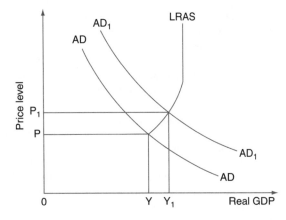

If the economy is producing at full capacity, a rise in AD will have no effect on output but will be purely inflationary. The figure below shows that the rise in AD forces up the price level to P1 but causes no change in output which remains at Y.

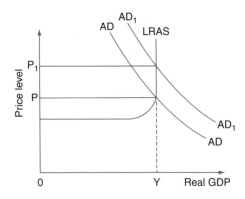

A rise in consumer expenditure may stimulate an increase in investment. This will further increase AD. It will also increase aggregate supply. In this case, output will rise. The effect on the price level will again depend on the initial level of economic activity but also on the relative size of the increases in aggregate demand and aggregate supply. The following figure shows the rise in AS matching the rise in AD and leaving the price level unchanged.

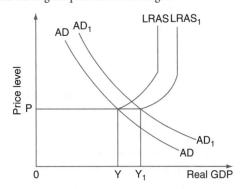

3. (a) Labour productivity may vary between countries for a variety of reasons. These include differences in the quantity and quality of education and training, levels of investment, adoption of new technologies, scientific innovation, quality and quantity of infrastructure and the organisation of production.

Countries, including Germany and Singapore, which have children receiving good quality education over a relatively high number of years and workers receiving regular and good quality training, usually have high labour productivity. A high proportion of their workers are skilled, occupationally and geographically mobile and able to work with complex equipment.

The more capital, workers work with, the higher their output tends to be. In recent years, the rise in net investment in China has been an important reason for the growth in the country's labour productivity.

As well as the quantity of investment, the quality of investment is also important. Countries which buy capital equipment which embodies new technology usually see an increase in output per worker hour. South Korea and other Asian countries have been adopting new technology rapidly in recent years and this has contributed to their productivity growth.

Scientific innovation can cause rapid increases in productivity. The silicon revolution in the US in the 1990s was the major driving force behind the remarkable increase in US productivity in that decade.

Good infrastructure can promote rises in labour productivity. It is generally thought that whilst India's labour productivity growth has been boosted by improved education and advances in IT, it has been held back by poor infrastructure.

Possibly, the easiest and most straightforward way of increasing labour productivity is improving the organisation of the factors of production. For instance, introducing shift work may enable workers to make better use of scarce capital equipment. Where people work such long hours that they get tired, cutting working hours may increase both labour productivity and total production.

(b) No measure of unemployment is perfect and whilst both the claimant count and the labour force survey measures have their strengths, they also both have some drawbacks.

A measure of unemployment seeks to include those who are without jobs but also are willing and able to work. The claimant count includes those who are receiving unemployment related benefits. This is a relatively cheap and quick measure as the information is collected when the benefits are paid out.

There are, however, questions about the reliability of the claimant count. Some of those included may not be unemployed whilst others who are unemployed are not included. Some people who are not actively seeking work, sometimes referred to as voluntarily unemployed, may claim benefit and some people who are employed in the informal (shadow) economy may also be illegally claiming benefit. It is generally thought that those who claim benefits who are not entitled to them, are more than outnumbered by those who are unemployed but are not allowed to claim unemployment benefits. These include, for instance, those who have recently given up a job and those with partners in work.

Differences in who is entitled to receive benefits, makes it difficult to make comparisons over time and between countries. The labour force survey is more suitable to make international comparisons. This is because it is based on the International Labour Organisation's (ILO's) definition of unemployment which is used in most countries. The labour force surveys asks a sample of the population if they are out of work and if they are actively seeking employment and are available for work in the next two weeks. It is thought that this method includes more of those who are unemployed.

Its perceived greater accuracy and its usefulness for international comparisons means that the labour force survey is usually the more favoured measure. It is, however, a relatively expensive measure. It takes time to collect the information and the measure may be subject to sampling problems. It is important that the sample is representative of the population and that answers are carefully interpreted.

Chapter 10

Answers to short questions

1. Deflation is a sustained fall in the general price level.

2. The two main causes of deflation are a fall in aggregate demand and a fall in the costs of production. The latter may benefit an economy but the former may cause a fall in real GDP.

3. An excessive growth of the money supply may give rise to demand-pull inflation.

4. To turn the Fisher equation into the Quantity Theory of Money, it has to be assumed that the velocity of circulation and output (transactions) are constant for a period of time.

5. Wage drift is the difference between an increase in workers' earnings and the increase in wage rates.

6. A rise in wages may not result in cost-push inflation if the rise is matched by an increase in labour productivity. In this case, wage costs per unit will not rise. It is also possible that other costs may have fallen.

7. A fall in the price of a currency may result in cost-push inflation as it will increase the price of imported raw materials. Such an increase may push up the costs of production if firms do not choose or are unable to switch to domestically produced raw materials.

8. A fall in the price of a currency may lead to demand-pull inflation as cheaper exports and more expensive imports can result in a rise in net exports and so aggregate demand. If aggregate demand grows faster than aggregate supply, the general price level will be pulled up.

9. A consumer boom is a high level of consumer expenditure and is likely to be associated with output and low unemployment.

10. A consumer boom may cause demand-pull inflation if the higher aggregate demand is not met by an equal increase in aggregate supply.

11. A fiscal boost is stimulus to the economy resulting from an expansionary fiscal policy, that is a rise in government spending and/or a cut in taxation.

12. A fiscal boost may result in inflation, if the government has underestimated the size of the multiplier or overestimated the size of a negative output gap and as a result increases government spending or cuts taxation by too much.

13. Tax revenue tends to rise during inflation because of fiscal drag – tax rates do not always rise in line with inflation. Governments are often net borrowers and if the rate of interest does not rise in line with inflation, the real burden of debt falls.

14. The only thing that must definitely occur as a result of inflation is a fall in the value of money. With higher prices each unit of money will purchase fewer goods and services.

15. Anticipated inflation is a rise in the general price level at the expected rate. In contrast, unanticipated rate is a rise in the general price level at a different rate to that expected.

16. Menu costs are significant when inflation is high and unanticipated.

17. Agflation is inflation caused by a rise in agricultural prices.

18. Cost-push inflation is likely to be associated with stagflation as increases in the costs of production may be accompanied by high unemployment and low economic growth. In contrast, demand-pull inflation is likely to be accompanied by falling unemployment and relatively high economic growth.

19. A trade deficit occurs when the expenditure on the imports of goods and services exceeds the revenue earned from the sale of exports and goods and services.

20. The current account and the financial account are linked in a number of ways. Foreign direct investment, portfolio investment

and other investment which appear in the financial account give rise to profit, interest and dividends which are recorded in the income section of the current account. A trade surplus may encourage multinational companies to set up in the country, foreigners to buy shares in domestic firms and foreign banks to lend to domestic firms and so lead to an inflow of foreign direct investment, portfolio investment and other investment.

21. One reason why a country's reserves of foreign currency may decline is that they may be used to finance a deficit on the balance of payments. Another reason is that they may be used to buy the currency to maintain a fixed exchange rate or to influence a floating exchange rate.

22. Special Drawing Rights (SDRs) are a form of international reserve asset that are issued by the International Monetary Fund and can be used to finance a balance of payments deficit or to support an exchange rate.

23. A merchandise trade deficit is a deficit in trade of goods (visible balance deficit). A rise in a country's real GDP may result in an increase in a merchandise trade deficit for two reasons. One is that as incomes rise, households may purchase more finished imported products and firms may buy more imported raw materials. The other reason is that domestic firms may divert products originally intended for the export market to the more buoyant home market.

24. There are a number of reasons why a country might experience a rise in its inflation rate but a fall in its merchandise trade deficit. These include other countries' inflation rates being higher (making the country's products relatively cheaper), a rise in incomes abroad, other countries removing trade restrictions and a rise in the quality of domestically produced products.

25. An increase in a trade deficit would, ceteris paribus, be expected to increase unemployment. A fall in export revenue relative to import expenditure would reduce aggregate demand and so may result in cyclical unemployment.

26. An increase in trade deficit might be accompanied by a rise in the current account surplus if an increase in a surplus on income and current transfers outweighs it.

27. A fixed exchange rate provides the most certainty for traders. This is because, at least for a period of time, the price of the currency will not change. This means that traders will know how much they will receive in their own currency when selling exports and how much they will have to pay for imports.

28. An exchange rate system determined by a mixture of market forces and state intervention is a managed float. The government may allow the exchange rate to float within certain margins but intervene if the exchange rate rises too high or falls too low.

29. The concept of purchasing power parity is associated with a floating exchange rate. In theory, a floating exchange rate will move to a value where the same amount of products can be purchased in the two currencies.

30. A revaluation will raise the price of exports in terms of foreign currency and lower the price of imports in terms of the domestic currency.

31. Hyperinflation will reduce the internal value of the currency. Each currency unit will buy fewer goods and services. It is also likely to reduce the external value of the currency. This is because a large rise in the price level is likely to make domestic products less price competitive and so demand for the currency is likely to fall.

32. The government of a country operating a fixed exchange rate may be under more pressure to reduce inflation than a government operating a floating exchange rate as its currency will not adjust to regain international competitiveness.

33. Two factors that influence the PED of the products produced by a country's firms are the extent to which they have substitutes and the extent to which they are seen as luxuries or necessities.

34. As the combined PED at – 0.7 is less than one, the Marshall-Lerner condition is not met and a depreciation will increase the country's trade deficit.

35. A country's trade balance may experience a reverse J-curve effect as a result of an appreciation or revaluation. Initially if the value of a currency rises, demand for exports and imports will be inelastic as households and firms will not have had time to adjust their purchase in response to the change in prices. With higher export revenue and lower import expenditure, any trade surplus will get larger. If, as time progresses, demand becomes elastic, the trade balance may move into deficit.

36. A run on a currency occurs when there is a sudden and large increase in the sales of the currency. Such an event can lead to a significant depreciation or force a government to devalue the currency.

37. Two reasons why someone may sell US dollars to the UAE's dirham is because she or he may want to go on holiday to the UAE or because she or he may want to buy property in the UAE.

38. The foreign exchange market is the market in which currencies are exchanged for each other. Dealers in foreign exchange operate throughout the world linked by modern technology.

39. Net exports is a component of aggregate demand. An increase in a country's current account deficit will reduce aggregate demand.

40. Two causes of a fall in a current account deficit, other than government policy, are a fall in the exchange rate resulting from market forces and a rise in incomes abroad.

Answers to revision activities

1. (a) Hyperinflation is an inflation rate above 50% whereas creeping inflation is a low rate of inflation.
   (b) Core inflation is the underlying trend inflation which excludes energy and food prices which can be volatile. In contrast, the headline rate of inflation is the one which receives the most publicity. This is usually the consumer price index which includes energy and food prices.
   (c) Inflation illusion is a situation where people confuse nominal and real values whereas an inflationary gap occurs when aggregate demand exceeds aggregate supply resulting in demand-pull inflation.
   (d) Stable inflation is a consistent rate of inflation with the price level rising at a similar rate. Accelerating inflation occurs when the inflation rate is increasing.

2. (a) False. The overall balance of payments always balances but the current account is more commonly in deficit or surplus.
   (b) False. Transactions in assets involve money leaving the country and as such they appear as a debit item in the balance of payments.

(c) True. The total value of debit items equals the total value of credit items.

(d) False. A government may not be too concerned about a short run, relatively low current account deficit or surplus. It may want to give priority to another objective. For instance, it may want to focus on reducing inflation and a current account surplus may be the result of this and a current account deficit may result from a government concentrating on reducing unemployment.

(e) False. The merchandise balance is the difference between the value of exports of goods and the value of imports of goods.

(f) True. Devaluation may not reduce a trade deficit. If PED for exports and imports is less than one, devaluation will increase a trade deficit.

3. (a) The trade balance is initially 400m pesos – 500m pesos = a trade deficit of 100m pesos.

(b) The trade balance changes to 600m pesos – 450m pesos = a trade surplus of 150m pesos.

(c) The change in the trade balance indicates that the combined price elasticities of demand for exports and imports is greater than one. This is because the fall in the price of the currency has moved the current account position from a deficit to a surplus.

Answers to multiple choice questions

1. C

Product	Weight		Price change (%)		Weighted price change (%)
Food	2/5	x	15%	=	6
Electricity	1/10	x	10%	=	1
Transport	1/5	x	20%	=	4
Entertainment	1/10	x	−10%	=	−1
Clothing	1/5	x	30%	=	6
					16

The overall rise in the cost of living (the inflation rate) was 16%.

2. C   A fall in the price of the currency will increase import prices including the prices of imported raw materials. More expensive raw materials will push up costs of production. A, B and D may cause demand-pull inflation.

3. B   Fiscal drag is the term given to people's incomes going into higher tax brackets because a government has not adjusted tax brackets in line with inflation. A is menu costs, C may occur but does not have a specific term and D is unlikely to happen as, in part due to fiscal drag, inflation tends to result in a redistribution of income from taxpayers to the government.

4. A   The Quantity Theory of Money, $MV = PT$, assumes the velocity of circulation (V) and real GDP (Y) are constant in the short run, so a fall in the stock money (M) will result in a fall in the price level (P).

5. B   Between January 2009 and January 2010, Pakistan experienced a lower rate of inflation. This meant that the price level was still increasing albeit at a slower rate. Higher prices will increase the cost of living. The internal purchasing power of the Pakistani rupee will have fallen, the Pakistani consumer price index will have increased. Costs of production are likely to have risen with inflation.

6. B   Inflation can enable firms to reduce the real cost of labour by raising wage rates by less than inflation. Lower real costs of production enable firms to survive or may increase their profits. A, C and D would all be disadvantages for firms.

7. B   A straightforward definition question. A is deflation, C is stable inflation and D is core inflation.

8. C   Insurance is included in the trade in services (invisible) section of the current account of the balance of payments, B the trade in goods balance of the current account and D the financial account.

9. D   The current account has a surplus of $30bn (trade balance = $20bn + net income −$20bn + current transfers $30bn). The financial account balance has a deficit of− $10 + −$20bn + −$10bn + $20bn = $20bn.

10. C   An increase in the exchange rate would raise export prices and lower import prices. If PED for exports and imports is greater than one, a current account deficit would increase. A may reduce imports and so decrease a country's current account deficit. B and D may increase the international competitiveness of the country's products and so again reduce the current account deficit.

11. C   A depreciation of the Pakistani rupee against the Chinese yuan would lower the price of Pakistan's exports to China and raise the price of Chinese imports to Pakistan. Pakistani's airlines are likely to gain extra passengers as their fares in yuans will be lower. Chinese car producers may lose sales as their prices will rise in terms of the Pakistani rupee, Chinese speculators holding Pakistani rupees will find that their holdings are worth less in terms of yuan and Pakistani students studying in China will be faced with higher living costs and tuition fees in terms of their home currency.

12. A   A rise in the rate of interest may encourage people to purchase the currency and so raise the exchange rate. B would result in an increase in the supply of the currency and so lower its value. C and D may reduce the international competitiveness of domestic firms and so lower export revenue and increase import expenditure. Such changes are likely to reduce demand for the currency, increase its supply and so put downward pressure on the price of the currency.

13. C   A trade weighted exchange rate is the price of a currency in terms of a basket of the currencies of the main countries, the country trades with.

14. A   With the price of the currency not changing, exporters will know how much they will earn and importers will know how much they will have to pay. B, C and D are advantages of a floating exchange rate.

15. A   If the value of Argentine imports increases, more Argentine pesos will have to be sold to buy the foreign currency needed to produce the imports. B would increase demand for the Argentinian pesos. C and D would decrease the demand for the Argentine pesos.

16. A   A decrease in foreign direct investment in Bangladesh would reduce demand for takas and so may result in a fall in the price of the taka. B, C and D could also lead to an increase in demand for the taka and so a rise in the value of the taka.

17. D   A fall in the price of a country's currency would lower export prices. To take advantage of the likely rise in the demand for

exports, it is important that the supply of exports is elastic. A and B would result in a deterioration in the trade balance and are not directly linked to a fall in the price of the currency. C would mean that the value of net exports would fall and so the trade balance would deteriorate.

18. B Egyptians taking more holidays abroad would increase the supply of Egyptian pound as Egyptians sell the currency to purchase the currencies of the destinations they are going to. Egyptian firms selling fewer products will result in a fall in demand for the Egyptian pound. An increase in supply of and a decrease in demand for the currency would push down its price.

19. B A devaluation might more than offset a rise in the inflation rate, making the country's products more price competitive. A, C and D would be likely to increase a current account deficit. A and D would increase imports and C reduce exports.

20. D A freely floating exchange rate is determined by the intersection of demand and supply. A and B involve government intervention and may be used to influence a managed exchange rate or to maintain a fixed exchange rate. C, again, involves government intervention and tax rate changes are used to influence aggregate demand.

Answers to data response questions

1. (a) (i) November (having to give fewer Kwacha to buy a US$ means that the value of the Kwacha has risen).
    (ii) At the start of November 4,300 Kwacha had to be sold to buy US$1. By the end of November, only 3,300 Kwacha had to be exchanged. So the Kwacha had appreciated by 1,000 that is by 1,000/4,300 x 100 = 23.26%.
   (b) Figure 10.2 suggests that Zambia uses a floating exchange rate, that is an exchange rate determined by the market forces of demand and supply. The evidence for this is the frequent changes in the exchange rate.
   (c) (i) An improved export performance would increase demand for the Kwacha. Those foreigners wishing to buy Zambian products will have to purchase Kwacha. The higher demand for the Kwacha will increase its price and result in an extension in the supply of the currency as shown in the figure below.

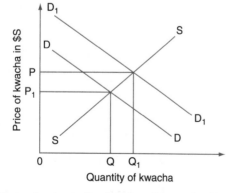

    (ii) A reduction in foreign debt will mean that Zambia will not have to sell as much of its currency to repay or service the debt. This will reduce the supply of the Kwacha, which will again push up the price of the currency and in this case will cause a contraction in demand for the currency as shown in the following figure.

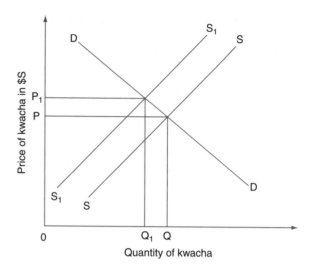

   (d) An appreciation of the Kwacha would lead to a rise in the price of Zambian exports and a fall in the price of Zambian imports. Such price changes will lead to an improvement in Zambia's terms of trade.
   (e) An appreciation of its exchange rate has both advantages and disadvantages for a country. It means that each export can be exchanged for more imports and that the foreign debt burden will be reduced. For instance, if the Kwacha appreciates from 5,000 to US$1 to 4,000 to US$1, a Zambian export priced at 20,000 Kwacha would before have bought four US imports of $1 each. With the appreciation, the Zambian export would now buy five US imports. If Zambia had to repay $10m of foreign debt, initially it would have to pay 50,000mKwacha. Now it would have to repay 40,000m Kwacha. An appreciation of the exchange rate might also increase confidence in the currency and as a result attract an inflow of portfolio investment and make it easier to obtain loans.

    In addition, an appreciation of the exchange rate can lower inflationary pressure. The price of finished imports bought by domestic consumers will fall, cheaper imported raw materials will reduce costs of production and lower import prices will put pressure on domestic firms to restrict price rises in order to remain competitive.

    On the other hand, an appreciation could result in a current account surplus turning into a deficit, or a current account deficit getting larger. This will occur if the combined elasticities of demand for exports and imports are greater than one. With elastic demand for exports and imports, net exports will decline. Net exports are a component of aggregate demand and with lower aggregate demand, unemployment may increase and economic growth may be slowed.

2. (a) (i) Two ways the Chinese government could intervene in the foreign exchange market to prevent the yuan rising in value against the US dollar are to sell yuan and lower the rate of interest.
    (ii) A currency is described as undervalued if the exchange rate does not reflect its purchasing power parity and if it gives rise to a current account surplus over time. For instance, if a basket of goods and services is priced at $100 in one country and 1,000 rupee in another country, the exchange rate would be expected to be $1 = 10 rupee. If the exchange rate was $1 = 15 rupee,

the rupee would be regarded as undervalued. A country which continues to experience a current account surplus, such as China in recent years, may be claimed to be operating an undervalued exchange rate.

(b) (i) More pence had to be given to buy US$1 over the period. Between 2007 and 2010, the value of the pound fell by 17/50 = 34%. This indicates that the value of the pound declined by rather more than 30%.

(ii) The slight fall in the value of the pound between 2007 and 2008 was accompanied by an increase in the current account deficit. Over the whole period, however, the fall in the value of the currency is accompanied by a decline in the current account deficit. A fall in the value of the pound would have made exports cheaper, in terms of foreign currency, and import prices higher, in terms of domestic prices. With elastic demand, export revenue will rise and import expenditure will fall, reducing a current account deficit.

(c) A depreciation in the exchange rate is likely to increase the rate of inflation for two main reasons. One is the higher price of imports. More expensive imports that the country's consumers buy will directly push up the country's consumer price index. A rise in the price of imported raw materials will push up costs of production and can result in cost-push inflation. Domestic firms may also be tempted to raise their prices now they are competing with more expensive imports.

The other reason why a depreciation may increase the rate of inflation is because it may result in a rise in net exports which will increase aggregate demand. If the economy was initially operating close to full capacity, it will be difficult to match the higher aggregate demand with higher aggregate supply and demand-pull inflation may occur.

There is a chance, however, that a depreciation in the exchange rate will not result in a rise in the inflation rate. If the country is an important market for the products of other countries, their firms may cut their prices in terms of their own currency, so that they remain at the same price when sold in the country. Even if the price of imported raw materials rises, it may not have much effect on the country's inflation rate if firms can switch to domestic producers. Higher aggregate demand may also not have much impact on the price level if there is considerable spare capacity in the economy.

3. (a) (i) Over the period 1980 to 2002, the US current account position moved from a small surplus to a deficit equivalent to approximately 4 ½ % of US GDP. For most of the period, the current account was in deficit with only three years, 1980, 1981 and 1991 in surplus.

(ii) Between 1992 and 2002, the US had a current account deficit. This might have been financed by borrowing, attracting financial investment or drawing on the country's reserves of foreign currency.

(b) If a surplus increases in size, it would be expected to increase the price of the currency. This is because if more is being spent on the country's exports, demand for the currency will increase. If less is spent on imports, the supply of the currency will fall. The following figure shows that an increase in demand for the currency combined with a fall in its supply will result in an appreciation of the exchange rate.

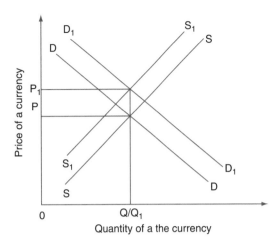

(c) (i) A lower price of the currency will reduce export prices and increase import prices. If the Marshall-Lerner condition is met, that is the price elasticity of demand for exports and imports is greater than one, the depreciation will reduce a current account deficit or increase a current account surplus.

(ii) The largest increases in the US$ which occurred between 1981 and 1985 and between 1996 and 2002 were largely accompanied by a rise in the current account deficit. In 1991, a fall in the value of the US$ coincided with the current account moving into surplus. There are, however, exceptions to this. For instance, between 1985 and 1987, the value of the US dollar fell but the current account deficit increased. Of course, there may be a time lag as suggested by the J-curve effect. It can take time for people to recognise that prices have changed and adjust their purchases. As a result, demand for exports and imports can initially be inelastic leading to a fall in the exchange causing a rising deficit. After a period of time, demand may become more elastic.

(d) There are both arguments for and against a government fixing the country's exchange rate. One argument is that a government can set it at a level to achieve one or more of its macroeconomic objectives. For instance, a government may want to set it at a low rate to improve its current account position and promote economic growth and employment. Alternatively, it may want to set it at a high rate in order to reduce inflationary pressure.

A fixed exchange rate also can discourage speculation and has the advantage of certainty. It can encourage trade as exporters and importers will know the price they will receive and the price they will pay. In addition, a fixed exchange rate can force a government to tackle inflation. This is because it will not be able to rely on a fall in the exchange rate to restore international competitiveness.

There are, however, disadvantages in operating a fixed exchange rate. One is that if it is not set at or close to the long run equilibrium level, it will come under market pressure to change its value. To maintain the price, the central bank will have to keep reserves of foreign currency and this will involve an opportunity cost. The central bank may also alter the rate of interest to keep the value at the fixed rate and this can have an adverse effect on some of the government's macroeconomic objectives. Indeed, a central bank may be restricted in its use of interest rate changes to influence domestic policy objectives because of concern about how a change in the interest rate could affect the exchange rate. A fixed exchange also does not, unlike a floating exchange rate, adjust automatically to offset a current account deficit.

Answers to essay questions

1. (a) Inflation is a sustained rise in the general price level. With prices rising, the value of money must fall. Each currency unit will buy fewer goods and services. There are two main causes of inflation. One is cost-push inflation. If costs of production rise, aggregate supply will decrease, pushing up the price level as shown below.

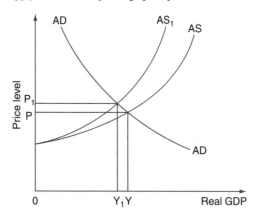

A common cause of cost-push inflation is wage rates rising faster than productivity. This can develop into a wage-price spiral, with higher wages pushing up prices and then workers demanding wage rises to maintain their real wage rates. Significant wage rises are more likely to occur when trade unions are strong and workers feel confident about retaining their jobs.

Another important cause of cost-push inflation is a rise in the price of raw materials. Such a rise might occur as a result of a fall in the exchange rate. Other causes of cost-push inflation are rises in profit margins and increases in indirect taxes.

Demand-pull inflation, in contrast to cost-push inflation, is caused by aggregate demand increasing more rapidly than aggregate supply. This often occurs when an economy is at or close to full employment. When the economy has little spare capacity, it is difficult to match higher aggregate demand with an extension in aggregate supply. The figure below shows an increase in aggregate demand from AD to AD1 pushing up the price level from P to P1.

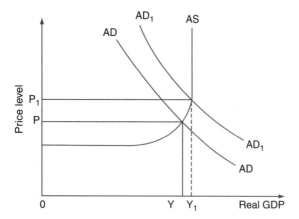

Demand-pull inflation may be caused by an increase in any of the components of aggregate demand. A consumer boom with a significant increase in consumer expenditure can lead to a shortage of both products and resources, resulting in a rise in the price level. An increase in net exports or government spending may also give rise to demand-pull inflation when the economy is operating close to maximum capacity. It is also possible that an increase in investment could give rise to demand-pull inflation in the short run, although in the long run an increase in net investment will generate extra capacity to meet the higher aggregate demand.

Although cost-push inflation and demand-pull inflation have different causes, they are linked and, in practice, it can be difficult to determine what was the cause of inflation. For example, rises in wages may coincide with a consumer boom and it may be difficult to decide which came first.

(b) A country experiencing inflation may encounter a balance of payments problem but this will not necessarily occur. Inflation will raise export prices and may make imports relatively cheaper. If demand for exports is elastic, a rise in their price will result in a fall in export revenue. Similarly, if demand for imports is elastic, a fall in their price will result in a rise in import expenditure. Such changes could result in an increase in a current account deficit.

Inflation may also discourage inflows into the financial account. Multinational companies may be reluctant to set up in a country experiencing high and fluctuating inflation. As well as foreign direct investment, inflation may also discourage portfolio investment and make it more difficult for firms and the government to borrow from other countries. People and institutions may be reluctant to buy the shares of firms and government bonds in a country experiencing a high rate of inflation. Banks may also be worried about lending to firms and a government in a country whose economic performance might be hindered by inflation.

Inflation, however, may not result in a balance of payments problem for a number of reasons. One is that the inflation rate may be lower than rival countries' inflation rates. In this case, the country's products will become more internationally competitive. A low and stable inflation may also encourage foreign direct investment and portfolio investment and may make it easier to obtain loans from overseas.

Even if inflation is higher than rival countries, the international price competitiveness of the country's product may not be lost if the upward pressure on the price of the country's exports is being offset by a fall in the value of the country's currency. Such a depreciation or devaluation will also raise import prices.

Inflation may be caused by a rise in demand for exports and so it may be accompanied by a decline in a current account deficit. It also has to be remembered that the trade balance is only part of the current account. The trade balance may move into deficit but be offset by a rise in income or current transfers.

In addition, if a country starts with a large current account surplus, inflation may reduce the surplus but such a reduction may not be considered that significant.

To decide whether a country experiencing inflation will always have a balance of payments problem, it would be necessary to know its rate, its stability, its cause and what is happening to other parts of the balance of payments.

2. (a) The exchange rate is the price of one currency in terms of another currency or currencies. If the currency is a freely floating one, its value is determined solely by the market forces of demand and supply. If it is a fixed exchange rate, the price is set by the government, or central bank acting on behalf of a government, or group of governments. The price of a managed exchange rate will be influenced by the government or central bank.

A floating exchange rate will fall if demand for it decreases and/or supply of it increases. There are a number of reasons why these changes may occur. If the price of the country's products are rising and demand for its exports is elastic, foreigners will spend less on the country's exports and so will buy less of the currency. Inflation can also result in an increase in the supply of the currency if it leads to a rise in import expenditure. This is because more of the currency will have to be sold to purchase foreign currency in order to try to buy imports. The effect of a loss in price competitiveness on the price of a currency is illustrated in this figure.

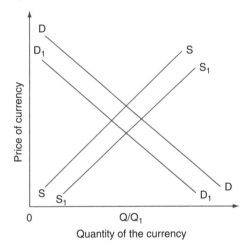

A relative fall in the quality or marketing of the country's products would have a similar effect.

The demand for and supply of currency is influenced not only by changes in demand for the country's products and foreign products but also by direct, portfolio and other investment flows and by speculation. If the country's interest rate falls, 'hot money' is likely to flow out of the country. Financial investors will sell the currency to buy the currencies of those countries which offer a higher rate of interest. If the economic performance of the country declines, some firms may be encouraged to move abroad. This would again result in an increase in the supply of the currency and a fall in its value.

A large part of the activity in the foreign exchange market is accounted for by speculation. If people believe that the value of a currency will fall in the near future, they will sell it now to prevent encountering a loss. Of course, their action can contribute to the very outcome they are expecting.

In the case of a fixed exchange rate or managed exchange rate, a government or central bank may lower the price either because the price cannot be sustained or because it is thought that a lower value will help improve the country's

macroeconomic performance. If there is widespread selling of the currency because it is thought to be overvalued, there may not be sufficient reserves of foreign currency to maintain the exchange rate and raising the rate of interest may also prove ineffective. A government may also encourage a central bank to move its exchange rate to a lower price to improve its current account position or stimulate economic growth and employment.

(b) A fall in a country's exchange rate is likely to influence its macroeconomic performance but the impact it will have is somewhat uncertain.

A fall in the exchange rate will reduce the price of the country's exports, in terms of foreign currency, and will raise the price of imports, in terms of the domestic currency. A fall in the price of exports should make them more price competitive and lead to higher demand. Whether revenue will rise or not depends on whether demand is elastic or inelastic. If demand is elastic, a fall in price will cause a greater percentage rise in demand and so result in an increase in revenue. A rise in import prices should lead to a fall in demand for imports. The effect on expenditure of imports will again depend on price elasticity of demand. If demand is elastic, import expenditure will fall.

The importance of elastic demand is emphasised in the Marshall-Lerner condition. This states that a fall in the value of a currency will lead to an improvement in the trade balance only if the sum of the price elasticities of demand for exports and imports is greater than one.

The J-curve effect suggests that a fall in the exchange rate may initially worsen the trade position before it improves it. This is shown in the following figure.

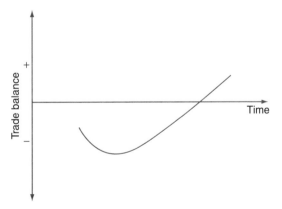

When the exchange rate first falls, there is not sufficient time for consumers to make much of an adjustment to the change in prices. It can take a while for households and firms to notice price changes, to change contracts and to find new suppliers. With inelastic demand, export revenue will fall and import expenditure will rise. Then as time progresses and buyers find alternative sources, demand is likely to become more elastic. This may cause export revenue to rise, import expenditure to fall and so the trade balance to improve.

Net exports are a component of aggregate demand and so a rise in net exports will increase aggregate demand. Such an increase can result in a rise in real GDP, if there is initially spare capacity in the economy. The following figure shows that a shift to the right in the aggregate demand curve will cause a rise in real GDP from Y to Y1.

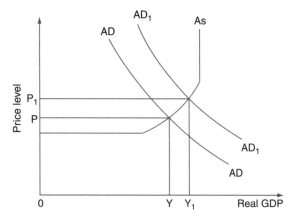

The higher output is likely to reduce cyclical unemployment. There is a risk, however, that the increase in aggregate demand may result in inflation. If the economy is operating close to full employment, the higher aggregate demand may lead to shortages and so push up the price level. Higher import prices also increase costs of production as imported raw materials become more expensive, directly raise the price of finished imported products and reduce the pressure on domestic firms to keep their prices low.

Whether a fall in its exchange rate will improve or worsen a country's macroeconomic performance will be influenced by the size of the fall, the price elasticities of demand for exports and imports, the level of spare capacity in the economy and how domestic firms respond.

3. (a) There are a number of reasons why a government may seek to avoid a current account deficit. One is that it will reduce aggregate demand. Net exports are a component of aggregate demand and if imports exceed exports, net exports will be making a negative contribution to the total demand for the country's products. As a result a current account deficit can slow down economic growth and can cause unemployment.

A current account deficit may indicate problems in the economy. It may arise due to poor quality products being produced or a relatively high inflation rate. A current account deficit can also put downward pressure on the country's exchange rate. A fall in the price of the currency may restore the current account to balance but there is a risk that domestic firms will become reliant on such declines to restore their price competitiveness. Such a tendency would lead to a significant fall in the price of the currency. A lower value of the currency may also give rise to inflationary pressure by raising the cost of raw materials and the price of a range of products bought by domestic consumers.

A current account deficit, especially if it is of a large size, may reduce confidence in the currency and the strength of the economy. This could result in an outflow of direct and portfolio investment.

A current account deficit also means that an economy is living beyond its means, in the sense that the country's population is enjoying more goods and services than it is producing. This situation can lead the country into debt and

there may be a concern that other countries will become reluctant to continue to lend to the country.

Most governments seek to achieve a current account balance in the long run. The extent to which it will be concerned about a current account deficit will be influenced by its cause, its size, its duration and whether it is offset by an inflow of direct and portfolio investment.

(b) One advantage claimed for the adoption of a floating exchange rate is that it will free a government to concentrate on domestic problems. Whilst, however, it is likely to give a government more influence over its domestic performance, it does not mean that a government can neglect the balance of payments.

In theory, under a floating exchange rate, the balance of payments or at least the current account will always move to an equilibrium position. This is because if, for instance, there is a current account deficit, the exchange rate will float downwards and, in theory at least, export revenue will rise and import expenditure will fall until an equilibrium is again restored. This will occur without any government action. So a government wishing to concentrate, for example, on raising employment may not regard the balance of payments as a constraint when it decides to increase demand. For even if a reflationary policy initially results in an increase in demand for imports, the resulting depreciation of the exchange rate will reduce the number of imports purchased.

It is also claimed that a floating exchange rate enables a government to have more control over its interest rate policy. With a fixed exchange rate, a government may have to use interest rate changes to maintain the exchange rate at the set value.

A floating exchange rate, however, whilst giving a government more opportunity to concentrate on internal problems will not in practice mean that it can neglect the balance of payments. This is because a floating exchange rate does not always guarantee a balance of payments equilibrium and because movements in the exchange rate will have an impact on the economy.

If there is a balance of payments deficit, the exchange rate will float downwards. The depreciation, however, will not restore a current account equilibrium if demand for exports and demand for imports is inelastic. This is because, in this case, a fall in the price of exports will result in a smaller percentage increase in demand for exports and hence, a fall in export revenue. Also, a rise in import prices will cause a smaller percentage fall in demand for imports and so expenditure on imports will rise.

A balance of payments deficit may also arise due to a net outflow from the financial account. Higher interest rates abroad, lower taxes and better economic prospects may cause direct and portfolio investment to leave the country.

A government may be concerned that a fall in the value of the exchange rate may give rise to inflationary pressures since the rising import prices may increase raw material costs and may stimulate wage demands. One disadvantage of a floating exchange rate is the ratchet effect. This is that when the exchange rate floats down workers, experiencing rises in the cost of living, press for wage rises but when the exchange rate floats upwards and the workers experience a reduction in the cost of living, they do not ask for a wage reduction. So the pressure over time is for wage rises.

A government may also use the exchange rate to stimulate economic growth and employment. Lowering the exchange rate may increase aggregate demand, raise output and create more jobs.

So although a floating exchange rate may be used as a policy measure rather than a policy objective, the government cannot ignore the balance of payments position nor can it presume that a floating exchange rate will guarantee a balance of payments equilibrium.

Chapter 11

Answers to short questions

1. Exchange controls are an expenditure switching policy measure. They are designed to encourage people to switch from imports to domestic products by restricting how much of the domestic currency can be changed into foreign currency to spend on imports and investment abroad.

2. Government subsidies can enable domestic firms to reduce their prices. This can encourage people to switch their expenditure from imports to domestic products.

3. Two factors that would restrict a central bank's ability to influence its exchange rate are a lack of reserves of foreign currency and the effect that a change in the interest rate will have on its macroeconomic objectives. A central bank will not be able to raise the value of its currency by buying the currency, if it does not have sufficient reserves of foreign currency to do so. Raising the rate of interest to create an increase in demand for the currency may have an adverse effect on economic growth and employment if it reduced aggregate demand.

4. A current account deficit may be self correcting if there is a floating exchange rate. In this case if, for instance, imports are exceeding exports of goods and services, there will be downward pressure on the exchange rate. If the exchange rate falls, export prices will decline and import prices will rise. If the Marshall-Lerner condition is met, this will lead to an improvement in the current account.

5. Expenditure dampening policies may reduce aggregate demand if they lower consumption and investment more than they increase net exports. If expenditure dampening policies do reduce aggregate demand, they will lower economic growth and employment. In contrast, expenditure switching policies are more likely to increase aggregate demand and so increase economic growth and employment.

6. Removing trade restrictions, such as tariffs and quotas, may reduce a current account surplus as it may result in more spent on imports. Removing tariffs is likely to reduce the price of imports and removing quotas will permit more imports into the country.

7. The imposition of import restrictions may increase the inflation rate by raising both the price of imported products and the price of domestically produced products. A tariff is a tax on import, and at least some of this tax may be passed on to the consumer in the form of a higher price. Observing that the price of imports has increased may encourage some domestic firms to increase their prices.

8. Two factors that would be likely to increase the effectiveness of expenditure switching policies are rising incomes abroad and a high price cross elasticity of demand between imports and domestically produced products.

9. Moving from a current account deficit to a current account surplus could create inflationary pressure as it would increase aggregate demand and bring more money into the economy.

10. A rise in the rate of interest may reduce spending on imports. It will discourage borrowing and encourage saving and so encourage spending on both imports and domestically produced products. There is a possibility, however, that it may increase expenditure on imports. This is because a higher rate of interest may attract hot money flows from abroad and drive up the exchange rate. A higher exchange rate will lower import prices and so may encourage people to switch from buying domestically produced products to buying imports.

11. It will rise, making Pakistan's products less internationally competitive. The real exchange rate is the nominal exchange rate X domestic price level/foreign level.

12. Expenditure switching policies would be more likely to increase aggregate demand. This is because they are designed to increase exports and reduce imports. In contrast, expenditure dampening policies are deflationary.

13. A firm may welcome a rise in its country's exchange rate if it is an importer or if it is an exporter of products which have inelastic demand.

14. Devaluation would lower the external value of the currency. So by avoiding devaluation, the government is seeking to maintain the external value of the currency. Adopting deflationary fiscal policy would imply that the government is trying to prevent the internal value of the currency falling.

15. A central bank may seek to reduce a current account deficit by attempting to reduce the exchange rate. It may try to do this by selling the currency and/or lowering the rate of interest. Selling the currency may increase the domestic money supply if some of the extra money comes back into the country. A rise in the money supply can cause an increase in aggregate demand. A cut in the rate of interest can also raise aggregate demand by encouraging an increase in consumer expenditure and investment. Higher aggregate demand can result in demand-pull inflation if there is a lack of spare capacity in the economy.

16. The expenditure dampening measure is to change the rate of interest and the expenditure switching measure is devaluation.

Answers to revision activities

1.

Sale of the currency by the central bank

↓

Increase in the supply of the currency

↓

Depreciation of the currency

↓

Fall in export prices and rise in import prices

↓

Rise in net exports

↓

Reduction in current account deficit ← → Rise in aggregate demand

↓

Increase in demand-pull inflation   Fall in unemployment

2.

Policy measure	Pairs of government macoeconomic policy objectives
A cut in the rate of interest	Reduce cost-push inflation Reduce unemployment
An increase in government spending on education	Increase economic growth Increase net exports
A rise in income tax	Reduce demand-pull inflation Reduce income inequality
The removal of tariffs	Reduce a current account surplus Reduce cost-push inflation

3. (a) A depreciation. A rise in the US rate of interest may encourage some Australians to place money in US financial institutions. In this case, they will sell Australian dollars to buy US dollars. The rise in the supply of Australian is likely to reduce its price.
   (b) An appreciation. If incomes rise in the European Union, demand for Australian exports is likely to rise. This will increase demand for the Australian dollar.
   (c) A depreciation. Demand for Australian exports will fall, resulting in a fall in demand for the Australian dollar. Australian demand for imports will increase, causing an increase in the supply of dollars.
   (d) A depreciation. Successful expenditure policies would result in lower spending by Americans, including lower spending on imports. Demand for Australian products would fall, causing a fall in demand for Australian dollars.
   (e) An appreciation. A large number of tourists would come to Australia to witness events at the global sporting event. This will increase demand for Australian dollars.
   (f) An appreciation. Multinational companies will demand Australian dollars to spend on setting up their overseas branches.

4. Internal: (b), (c) and (e)
   External: (a) and (d)

5. Decrease: (b), (c) and (d)
   Increase: (a) and (e)

Answers to multiple choice questions

1. C   An increase in the rate of interest may encourage foreigners to put money into the country's financial institutions. This will increase demand for the currency, driving up the exchange rate. A higher interest rate would increase savings but reduce investment. It would also reduce aggregate demand and so increase unemployment.

2. A   A devaluation of the currency would reduce export prices and raise import prices. If demand for exports and imports is elastic, this will result in an improvement in the balance of trade position. B, C and D would all increase spending on imports and might divert products from the export market to the domestic market and so would tend to increase a balance of trade deficit.

3. A   Moving to a deficit would result in money leaving the country. A trade deficit would tend to increase unemployment, reduce the exchange rate and reduce real GDP.

4. B   A government subsidy to domestic producers would reduce the price of their products and so may encourage both domestic consumers and foreigners to buy more domestically produced products and fewer foreign produced products.

5. D   By lowering export prices and raising import prices, devaluation will lower the terms of trade. Devaluation may raise employment. It may be caused by a fall in the rate of interest but is unlikely to cause it. If devaluation stimulates economic activity, it may increase a budget surplus.

6. B   A reduction in government spending may lower aggregate demand or at least lower the growth in aggregate demand. This may reduce inflationary pressure. It may also lower expenditure on imports and put pressure on domestic firms to increase their exports.

7. B   An expenditure dampening policy would reduce consumer expenditure. With consumers spending less, imports are likely to fall and domestic firms may put more effort into selling products to other countries.

8. C   An upward revaluation of the currency would raise export prices and lower import prices. This will reduce a current account surplus if demand for exports and imports are elastic. Lower import prices will reduce the prices of some finished products produced, and may lower the prices of some domestically produced products if they use imported raw materials. Higher export prices do not affect the country's inflation rate as the country's population does not buy exports.

9. D   An increase in income tax would reduce the country's inflation rate by reducing consumer expenditure and so aggregate demand. It may also reduce a current account deficit if the fall in consumer expenditure reduces spending on imports. A and B would reduce a current account deficit but would increase inflation. C might reduce the inflation rate but may also increase a current account deficit by raising export prices and lowering import prices.

10. B   Lower government spending and a higher rate of interest would be likely to reduce aggregate demand. This might lower the inflationary pressure which might result from a current account surplus.

Answers to data response questions

1. (a) Both a customs union and a free trade area do not have restrictions on the movement of products between member countries. A customs union requires member countries to impose the same tariffs on the imports from non-members. A free trade area, however, allows member countries themselves to decide whether they want to impose tariffs on imports from non-members and, if so, what the tariffs should be.
   (b) (i) The value of the Mexican currency fluctuated over the period. It rose against the US dollar from the start of 1997 through 1998 and then fell. Over the whole period, it rose by approximately 15 index points. In contrast, the value of the Argentine currency remained fixed throughout the period.
       (ii) The value of the Mexican currency may have risen because the price competitiveness of the country's products may have increased. A lower rate of inflation in Mexico may have increased demand for Mexico's exports and reduced Mexico's demand for imports. These changes would have increased demand for Mexico's currency whilst decreasing the supply.

(iii) The fixed level of the Argentine exchange rate may have been achieved by its central bank buying and selling the currency to avoid fluctuations. The central bank may also have changed its interest rate to influence the demand and supply of the currency in order to offset market fluctuations.

(c) (i) Mexico experienced an inflation rate of 50% from the start of 1997 to the end of 1999. Argentina experienced a slight fall in its price level over the period, in other words, deflation.

(ii) Mexico's high rate of inflation would be expected to cause a fall in the exchange rate. With prices rising rapidly in Mexico there is likely to have been a fall in Mexico's exports and a rise in Mexico's imports. This would have reduced demand for the currency and increased its supply. The impact of the high inflation may have contributed to the fall in the peso in 1999.

(d) The devaluation of a country's exchange rate may improve the country's balance of trade position. If demand for exports and imports is price elastic, and so the Marshall-Lerner condition is met, export revenue will rise and import expenditure will fall.

If, however, demand is price inelastic, a devaluation will worsen the balance of trade position. It is also possible that whilst devaluation may cause demand for exports to rise by a greater percentage than the fall in their price, it may be difficult to increase exports if the economy is working at full capacity. The effect of the devaluation may also be offset by a fall in income abroad, the imposition of trade restrictions or other countries devaluing.

2. (a) (i) Balance on services
(ii) Balance on goods

(b) The current account balance was 686.7m emalangeni in 2003, −652.7m in 2005 and −461.8m in 2007. So the balance went from a surplus to a deficit. The deficit fell slightly from 2005 to 2007.

(c) (i) Comparative advantage arises when a country can produce a product at a lower opportunity cost than another.

(ii) The Central Bank's report suggests that Swaziland has a comparative advantage in agricultural products, processed agricultural products and relatively low valued-added manufactured products. Countries usually have a comparative advantage in agricultural products when they have fertile agricultural land and in low valued manufactured products when they have a supply of low skilled, low waged labour.

(d) For a depreciation to succeed in increasing a country's export revenue, it is necessary for demand for exports and imports to be price elastic, so that export revenue rises and import expenditure falls. It is also necessary that the supply of exports is elastic, so that advantage can be taken of higher demand. In addition, it is important that other countries do not devalue their currencies and that they do not improve import restrictions on the country's products.

(e) Swaziland's use of tariffs may have been one of the reasons why the EU removed its ban on Swaziland's beef exports. Tariffs also raise revenue for the government and may be used to protect infant, strategic or declining industries. For instance, Swaziland may want to build up its meat products industry but the industry may not yet be large enough to exploit the available economies of scale. In the short term,

tariffs may also turn the country's deficit into a surplus. In the longer term, however, it may not reduce imports but also reduce exports if the EU retaliates and a trade war develops. Swaziland may also be reluctant to impose tariffs on EU products if the EU imports put competitive pressure on domestic firms to keep prices low or if the imports lack domestic substitutes.

3. (a) A country might open its economy to free international trade by removing tariffs and quotas on imports.

(b) (i) Japan's comparative advantage appears to lie in high technology-intensive products. This suggests that Japan benefits from high levels of investment and high skilled workers. In contrast, at the time, China seemed to have a comparative advantage in making lower quality manufactured products due to its low productivity and low wage labour.

(ii) The total value of trade was greater for Japan than China in 2000 −$802bn compared to $465bn. As a percentage, however, international trade was more important for China as it accounted for 42% of its GDP whereas for Japan it was 16%.

Both countries experienced a current account surplus, and so in both countries, international trade made a positive net contribution to aggregate demand. Japan had a larger surplus as a percentage of GDP than China.

(c) Japan experienced a very low rate of inflation over the previous ten years and actually experienced deflation in 2000. China's inflation rate over the previous ten years was noticeably higher than Japan's – 6.3% points. In 2000, it experienced inflation but it was a very low rate.

(d) China presents both a threat and an opportunity to Japan. In 2000, it was producing standard models of TVs and cars and as the economy develops it may move into high quality models and compete with Japan. China may also start to compete with Japan in other products and may capture some of its exports markets. Japanese multinational firms have set up branches in China and more Japanese firms may decide to relocate production to China. This may have an adverse effect on Japanese employment and economic growth.

There is the possibility, however, that if Japanese firms respond to the increased competition from China by raising their efficiency, Japan may benefit from the economic development of China. A growing Chinese economy will purchase more imports. As the skills of Chinese workers and their productivity rises, the country may provide Japan with cheaper imports of raw materials. This would reduce Japanese firms' costs of production. It may also enable branches of Japanese multinational companies, based in China, to send greater profits back to Japan.

Answers to essay questions

1. (a) A current account deficit means that the value of debit items on the current account exceeds the value of credit items. This means that there has been a net outflow of money on the current account.

The current account has four parts. These are trade in goods, trade in services, income and current transfers. Trade in goods is also sometimes known as the visible balance or the merchandise balance. It is calculated by deducting the value of imports of goods from the value of exports of goods. Trade in services is also referred to as the invisible balance. This part covers, for example, transportation, travel, communications, construction, finance and education. A trade in services surplus means that the income earned from selling services to other countries would exceed expenditure on services bought from other countries. Income covers mainly investment income and includes net flows of dividends, profits and interest. Current transfers include food aid, subscriptions to international organisations and workers' remittances.

A current account deficit might result from a deficit on all four parts or a deficit on one or some parts which is greater than a surplus on one or more parts. For instance, a trade in goods deficit may be larger than surpluses on the other three parts.

(b) Imposing tariffs to correct a current account deficit is an expenditure switching measure. It is designed to encourage people to switch their spending from imports to domestically produced products.

A tariff is a tax on imports. To cover the tax, foreign producers are likely to raise the price of imports. If the tariffs are large enough, the price of imports may fall. This may be sufficient to correct a current account deficit. Imposing tariffs may also permit infant industries to grow which in the longer run may reduce imports and increase exports. Tariffs may also be used to control dumping from foreign producers. There is a risk, however, that tariffs may provoke retaliation and lead to a price war, reducing not only imports but also exports.

If, however, the domestically produced substitutes are of a poorer quality, people may not buy them even if they are now cheaper than imports. It is also possible that foreign producers may absorb most or all of the tax and so imports may still remain price competitive. The imposition of tariffs will also not be an effective policy measure, if the current account deficit has arisen due to a net outflow of income and/or current transfers.

Even if the imposition of tariffs does reduce a deficit in the short run, it is possible that it may not be a long term solution. If the deficit has been caused by high inflation or an overvalued exchange rate, the problem is likely to reappear. Indeed, tariffs may make the situation worse in the long run. This is because they may reduce the competitive pressure on domestic firms to keep their prices down and to raise the quality of what they produce.

The government of a country which is a member of a trading bloc will be restricted in its ability to impose tariffs. Any country, whether a member of a trading bloc or not, has to consider what has caused the deficit, whether imposing tariffs is likely to impose the situation and what is the risk of retaliation.

2. (a) A devaluation occurs when a government changes its fixed exchange rate to a lower fixed exchange rate. It will reduce the price of exports in terms of foreign currency and raise the price of imports in terms of the domestic currency.

For a devaluation to succeed in reducing a trade deficit, it is necessary for the combined elasticities of demand for exports and imports to be greater than one. This is known as the Marshall-Lerner condition. If the Marshall-Lerner condition is met, there should be a fall in the gap between import expenditure and export revenue. If, however, demand

is price inelastic, reducing export prices and raising import prices would increase a trade deficit.

The J-curve effect suggests that the price elasticity of demand for exports and imports and that a depreciation may make the trade deficit larger before it improves it. In the short term, demand for exports and imports is usually price inelastic. This is because it takes time for people to recognise that prices have changed and to change contracts. With inelastic demand, a devaluation will reduce export revenue and raise import expenditure. In the longer term, when people have had time to realise the price changes and to adjust their purchases, demand may become price elastic and the trade position may improve. The following figure shows that the country starts with a trade deficit. This deficit initially increases before it moves into a surplus, giving the curve its J-shape.

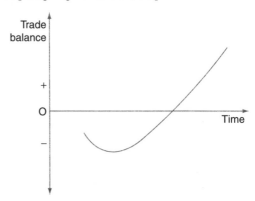

(b) Some expenditure dampening policies may reduce a trade deficit in the short run. For example, if increasing income tax reduces consumer expenditure, demand for imports is likely to fall. Lower demand for domestically produced products may also encourage domestic firms to increase their efforts in searching out new export markets. In addition, lower consumer expenditure may reduce inflation and make domestically produced products more price competitive.

In the longer run, however, if the trade deficit has been caused by a lack of quality competitiveness or inflationary pressure due to higher costs of production, any reduction in a trade deficit may be short lived. As consumer expenditure returns to its previous levels, imports will again increase.

There is also the risk that some expenditure dampening policies may actually increase a trade deficit. A rise in the rate of interest may reduce spending on imports by lowering consumer expenditure. It may, however, also increase the value of the currency by attracting hot money flows into the country and so raising demand for the currency. The appreciation of the exchange rate will increase export prices, which may reduce export revenue, and lower import prices, which may raise import expenditure. In addition, a higher interest rate may discourage investment which, in the long run, may reduce both the price and quality competitiveness of domestic firms.

A cut in some forms of government spending may also reduce the price and quality competitiveness of domestic firms. For example, a reduction in government spending on education and training may lower labour productivity and cuts in government spending on investment may reduce product development and raise costs of production.

Although expenditure dampening measures may reduce a trade deficit in the short run, they may have adverse effects on other macroeconomic objectives. In the long run, other policy measures, including education and training may be more effective in reducing a trade deficit.

3. (a) An appreciation of a currency means that its exchange rate rises with each unit of the currency purchasing more of another currency or currencies. It will result in a rise in export prices in terms of foreign currencies and a fall in import prices in terms of the domestic currency.

Domestic citizens do not buy their country's exports and exports are not included in the consumer price index, an important measure of inflation. The citizens do buy imports and lower import prices are likely to reduce inflationary pressure in the country. This is for three main reasons. One is that some of the products consumers buy will fall in price. The more imports they do buy, the larger this impact will be.

The price of imported raw materials will also fall. This will reduce domestic firms' costs of production. This may enable them to reduce their prices and possibly increase their profit levels.

In addition, lower import prices will put pressure on domestic firms that produce substitute products to avoid price rises. So both directly and indirectly an appreciation can lower a country's inflation rate. The extent of its influence will depend on the size of the appreciation and what proportion of raw materials and products purchased by consumers are imported.

(b) A rise in an economy's inflation rate may result in a depreciation of the floating exchange rate. If the economy's price level is rising more rapidly, this may make its products less internationally competitive. This may result in a fall in demand for its exports and a rise in demand for imports. Demand for the currency will fall whilst the supply of the currency will rise. As shown here, these changes will cause the value of the currency to fall.

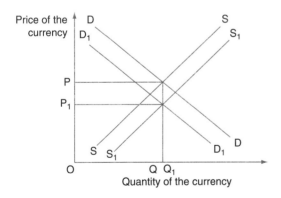

A rise in an economy's inflation rate may put downward pressure on a fixed exchange rate but may not cause the exchange rate to change. This is because a central bank may buy up the currency or raise the rate of interest to offset the downward pressure.

Even in the case of a floating exchange rate, a rise in an economy's inflation rate will not necessarily cause a depreciation in the exchange rate. Other countries may be experiencing even higher rates. It is also possible that despite a high inflation rate, other countries will still want to purchase the country's exports. This may be because incomes abroad are increasing or the quality of the country's products is improving.

As well as trade flows, there are other influences on the exchange rate. An economy may currently be experiencing a rise in its inflation rate, but if its economic performance is forecast to improve, speculators may increase demand for the currency. In addition, foreign multinational companies may be encouraged to set up in the country if consumer expenditure is rising. This would again increase demand for the currency.

For an economy operating a floating exchange rate, a rise in an economy's inflation rate might be expected to cause depreciation in the external value of the currency. There are a number of reasons, however, why this might not be the case.

Chapter 12

Answers to short questions

1. Economic inefficiency occurs when productive efficiency and allocative efficiency are not achieved. Output is produced at a higher cost than possible and does not fully reflect consumer demand.

2. Productive inefficiency is illustrated on a production possibility curve diagram by a production point inside the curve as illustrated here.

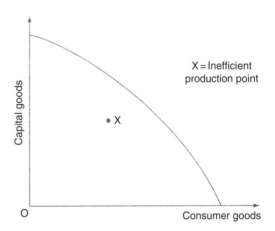

3. If output is at a point where price is below marginal cost, too many resources are being devoted to producing the product. The cost of the last unit exceeds the value consumers place on it.

4. Achieving economic efficiency will not solve the economic problem. Wants are always likely to exceed resources. The gap, however, will be lower than if there is economic inefficiency.

5. The existence of long run shortages in a market indicates that a market is inefficient. Devoting more resources to producing the product would increase welfare.

6. International trade may promote economic efficiency as it can permit countries to specialise in producing those products it is better at making. This way it can raise output and lower production costs.

7. The existence of unemployed resources means that national output is lower than possible. The country will be producing below its productive potential and will be productively inefficient.

8. In the case of merit goods, allocative efficiency is not achieved. Not enough resources are allocated to their production. They are under-consumed and so under-produced.

9. A tax on a demerit good will reduce efficiency if it is set so high that it moves the market from over-consumption to even greater under-consumption.

10. It is difficult to achieve allocative efficiency in the case of public goods as it is difficult to assess how much consumers value the products. This is because they do not have to reveal their preferences.

Answers to revision activities

1. (a) A reduction in unemployment would increase productive efficiency. More use would be made of factors of production and output would increase.
   (b) Reallocating resources from producing demerit goods to producing merit goods would increase allocative efficiency. This is because resources would be moved from producing products that are overproduced to those that are under-produced.
   (c) A reduction in surpluses would increase allocative efficiency. Resources would not be wasted.
   (d) A reduction in labour productivity would reduce productive efficiency. Costs per unit would be driven up above the minimum level.
   (e) A switch from producing less popular to more popular products will increase allocative efficiency. Producers will be responding more effectively to consumer demand.
   (f) Organisational slack can also be referred to as X inefficiency. It arises when firms do not minimise average costs due to a lack of competitive pressure. It may involve managers permitting some spare capacity to exist. If too much organisational slack exists, a reduction may increase both productive and allocative efficiency. If, however, all organisational slack is eliminated, it may become difficult for firms to adapt to market conditions and so both allocative and productive efficiency may be reduced.

2. (a) False. A shift to the right in the production possibility curve shows that the economy is capable of producing more. It does not in itself, however, indicate an increase in productive efficiency. To assess whether productive efficiency is being achieved or not, it would be necessary to assess where the production point is relative to the production possibility curve.
   (b) True. If average revenue is above marginal cost, not enough resources are being allocated to producing the product. Consumers are valuing the product more than it is costing to produce it – there is a welfare loss.
   (c) True. When allocative efficiency is achieved, consumers are paying an amount for the last unit consumed which matches society's valuation of the resources used to make that unit.
   (d) False. To achieve economic efficiency, marginal social cost should equal marginal social benefit.
   (e) True. Information failure can result in people under-consuming merit goods and over-consuming demerit goods. This causes too few resources being devoted to consumer goods and too many resources being devoted to demerit goods.

Answers to multiple choice questions

1. C   Allocative efficiency is achieved where the value consumers place on a product (reflected in the price they are prepared to pay) is equal to the cost of producing the last unit.

2. B   A more efficient allocation of resources will enable an economy to produce more goods and services with existing resources.

3. B   To be productively efficient an economy needs to be making full use of its resources. The economic problem cannot be solved. Macroeconomic equilibrium may occur at any level of economic activity. If an economy can produce more of one good without producing less of another good, it is inefficient.

4. C   Output C occurs where average revenue (price) equals marginal cost. At this point allocative efficiency is achieved.

5. D   If a tax is imposed which is equivalent to external costs, output should occur where marginal social cost equals marginal social benefit. Fewer resources will be devoted to the product. Output will decrease and demand will contract rather than decrease.

6. C   Points X and Y are both productively efficient as they both occur on the production possibility curve. To know what effect the movement along the production possibility curve has on allocative efficiency, it would be necessary to have information on consumers' preferences.

7. A   International trade can allow countries to specialise which, in turn, can increase output and reduce costs. B and D would reduce efficiency and C does not directly affect efficiency.

8. A   An improvement in the state of technology would increase the amount that could be produced with existing resources. B, C and D would not increase output and, indeed, D would reduce productive capacity and may lower output in the longer term.

9. C   Producing inside a production possibility curve (PPC) means that not all resources are employed. Making use of unemployed resources means that more can be made of both products. It is not necessary to switch resources from making less of one product to making more of another product. A PPC does not provide information about the distribution of income or prices.

10. B   A Pareto improvement occurs when a change in the use of resources causes at least one person to be better off without causing any one else to be worse off. A movement from X to B increases person R's utility, whilst leaving person T's utility unchanged. A movement from X to A reduces both people's utility. A movement from X to C increases person R's utility but reduces person T's utility. Finally, a movement from X to D leaves person R's utility unchanged but reduces person T's utility.

11. D   The tax on the substitute good might initially have meant that its price exceeded its marginal cost and that it was under-consumed. The introduction of a tax on a rival product may increase the demand for the substitute moving consumption closer to the efficient level. A tax on a complementary good may also mean that this product was initially under-consumed. This time, however, the introduction of a tax on a product which is in joint demand will reduce demand for the complementary good. A and D would also increase inefficiency as they would move price away from marginal cost.

12. D Producer surplus rises by PPXUV whilst consumer surplus falls by PPXTV, giving a net benefit of TVU. The cost to the government is PX multiplied by quantity purchased which is YZ – giving a total cost of YTUZ. So the net loss to society is YTUZ – TVU which is equal to YTVUZ.

Answers to data response question

1. (a) In the short run, firms are regarded as being productively efficient when they produce any given output at minimum cost. In the long run, productive efficiency is achieved by producing at the lowest point on the lowest average costs curve. To be allocatively efficient, firms have to produce where price equals marginal cost.

   (b) The information mentions that a lack of competition reduces the pressure on firms to keep their costs low. This indicates that at least some Latin American firms might not have been productively efficient. It also mentions that the lack of competition reduces pressure on firms to 'respond quickly and fully to changes in consumer demand'. This implies that firms may be allocatively inefficient – not producing the right products in the right quantities.

   (c) Latin American governments could increase spending on education and training and could provide investment subsidies to increase efficiency. A more educated and better trained labour force should have higher productivity. This should reduce firms' costs of production. A more educated and better trained labour force should also be more geographically and occupationally mobile, should adapt more quickly and fully to changes in consumer demand and so increase efficiency.

   Investment subsidies should encourage firms to undertake more investment. New investment also often embodies advanced technology. Having more and better quality capital equipment should cut costs of production and may increase flexibility of production. If this is the case, again productive efficiency and allocative efficiency should increase.

   (d) (i) The information in Table 1 is inconclusive. Venezuela does have both the highest inflation rate and the highest unemployment rate. Argentina, however, which has the second highest inflation rate has only the fourth highest unemployment rate. Chile has the lowest inflation rate but the second highest unemployment rate.

   (ii) A high unemployment rate is an indication of productive inefficiency. When an economy has a high unemployment rate, it will be producing inside its production possibility curve. With unemployed resources, actual output is below potential output.

   A high inflation rate may indicate inefficiency but this is not necessarily the case. Producers becoming more productively inefficient can contribute to cost-push inflation. This type of inflation, however, may also be caused by external shocks. Domestic producers may be relatively efficient but a rise in the price of imported fuel may drive up costs. In addition, inflation may be of a demand-pull nature. Producers may be producing the right products in the right quantities and at the lowest possible cost. If, however, aggregate demand exceeds the maximum output the country can produce, inflation will occur.

Answers to essay questions

1. Economic efficiency occurs when resources are used in a way that maximises economic welfare. It is achieved when both allocative efficiency and productive efficiency are attained. Allocative efficiency occurs when the right products are produced in the right quantities. Firms produce where price equals marginal cost and, in the wider sense, where marginal social benefit equals marginal social cost.

   Productive efficiency is achieved when firms produce at the lowest possible average cost and the economy makes full use of its resources.

   Market forces can provide both an incentive (carrot) and a threat (stick) for firms to be efficient. Those firms which make the products consumers want at the lowest cost may earn supernormal profits at least in the short run. In contrast, those firms which fail to be allocatively efficient may be driven out of the market.

   There are, however, a number of reasons why market forces may not achieve economic efficiency and so why government intervention may be needed. Private sectors, monopolies and oligopolies, pursuing maximum profit, will produce where price is greater than marginal cost.

   Private sector firms are also likely to take into account in making their production choices. They are likely to base their decisions on private costs and benefits. For example, private sector farms may cut back on costs by disposing of animal waste in rivers and private sector factories may emit greenhouse gases which contribute to climate change.

   Private sector firms will also fail to produce public goods, will under-produce merit goods and will over-produce demerit goods. They will not make public goods as they will have no financial incentive to charge directly for them. This is because they will not be able to exclude free riders. They will not devote sufficient resources to producing merit goods as they will be under-consumed. In contrast, too many resources will be devoted to demerit goods as these will be under-consumed.

   A lack of factor mobility may also result in shortages and surpluses in markets. Labour market failure may occur leading to unemployment.

   Government intervention can seek to regulate private sector monopolies and oligopolies. It can also try to convert externalities into private costs and benefits, to encourage the consumption and production of merit goods and discourage the consumption and production of demerit goods. Government policies, including educational policies, may promote an increase in labour mobility and expansionary fiscal and monetary policies may be used to increase aggregate demand and so lower unemployment.

   There is, however, no guarantee that government intervention will reduce inefficiency. Government failure may occur, for instance, if government intervention is based on inaccurate information. There are disagreements, for instance, about the causes of climate change and whether government intervention is needed.

2. (a) Efficiency in the use of resources covers both allocative efficiency and productive efficiency. The following figure shows the allocatively efficient output is X. At this level of output the value people place on the last product consumed is equal to the cost of that last unit. The right quantity of output is produced.

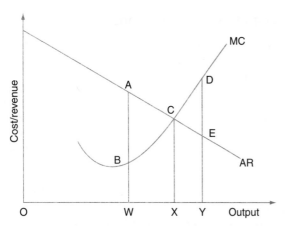

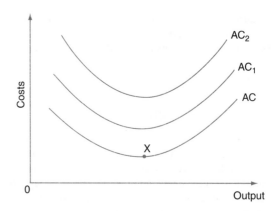

An output of W would be allocatively inefficient. This is because consumers value the product more than it is costing to make it. There is an insufficient quantity of resources devoted to making the product and there is a welfare loss represented by the area ABC. Mutually beneficial transactions are being forgone. An output of Y would also be allocatively inefficient. In this case, the cost of producing the product exceeds the utility consumers gained. There is overproduction and a welfare loss of CDE is created.

Productive efficiency occurs when firms are producing the maximum output for a given quantity of resources and producing that output at the lowest possible cost. This can also be viewed as producing a given level of output with the minimum quantity of resources.

Productive efficiency in an economy takes place when production occurs on the production possibility curve as illustrated by point Z in the following figure.

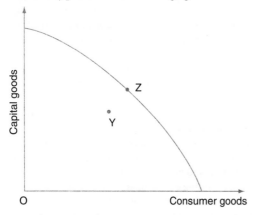

If the economy is producing at point Y, there are unemployed resources and some industries may be losing more resources than necessary for the output they are making. By employing all resources and by ensuring all industries are productively efficient, the economy's output can rise to Z. At production point Z, it will not be possible to make more capital goods without making fewer consumer goods and vice versa.

For an individual firm, productive efficiency occurs when it produces any given output at the lowest possible average cost. Productive efficiency is sometimes defined as producing at the lowest point on the lowest average cost curve. This is shown as point X in the following figure.

(b) The government's decision to spend a large amount of money and resources to help with the reconstruction of the area will have implications for taxation or government spending. The government may raise tax rates or widen the tax base to finance the extra expenditure. Alternatively, the government may switch spending from, for instance, defence or education to the restoration efforts. Either way, there will be an opportunity cost. Taxpayers may have to forgo some consumer expenditure or saving, or some government departments may have to sacrifice some of their projects.

The government might also decide to borrow money to finance their expenditure on the restoration. An increase in government borrowing may push up the rate of interest. This may result in crowding out with private sector firms reducing their investment because of the higher interest rate and possibly shortage of loanable funds.

Extra government spending, however, may raise real gross domestic product (GDP). Government spending is an injection into the circular flow and, ceteris paribus, will cause a multiple increase in real GDP. The following figure shows a rise in government spending will cause a rise in aggregate demand from AD to AD1. If the economy is initially operating below full capacity, this can result in a rise in both output and the price level.

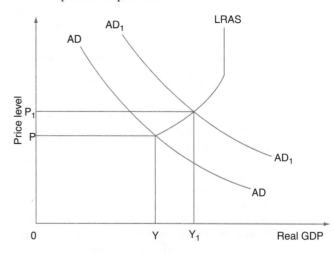

Not all of the extra government spending may be injected into the domestic economy. Some may be spent on imports. In this case, any trade deficit may increase. The huge floods would themselves have been likely to reduce exports. Some resources would have been destroyed and damaged which would have lowered the output of a range of products, including agricultural goods.

As well as considering the effects on the macroeconomy, the government would have to decide how to spend the money and allocate resources. The government may directly provide goods and services in the region or may decide to subsidise private sector firms to provide the products. In making its decision, it will have to consider the relative efficiency of the private sector and public sector.

Chapter 13

Answers to short questions

1. A change in the price of a product or a change in the marginal utility gained from a product could cause consumers to change the pattern of their spending.

2. A person would be unlikely to consume a quantity of a product where total utility is falling as this would mean that marginal utility is negative. Consuming the last unit would have reduced total satisfaction which would be irrational.

3. If the price of one of the two products falls, the budget line will rotate outwards. This shows that a larger amount of the product can be purchased with a given level of income.

4. A rise in marginal product and a rise in the price of the product.

5. More capital-intensive methods of production and improvements in the capital equipment used have raised agricultural output.

6. Labour is not demanded for its own sake but for what it can produce. So the amount of labour demanded depends on the demand for the products it can produce.

7. One reason why demand for nurses has increased in most countries in recent years is due to increase in life expectancy. With people living longer, demand for health care has increased. Another reason why demand for health care has increased and so demand for nurses has also increased is a rise in income. As people get richer they demand more and better health care.

8. Three reasons why people may switch from a higher paid to a lower paid occupation are that they may believe they will gain more job satisfaction from it, it may provide more job security and it may provide a better pension.

9. A closed shop occurs when to work in a job, people have to belong to a trade union.

10. Raising educational standards should increase workers' skills and mobility. This is likely to mean that there will be a range of well-paid jobs they can do. The gap between what they are earning now and what they could earn in their next best paid job should narrow.

11. Fixed costs are costs which do not vary with output in the short run. In contrast, variable costs rise as output rises. In the long run all costs are variable costs as this is the time period when the quantity of all factors of production can be altered.

12. Financial economies of scale, managerial economies of scale, staff facilities economies of scale and, of course, risk bearing economies of scale.

13. External economies of scale are sometimes referred to as economies of concentration because they arise when the firms in the industry are concentrated in one area. For instance, colleges are more likely to run specialist courses for car mechanics if there are a number of car firms nearby.

14. Among the factors that may prevent a firm from expanding are demand for a product not increasing, a lack of finance and difficulty in attracting factors of production.

15. A sole trader can use retained profits or borrow to finance investment. A large public limited company can also use retained profits and borrow but it can also sell shares on the stock exchange.

16. A perfectly competitive firm's marginal revenue and average revenue are equal and both will remain constant if the firm increases its output. In contrast, a firm operating under conditions of monopolistic competition, a firm's average revenue will exceed its marginal revenue. Average revenue and marginal revenue will also fall with output.

17. A firm might stay in an industry, despite making a loss, as long as it can cover its variable costs and believes that in the longer term it will return to profitability.

18. Predatory pricing is designed to drive out rivals from an industry whereas limit pricing is intended to prevent new firms coming into the industry.

19. A firm may not seek to maximise profit either because it cannot determine what the profit maximisation level of output is or because it wants to pursue another objective such as sales revenue maximisation.

20. A number of market imperfections may exist at any one time and seeking to correct one market failure may increase another market failure. The second best theory argues that governments should seek to minimise the overall adverse effects on efficiency of any government policy measure designed to reduce market failure in one market.

21. Perfect competition is the highest level of competition possible but it is not perfect. Whilst it offers consumers a choice of producers, it does not provide a choice of differentiated products. In addition, although firms will be allocatively efficient in the sense of producing where price equals marginal cost, they will not necessarily produce where marginal social benefit equals marginal social cost.

22. A monopsony is a single buyer and an oligopsony is one of a few dominant buyers.

23. The kinked demand curve suggests that prices will be 'sticky' under conditions of oligopoly. This means that they will not change on a frequent basis as firms will think that either raising or lowering price will lose them revenue.

24. A bilateral monopoly occurs in a market when there is a single buyer and a single seller – i.e. a combination of monopsony and monopoly.

25. Two possible barriers to exit from an industry are long term contracts and sunk costs. A long term contract will mean that a firm may have to continue to provide a good or service for at least a year, for example. Sunk costs are costs which are not recoverable should the firm leave the industry. Spending on market research and the cost of firm specific equipment are examples of sunk costs.

26. The theory of contestable markets suggests that privatisation may make a market efficient even if the state owned enterprise is sold off to one firm provided there is free entry into and exit from the market. In such a situation, potential competition may be sufficient to keep the market allocatively and productively efficient.

27. A natural monopoly is an industry where economies of scale are so significant that average costs would be higher if there were to be more than one producer.

28. The three conditions for a firm to engage in price discrimination are that it must have the market power to set price, there must be no seepage between the two markets and the price elasticity of demand must be different in the two markets.

29. The prisoners' dilemma can be applied to emphasise the interdependence of oligopolists. It suggests that firms consider how their rivals will respond to their actions.

30. In monopolistic competition, the profit maximising output occurs at a point below where average costs are minimised.

Answers to revision activities

1.

Output	Total cost	Fixed cost	Variable cost	Average cost	Average variable cost	Average fixed cost	Marginal cost	Marginal variable cost
0	100	100	–	–	–	–	–	–
1	150	100	50	150	50	100	50	50
2	180	100	80	90	40	50	30	30
3	201	100	101	67	33.67	33.33	21	21
4	220	100	120	55	30	25	19	19
5	280	100	180	56	36	20	60	60
6	360	100	260	60	43.33	16.67	80	80

2. (a) True. A fall in the price of a product, for instance, will increase people's purchasing power (income effect) and will encourage them to switch from substitutes to this product (substitution effect).

(b) False. Average fixed costs always fall with output. Average variable costs, however, tend to fall at first and then rise.

(c) True. Demand for skilled labour is usually more inelastic than demand for unskilled labour. This is because it is more difficult to replace skilled labour with machinery and with new workers.

(d) False. If the supply of labour is perfectly elastic, all of the workers' earnings will be transfer earnings.

(e) False. Although seeking to raise wage rates is a major function of trade unions, it will not pursue this aim at all times. During a recession, trade unions are more likely to pursue job security for their members.

(f) True. One reason why small firms survive is that they cater for a market which has a relatively small demand. Larger firms may not find it profitable to make products for such a market.

(g) True. If marginal revenue is zero and PED is unitary, selling one more unit will not add to total revenue. Total revenue will be at its highest level.

(h) True. A perfectly competitive firm is a price taker. A change in the quantity it sells would be too insignificant to influence price. What determines the amount it sells at the market price is the relationship between that price and its marginal cost (MR = MC). In contrast, a monopolist is a price maker. To encourage people to buy more of the product, it would have to lower price.

(i) False. It is true that a perfectly competitive market is a contestable market as both lack barriers to entry and exit. It does not, however, follow that a contestable market is a perfectly competitive market. It is possible for other market structures to be contestable markets. What is significant in the case of contestable markets is not the actual competition in the market but the potential competition.

Comparison of market structures				
Feature	Perfect competition	Monopolistic competition	Oligopoly	Pure Monopoly
No. of firms in the market	Very many	Many	Dominated by a few large firms	One
Market concentration ratio	Very low	Low	High	100%
Barriers to entry and exit	None	None or low	High	Very high
Type of product produced	Homogeneous	Differentiated	Differentiated and homogeneous	Unique
Influence on price	Price taker	Price maker	Price maker	Price maker
Ability to earn supernormal profit in the long run	No	No	Yes	Yes

Answers to multiple choice questions

1. B   Total utility rises when marginal utility is positive. It reaches its peak when marginal utility is zero and would decline if marginal utility became negative. A zero marginal utility does not indicate that the product is free and the consumer is in equilibrium when the marginal utility of each product divided by its price is equal.

2. D   The person is receiving a higher utility per $ spent on product Y than on products X and Z. The marginal utility per $ spent on product Y is 8 whereas it is 5 each on the other two products.

3. D   This is the paradox of value. Whilst food has a high total but low marginal utility, the reverse is true for diamonds.

4. A   The consumer would be able to buy less of good X. If the price of good Y falls relatively more than the fall in income, the consumer might be able to buy more of the good.

5. B   The total cost when output is zero is 20. The total cost and average cost of the different units is shown below.

Units of X	Total cost	Average cost
1	60	60
2	90	45
3	120	40
4	180	45
5	260	52

6. D   At 5 units of output, the average total cost is $110, so the total cost is $550. At 6 units, the average total cost is $98 and so the total cost is $588. Therefore, the marginal cost is $588 − $550 = $38.

7. C   The firm's average fixed cost (AFC) is $2 ($10 − $8). AFC is total fixed cost/output. In this case, $2 = $8,000/output. So output is 4,000 units.

8. A   A definition question. B is decreasing returns to scale. There are no specific terms for C and D.

9. B   Diminishing returns set in when successive units of a variable factor of production are applied to a fixed factor.

10. C   Between 450 and 600 tonnes, output is rising more slowly than previously.

11. C   A shift to the right of the demand curve for labour indicates an increase in the demand for labour. An increase in labour productivity would increase the return from employing workers. A and B would cause a decrease in demand for labour and D would result in a contraction in demand for labour.

12. D   Employing the third worker causes total output to rise by 40 units.

13. A   The marginal product received per $ spent on factor Y is 7 whereas it is 5 for factor Z. The higher return from factor Y suggests more of it should be employed and less of factor Z.

14. B   A backward sloping supply curve shows that at first a rise in wages will encourage workers to work longer hours. At this stage the substitution effect is dominant. Between W and W1 the income effect becomes dominant and workers, in effect, buy more leisure time.

15. C   The successful negotiation will set a wage floor of W1. Workers will supply their labour for a wage of W1 or above. Employment will fall from Y to X, the wage bill will fall to OW1X and the total economic rent earned will decline.

16. A   The number of workers employed is determined where the marginal cost of labour (MCL) equals the demand for labour (marginal revenue product of labour MRPL). The wage rate is then found from the average cost of labour curve below where MCL crosses the MRPL curve.

17. B   The army is likely to be the only buyer of the labour of soldiers. A national trade union may have monopoly power in the sale of labour. The extent to which workers who possess skills in high demand may have some degree of monopoly power will depend on whether they negotiate collectively. The sole seller of a particular brand of orange juice is likely to be competing to buy the services of workers with other firms producing orange juice.

18. A   The initial economic rent earned is MNT. It then rises to MPS – an increase of NPST.

19. A   A national minimum wage will increase employment if it results in a rise in spending and so a need for more workers. If the national minimum wage is set below the equilibrium level, it will have no effect. The elasticities of demand for and supply of labour will influence the extent to which unemployment might result from the introduction of the national minimum wage.

20. A   If average cost (AC) is falling, marginal cost (MC) will be lower as shown in the figure here.

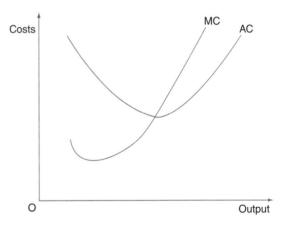

21. D   Marginal cost is zero. This means that profit will be maximised where marginal revenue is zero and this occurs where total revenue is at a maximum.

22. C   Wages increased by 30% and output per hour by 15%, so labour costs rose. Hours worked fell by 5% but output per hour increased by 15%, so total output rose.

23. B   If a firm can cover the direct costs of production and possibly make some contribution to fixed costs, it may stay in business if it also thinks conditions will improve in the future. It may also be able to postpone paying some fixed costs but will have to pay now for raw materials, for example.

24. A   The supply curve shows the different quantities which will be supplied at different prices. A profit maximising firm will produce where marginal cost equals marginal revenue. Under conditions of perfect competition, average revenue equals marginal revenue. So the supply curve can be plotted from

where the average revenue and marginal revenue line cuts the marginal cost curve. In the short run, a firm will continue in production if it can cover its variable costs and therefore, its supply curve is based on its marginal cost curve above the average variable cost curve.

25. D Monopolistically competitive firms make slightly different products. In the long run, firms produce where average revenue equals average cost and so makes only normal profit. There are no or only low barriers to entry and exit.

26. A The firm will produce where MC = MR. This is where output is M. Average cost is F and total cost is OFKM.

27. B Sales maximisation occurs at a higher level of output than profit maximisation. To increase sales, the monopolist will have to lower prices. To maximise sales, a firm is likely to increase sales until average revenue equals average cost. Increasing output past this point would be unlikely as it would involve making a loss.

28. B At Q supernormal profit is GPJN. When the firm cheats, it will produce where MC = MR and earn supernormal profit of PKLH. So there is extra supernormal profit of JKLM minus the loss of HMNG supernormal profit.

29. B The profit maximising output is where MC = MR, which is the case at V. The allocatively efficient output is where MC = P (AR) – which, in this case, is at Y.

30. A A contestable market is one in which there are no barriers to entry and exit including no sunk costs. B and D would mean that a market is not contestable. C, in itself, does not make a market contestable or incontestable.

Answers to data response questions

1. (a) The North Sea oil industry contributes to UK national income in a variety of ways. Its output contributes to the country's gross domestic product and it generates employment. It also contributes to net exports, making a positive contribution to the country's current account position.

   (b) The information provided is not conclusive on whether the North Sea oil industry is declining or expanding. There is some evidence to suggest that it is declining. Figure 13.18 shows a decline in production in terms of the number of barrels a day. The government was also trying to encourage smaller and more enterprising firms in order to protect the 250,000 jobs.

      The information does, also, however, provide some evidence to suggest that the industry may be expanding. Investment increased in 2006 and it is mentioned that the 'oil companies made huge profits.' Technological advances, innovation and the possible discovery of new oil fields may enable the industry to expand in the future.

   (c) Figure 13.19 shows that the price of oil is rising so even though less oil is being produced from the North Sea, revenue may be rising. Higher profits would provide the incentive and finance for investment. The new licences given out by the government and the greater access to pipelines might also encourage mall firms to invest.

   (d) Small firms may exist because they are catering for a niche market. This clearly does not apply in the case of oil which is a largely homogeneous product demanded in large quantities. Oil firms also do not provide a personal service. In addition, the industry would appear to be one in which economies of scale are likely to be significant.

      Advances in technology, however, may be reducing the cost advantages of large oil companies which may explain the increase in the number of small firms drilling for oil. Government support, which is mentioned in the information, is a major reason for the existence of small firms. To get established, firms new to an industry may need some help. New firms entering a profitable industry is a reason given by economists for the existence of small firms. A firm may start off small and then grow.

      It is not uncommon for a profitable market to have new small firms, especially when those firms receive government support. It is, however, perhaps surprising that there is an increase in the number of small firms in a market in which the benefits of producing and investing on a large scale seem so significant.

2. (a) The market for food retailing has become more concentrated. In the past, there were a high number of small firms in the market, essentially a monopolistically competitive market. It is now dominated by a few supermarkets with 75% of sales being made in supermarkets. The market has become an oligopoly although it can also be viewed as a legal monopoly as Tesco has a 30% share of supermarket sales.

   (b) The article provides conflicting information on whether the consumer is sovereign in food retailing. It mentions that 'supermarkets gave the customers what they wanted'. It also refers to the wide choice and low prices that supermarkets offer. Nevertheless, it seems that supermarkets are, to a certain extent, the decision makers. For instance, it mentions that they have introduced a store loyalty card but it does not suggest that this was in response to consumer demand.

   (c) There are a number of objectives a firm may pursue. The best known is profit maximisation with a firm trying to earn as much profit as possible to reward the owners of the business. Another objective is sales maximisation. This may permit a firm to gain a larger share of the market which will increase its market power. In addition, a firm might decide to engage in profit satisficing, This involves achieving a level of profit which will satisfy the shareholders but may also allow a firm to pursue other objectives including, for instance, becoming more environmentally friendly or expand into new areas.

   (d) There are some reasons why the size of a firm may not be a cause of concern. A large firm may benefit consumers with low prices because of economies of scale and high quality of products because of the availability of finance to spend on research and development and innovation. The information provided mentions that the large supermarkets are providing wide choice, reasonable prices and 24 hours opening.

      The information does, however, also mention that the supermarkets earn large profits and have driven some small shops out of business and have been criticised for damaging the environment. A firm that comes to dominate its market may push up price significantly above marginal cost, it may become complacent and the lack of competitive pressure may also mean that the quality of products may not remain high. A large share of the market may also mean that a supermarket gains monopsony power relative to its suppliers. This may result in it driving down the price it pays its suppliers which may have an effect on their ability to invest and innovate.

In assessing whether the size of a firm should be a source of concern or not, it would be useful to consider the level of actual and potential competition in the market. A large firm in a contestable market may not be a source of concern whereas a monopoly operating in a market with high barriers to entry and exit may be.

3. (a) The information indicates that Coca-Cola is not a pure monopoly. It does, however, indicate that it had a dominant monopoly position in 2011 as it had a market share in excess of 40%.

(b) A price war may be risky as it can result in all the firms that engage in it losing out. There are circumstances, however, when a firm may benefit from engaging in a price war. If a firm has lower costs, more retained profits than its rivals or can cross subsidise its products, it may be able to drive its rivals out of business by reducing its price. If it does succeed in forcing its rivals out of the market, it could then increase its price.

(c) There is some evidence in the information that Pepsi-Cola is becoming more allocatively efficient. Allocative efficiency occurs when resources are allocated in a way that maximises consumer satisfaction. Resources are devoted, in the right quantities, to producing what consumers are willing and able to buy. The information mentions that consumer demand is switching away from making fizzy drinks to making healthier drinks. Pepsi-Cola is seeking to produce a higher proportion of healthier drinks in order to respond to a change in demand.

(d) It is uncertain whether increasing spending on advertising will increase a firm's profits. It will do so if it increases revenue by more than costs. There is a chance, however, that it may not result in a significant rise in demand if the advertising campaign is not popular. There is also the possibility that one firm's increase in expenditure on advertising may be less than the rise in a rival's advertising expenditure.

(e) Pepsi-Cola's approach to diversification is to diversify not only into other soft drinks but also into food. Pepsi-Cola has a wider product range than Coca-Cola. This has some advantages. It means that its risks are spread. If, for instance, demand for fruit drinks declines, the impact on the company's profit may not be very significant and resources could be shifted from fruit juices to, for example, breakfast cereals. Wide diversification may also enable workers to develop a range of skills, will permit cross fertilisation of ideas and may provide savings in advertising costs since advertising one product may promote the whole company.

On the other hand, adopting Coca-Cola's approach to specialising in soft drinks, has a number of potential advantages. It may be easier to manage and co-ordinate a more focused range of products. The firm may also be able to build up expertise and may gain a good reputation in producing soft drinks.

Answers to essay questions

1. (a) According to utility theory, consumers would change their spending patters if the satisfaction they gain from the products or the prices of products alter. For instance, initially a consumer may be allocating her spending between three products, X, Y and Z such that:

$$\frac{\text{Marginal utility of X}}{\text{Price of X}} = \frac{\text{MU of Y}}{\text{P of Y}} = \frac{\text{MU of Z}}{\text{P of Z}}$$

$$\frac{20}{4} = \frac{35}{7} = \frac{15}{3}$$

She is maximising her total utility, as changing her purchases cannot increase her satisfaction. She is gaining the same marginal utility per dollar spent i.e. 5 units. If, however, the price of product X fell to $2, she would not be maximising her utility if she did not alter her purchases.

$$\frac{20}{2} > \frac{35}{7} = \frac{15}{3}$$

She is gaining more satisfaction per dollar spent from X. She will reallocate some of her spending from Y to Z and Z to X. As she consumes more of X the marginal utility she gains from it will fall. As she consumes less of Y and Z, the marginal utility she gains from them will rise until she is again in equilibrium with her purchases.

$$\frac{12}{2} = \frac{42}{7} = \frac{18}{3}$$

A change in the satisfaction she gains from one or more of the products she buys would also cause her to alter her spending pattern. If, from the above position, she gained more satisfaction from product Y than before, perhaps because of a rise in its quality, she would again be encouraged to alter her spending pattern. For instance, if the marginal utility of Y rose to 56 she would gain 8 units of marginal utility per $ spent but only 6 from X and Z. This would encourage her to buy more of Y and less of X and Z.

(b) Reducing the scale of production will reduce total costs of production. With less being produced, less resources will be employed and so total costs will be lower. What is more uncertain is what will happen to average costs of production. If economies of scale are being experienced, a fall in output will raise average costs of production. For instance, less discount might be given on the purchase of raw materials if less are purchased. Few specialist workers may be employed, banks may charge high interest rates and may be more reluctant to lend to smaller firms and capital equipment might be used less efficiently.

A smaller industry may also mean that less advantage can be taken of external economies of scale. With fewer potential students, colleges might decide to stop running courses. Specialist services provided by financial institutions might be withdrawn and specialist markets may be closed.

It is, however, possible that a fall in the scale of production may lower average costs if the firms and/or the industry was initially too large. Smaller firms may have less managerial problems. With fewer managers, decisions might be made more quickly. With fewer workers and closer contact between workers and managers, industrial relations might be better.

With a smaller industry, external diseconomies of scale may be reduced. Firms may find that they have to pay less for factors of production and may have reduced transport costs due to less congested roads and less demand for rail services.

2. Inequalities in wage rates in an economy occur for a number of reasons. Market forces explain why some workers are paid more than other workers. Workers whose skills are in high demand and in short supply are likely to be paid more than those whose skills are demanded less and which are in greater supply. Marginal revenue productivity theory suggests that demand for labour will be high if labour productivity is high and/or the product provided is in high demand. For example, top lawyers are highly skilled and their services are in high demand. The supply of top lawyers is also limited, with not many people possessing the necessary skills, qualifications and experience. In contrast, the supply of cleaners in many countries is high relative to demand for their services. The following figure shows that the wage rate of top lawyers is significantly higher than that of cleaners.

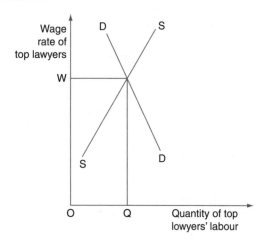

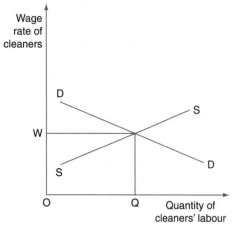

It is interesting to note the differences in the elasticities of demand and supply in the two labour markets. Most people could work as cleaners and so supply is elastic whereas not many people have the skills, qualifications and experience to be top lawyers. Demand for cleaners is more elastic as it is easier to reduce the number of cleaners required by introducing the use of more capital equipment.

Some individual workers are so skilful in a particular field and/or their services are in such high demand, that they can enjoy considerable economic rent. For instance, whilst there are many low paid actors, a few top actors earn considerably more than in their next best paid job. This is illustrated in the figure below.

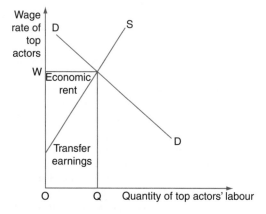

Whilst the need for particular skills or qualifications can act as a barrier to stop workers moving from low paid to high paid jobs, other factors can keep people in low paid jobs. They may not want to move from one region of the country to another region, perhaps because of family ties. Workers also base their decisions as to what jobs to do not just on the wage rate. They take into account, for instance, promotion chances, job security and working hours. Some nurses, for example, might earn more in other jobs but stay working as nurses because of the job satisfaction they gain.

Wage rates are also not determined just by demand and supply. Governments and trade unions may influence the wages and so the extent to which inequalities in wage rates occur. The gap between the highest paid and the lowest paid workers may be less in countries which operate a national minimum wage although the effects of national minimum wage legislation are somewhat uncertain. Governments also influence wage rates through the wages they pay public sector workers and the education and training they provide. The more educated and the better trained workers are, the greater their earning potential.

Trade unions engage in collective bargaining with employers. They seek to raise the wage rates of their members as well as trying to achieve other benefits for their members. How influential trade unions are in determining wage rates depend on a number of factors. These include what proportion of workers the union represents, whether the members are prepared to take industrial action, the buying power employers have and the relative bargaining strength and skills of the trade unions and employers.

Wage negotiations may take place under conditions of bilateral monopoly. This occurs when a monopolist trade union, representing everyone in a particular group of workers, and a monoponsonist employer, the sole buyer of a particular group of workers. The following figure shows such a market.

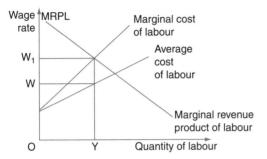

The employer will want a wage rate of W but the trade union may be able to raise the wage rate to W1. If the trade union is successful, it will not cause a loss of jobs as the new supply curve of labour will become W1XY. If, however, the employer operates in a competitive market, a trade union pushing up the wage rate above the equilibrium level may cause unemployment. The figure below shows that raising the wage rate from W to W1, in this case, causes the quantity of labour employed to fall from Q to Q1.

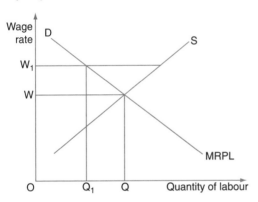

One group of workers may be paid less than another group if there is discrimination. If, for example, employers in a country think that female workers are less productive than male workers, they may pay them less. The following figure shows that the wage rate of female workers and their employment will be lower if there is discrimination.

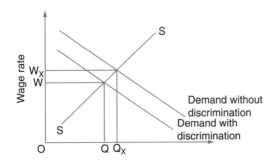

Workers and jobs differ and so demand for and supply of labour vary. Market forces, however, do not explain all of the wage

differentials in an economy. Institutional factors and government policy also play a role. In addition, in some cases, discrimination may occur.

3. (a) Monopoly and monopolistic competition are two types of market structure. In both cases, each firm in the market is a price maker. A rise in its output will lower price. As a result, its average revenue will exceed its marginal revenue and both will decline with output.

A monopoly and a monopolistically competitive firm will be likely to produce where neither allocative nor productive efficiency is achieved. In the private sector, a monopoly and a monopolistically competitive firm are likely to be profit maximisers and will produce where marginal cost equals marginal revenue, but also where price exceeds marginal cost and where average cost is greater than marginal cost.

In a pure monopoly, however, there is only one firm in the industry whereas in monopolistic competition there is a relatively high number of firms. This means that in monopoly there is a high degree of market concentration. Indeed, in the case of a pure monopoly there will be a one firm market concentration ratio of 100%. Even if a monopoly is defined in terms of a firm with a share of 25% or more of the market or a 40% plus share, one or two firms are likely to account for a large share of the market. On the other hand, in monopolistic competition the firms are small relative to the size of the industry. As a result the market concentration ratio is low.

In a pure monopoly, there is a unique product with only one firm producing the product. In monopolistic competition, consumers can enjoy variations in the product. This is because the firms produce slightly differentiated products.

A major difference between the two types of market structure is that whilst there are high barriers to entry and exit in the case of monopoly, there are no or low barriers to entry and exit in the case of monopolistic competition. This difference explains why a monopoly can protect any supernormal profit in the long run but a monopolistically competitive firm cannot. If a monopolistically competitive firm does earn supernormal profit in the short run, new firms will enter the market and compete it away. The diagram here shows shows how the long run equilibrium output of a monopolistically competitive firm which is producing where MC = MR and where it is earning only normal profit.

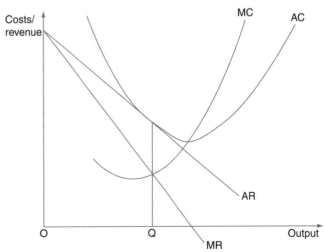

(b) A monopoly may disadvantage consumers in a number of ways. It may mean that consumers have a limited choice of products. They may also have to pay a high price. This is because a monopoly may restrict output below the efficient level in order to drive up price.

There is another reason why price may be higher under conditions of monopoly than under a more competitive market structure. This is because average costs of production may be higher under monopoly. Higher costs may result from x-inefficiency or diseconomies of scale. X-inefficiency refers to a lack of drive to keep costs low. Without the threat of competition, a monopoly may not spend time and effort searching for the cheapest raw materials, its managers may take long breaks and it may employ out of date equipment and too many workers. A monopoly may also experience diseconomies of scale if it grows too large. The firm may become difficult to manage and there may be more industrial disputes.

Not all monopolies, however, are large and those that are may experience economies of scale. A monopoly may earn supernormal profit but if its average cost is significantly lower than would be the case under more competitive conditions, price might also be lower. The existence of supernormal profit may also allow a monopoly to spend more on research and development and innovation. This means that it is possible that quality may be higher under conditions of monopoly. A monopoly may also seek to improve the quality of its existing products and develop new products in order to strengthen barriers to entry. In addition, a monopoly may choose to produce a range of products so that whilst consumers may not have a choice of producers, they may have a choice of products.

In the case of a natural monopoly, long run average costs will be lower if one firm controls the market than if a number of firms supply the market. One firm operating in a natural monopoly market allows economies of scale to be exploited fully and for wasteful duplication to be avoided. A state run monopoly may also benefit consumers if it bases its production and pricing decisions on social costs and social benefits.

There is the possibility that a monopoly may disadvantage consumers by limiting choice, reducing quality, restricting output and charging a high price and so reducing consumer surplus. The outcome, however, is uncertain as it is possible that a monopoly may result in lower average costs, lower prices and better quality.

4. (a) Price discrimination involves charging different groups of consumers different prices for the same product. For price discrimination to occur, three conditions have to be met. These are that the firm must be a price maker, so that it can set different prices in the different markets. The firm must be able to keep the markets separate, so that people cannot buy the product in the cheaper market to resell in the more expensive market. Markets may be kept separate in terms of time, geographical area and the age of the consumer etc. The last condition is that the price elasticity of demand is different in the different markets. If it was the same, the profit maximising price would be the same in the two markets. The following figure shows that the price will be higher in the market which has more inelastic demand.

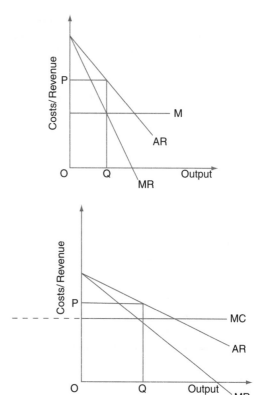

There are different degrees of price elasticity of demand. First degree price discrimination is when a different price is charged for each unit sold. This could eliminate consumer surplus. It is very difficult to implement as it requires accurate information about the willingness and ability of all consumers to pay different prices. Second degree price discrimination occurs when one price is charged for some units of the product and a different price is charged for further units. Third degree price discrimination if the standard form of price discrimination. It involves selling the same product to different consumers at different prices.

The evidence does suggest that third price discrimination was occurring in the case of the sale of tickets on London trains. Consumers were being divided into different markets according to time of purchase, time of travel and age.

(b) The output and pricing policy of a firm will be influenced by the type of market structure in which it operates. A perfectly competitive firm is a price taker. It accepts the market price and produces where marginal cost equals marginal revenue as shown here.

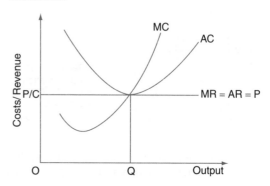

Its output will change if market forces change the price. A rise in demand will push up the price and cause the firm to supply more. It will also enable it to make supernormal profit. This, however, is only a short term situation as the lack of barriers to entry and exit will enable new firms to enter the industry and compete away the supernormal profit.

A monopolistically competitive firm can also earn supernormal profit only in the short term. It is, however, a price maker. To sell more, it has to lower price and so its average revenue exceeds marginal revenue. It will again produce where marginal revenue equals marginal cost as shown in the following figure.

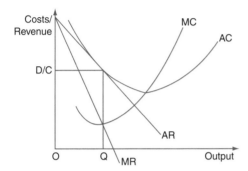

The output and pricing policy is more complex under conditions of oligopoly. An oligopolist may set price below the profit maximising level in order to drive out current or potential competitors. It may also collude with other firms to drive up their joint profit or may decide to follow the price set by a price leader. In making its output and pricing decisions, an oligopolist will take into account how its rivals will react.

A monopolist may make a range of output and pricing decisions. For example, a state owned enterprise may seek to achieve allocative efficiency and so produce where marginal cost equals price as shown by output C in the following figure. It might also seek to maximise sales revenue which is achieved where marginal revenue is zero. This achieved at an output of B.

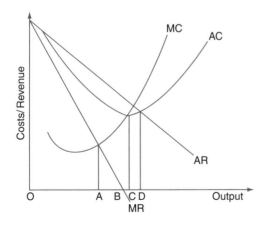

Alternatively, it could seek to maximise sales whilst still making normal profit – output D. The traditional decision, of course, is to produce where marginal cost equals marginal revenue in order to maximise profit – output A. If

a monopolist decides on a given output, it will have to accept the market price. It can determine output or price but not both.

Under conditions of perfect competition, a firm's output and price are determined by the market. In contrast, a firm producing under conditions of monopolistic competition, oligopoly or monopoly can influence price or output.

Chapter 14

Answers to short questions

1. Asymmetric information is unbalanced information. One of the parties to a transaction has more information than the other party. For example, a dentist is more informed about the patient requires than the patient.

2. Factor immobility leads to market failure because it stops markets making full adjustment to changes in demand. If there is difficulty moving resources from making less popular products to more popular products, there will be surpluses and shortages.

3. The more inelastic demand is, the greater the size of a deadweight loss.

4. Price discrimination usually reduces consumer surplus and increases producer surplus. It pushes up the price paid by some consumers and so reduces the surplus they receive whilst increasing the surplus gained by producers.

5. The government should not necessarily prevent the consumption of a demerit good unless it is very harmful to those consuming it or to third parties. It is more likely to reduce its consumption to the allocatively efficient level.

6. A pure monopoly is a single seller. The firm has a 100% of the market. This contrasts with a legal monopoly (25% or larger share of the market) and a dominant monopoly (40% or greater share of the market).

7. Governments seek to promote efficiency to increase living standards. An increase in efficiency should raise output and mean that the output more fully reflects consumer demand.

8. A perfectly equal distribution of income would not be equitable. This is because people have different needs. A disabled person who needs specialised equipment and medicines would need more income than a fit person.

9. There may be regulation requiring consumption of a product example, the legal requirement in many countries that children have to go to school. Regulation may also ban the consumption of a product, for instance, making it illegal to buy and sell alcohol. In addition, regulation may set standards. This may involve, for instance, requiring restaurants to prepare food in hygienic conditions and drivers to have their cars checked for road worthiness on a regular basis.

10. The sale of cigarettes to children is banned by many governments whilst adults are allowed to buy cigarettes because it is thought that information failure is more significant in the case of children.

11. Income is a flow whereas wealth is a stock. Income is a payment received on a regular basis. Income may be received in the form of

wages, dividends, interest, profits and state payments. Wealth is a stock of assets and may include property, vehicles, savings etc. A person may survive with zero wealth but not zero income.

12. One argument for a tax on wealth is that it may reduce income inequality. Wealth generates income and a reduction in the wealth of the rich would reduce their earning capacity. An argument against a wealth tax is that it may discourage saving and enterprise.

13. An argument for universal benefits is that it is a simple system to operate. An argument against is that it is an expensive system with some receiving the benefit who do not really need it.

14. Tradable pollution permits are licences which allow firms to pollute up to a certain level which can be sold if firms pollute less than their permits allow.

15. A government's ability to raise the rate of corporation tax may be restricted by other countries' corporation tax rates. A government may be worried that if it raises its corporation tax rate above other countries' tax rates, firms may move abroad.

16. An increase in unemployment benefit may increase income inequality if it encourages some people to become voluntarily unemployed.

17. A rise in income tax rates may reduce tax revenue if it causes some people to work fewer hours, some to withdraw from the labour force and some to move into the informal economy.

18. A government may decide to regulate a privatised industry if it believes it is abusing its market power.

19. This is a system which integrates income tax and benefits. There would be one transaction between taxpayers and the government. People earning above a set level would pay tax and those earning below the level will receive a payment from the government (negative tax). Advocates suggest that such a system is simpler and cheaper to administer than separate tax and benefits system. It is also claimed that it could be used to eliminate the poverty trap.

20. A privatised firm and a nationalised industry can use both retained profits and can borrow. A privatised firm, but not a nationalised industry, can also issue shares.

Answers to revision activities

1. (a) EAI   (b) BAD
   (c) EBDI   (d) GCDH
   (e) EBCF   (f) FCG and HDI

2. (a) Energy security means that a country has sufficient supplies of its own. It is not dependent on other countries for its energy supplies.
   (b) A first party is either an energy firm or a consumer of energy. A third party who might suffer from the government's proposed scheme is a member of the Ashanikas tribe.
   (c) The imposition of an indirect tax of PYP1 per unit will cause the supply curve to shift to the left as shown in the figure below. This will raise price from P to P1 and cause demand to contract from Q to Q1. The price the producer receives falls to PY.

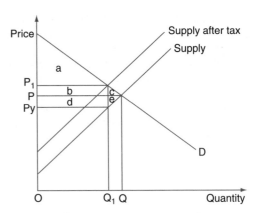

Consumer surplus falls by bc amount and producer surplus by de amount. The government receives tax revenue of bd amount and so there is a deadweight loss of ce.

3. (a) (ii) Unemployment benefits are designed to provide the unemployed with a basic income.
   (b) (i) A tax on cigarettes is designed to discourage the consumption of a demerit good and move the market towards the allocatively efficient output.
   (c) (i) One of the motives behind privatisation is to put market pressure on firms to respond to consumer demand and to keep costs down.
   (d) (iii) Providing state education free helps the poor and helps increase labour productivity.
   (e) (i) A cut in the top rate of tax would benefit the well paid and would be designed to increase incentives.
   (f) (iii) A subsidy to providers of public transport will help the poor and will increase the output of a product which was previously under-consumed.

Answers to multiple choice questions

1. D   Street lighting would not be provided by the private sector as it is a public good. A, B and C all indicate a market working efficiently.

2. B   Market forces will not provide public goods as it would not be possible to charge directly for them. A and D would enable markets to work more efficiently and merit goods are under – provided rather than over-provided.

3. C   The monopolisation of the industry causes consumer surplus to fall by RSUW. Of this, RSUV is converted into producer surplus and UVW is lost.

4. C   An increase in the exploitation of economies of scale would lower average cost and consumers might benefit from lower prices. A, B and D may all be causes for concern.

5. B   Firms may be reluctant to train their workers for fear that once trained, they will move to another firm and that firm will reap the benefits of the training. A, C and D would encourage firms to provide training and would not necessitate government intervention.

6. C   The socially optimum level of pollution is achieved where the marginal social cost and marginal social benefit of pollution reduction are equal. It would not be efficient to have zero pollution if the allocation of resources devoted to achieving it exceeds the benefits people receive from its elimination.

7. D   The socially optimum output is achieved where marginal social cost equals marginal social benefit. To encourage people to buy this quantity, the price would have to fall to X and this would require a subsidy of XZ per unit.

8. D   If it is expensive to investigate monopolies, a ban might be used. A, B and C will not necessarily occur.

9. C   In the case of a regressive tax, the poor pay a greater percentage of their income in tax than the rich. The marginal rate of tax will fall as income rises, dragging down the average rate of tax. In terms of the actual amount paid, the poor are likely to pay less than the rich.

10. C   A poll tax is the most regressive tax as it is a tax per head of the population. Everyone pays the same amount regardless of income and personal circumstances.

11. D

Price ($)	Original quantity demanded	Quantity supplied	Quantity demanded after tax
10	20	1280	600
9	60	1000	400
8	150	850	150
7	260	600	50
6	400	400	
5	600	150	
4	900	50	

After the tax, if a firm sells a product for $10, it will only receive $7 as the $3 will have to be passed to the government. So the firm would now supply at $10 what before it had supplied at $7 and so on. The price rises by $2, so consumers are bearing $2/$3 of the tax and producers are bearing the remaining third.

12. B   Moving towards greater reliance on indirect taxes is a regressive move as indirect taxes tend to take a higher proportion of the income of the poor. Lowering income tax may encourage people to work longer hours and accept promotion.

13. D   As income rises the tax payment is rising but at a decreasing rate. This means that the rich will be paying proportionately less and the poor proportionately more.

14. A   The poor spend a higher proportion of their income than the rich. Redistributing income from the rich to the poor would be expected to raise spending and reducing saving. If the mpm of households is higher at high levels of income, a redistribution of income to the poor would reduce a current account deficit. If progressive taxes act as a disincentive to work and effort, a redistribution of income to the poor would reduce real GDP.

15. B   There is a risk that a regulatory agency may get too close to the producers it is regulating. If this occurs, it may protect the interests of the producers.

16. C   The allocatively efficient output is achieved where marginal cost equals price (average revenue). In the case of a natural monopoly, such as illustrated in Figure 14.7, this may be where average cost is still falling and a loss would be made as average cost is greater than average revenue.

17. A   Failure to take into account social benefits may result in allocative inefficiency. B, C and D would all increase the chances of government intervention increasing efficiency.

18. C   If the privatised industry becomes more responsive to changes in demand, it is more likely to be allocatively efficient. A, B and D would all increase efficiency.

19. C   The following figure shows that a maximum price on the rent of private accommodation will cause the quantity of rented accommodation to fall from Q to QX. The lower price will encourage more people to seek to rent but the supply will fall and a shortage will be created with demand (QY) exceeding supply (QX). Producer surplus will fall and the PED and PES will be unaffected.

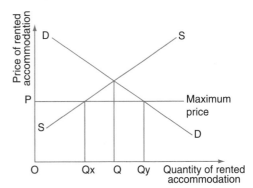

20. B   If demand for the product is perfectly price inelastic, a subsidy will not alter the amount consumed and so will not increase allocative efficiency. The following figure shows the subsidy shifting the supply curve to the right and the quantity purchased remaining unchanged when demand is perfectly inelastic.

In contrast, when demand is price elastic, the quantity bought will increase. C would move the market towards allocative efficiency. So would D even if it would fail to reach the allocatively efficient level.

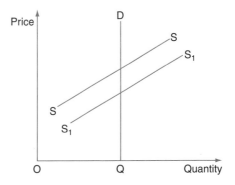

Answers to data response questions

1. (a) Industrialisation involves moving resources from agriculture into manufacturing.

   (b) There are a number of economic costs that Odisha is experiencing as a result of industrialisation. One is the damage to the environment which includes soil erosion, air and water pollution and loss of wildlife. Another cost is people being forced to move to new homes. In addition, the damage to the environment may reduce the fertility of land and reduce farmers' incomes.

(c) It is debatable whether the development of the steel plant is likely to benefit the workers in Odisha. On the one hand, it is creating jobs which are better paid than agricultural work. Those who are employed in the steel industry will experience a rise in income and they may gain transferable skills. In addition, the employment and wages will not be stopped by bad weather.

On the other hand, however, the factory is not labour-intensive and not many people will be employed. Some people may no longer be able to work on the land due to soil erosion and some may have lost their land due to the building of the steel industry. Not all of those who have lost their jobs in agriculture will gain jobs in the steel industry. Some of those who have gained jobs may have lost their homes and may have had to move away from their community.

(d) There is some evidence to suggest that the standard of living in Odisha is low. The information mentions that 48% of the population lives below the poverty line, that its literacy level is below the national average, that there is a high rate of infectious disease and malnutrition. It does, however, also mention that the area is attracting big industrial companies. Whilst the steel industry is not creating many more jobs, the other industries might be raising overall employment and increasing income levels.

More information would be needed to come to a firm conclusion on what is happening to the standard of living in Odisha. It would be useful to have, for instance, GDP per head figures, HDI figures for the area and levels of crime in the area.

2. (a) A negative externality is a harmful effect imposed on a third party. Two negative externalities created by the Deepwater Horizon disaster are the damage to the local shrimp industry and the damage to wildlife. Those in the shrimp industry and wildlife are not involved in the decisions of the oil industry. Shrimp fishermen lost income and a high number of mammals, birds and fish were killed.

(b) The existence of negative externalities does not necessarily mean that too many resources are devoted to oil production. This is because oil production may also generate positive externalities.

To decide whether too many resources are devoted to oil production, it would be necessary to consider the relationship between marginal social cost and marginal social benefit.

(c) The disaster is likely to have reduced BP's profits in the short run. This is because the firm had to pay out a large amount in compensation and because the adverse publicity is likely to have reduced demand for its oil.

(d) There are both arguments for and against a government allowing foreign companies to drill for oil off its coast. Oil companies are likely to employ some workers from the country. The government can also charge for the right to drill in its waters and will gain some tax revenue. The oil companies may also buy equipment and supplies from local companies and may sell oil to the country at a reduced rate.

The drilling for oil may, however, cause environmental damage as in the Deepwater Horizon disaster. Foreign companies may also generate less employment for the country than any domestic oil companies and may pay less corporate tax if some of the profits earned are returned home. In addition, they may extract the oil at a fast rate whilst the government may want to conserve oil supplies to ensure energy security.

3. (a) Deregulation involves the removal of legal restrictions on firms usually with the intention of increasing competition in markets.

(b) The information largely suggests that taxi firms operate under conditions of monopolistic competition. It indicates that there is a low market concentration ratio as the industry has a large number of small firms, which is a key feature of monopolistic competition. It also mentions that the firms produce a slightly differentiated product and that it is possible to enter the market with one or two vehicles.

The information does, however, mention that regulation can create a barrier to entry. It would be expected that monopolistic competition would either have no barriers to entry and exit or only low barriers. In order to judge whether any remaining regulation in this case is consistent with monopolistic competition, it would be necessary to assess how strong a barrier it creates.

(c) A more contestable market means that it easier to enter and exit a market and that firms that come into the market will not be at a disadvantage compared to incumbent firms. A more contestable market will increase potential but not necessarily actual competition. If supernormal profits are being made new firms will come into the market. Indeed, there may be hit and run competition. Firms may enter and then leave when the circumstances that gave rise to the supernormal profits disappear.

At any one time, however, a more contestable market does not necessarily mean that there will be more firms in the market. Indeed, a contestable market may consist of only one firm. What will provide the competitive pressure will be the threat from outside the market.

(d) Economic efficiency is achieved when both productive efficiency and allocative efficiency are achieved. Deregulation may increase economic efficiency in the taxi market. Removing restrictions on the number of firms that can operate in the market will be likely to increase competition. Greater competitive pressure can make firms more sensitive to changes in consumer demand and more determined to keep costs low. It can also push down fares and may encourage firms to innovate to increase the quality competitiveness. In addition, more firms in the market can reduce waiting time for passengers.

There is, however, no guarantee that deregulation will increase economic efficiency. Any increase in the number of firms in the market may cause congestion in city centres and tourist spots. Removing controls on the geographical coverage of taxi firms may result in more remote areas not being covered. Allowing taxi firms to charge whatever fares they want may also cause problems. Fares to destinations without a return fare may rise significantly. Consumers may not have the time or confidence to find the lowest fare, especially if they have to hail down taxis in the street.

Answers to essay questions

1. (a) A market is in equilibrium when demand and supply are equal and there is no pressure for the price and the quantity traded to change. An efficient market equilibrium would mean that the equilibrium is occurring at a point where economic efficiency is achieved. Economic efficiency is attained when both allocative and productive efficiency are achieved.

Allocative efficiency means that resources are allocated in a way that maximises utility. Producers make the products consumers demand in quantities where demand and supply are equal and there are no surpluses or shortages. This desirable outcome is achieved where price (reflecting consumer utility) equals marginal cost.

Productive efficiency is achieved when a given level of output is produced at minimum cost. In the long run, this can be taken to be at the lowest point on the lowest average cost curve.

In theory, perfectly competitive firms are always allocatively efficient. This is because they produce where marginal cost equals marginal revenue and as marginal revenue equals average revenue, it is where marginal cost equals average revenue (price). They also produce at the lowest possible average cost and, in the long run, at the lowest point on the average cost curve. The following figure shows the long run output position of a perfectly competitive firm with it producing where marginal cost equals marginal revenue and marginal cost equals average cost.

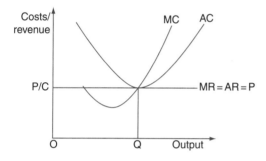

In contrast, firms that are price makers may not be allocatively efficient as they may restrict output to drive up price above marginal cost. They may also fail to produce at the lowest point on the average cost curve. The figure below shows a monopolist producing where price exceeds marginal cost and average cost exceeds marginal cost.

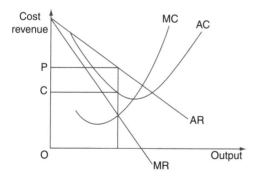

Of course, it is possible that a monopolist state owned enterprise may be instructed to produce where marginal cost equals average revenue and so may be allocatively efficient. It is also possible that average costs may be lower under conditions of oligopoly and monopoly if economies of scale are significant.

(b) There are a number of reasons why the market mechanism might fail in the allocation of resources. One is that there may be information failure about the benefits and costs of consuming a product. In the case of merit goods, too few resources are devoted to the production of the product. This is because consumers do not fully appreciate the benefits they gain from consuming the product. As a result, the products are under-consumed and so under-produced. Demerit goods, in contrast, will have too many resources allocated to them. Consumers do not recognise the full costs of consuming the product and so their demand is above the allocatively efficient level.

The market mechanism reflects consumer demand and producer supply. When making their decisions on what to buy and what to produce, consumers and producers usually only take into account private costs and benefits. This means that the existence of positive and negative externalities can result in inefficiency. The following figure shows that the presence of positive externalities will result in too few resources being devoted to the output of the product.

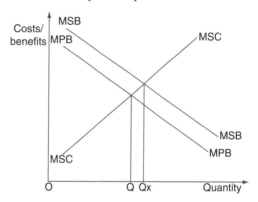

The allocatively efficient output is QX since this is where marginal social benefit equals marginal social cost. Consumers, ignoring the external benefits arising from consuming the product, however, will only demand Q amount and this is the quantity producers will provide.

There is likely to be an even more significant failure in the case of public goods, i.e., goods which are both non-rival and non-excludable. As people can act as free riders, it is very difficult to make consumption dependent on payment. This problem may discourage private sector firms from providing public goods.

Private sector firms may also use any market power they possess to restrict output below the allocatively efficient level. They may do this in order to drive up price and and increase their profits. A lack of competitive pressure may also mean that firms do not strive to keep their costs low.

Firms may want to respond to consumer demand in an efficient manner, but a lack of factor mobility may prevent them doing so. Workers, for instance, may lack the skills necessary to move from declining to expanding industries.

To work efficiently, the market mechanism needs perfect information, the absence of externalities and possibly perfect product and factor markets.

2. (a) Negative externalities are the harmful effects on third parties arising from the production or consumption by others. Those suffering from these adverse effects are not compensated through the market. So those who create these negative externalities by their production or consumption activities do not pay those who suffer as a result of their activities.

When private sector firms decide on their output and price they take into account private costs and benefits. For example, a chemical firm will consider its wage, fuel, advertising and other production costs and the revenue it will receive from selling the chemicals it produces. Without government intervention, it is unlikely to consider the external benefits, for example employment generated in the local area and the external costs such as employment generated in the local area, and the visual, air and noise pollution, it creates.

The following figure shows a firm seeking to maximise profits produces where marginal private cost equals marginal private benefit.

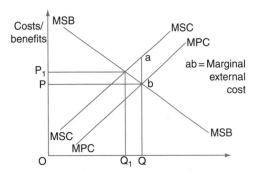

The output of Q is above the allocatively optimum level of QX. The price of P also fails to reflect the social cost and is below the allocatively efficient price of PX.

(b) There are a number of measures which a government could introduce to reduce pollution. One measure is to seek to internalise the external cost of polluting by taxation. This means seeking to change pollution from an external cost into a private cost so that social cost equals private cost. To achieve this, the government would have to impose a tax on polluting firms which is equal to marginal external cost. The revenue raised could be used to treat pollution or compensate suffers.

However, in practice, it is difficult to estimate external costs as they do not go through the market. There is the risk that a tax may add to inflationary pressure, especially if the demand for the products produced by the polluting firms is inelastic. A tax may also reduce international competitiveness and place a greater burden on the poor than on the rich.

An alternative approach is to regulate. Legislation could be passed which sets rules for the maximum permitted level of pollution emissions from different industries. As with a tax, this measure should internalise the external cost and shift the supply curve to the left. Regulation is more common than imposing a pollution tax but regulation also has its drawbacks. As with a tax, it is difficult to determine the 'right' level of pollution. Levels of pollution have to be checked and this involves a cost. Regulation also does not compensate the victims and does not provide firms with an incentive to reduce pollution below the maximum level.

One measure which does build in an incentive to reduce pollution is tradable pollution permits. This involves issuing permits to firms which allow them to pollute up to a certain level. Firms that pollute more have to buy additional permits. Those that pollute less can sell some of their permits. This should mean that high polluting firms will experience a rise in costs which should reduce their output whilst low polluting

firms will gain extra revenue, encouraging them to expand. It is important that the level of pollution set reflects an efficient level. If, for instance, it is set too high, no firm would have to purchase additional permits. Money would have to be spent monitoring firms and, in practice, it can be difficult to determine where some pollution has come from. The measure is also more likely to be successful if it is introduced on a global scale as otherwise, firms may relocate to countries not operating tradable permits.

A measure that has become popular in some countries, including the US, is extending property rights. By giving local inhabitants ownership rights over, for instance, fresh air, rivers and the sea, it makes it easier for people to take legal action against a firm that pollutes. People could be given the right to sell pollution opportunities to firms. The allocatively efficient level of pollution might be achieved if they sell these rights at a price equivalent to marginal external cost. There may, however, be disagreements over ownership and it may be expensive for a firm to negotiate with a high number of people if the pollution is widespread.

A government might also decide to subsidise the installation of equipment which will generate less pollution. This will involve government expenditure. This might involve reducing spending on other areas. It might also involve higher taxes which will involve a transfer of income from tax payers to the firms which cause the pollution.

In the case of all the measures discussed, there is the problem of measuring external costs accurately. Each measure has a number of disadvantages and advantages.

3. (a) Economic efficiency occurs when both productive efficiency and allocative efficiency are achieved. In such a situation, resources are being fully used to produce, at the lowest possible cost, the products which consumers are demanding. Market forces may not always lead to economic efficiency. In situations where there is market failure, there is the possibility that government intervention may improve economic efficiency although there is also the risk of government failure. Government intervention may take the form of regulation or state provision. Regulation covers legally enforced requirements or standards made by the government of a country. State provision involves the government providing products to consumers.

One situation where market failure occurs is when there are negative externalities. Households and firms base their consumption and production decisions on private costs and benefits. If there are negative externalities, overconsumption and overproduction of products will occur. In this case, a government might seek to restrict the consumption and or output of the products involved. For example, a government may place limits on the amount of pollution firms are allowed to emit. Such a limit will increase economic efficiency if the government can measure the marginal external cost accurately and set the limit appropriately. A government might also decide to produce the product itself so that its output level is based on social costs and benefits.

A government may also decide to produce merit goods and provide them free at the point of delivery. Merit goods not only generate external benefits but also greater private benefits than their consumers realise. An alternative or additional measure is for a government to pass a law requiring, for instance, children to attend school and people to have vaccinations against certain diseases.

There is a strong case for government provision of public goods. This is because, if left to market forces, these types of products would not be produced. Market forces do not provide an incentive for private sector firms to produce public goods as their characteristic of non-excludability makes it impossible to charge directly for them. A government may decide to produce a public good itself or pay a private sector firm to produce it.

A government may intervene in an economy if it thinks that the private sector lacks sufficient finance to undertake the necessary investment in an industry. Efficiency may also be increased if a government decides to produce a product made under conditions of natural monopoly. In such a situation, average cost will be lowest with one producer. A private sector firm may not produce where price equals marginal cost. This is both because such an output may result in the firm making a loss and because it may be tempted to exploit its monopoly power by restricting its output to ensure price exceeds marginal cost. A state owned enterprise may have both the finance and the motivation to achieve allocative efficiency.

As well as providing products produced under conditions of a natural monopoly, a government may decide to regulate private sector monopolies. This may take the form, for instance, of placing limits on the amount by which the monopolies can raise price. In addition, a government may employ regulation in order to create the market failure which occurs in the case of demerit goods. These goods are over-consumed and so overproduced. A government may ban their consumption or production completely or may place age restriction on those who can consume the product.

Regulation and state provision have a number of advantages. They are both backed by the force of law. For instance, those who do not obey government regulation may be fined or even imprisoned and private sector firms can be stopped from competing with state-owned enterprises. Regulation may also be relatively easy to understand and state provision can take into account social costs and benefits.

These forms of government intervention, however, may have a number of disadvantages which may lead to government failure. For example, it may be difficult for a government to measure external costs and benefits and so there is no guarantee that state intervention will move output to a more efficient level. A government may also find it difficult to estimate the extent of information failure in the case of merit and demerit goods and the efficient level of output when it decides to provide products itself. Regulation involves costs of enforcing and monitoring and state provision obviously involves a cost. In addition, there is the risk that stated owned enterprises may be inefficient due to a lack of both competition and the profit motive.

Chapter 15

Answers to short questions

1. Replacement investment is investment undertaken to replace worn out and obsolete capital equipment.
2. A country's informal economy may fall if tax rates are reduced and if the rules and regulations that firms have to comply with are reduced.
3. Gross National Happiness is a measure of the quality of life developed in Bhutan. It takes into account a variety of influences on happiness including culture, the environment as well as the quality of products consumed.
4. It is difficult to measure the money supply as it is difficult to decide what to include in any measure. The items that are used as money can change relatively quickly.
5. The government's budget position is the relationship between its spending and revenue. If a government spends more than it receives in revenue, it has a budget deficit. To finance this, it will have to borrow. The national debt is the accumulation of government borrowing that has built up over time.
6. A closed economy is an economy which does not engage in international trade. In such an economy, the injections are investment and government spending and the withdrawals are savings and taxation.
7. Keynesians think that there may be large scale unemployment which will continue without government intervention. In contrast, monetarists think that in the long run, unemployment will return to the natural rate.
8. The average propensity to consume is the proportion of total income that people spend. The marginal propensity to consume is the proportion of extra income that people spend.
9. An increase in net exports will increase aggregate expenditure and increase money GDP by a multiple amount.
10. An increase in aggregate expenditure would reduce a deflationary gap.
11. An injection causes a multiplier effect as extra spending generates income. Some of this income is spent which, in turn, raises income and so on until the injection is matched by withdrawals.
12. The paradox of thrift suggests that a decision by people to save more can, in the longer term, result in them saving less. This is because an increase in saving means a fall in spending. If people spend less, firms may produce less and incomes may fall. With lower incomes people will not be able to save as much.
13. A rise in the marginal rate of tax will increase the size of the multiplier. Less of extra spending caused by an injection will leak out of the circular flow. For example, if initially mps is 0.2, mrt is 0.1 and mpm is 0.1, the multiplier would be $1/0.4 = 2\frac{1}{2}$. If mrt then increased to 0.2, the multiplier would fall to $1/0.5 = 2$.
14. Firms may buy more capital equipment despite a rise in the rate of interest if they expect the yield will exceed the cost of the investment. During a consumer boom, firms may expect to earn more profit from expanding, even if they are paying a higher rate of interest.
15. If there is a rise in liquid assets of $20 million with a credit multiplier of 20, bank loans will rise to $380 million ($400m−$20m).
16. Banks may lend less than their liquidity ratios permit due to a shortage of credit worthy borrowers.
17. The velocity of circulation is the number of times, on average, money is spent. It is calculated by dividing GDP by the money supply. For instance, if GDP is $80bn and the money supply is $20bn, the velocity of circulation is 4.
18. An increase in the money supply may cause an increase in output as the resulting rise in spending may encourage firms to increase their output rather than prices.
19. Active money balances covers money held for transactions and precautionary motives. In contrast, idle balances are held for speculative purposes.
20. The liquidity trap occurs when the rate of interest is so low (and the price of government securities is so high) that it becomes impossible to lower it further. A central bank may increase the money supply in a bid to reduce the rate of interest but if people

think it will rise in the future and so the price of government securities will fall, they will hold all the extra money. This means that the rate of interest will remain unchanged.

Answers to revision activities

1. (a) National income = injection × multiplier.
        Investment is the injection. The multiplier is 1/mps = 1/0.25 = 4.
        So national income = $50bn × 4 = $200bn.
    (b) (i) mrt = 0.2 and mps = 0.25 × 0.8 = 0.2. k = 1/0.2 + 0.2 = 1/0.4 = 2 ½.
        (ii) I + G = $50bn + $70bn = $120bn.
              National income = $120bn × 2 ½ = $300bn.
        (iii) Tax revenue = 0.2 × $300bn = $60bn.
               Budget balance = tax revenue – government spending = $60bn – $70bn = –$10bn.
               There is a budget deficit of $10bn.
    (c) (i) mrt is 0.2, mps is 0.2 and mpm is 1/8 of 0.8 = 0.1.
              So k = 1/0.2 + 0.2 + 0.1 = 1/0.5 = 2.
        (ii) NY = J × k = I + G + X × k = $50bn + $70bn + $40bn × 2 = $320bn.
        (iii) The trade balance = X – M
               = $40bn – 0.1 × $320bn
               = $40bn – $32bn
               = $8bn.
               There is a trade surplus of $8bn.
        (iv) The budget position = T – G
               = 0.2 × $320bn – $70bn
               = $64bn – $70bn
               = –$6bn.
               There is a budget deficit of $6bn.

2.

	Keynesians	Monetarists
View on market failure	Significant	Not very significant
View on government failure	Not very significant	Significant
View on Quantity Theory	Reject	Support
Cause of inflation	May be demand-pull or cost-push	Excessive growth of the money supply
Main cause of unemployment	Frictional, structural and cyclical	Frictional and structural
Effect of government borrowing	Crowding in	Crowding out
Shape of LRAS curve	Horizontal, then upward sloping and then vertical	Vertical
Macroeconomic policy	Favour demand management	Favour supply-side policies
Government intervention	Needed to ensure the smooth running of the economy	To be kept to a minimum. Main responsibilities = remove market imperfections and keep inflation low.

3. (a) True. Current accounts are included in both measures. In contrast, deposit (time) accounts are only included in broad measures.
    (b) True. By engaging in credit creation, banks create more accounts than they have cash. This practice will not cause a problem as long as bank customers believe they could get the cash out of their accounts – which in fact they could not do.
    (c) True. The more liquid a bank's assets are, the less profitable they are. The bank's most liquid asset is cash which does not earn any money whereas its advances (bank loans) are profitable but not liquid.
    (d) True. Both sides represent total spending on goods and services.
    (e) False. A budget deficit may increase the money supply but it may not do so. If the government finances it by borrowing from the banking sector or from abroad, it will do so. If, however, it finances it by selling government securities to the non-bank sector, it will just be using existing money.
    (f) False. A credit crunch involves a shortage of bank loans which can lead to a recession.

Answers to multiple choice questions

1. B   The difference between GDP and NDP is depreciation. Net property income from abroad is the difference between a national and a domestic income measure. Net exports and consumer expenditure are included in both national and domestic measures.

2. D   Real GDP in 2012 was $30bn x 100/125 = $24bn. Real GDP in 2002 per head was $20bn/20m = $1,000. In 2012, it was $24bn/22m = $1,090.91.

3. A   A fall in the average number of hours worked would mean that people are enjoying more leisure time and this is likely to have increased people's standard of living. B and D are taken into account in national income per head measured at constant prices. C may mean that total output has risen by less than the official figures suggest.

4. A

Disposable income ($billion)	Consumer expenditure ($billion)	Average propensity to consume	Marginal propensity to consume
100	120	1.2	–
200	200	1.0	0.8
300	270	0.9	0.7
400	320	0.8	0.5
500	350	0.7	0.3

5. B   A definition question. The monetary base forms the basis of bank credit.

6. C   An increase in the budget surplus may result in more money being taken out of circulation. A and B may result in an increase in bank lending and so an increase in the money supply. D would mean more money would be entering the economy.

7. D   The sale of treasury bills to the banking system would increase the liquid assets of the commercial banks, permitting them to lend more. A, B and C would all be making use of existing money.

8. B   Monetarists believe that inflation is caused by an excessive growth of the money supply. They do not think that an increase in the money supply will increase productive capacity. They also

think the government may be able to reduce unemployment by a variety of supply side measures including the gap between paid employment and unemployment benefit. Keynesians think the government can reduce cyclical unemployment by increasing aggregate demand.

9. A   Keynesians favour increasing aggregate demand to reduce unemployment. B contains a supply side policy. C contains two monetarist policy measures and a rise in the exchange rate is not likely to be favoured by either group as a measure to reduce unemployment. D also contains two monetarist policy measures and one measure, a rise in the rate of interest, which again would not tend to be favoured by either Keynesians or monetarists as solution to unemployment.

10. D   The crowding out view suggests that a rise in government borrowing will push up the rate of interest and reduce the quantity of loanable funds available to private sector firms.

11. A   Aggregate expenditure would need to cut the 45 degree line at R to ensure full employment. There is a gap in expenditure of RS.

12. B   Aggregate expenditure consists of C + I + G + (X – M). The difference between aggregate expenditure and C + I + (X – M) is G.

13. C   In a closed economy with a government sector, the injections will be I + G and the leakages will be S + T. The equilibrium level of income is where injections equal leakages.

14. B   Investment is an injection into the circular flow. A and D are leakages and C is part of the circular flow.

15. C $k = 1/0.1 + 0.005 = 0.05 = 5$.

16. B   In this case, mps is 0.1, mrt is 0.1 and mpm is 0.2. So the multiplier is $1/0.4 = 2 \frac{1}{2}$. To raise national income by the $50bn desired, the government would have to increase its spending by $50bn/2 \frac{1}{2}$ = $20bn.

17. C   A definition question. A is a partial explanation of the multiplier. In terms of B and D, actual saving will always equal actual investment.

18. B   The Loanable Funds Theory argues that the rate of interest is determined by the demand for and supply of loanable funds. The supply of loanable funds comes from savings. A decrease in the level of savings would cause a rise in the rate of interest as shown here.

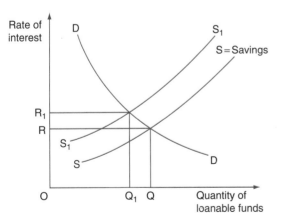

19. A   If people expect the price of government bonds to fall, they will want to hold money now. This is because they will not want to buy bonds as they will anticipate making a loss and will not be forgoing much interest. D is concerned with the precautionary motive and

B with the transactions motive. An expectation that the rate of interest will fall C, would encourage people to reduce their holdings of money. They would want to buy bonds, expecting the price of bonds to rise.

20. C   A rise in bank lending would increase the money supply. Liquidity preference will increase as a result of more money being demanded for transactions purposes.

21. B   In a closed economy, Y = C + I + G. In this case, C = 0.75 x $400m = $300m. C + G = $340m, so investment must be $60m. $60m = $(70 – 2 x 5) m.

22. D   The multiplier is $1/0.1 + 0.05 + 0.1 = 1/0.25 = 4$. National income will increase by $200m x 4 = $800m. Consumer expenditure will rise by 80% of $800m = $640m. Note of this $40m will be spent on imports and $600m on domestic products.

23. D   Consumer expenditure is MN. Planned investment is PY. Firms make MP amount of consumer goods but only MN is sold. This means there are unsold stocks of NP. Actual investment consists of planned investment and changes in stocks, which in this case is PY + NP = NY.

24. B   If interest rates decrease, it means that the price of bonds has increased. People will want to hold money now as they will expect the price of bonds to fall in the future. They will also sacrifice less by holding money. If the price level increases, people will want to hold money for transaction purposes. Real income remaining constant will have a neutral effect. So with two of the influencing increasing demand for money will increase. In the case of A, the decrease in interest rates will increase demand for money but a fall in real income will reduce the transactions demand for money – so the effect is uncertain. C – all of the factors would reduce demand for money. D – the increase in interest rates would reduce demand for money, a constant price level is neutral and the increase in real income would increase demand for money. This means that the effect is again uncertain.

Answers to data response questions

1. (a) Investment and government spending are not mentioned in the first paragraph of the extract.
   (b) Between 1st July and 1st October 2007, the US economy grew at a rate equivalent to a 4% annual rate, so in three months it grew by 4%/4 = 1%.
   (c) Figure 15.15 shows that over the time period more dollars had to be given to buy a euro. For instance, in 2002, it cost approximately $0.9 to buy 1 euro but by 2007, this had risen to $1.4. This means that the value of the dollar fell over the period – dollars were able to buy fewer euros.
   (d) A reduction in the rate of interest would be designed to encourage an increase in consumer spending and investment. A lower interest rate would reduce the cost of borrowing. This may encourage more people to borrow to buy houses. Higher demand for houses would encourage house building firms to supply more houses. People, however, may not borrow more if they are worried about future economic prospects or if they expect the rate of interest to rise fairly soon. In addition, if the rate of interest is cut from a very low rate of interest such as 0.75% to 0.5% it may have little effect.
   (e) The data does seem to support the view that the US economy was facing conflicting policy objectives. The Federal Reserve was considering cutting the rate of interest. Such a policy

measure may stimulate the housing market. A more buoyant housing market may reduce unemployment and increase the economic growth rate, two of the main government macroeconomic objectives.

A cut in the rate of interest may also mean an outflow of hot money flows which reduces the exchange rate. If the combined elasticity of demand for exports and imports is greater than one, this would improve a current account deficit which is another government macroeconomic objective.

A lower rate of interest could, however, make it more difficult for the government to achieve the objective of a low and stable inflation rate. If it stimulates higher consumer spending and the economy would experience demand-pull inflation. There may also be cost-push inflation with the falling exchange rate pushing up the price of imported raw materials, including imported oil.

2. (a) Economic growth is usually measured by increases in real gross domestic product (GDP).

(b) A deflationary gap occurs when there is not sufficient aggregate expenditure to achieve full employment. Injecting government spending into the circular flow will cause a multiple increase in GDP. If the multiplier has been calculated accurately and sufficient government spending has been injected, a deflationary gap may be reduced. For instance, if there is a deflationary gap of $20bn and the multiplier is 4, the government would have to increase its spending by $5bn. In the following figure, there would be a deflationary gap of ab unless government spending rises to G1.

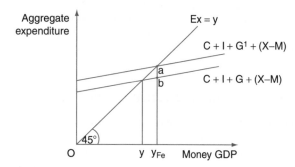

(c) (i) A large budget deficit will mean that the government will have to borrow a considerable amount to finance the gap between its spending and its revenue. The high demand for funds may result in a high interest rate.

(ii) The information in Table 1 does not really confirm this relationship. UK had the largest relative budget deficit but only the third highest interest rate. The US had the second largest deficit but the lowest interest rate. The country with the highest interest rate, Brazil, had the second to lowest budget deficit. Of course, only a small number of countries are shown. It is also possible that a large budget deficit may be associated with a low interest rate when both high government spending and a low interest rate are used to stimulate economic activity.

(d) Liquidity ratios are the percentages of liquid assets that commercial banks have to keep relative to their liabilities (deposits). For instance, a liquidity ratio of 10% would mean that a bank with deposits of $200m would have to keep liquid

assets of $20m. If a central bank raises a formal liquidity ratio requirement from 10% to 12 ½ %, this would mean that $20m of liquid assets would only be able to support deposits of $160m. This would reduce the amount the bank could lend from $180 m ($200m – $20m) to $140 m ($160m – $20m). A reduction in bank loans (advances) could reduce consumer spending.

It is possible, however, that whilst a bank may have been able to lend $180m, demand for loans might have only been, for instance, $120m. In this case, the increase in the liquidity ratio would have no effect. Even if demand for loans is $180m or more, an increase in the liquidity ratio, may not reduce bank lending if banks can arrange loans through their foreign branches. Consumer demand may also increase, even if loans are being reduced, if income is rising or consumers are becoming more confident.

3. (a) A fall in capital stock would be expected to reduce aggregate supply. This is because there will be a reduction in resources and so the economy would be capable of producing less.

(b) (i) It cannot be concluded that the output of Pakistan's economy was $163bn in 2011. This is because although its GDP as measured by C + I + G + (X – M) was $163bn, the information states that there was informal economic activity. Such undeclared output is not recorded in GDP.

(ii) Domestic demand consists of C + I + G. In India in 2011 this amounted to $1,432bn.

(c) The size of the multiplier varies between countries because of the tendency to withdraw extra income from the circular flow. The marginal propensity to save, the marginal rate of taxation and the marginal propensity to import differ between countries. If people in a country save a higher proportion of their income, are taxed more and spend a higher proportion of their income on imports than in another country, the size of the multiplier will be smaller. Less of any extra income will spent on the economy's products. For example, if one country's mps, mrt and mpm are 0.2, 0.2 and 0.2, the multiplier will 1/0.6 = 1.67. In contrast, if another country has a mps, mrt and mpm of 0.1, 0.05 and 0.05, its multiplier will be 1/0.2 = 5.

(d) An output gap occurs when an economy is not producing at full capacity. A negative output exists when an economy's output is below its potential output.

(e) Consumer expenditure may increase in absolute terms if disposable income rises in Pakistan. Of course, there is the possibility that as income rises, the average and marginal propensities to consume may fall. This is because as people get richer they are more able to save.

Consumer expenditure might increase even if income does not rise if people become more confident about the future or if the rate of interest falls. In the latter case, the reward from saving will fall and it will become cheaper to borrow to buy items such as cars. In addition, a rise in wealth may increase consumer expenditure. If, for instance, the value of housing and/or shares increases, people may spend more as they will feel richer and will have more collateral to borrow against.

If, however, income falls, confidence declines, the rate of interest rises or the value of wealth decreases, it would be expected that consumer expenditure would decline.

Answers to essay questions

1. (a) Households and firms demand money when they decide to hold some of their wealth in a money form. Demand for money is called liquidity preference. There are three main motives for demanding money. One is the transactions demand for money. This is money kept to make everyday purchases. Households keep some of their money in cash, notes and current bank accounts in order to buy goods and services. Firms also keep some money in a liquid form in order to, for instance, pay for raw materials and pay wages.

Most households and firms keep more money than they think they will need for transaction purposes. This is because they want to be able to meet unexpected expenses and take advantage of unexpected bargains. This motive is referred to as the precautionary motive.

Together the transactions and precautionary motives are known as active balances. This is because the money held is likely to be spent in the near future. Both motives tend to be relatively interest inelastic. This means that they are not significantly affected by a change in the rate of interest. A rise in the rate of interest, for instance, would not cause households and firms to cut back on the money they keep in a liquid form. The following figure shows the inelastic demand for active balances.

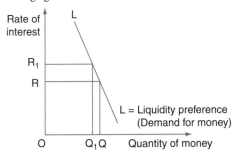

The third motive for holding money is the speculative motive. Money may be held to avoid making a loss on financial assets and to be ready to take advantage of changes in the price of government bonds (securities) and the rate of interest. If the price of government bonds is high and the rate of interest paid on them is low, households and firms may want to hold money for two related reasons. They are likely to expect the price of bonds to fall and so they will not want to make a loss by buying bonds. They will also not be forgoing much as interest is low. In contrast, if the price of government bonds is low and the rate of interest is high, households and firms will want to minimise the amount of money they hold for speculative purposes.

The speculative motive is also referred to as the demand for idle balances. The figure below shows the demand for idle balances.

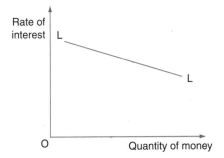

Together the demand for active balances and idle balances make up the demand for money as shown here.

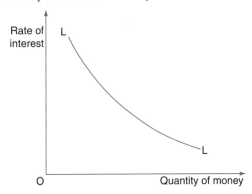

(b) An increase in the supply of money will be expected to reduce the rate of interest and increase national income.

A rise in the money supply will increase the money balances people hold. Some of these balances are likely to be used to purchase government bonds. An increase in demand for government bonds will raise their price and lower the rate of interest. The following figure shows an increase in the money supply from MS to MS1 causing the rate of interest to fall from R to R1.

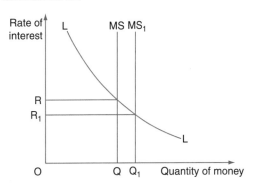

If the rate of interest is initially very low, an increase in the money supply may have no effect on the rate of interest. This is referred to as the liquidity trap. The figure below shows that at the rate of interest R, demand for money becomes perfectly elastic. With the rate of interest being so low, the price of government bonds will be very high. People will expect the price of government bonds to fall and they will not be sacrificing much interest. As a result all of the extra money will be held.

In most cases, however, an increase in the rate of interest is likely to reduce the rate of interest. This, in turn, would be expected to increase consumer expenditure and investment. People will probably save less, borrow more and so spend more. Firms will be encouraged to invest more as it will be cheaper to borrow and they will be expecting consumer expenditure to rise.

There is a possibility that a fall in the rate of interest will not increase consumer expenditure and investment. People may not spend more and firms may not increase their investment if they are not confident about the future. They may also think that the fall in the rate of interest may not last.

If the rate of interest does increase consumer expenditure and investment, this will increase aggregate expenditure and increase it by a multiple amount. The size of the increase will depend on the initial injection and the size of the multiplier. The following figure shows national income rising as a result of an increase in aggregate expenditure.

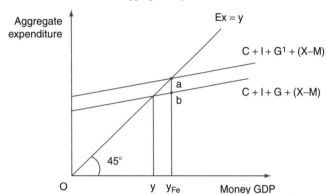

2. (a) A free market closed economy is an economy without a government sector and without an international trade sector. In such an economy, there are only two sectors. These are households and firms. This means there is one injection into the circular flow of income, which is investment, and one leakage (withdrawal) from the circular flow, which is savings.

The size of the multiplier is influenced by how much of extra income is spent on domestically produced products. The more that is passed on in the circular flow, the larger will be the multiplier. This means that the size of the multiplier varies inversely with the tendency for extra income to be withdrawn from the circular flow – the marginal propensity to withdraw. It is calculated by 1/marginal propensity to withdraw. In the case of a two sector economy, this is 1 divided by the marginal propensity to save (mps). For example, if people save $20 out of an increase in income of $100, the mps will be 0.2 and the multiplier will be 1/0.2 = 5.

In a mixed economy with foreign trade, there will four rather than two sectors. These are households, firms, the government and international trade. There are now three injections which are investment, government spending and exports and three leakages which are saving, taxation and imports. The formula for the multiplier is:

$$\frac{1}{\text{marginal propensity to save (mps)} + \text{marginal rate of tax (mrt)} + \text{marginal propensity to import (mpm)}}$$

If mps is 0.2, mrt is 0.1 and mpm is 0.2, the multiplier is 1/0.5 = 2.

The size of the multiplier is likely to be lower in a mixed economy than in a free-market, closed economy as there are more leakages. There is, however, a greater chance that national income will change as there are more injections and leakages.

(b) An injection into the circular flow will cause national income to rise by a multiple amount. This is because extra spending will result in further spending and so on. For example, a rise in government spending will increase demand and output.

The higher demand and output will generate more income which will increase spending further. The multiplier process will continue until the injection is matched by an equal rise in leakages.

If, for instance, government spending rises by $30m and the multiplier is 3, national income will increase by $90m. Leakages will rise by $30m to match the injection of $30m.

If an economy is initially operating below full employment, a rise in national income would be expected to cause unemployment to fall. Figure ? shows an increase in government spending causing national income to rise, closing a deflationary gap and resulting in full employment.

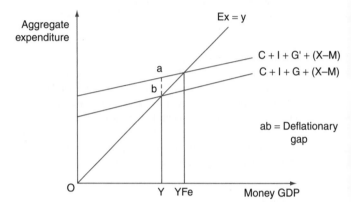

There is a risk that an injection may take national income beyond the full employment level. For example, a government may underestimate the size of the multiplier and may increase national income by too much. In this case, an inflationary gap will be created as shown in the figure below.

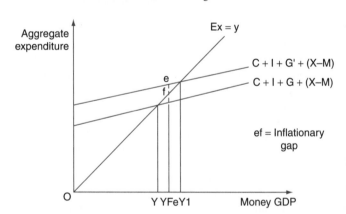

3. To decide whether people in the Netherlands enjoy living standards four times as great as that of people in Chile, more information than the GDP per head would be needed.

It would be necessary to know real GDP per head as the difference in GDP per head would be exaggerated if the Netherlands experienced higher inflation than Chile and would have been understated if Chile's price level had risen by a greater percentage.

It would also be useful to know what type of products is being produced. A country may have a high output relative to its population but if it devotes a significant proportion of its resources to its armed forces, its population may not enjoy many consumer goods and services. A country's output might consist of a high proportion of capital goods. This might mean higher living standards in the future, but relatively low living standards now.

It is also important to check whether the GDP per head figures have been compared using purchasing power parity, that is an exchange rate based on the cost of a given basket of products. This would prevent a misleading impression being given of the gap by the comparison being made when there has been a significant change in the market exchange rate which does not reflect the internal purchasing power of the currencies.

GDP per head takes into account differences in population size but it does not provide information about the distribution of income. If income is very unevenly distributed in a country with a higher GDP per head, only a small proportion of the population may enjoy higher material living standards than those in a country with a lower but more evenly distributed GDP per head.

Income plays a key role in determining the living standards people enjoy, but it is not the only determinant. A number of other measures of economic welfare take into account not only income per head but also other indicators. For instance, the Human Development Index (HDI) also considers the education people experience and the life expectancy they enjoy. A good quality education increases a person's career choices, earning capacity and often their interests. Higher life expectancy indicates better quality health care. The Index of Sustainable Economic Welfare (ISEW) includes in its measure income inequality, environmental damage and depletion of environmental assets.

It is difficult for a measure of a country's income to take into account all economic activity since not all such activity is declared. A country may have a smaller real GDP per head than another country but if it has a larger informal economy, depending on its composition, its citizens may enjoy higher living standards. The quality of the products produced and differences in working hours and working conditions also have to be considered.

In addition, cultures and climates vary between countries. People living in a non-materialistic culture may be satisfied with fewer goods and services. People living in cold climates may have to spend more on heating than people living in warmer climates, just to experience the same standard of living.

GDP per head figures give some indication of living standards but other factors have to be considered to gain a fuller picture of the quality of people's lives.

Chapter 16

Answers to short questions

1. A country may have a higher GNI than another country but a lower value on the HDI due to lower life expectancy and people spending less time in education.

2. Countries with a low income per head tend to have a high birth rate for a number of reasons. The countries often have a high infant mortality rate and so some people have a relatively high number of children because they do not expect them all to survive. In the absence of a welfare system, people have children to provide for them in old age. In addition, women receive less education and are likely to marry younger and so have more children. There may also be a lack of knowledge and availability of contraception.

3. If a rich country provides aid to a poor country, it may enable the poor country to grow. As the poor country experiences a rise in income, it may buy more products from the rich country. This would increase the rich country's exports and contribute to economic growth.

4. The Malthusian theory of population states that whilst agricultural output grows arithmetically, population has a tendency to grow geometrically. In other words, population growth tends to outstrip the growth of food supplies. The theory assumes that the supply of land is fixed and that as more workers are employed on the land, agricultural output would increase but at a diminishing rate.

5. A natural increase in population occurs when the birth rate exceeds the death rate.

6. If population grows above the optimum level it may hinder economic development. This is because if output per head falls, people's material living standards may decline and less tax revenue per head may be raised to spend on education and health care.

7. A supply constraint means that a lack of aggregate supply is limiting economic growth. When an economy is operating at full capacity, it will not be able to produce more unless there is an increase in aggregate supply. In contrast, a demand constraint occurs when a failure of aggregate demand to increase stops economic growth taking place.

8. Net investment is very significant in generating economic growth because it both increases aggregate demand and increases aggregate supply to meet the higher aggregate demand.

9. Zero net investment may be associated with economic growth if replacement investment incorporates advanced technology. With a better quality of investment, aggregate supply will increase.

10. External debt may hinder economic growth for a number of reasons. One is that some of the export revenue earned may have to be spent on servicing the debt rather than, for instance, buying imported capital goods to expand output. Another reason is that foreign financial institutions may be reluctant to lend to a country which has a high level of debt. This may restrict spending on, for instance, the country's infrastructure.

11. Cutting down rainforests could contribute to global warming as rainforests absorb carbon dioxide. It may also restrict future economic growth by depleting a natural resource.

12. Voluntary unemployment occurs when unemployed people refuse to accept the jobs on offer.

13. Frictional unemployment might be reduced by providing more information about job vacancies and by cutting unemployment benefit. More information may enable people to switch jobs more quickly. Cutting benefit may reduce search unemployment with people becoming more willing to accept the first or second job on offer.

14. Unemployment might increase despite a rise in real GDP, if the rise in real GDP is below the increase in potential output. With improved education, training and advances in technology, an economy may be able to produce more with fewer workers. If aggregate demand rises but at a slower rate than productive potential, there is likely to be a rise in unemployment.

15. Advances in technology may create jobs by increasing productivity and so raising international competitiveness. This may increase aggregate demand and real GDP and so may create jobs. In addition, advances in technology can lead to the development of new products, for example the ipad, and so can generate demand and jobs.

16. Discouraged workers are unemployed people who drop out of the labour force because they give up hope of finding a job.

17. The internal value of money may decline but its external value rises if the country experiences a lower inflation rate than its main rivals. The inflation will reduce the internal value. Its lower rate, however, may increase the country's international price competitiveness. This could increase demand for its currency and so raise its external value.

18. The slope of the traditional Phillips curve is steeper at higher levels of inflation and less steep at lower levels of inflation. This suggests that as unemployment falls, it takes larger increases in unemployment to reduce inflation further.

19. The long run Phillips curve is vertical at the natural rate of unemployment. It shows the relationship between the inflation rate and unemployment rate when the actual and expected inflation rates are equal.

20. A horizontal Phillips curve suggests that it is possible to reduce the rate of inflation without increasing the unemployment rate. This may occur if advances in technology and international competitiveness keep inflationary pressures in check despite the increase in aggregate demand which is likely to accompany falling unemployment.

Answers to revision activities

1. (a) The following figure shows actual economic growth using a production possibility curve diagram.

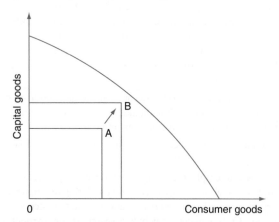

Moving from production point A to production point B, by employing previously unemployed resources, raises national output.

The figure below also shows actual economic growth, this time using an aggregate demand and aggregate supply diagram.

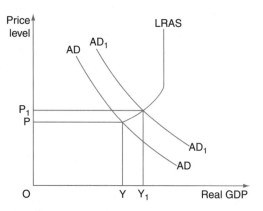

The increase in aggregate demand form AD to AD1 results in a rise in real GDP from Y to Y1.

(b) The following figure shows potential economic growth using a production possibility curve diagram.

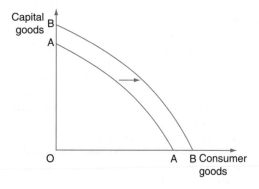

The shift to the right in the PPC from AA to BB shows that the economy is capable of producing more goods and services.

The following figure also shows a rise in productive capacity, this time illustrated by a shift to the right in the aggregate supply curve.

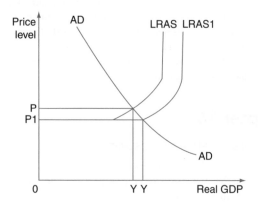

2.

3. (a) Cyclical   (b) Structural   (c) Frictional
   (d) Cyclical   (e) Structural   (f) Structural

Characteristic	Developed country	Developing country
GDP per head	High	Low
Population growth	Low	High
Proportion of population employed in agriculture	Low	High
Urbanisation	High	Low
HDI ranking	High	Low
Foreign debt as % of GDP	Low	High
Energy consumption	High	Low
Enrolment in tertiary education	High	Low
Life expectancy	High	Low
Population per doctor	Low	High
Labour productivity	High	Low
Savings ratio	High	Low

Answers to multiple choice questions

1. B Developing countries tend to have a large primary sector, with a relatively high proportion of its GDP and employment accounted for by agriculture, mining and forestry. Developing countries often have large population growth rather than large populations. They also tend to have infant mortality rates and a high rate of rural – urban migration.

2. C The Human Development Index takes into account, Gross National Income per head, life expectancy at birth and years of schooling. It does not include A, B and D.

3. C India's current account would have benefited from $6m coming into the income section. The financial account covers direct investment, portfolio investment and bank loans. The $5m borrowed from Indian banks would come out of India's financial account. This would mean that the net currency flow would have been $6m - $5m = $1m.

4. C Country C has the second highest GNI per head, by far the longest life expectancy and the largest time spent in education.

5. D The inelastic supply which primary products tend to have makes it difficult to adjust supply to changes in market conditions. Primary products tend to have low income elasticity of demand and have a relatively low price elasticity of demand. A low cross elasticity of demand between primary products and artificial substitutes would be an advantage for developing countries.

6. D A definition question. An optimum population is the population which maximises output per head and so allows income per head to be maximised.

7. C A slightly different way of defining the optimum population.

8. A Life expectancy is included in the HDI but not real GDP per head. Both HDI and real GDP per head are adjusted for inflation. Neither measure takes into account pollution or working hours.

9. D If an economy is producing below full capacity, an increase in aggregate demand could increase real GDP despite a decrease in productive capacity. This is illustrated in the following figure.

   B would cause a decrease in aggregate supply but also a decrease in aggregate demand and so a fall in real GDP. A and D would reduce aggregate demand and real GDP.

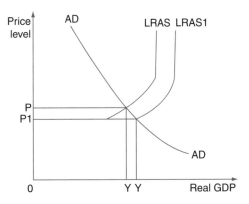

10. D Long run economic growth increases productive capacity and this is illustrated by a shift of the long run aggregate supply curve.

11. D The closure of bookshops and the resulting unemployment of bookshop assistants is affecting one industry. It has resulted from a change in supply conditions – a structural change. A and C are types of frictional unemployment which arises when people are in between jobs. B is caused by a general lack of demand for all products.

12. A The natural rate of unemployment is the rate of unemployment which is consistant with a stable rate of inflation. This occurs when the aggregate labour market is in equilibrium with no upward pressure on wages and the price level.

13. A An increase in the gap between paid employment and unemployment benefit may encourage the unemployed to seek work more actively. C may cause cyclical unemployment. D might reduce cyclical unemployment but not the natural rate. B is a consequence rather than a cause of a reduction in an economy's natural rate of unemployment.

14. D Potential output is forgone. B and C are not sacrificed – they are experienced. More or fewer imports may be purchased as a consequence of unemployment. If imports are being purchased they are not being forgone.

15. A With more people being unemployed, the government would spend more on unemployment benefit. It will also be receiving less direct and indirect tax revenue as income and spending will fall.

16. A In a fully employed economy, any cause of an increase in aggregate demand, not matched by an increase in aggregate supply would cause inflation. Increased government spending on pensions would raise aggregate demand. B, C and D would reduce inflation. This is because B would increase aggregate supply and C and D would reduce aggregate demand.

17. B An increase in potential output raises the productive capacity. If real GDP does not rise in line, there will be unemployed resources.

18. C Exports are injections into the circular flow of income. An increase in injections or a fall in leakages would increase the circular flow. A and B are reductions in injections and D is an increase in a leakage.

19. D Increasing government spending would increase aggregate demand. This would reduce unemployment as firms will expand to produce the extra products being demanded. Higher aggregate demand, however, would increase demand-pull inflation. More government spending will raise GDP by a multiple amount. This will increase consumer expenditure, some of which will be spent on imports. This will increase a current account deficit and reduce the value of the currency.

20. A Inflation reduces the internal value of money as each unit of money will be able to buy less. A high and accelerating inflation rate would also be likely to reduce demand for the country's exports. This, in turn, would reduce demand for the currency which would reduce the exchange rate and so the external value of money.

Answers to data response questions

1. (a) A developing country is one which has a variety of characteristics including low income per head and, by definition, low development. There is a range of evidence in Article 1 to suggest that The Gambia is a developing country. It mentions that it was ranked low in the HDI, more than half of the population lives on less than one US dollar per day, there is subsistence farming and there is a relatively large informal sector.

   (b) As well as employing workers directly, the tourism industry creates work for people employed by firms that supply the needs for the industry. For instance, an expanding tourism industry will cause more workers to be employed by coach companies, taxi firms, firms supplying food to hotels and firms providing entertainment for tourists.

   (c) There are a number of reasons why hotel workers might be lowly paid. They are low skilled and have low marginal revenue productivity. Hotel workers often suffer from seasonal unemployment, being out of work during the off-peak season, and this reduces their earning potential. Many hotel workers are not in trade union and so have relatively weak bargaining power against their employers, some of whom are powerful multinational companies.

   (d) The evidence is mixed on whether tourism exploits resources and is of little benefit. The second article suggests that tourism does not treat domestic workers well. It mentions that they are low paid and that the high paid jobs go to foreign workers. It also implies that 90% of the income earned from Tanzania's tourist industry leaves the country.

   The first article is rather more positive about the contribution that tourism makes to an economy. It mentions that the industry accounts for 7.8% of the GDP of The Gambia and creates jobs for 11,000 people. It does suggest, however, that tour operators abuse their market power to drive down the prices that they pay the hotels.

   To decide whether tourism exploits resources and is harmful, it would be useful to have more information. For example, is the industry paying workers a wage equivalent to their marginal revenue product? In addition, is it damaging the environment to make a short term gain by, for instance, using up scarce resources of water in swimming pools and bottled water?

2. (a) It rose by 1.8% points to 15%.

   (b) Unemployment and poverty are usually directly linked. A rise in unemployment would be expected to increase poverty. Most people experience a fall in income when they lose their job. The longer they are unemployed, the more financial difficulties they may experience.

   (c) There are a pieces of evidence to suggest that USA is a developed country. Although there is a significant proportion of the population living in poverty, the poverty line is set at a relatively high rate. In many countries, an increase of, for instance, $20,000 for a family of four would be considered to be a high income but in the US it would result in the family being classified as poor. Being ranked 4th in the Human Development Index (HDI) is strong indicator that the USA is a developed country. The HDI takes into account GDP per head, education and life expectancy. An average income of $52,029 is high. Long life expectancy, such as the 79.6 years that USA enjoys, is a sign of a developed country. A high percentage of the labour force employed in the tertiary sector also provides evidence that the US is a developed country.

   (d) The children of the poor tend to become poor adults because they are likely to have grown up in poor housing and may have only a few years of education. The resulting poor health and limited skills and qualifications reduces their earning capacity. They may also have lower expectations which will also limit the types of jobs they apply for.

   (e) Achieving the macroeconomic objectives of full employment and steady and sustainable economic growth is likely to reduce absolute poverty but not necessarily relative poverty. With more people in work, more people may have access to basic necessities. Economic growth may result in a rise in income per head. It should also increase tax revenue and some of this may be spent on education, health care and benefits which may reduce absolute poverty.

   Higher employment and economic growth may, however, be accompanied by a widening of the gap between those receiving a high income and those receiving a low income. Successful entrepreneurs and workers with skills in high demand may experience significant increases in their incomes during prosperous times. More people may be in work but some may be in low paid jobs and the standard of living that they are able to enjoy may be significantly lower than that of the highest paid. To reduce relative poverty, a government may have to pursue another macroeconomic growth, that is a redistribution of income.

3. (a) A low corporate tax rate and a low cost but educated workforce. Also, absence of exchange controls, financial hub of the Southern African Development Community, absence of exchange controls, adoption of free market principles.

   (b) There is evidence to suggest that Botswana is a developing country. Its exports are dominated by one primary product, diamonds, which also accounts for nearly half of its output. Wage rates are low in the country, there is a lack of usable water and life expectancy is low.

   (c) It is a pure monopoly. The article mentions there is only one company, Debswana. The company seems to enjoy high profits which may be supernormal profits. Government involvement in the company may indicate that the government considers it necessary to ensure that it does not abuse its market position.

   (d) A commitment to free market principles means that the government is allowing the free market forces of demand and supply to allocate resources. The price mechanism would be allowed to act as a signal to producers of what consumers are demanding. The profit incentive would encourage them to respond by shifting resources away from unpopular to popular products. The article suggests that the Botswana economy is making some commitment to free market principles. It is operating a low corporate tax and puts no restrictions on profits out of the country. It does, however, mention that the government owns half of Debswana. State ownership of firms is more a feature of a planned or a mixed economy.

(e) There are a number of advantages that a developing country may gain from becoming more diversified. Building up new industries will create more products for consumers to enjoy. It may also increase employment and the jobs may be better paid and may develop skills to a greater extent.

A more diversified economy may also allow advantage to be taken between industries. The diamond industry could provide diamonds for a domestic jewellery industry which, in turn, can provide products for domestic jewellery sellers. Agricultural producers can also provide products for domestic food processing firms and provide demand for manufactured products including tractors.

A more developed economy also reduces the risk of its output and exports being seriously affected by a fall in demand or supply problems. With a greater range of products being produced, it is likely that if demand for some is declining, demand for others will be increasing.

An economy needs to be not too narrowly focused but also to have some specialisms. It also has to recognise that the structure of its economy also has to be adaptable to changing economic conditions.

Answers to essay questions

1. (a) Economic growth is concerned with output and the ability to produce output whereas economic development is concerned with welfare.

Actual economic growth occurs when the output of an economy grows. With higher output comes higher income and higher expenditure. Potential economic growth enables an economy to continue to produce more products and so earn higher incomes. It is achieved when the productive capacity of the economy increases due to a higher quantity and/or quality of resources.

Economic growth may lead to an increase in economic development but, depending on how it is achieved, it may reduce economic development. A country will experience economic development if its population achieves an improvement in their economic welfare. It is possible that a rise in real GDP, especially if it is evenly distributed, may improve the quality of people's lives. People will be able to enjoy more goods and services and this may be particularly important in the case of people lacking basic necessities. Higher income will also be expected to raise tax revenue, some of which may be spent on education and health care – key influences on economic welfare. A high GDP per head is one of the main indicators of a developed country. High incomes also enable more saving which, in turn, can increase investment and lead to further economic growth.

Economic growth, however, may come at the cost of economic development if it is accompanied by a rise in pollution, longer working hours and worse working conditions. Economic growth also involves change, with workers often having to develop new skills and some may find this stressful.

(b) Multinational companies (MNCs) often contribute to the economic growth of developing countries, at least in the short term. Their impact on economic development is, however, more debatable.

A MNC is one which produces in more than one country. It has its headquarters in one country but may have plants in a number of countries.

When setting up in another country, MNCs usually add to that country's output unless they replace domestic firms. MNCs' output may increase more rapidly than that of domestic firms and they tend to export a relatively high proportion of their output. They may also promote economic growth in a developing country by providing workers with transferrable skills, introducing new technology and working methods and sometimes building transport links.

Their contribution to the economic growth of a particular developing country may be short lived if, after a period of time, they decide to relocate their production to an economy with lower costs or a larger market. MNCs may be less committed to the country than domestic firms. It is also possible that the MNCs may deplete some of the country's natural resources, thereby reducing the country's ability to grow in the future.

MNCs may promote economic development by providing employment, higher wages, training and better working conditions. It may also raise economic welfare by introducing new products. The more skilled are the jobs on offer, the more economic development there is likely to be. In some cases, MNCS may offer mainly low skilled jobs which may not allow workers to make much progress. For instance, foreign owned hotels may employ managers from their own countries whilst employing waiters and cleaners from the developing country.

MNCs may pay workers less in a developing country than it pays workers in developed countries. If, however, it pays them more than the wages in the country, it may nevertheless promote economic development. Generic training is likely to be more beneficial than firm specific training as the workers will be able to make use of it in other jobs.

The working conditions provided by MNCs may not be as good as they provide in their home countries but again may be better than in the developing countries.

They may introduce products which help to make life easier and promote health care. Some of the products they produce and sell in developing countries, however, may be demerit goods and may clash with the culture of the countries. MNCs may also cause environmental damage and may take risks in terms of health and safety to save costs. They may also put pressure on national governments to pursue policies beneficial to them but not necessarily to the country's economic development.

The effect MNCs have on developing countries varies across the world and really needs to be considered on a case by case basis.

2. (a) The natural rate of unemployment is the unemployment which exists when the labour market is in equilibrium with all those wanting to work at the going wage rate are able to find a job. As there is no shortage of workers there is no upward pressure on wages and so no upward pressure on the price level. The following figure shows the labour market in equilibrium at a wage rate of W. The aggregate labour force, however, is greater than those willing and able to work at the wage rate on offer. The natural rate of unemployment is Q – Q1.

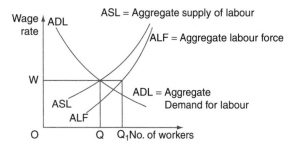

To reduce the natural rate of unemployment a government would implement supply side measures. One such measure would be to cut unemployment benefit. The intention behind this measure is to encourage the unemployed to look more actively for work and so reduce the search unemployment component of the natural rate.

A government might also reduce the gap between paid employment and benefits by reducing income tax. Such a reduction would increase disposable income and so would provide a financial incentive for the unemployed to take the jobs on offer.

In some countries, it might be thought that trade unions are pushing wage rates above their equilibrium levels and engaging in restrictive practices. These actions may be discouraging some employers from taking workers on and so a government may decide to pass laws to reduce trade union power.

The natural rate of unemployment may be reduced by increasing the occupational and geographical mobility of labour. The main way to increase occupational mobility is to promote education and training. This might be achieved, for example, by giving tax incentives to firms which undertake training. To increase geographical mobility a government might remove any restrictions on building homes in areas of the countries where industries are expanding so as to reduce the cost of housing there.

So there are a number of measures a government might take to reduce the natural rate of unemployment. The concept is, however, a somewhat controversial one and even its supporters admit, it can be difficult to determine what is the natural rate at any one time.

(b) According to the traditional Phillips curve, a decrease in unemployment will cause inflation. In the following figure, a fall in unemployment from 6% to 4% causes an increase in the inflation rate from 4% to 7%.

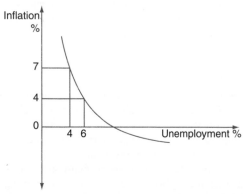

When unemployment falls, aggregate demand increases, firms compete more for labour and trade unions become more powerful. These forces can push up the price level.

The expectations-augmented Phillips curve suggests that attempts to reduce unemployment below the natural rate will cause inflation. It also suggests that, in the long run, unemployment will return to the natural rate but with a higher inflation rate. In the figure that follows, the economy is initially operating at the natural rate of unemployment of, in this case, 6% and on the short run Phillips curve (SPC). A government attempt to reduce unemployment succeeds in the short run in lowering unemployment to 3%. It does, however, also cause inflation of 5%. When some workers realise that with inflation the real wages they are being paid have not risen, they leave employment. Some firms will also realise that their real profits have not gone up and so will reduce the number of workers they employ. In this case, unemployment returns to 6%. Having experienced an inflation rate of 5% workers and firms will anticipate that inflation will continue. As a result they will behave in a way, for instance by asking for wage rises and by increasing prices, that will cause inflation to continue. The economy will move on to a higher short run Phillips curve (SPC1). Now any further attempt to reduce unemployment below 6% will generate even higher unemployment.

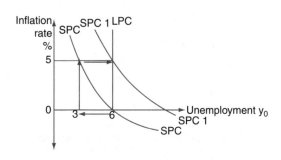

It is, however, possible to argue that a fall in unemployment will not necessarily cause inflation. Unemployment may decrease because workers become more educated or better trained. In this case, the higher aggregate demand which will result from more people being in work may be matched by the higher aggregate supply resulting from the increase in labour productivity. Aggregate supply may also be increased by advances in technology which again may offset any inflationary pressure arising from lower unemployment. In addition, increased global competition may make it difficult for firms to raise their prices even if domestic demand is increasing. It is possible that the nature of the relationship between unemployment and inflation may be changing in some countries. The following figure shows a new type of Phillips curve with a reduction in unemployment from 6% to 3% having no effect on the inflation rate which remains at 4%.

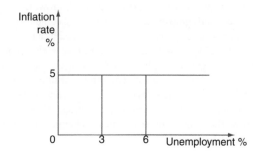

There are reasons to think that a fall in unemployment may contribute to inflation by raising both aggregate demand and costs of production but the relationship is not clear cut.

3. (a) There are a number of indicators which can be examined to determine whether a country should be classified as developed or developing. It does, however, have to be remembered that no country's characteristics will fit entirely into one of these two categories.

One of the key indicators is real GDP per head. A developed country would be expected to have a high GDP per head, giving its citizens the ability to consume a high number of goods and services. A developing country would be likely to have a low GDP per head. This does not mean that everyone in the country will be poor. Indeed, a developing country may contain a number of rich people but, on average, income will be low.

A higher proportion of the labour force would be expected to be employed in the primary sector and it would be anticipated that this sector would make a larger contribution to GDP in a developing country than in developed country. In contrast, the tertiary sector would usually make a larger contribution to employment and output in a developed country.

Primary products may also form a high proportion of the exports of developing countries. There may also be a narrow range of products forming those exports.

The labour force in developing countries usually has a lower productivity than that in developed countries. With lower income, the amount spent on education and training per head is likely to be lower, leading to lower skilled workers and hence lower productivity. In turn, lower productivity can result in lower income which can keep saving and investment low. This vicious circle of poverty is illustrated in the figure here.

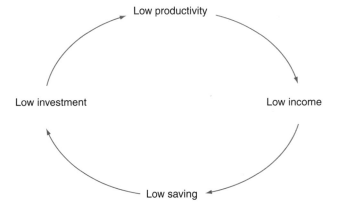

One of the reasons why spending on education per head may be low is because of high population growth. Developing countries tend to have higher population growth than developed countries. Their birth rates and death rates tend to be higher in developing countries, giving them a pyramid shaped population pyramid as shown here.

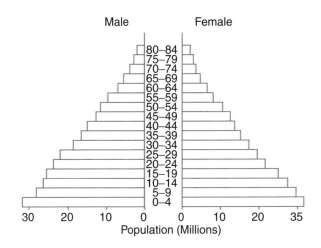

This contrasts with the typical population pyramid of a developed country which has a narrower base and wider apex as shown in the following figure.

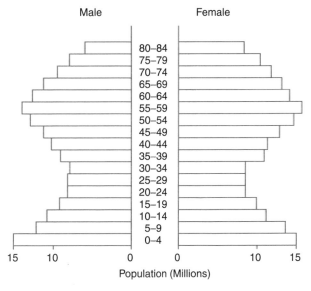

The population structure of developing countries leads to a higher dependency ratio. This means that a high proportion of economically inactive dependants rely on a relatively small proportion of economically active workers to supply them with goods and services.

Migration patterns may also be examined to decide whether a country should be classified as a developing country or developed country. Developing countries tend to experience net emigration as some people leave the country in search of employment and higher incomes abroad. Many are also experiencing rural to urban migration as people leave the countryside to move to towns and cities again in search of employment and higher income. In contrast, many developed countries experience net immigration and have already achieved a relatively high level of urbanisation.

Countries might also be classified according to a composite measure of living standards such as the Human Development Index. In practice, developing countries have differences as well as similarities. These differences include, for instance, disparities in factor endowment, the inequality of income and economic growth rates. Developed countries also exhibit a range of differences.

(b) The 'old' approach referred to in the question refers to fair trade. This is trade on equal terms that is without any restrictions and without any country seeking to gain an unfair advantage. The restrictions may be in the form of, for instance, tariffs and quotas and an unfair advantage may be sought by giving subsidies to domestic producers.

Allowing countries to trade on equal terms should allow them to concentrate on producing those products in which they have a comparative advantage. If countries can specialise in those products that they are best at producing, global output should increase which will raise global living standards. Resources should be used in an allocatively efficient way, meaning that they should not be wasted. Of course, it is not always easy to identify where comparative advantage lies. Transport costs may also offset comparative advantage.

The 'new' approach emphasises that resources may be wasted transporting products which could be produced at home. It also mentions an external cost, pollution, which arises from transporting products around the world. The existence of external costs may result in production occurring where marginal social cost being greater than marginal social benefit and so may create allocative inefficiency.

If some of the products that are currently produced in developing countries and are exported to developed countries were to be produced in the developed countries, there is no guarantee that a better use of resources would be achieved. It may take considerably more resources to produce the products in developed countries if their factor endowment does not favour their production. For example, a country may not have a suitable climate or sufficiently fertile land to grow a particular crop. This may mean that the crop would have to be grown under expensive artificial conditions.

In addition, there is no guarantee that relying on domestic production will reduce transport costs as significantly as it might first appear. Local producers may make more individual journeys and may not be able to take advantage of economies of scale.

Transport costs do have to be taken into account but if they do not offset the comparative advantage experienced by the developing countries in primary products, both developing countries and developed countries would lose from adopting the new approach. Both could make better use of resources by engaging in fair trade.

Chapter 17

Answers to short questions

1. Governments seek to keep unemployment low in order to protect people's living standards and encourage economic growth.
2. Every year most countries' ability to produce products with the same number of workers increases because of improvements in education and training and because of advances in technology. If output rises by less than potential output, firms are likely to dismiss some of their workers.
3. Price stability does not mean the price level remaining unchanged. It means the price level rising at a low and steady rate.
4. A positive output gap occurs when output is higher than potential output. This is an unsustainable output in the long run and will be accompanied by a rise in the price level. The higher the positive output gap, the higher the rate of inflation is likely to be.
5. Both a recession and a depression involve a fall in real GDP. Whilst a recession involves a decline in the economy's output over a period of six months or more, a depression is usually taken to mean a decline in the country's output over a period of three years or longer or a decline that exceeds 10%.
6. There are four main stages to the economic cycle. These are the upturn, the boom, the downturn and the recession.
7. A current account deficit which has resulted from the import of raw materials may lead to a current account surplus if the raw materials are converted into products, some of which are exported and some of which replace imports.
8. It is more useful to compare current account balances as a percentage of GDP than in absolute terms because this allows for the difference in size of economies. A country may have a larger absolute current account deficit than another country but if its economy is much larger, it may be of less significance.
9. Financial flows from developed to developing countries include commercial loans, workers' remittances and foreign aid.
10. A government of a developed country could open up its markets to firms in developing countries in a number of ways including removing tariffs and quotas on imports and stopping subsidies given to domestic producers.
11. The Prebisch-Singer hypothesis suggest that the terms of trade of developing countries that specialise in producing and exporting primary products tend to deteriorate.
12. A multinational company may set up in a developing country because the country may have a supply of labour available at relatively low wages and may have a growing market for the firm's products.
13. One disadvantage of a government defaulting on its overseas loans is that it may be difficult for it to obtain loans in the future.
14. Workers' remittances may increase if more of the country's population go abroad to work or if the same number of people working abroad earn more.
15. The 'canons of taxation' mean the principles of taxation – in other words, what qualities taxes should possess.
16. A poll tax is a tax per head of the population. This type of tax is sometimes referred to as a neutral tax as it does not influence what people buy, where they live, what jobs they do and how many hours they work. It is, however, a regressive tax as it takes a larger percentage of the income of the poor.
17. A flat tax is a tax which has one rate of tax. This can apply to income tax, corporation tax or the whole range of taxes.
18. The Laffer curve shows that tax revenue is zero when the tax rate is zero, then rises and then falls reaching zero when the tax rate is 100 per cent. It suggests that a cut in a high tax rate would increase tax revenue as it would increase the incentive to work.
19. A decrease in commercial banks' liquidity ratios would increase their ability to lend but they may not be able to do so if people do not want to borrow.
20. A cut in a central bank's interest rate may not increase bank lending if commercial banks do not pass on the cut to their customers.

21. 'Quantitative easing' is increasing the money supply through expansionary open market operations.

22. One way in which fiscal policy and monetary policy are connected is through the budget position. An expansionary fiscal policy may result in a budget deficit and this budget deficit may be financed by increasing the money supply.

23. Globalisation is the creation of a global economy as a result of:

    ✓ the increased mobility of products, capital and labour, arising from lower transport costs and advances in technology and the removal of some trade restrictions

    ✓ the development of common tastes, encouraged by an increase in travel and greater exposure to the same products through films, TV and websites.

24. There is an argument for imposing tariffs on products produced by industries that use child labour as such a practice can be regarded as unfair competition and socially undesirable. There is, however, a risk that imposing trade restrictions may increase poverty in the countries and increase the use of child labour.

25. Relaxing immigration controls may be regarded as a supply side measure as it may be intended to increase the size of the labour force in order to increase productive capacity.

26. Whether cutting income tax is a fiscal policy measure or a supply side measure is determined by whether its prime objective is to increase aggregate demand or aggregate supply. If it is the former, it is a fiscal policy measure whereas if it is the latter, it is a supply side policy measure.

27. Globalisation may increase tax competition. Governments may think they have to lower their spending on education and other areas so that they can reduce tax rates. They may consider that lower tax rates are needed to attract and keep MNCs in the country.

28. Globalisation may put upward pressure on governments to increase their spending on education in order to raise labour productivity and so remain internationally competitive and to attract and keep multinational companies in the country. Of course, this pressure may conflict with the pressure arising from the need to be tax competitive.

29. A conventional Phillips curve would shift to the left if any given rate of unemployment is accompanied by a lower inflation rate. This could occur if workers press for lower wage rises when unemployment falls. This may be the result of reduced expectations of inflation.

30. The expectations augmented Phillips curve suggests that whilst it may be possible to reduce unemployment in the short run by increasing aggregate demand, in the long run, unemployment will reduce to the natural rate of unemployment but at a higher inflation rate.

31. The vicious circle of poverty is the tendency of poverty to lead to low levels of saving which, in turn, results in low levels of investment. The low levels of investment give rise to low levels of productivity which then results in poverty.

32. Among the reasons why income may become less evenly distributed are a rise in unemployment, a fall in the minimum wage and an increase in the pay of chief executives.

33. A government may want its country's exchange rate to fall in value in order to reduce a current account deficit and to raise employment.

34. A government may move from operating a fixed to a floating exchange rate in order to concentrate on other policy objectives and to avoid the need to keep significant levels of foreign reserves.

35. An increase in unemployment may reduce a current account deficit in the short run for two main reasons. One is that the resulting fall in income is likely to result in a fall in demand for imports. The other reason is that lower income at home may force domestic firms to compete more aggressively in export markets and they may be successful.

36. It is not possible for a government to reduce unemployment to zero. This is because some workers will always be in between jobs – taking some time between leaving one job and taking up another job. There may also be some workers who lack the necessary skills or commitment to find employment.

37. An economy's trend growth rate is the expected percentage change in its potential output. In most cases, the productive capacity of an economy would be expected to grow because of advances in technology and improvements in education and training.

38. A government may seek to reduce the economy's economic growth rate if it thinks that the growth rate is unsustainable. A rapid increase in aggregate demand which is not matched by an equal increase in aggregate supply may lead to higher output in the short run but also inflation and possible balance of payments problems. A government may try to reduce the growth in aggregate demand so that it equals the growth in aggregate supply.

39. A counter cyclical policy measure is one designed to work against the economic cycle and smooth out fluctuations in economic activity. When the economy is going into a downturn, a counter cyclical policy measure would seek to increase aggregate demand. In contrast, during an upturn a counter cyclical policy might try to reduce the growth in aggregate demand.

40. One benefit of inflation targeting is that it can reduce inflation by lowering inflationary expectations. If people are convinced by a central bank, being set an inflation target, that it will keep inflation low, they will act in a way that causes inflation to be low. Consumers will not rush their purchases, workers will not push for wage rises and firms will not push up their prices.

Answers to revision activities

1. (a) $100bn ($20bn + $10bn = $30bn).
   (b) A surplus of $35bn ($75bn - $20bn + $25bn).
   (c) As GDP rises a government would gain more tax revenue to spend as people will earn and spend more, and so both direct and indirect tax revenue will rise. In addition, as income rises people demand a higher quality of publically financed services. Indeed, demand for education, health care and roads, for example, is income elastic.

2. (a) (i) Spain's budget position moved from a surplus to a deficit because the economy went into recession which reduced tax revenue and increased government spending on benefits and because the Spanish government introduced a large public works project.
       (ii) Mr Zapatero was implying that he had expected that automatic stabilisers would have offset the recession. Automatic stabilisers should reduce economic fluctuations but may not be sufficient, by themselves, to avoid a recession. Planning to cut a budget deficit by raising taxes may work in the short run. There is a risk, however, that in the longer run higher taxes may reduce economic activity and so lower tax revenue.

(b) (i) A fiscal deficit is another term for a budget deficit. So when there is a fiscal deficit, it means that government spending exceeds tax revenue.

(ii) One argument in favour of privatisation is that it will raise revenue for the government in the short run. The effect on government revenue in the long run will depend on whether the state owned enterprises were profitable or not.

Another argument touched on in the extract is to increase efficiency. Operating in the private sector provides both a 'carrot' and a 'stick' to be allocatively and productively efficient. Firms that are efficient may earn high profits whilst those which are not may go out of business.

(c) (i) The increase in wealth inequality would have increased income inequality. Wealth gives rise to income in the form of interest, profit and dividends.

(ii) An increase in state benefits would reduce income inequality. It would raise the incomes of vulnerable groups such as the sick, the old and the unemployed and would narrow the gap between their incomes and the incomes of richer people in the country. Other government policy measures include increasing the top rates of progressive taxes, education and training.

(d) (i) Devaluation may promote economic growth as it may result in a rise in net exports and so a rise in aggregate demand. If the economy has spare capacity, higher aggregate demand will lead to a rise in real GDP.

(ii) Devaluation may increase a country's inflation rate as it will increase the price of imported raw materials which may push up domestic costs of production and will increase the price of finished raw materials purchased by domestic consumers. It will also reduce the pressure on domestic firms to keep their prices low as rival imported products will be more expensive.

3. (a) The amount of tax paid would be $5,200 (0% on first $5,000 of income, 20% on income $5,000–$15,000 = $4,000 and 40% on income $15,000–$18,000 = $1,200).
The average tax rate (art) is $5,200/$18,000 x 100 = 28.89%.
The marginal tax rate (mrt) is 40%.

(b) The tax system is progressive. Tax rates rise as income rises and the mrt exceeds the art.

(c) The person would pay $18,000 x 15/100 = $2,700. So she or he would pay $2,500 less.

Answers to multiple choice questions

1. C   An economic boom would increase income and so lead to higher direct tax revenue. It would also increase spending and so result in more indirect tax revenue.

2. C   A regressive tax takes a larger percentage of the income of the poor. The rich are likely to pay more in absolute amount but will pay a lower percentage than the poor.

3. A   Devaluation would reduce the price of exports and increase the price of imports. These price changes could result in an increase in net exports. Such a change would improve the current account position and the resulting higher aggregate demand would reduce cyclical unemployment.

4. C   An automatic stabiliser is a form of government spending or tax revenue which changes, without an alteration of government policy, to dampen down economic fluctuations. Income tax and sales tax revenue will automatically rise whilst unemployment benefit will automatically fall during an economic boom. Higher tax revenue and lower government spending will reduce the growth of aggregate demand. In contrast, the amount paid out in a state retirement pension is influenced by the age of the population rather than economic activity.

5. A   Deregulation, the removal of laws and rules, is designed to improve the efficiency of markets and so increase aggregate supply. B would be designed to increase aggregate demand and C and D to reduce aggregate demand.

6. D   The traditional Phillips curve shows an inverse relationship between unemployment and money wages, although money wages are usually taken to be an indicator of inflation. The relationship between unemployment and inflation is inverse rather than direct.

7. B   A reduction in unemployment benefit may encourage those made redundant to seek work more quickly and so may reduce unemployment.

8. A   Some of the income taxed may have been saved and so one leakage will replace another leakage. For instance, a government spends an extra $100m and raises taxes by a $100m. If the savings ratio is 20%, the rise in taxes will reduce private sector spending by $80m and so there will be a net injection of spending of $20m. B would reduce the impact on aggregate demand. C is not true for most countries and is anyway concerned with absolute amounts rather than changes. D is true but does not explain why a rise in government spending matched by an equal increase in tax revenue is likely to increase aggregate demand.

9. C   If inflationary expectations are at 5%, the economy is operating on the second short run Phillips curve. To remove inflation from the system, unemployment would have to rise to the point where the short run Phillips curve cuts the horizontal axis. In this case, this would be at an unemployment rate of 13%.

10. A   A straightforward definition question. Cyclical unemployment is also sometimes known as demand-deficient unemployment as it arises from aggregate demand being too low to provide sufficient jobs for the labour force.

11. C   A deflationary monetary policy measure is a measure designed to reduce aggregate demand by reducing the money supply/growth of the money supply and/or raising the rate of interest. A sale of government securities by the central bank would take money out of circulation and, by lowering the price of government securities, raises the rate of interest. A is an expansionary monetary policy measure, B is a deflationary fiscal policy measure and D is a supply side policy measure.

12. B   A regressive tax system will take a smaller percentage of the income of the rich than of the poor. This may encourage some people to work longer hours or to take promotion. It is also likely to result in a more uneven distribution of income, a fall in the rate of taxation on higher income groups and a rise in savings as the rich tend to save a higher proportion of their income than the poor.

13. A   A rise in taxation and a cut in government spending would reduce household spending. Lower spending may reduce demand-pull inflation and, by reducing spending on imports, may lower a current account deficit. Lower household spending would raise unemployment and may increase a current account surplus.

14. C   The natural rate of unemployment is the rate of unemployment

which exists when the labour market is in equilibrium and the rate of inflation is stable. An increase in unemployment benefit may put less pressure on the unemployed to find jobs. It may also push up wages as employers will have to make employment more attractive. A and D would tend to reduce the natural rate of unemployment as they would make work more financially rewarding and would make labour more mobile. B may reduce cyclical unemployment.

15. A   A rise in the rate of interest may reduce consumer expenditure and investment and so reduce demand-pull inflation. It will, however, push up the costs of production of firms which have borrowed in the past and so may raise cost-push inflation. B might increase demand-pull inflation in the short run but reduce cost-push inflation in the long run. C and D may reduce cost-push inflation.

16. A   Borrowing from the central bank is sometimes referred to as 'resorting to the printing press' and will increase the money supply. B, C and D involve making use of existing money.

17. C   A fiscal policy measure is one designed to influence aggregate demand whereas a supply side policy measure is one intended to influence aggregate supply.

18. C   A high marginal propensity to save would reduce the size of the multiplier and so reduce the impact of expansionary fiscal policy. A, B and D would all tend to increase the effectiveness of expansionary fiscal policy.

19. A   A decrease in income tax rates may increase spending and the incentive to work. Both effects may result in a rise in real GDP and employment. An increase in real GDP and employment may increase tax revenue and reduce government spending on unemployment benefits and so may reduce a budget deficit. A decrease in income tax rates is likely to reduce tax evasion. Fiscal drag may arise due to inflation and a change in income tax rates does not affect the tax base.

20. C   A definition question. The incidence of taxation relates to which group or sector a tax falls on. For instance, the incidence of an indirect tax may be on both producers and consumers.

Answers to data response questions

1. (a) Economic growth involves an increase in real gross domestic product (GDP) in the short run and an increase in productive capacity in the long run. If there is spare capacity in the economy, an increase in aggregate demand can result in more goods and services being produced. In the longer run, to continue to increase output, the quantity and/or quality of resources have to be increased.

   (b) Developed countries may help to stimulate economic growth in developing countries in a number of ways. Multinational companies (MNCs) from developed countries set up in developing countries and contribute directly to the countries' economic growth rates. They may also contribute indirectly to economic growth by training workers and financing infrastructure projects.

   In addition, developed countries can provide aid to developing countries. Such aid may increase the quantity and quality of the countries' resources if the aid is used, for instance, to improve education and increase investment.

   Developing countries, in turn, contribute to the economic growth of developed countries by providing markets for their products and resources for their industries. As developing countries grow, the higher incomes of their people result in developed countries' exports rising. The raw materials purchased from developing countries are turned into finished products and sold at home and abroad and labour from developing countries can be employed in their own countries by MNCs or as migrant workers in developed countries.

   (c) Figure 17.2 shows that income inequality is unevenly distributed with the richest 20% of the population receiving 82.7% of total global income and the remaining 80% receiving only 17.3%. It cannot, however, be concluded from the figure that almost half of the world's population still lives on less than US$2 a day. It does not show the actual income earned. Whilst the poorest 20%, for instance, receive only 1.4% of income, it is not indicated how much that is per head of population. It cannot be calculated from the data what is the share of the poorest 50% or, indeed, what the average income is. Figure 17.2 also does not show global income distribution over time and so it cannot be concluded that half of the world's population continues to live in poverty.

   (d) Economic growth has the potential to reduce absolute and relative poverty. Increases in the country's output can result in rising employment and rising income. The extract mentions that many countries in Africa experience severe poverty and that there has been an increase in the gap between the rich and the poor within countries and between rich and poor countries.

   The extract, however, does not provide evidence on how economic growth has affected poverty. No data is given on economic growth and poverty over time. Economic growth may have led to a decline in poverty. Even if data indicated that poverty continued to exist during a period of economic growth, it is possible that poverty might have been higher without economic growth. In addition, the extract does not provide any evidence on the cost of living. Incomes might be low but if living costs fall, poverty could decline.

2. (a) Gross domestic product is the output produced in a country in a given time period. The money sent to Kenya by Kenyans has been earned as a result of producing products in other countries and, as such, would not be included in Kenya's GDP.

   (b) There is a range of reasons why people from developing countries might go to work in developed countries. One possible reason is the greater availability of work. In their home countries there may be a shortage of workers with their skills. Pay may be higher in developed countries which may enable the migrant workers to save some of their income for the future and to send some home to improve the working conditions of their relatives. In addition, there may be a greater variety of jobs, better promotion prospects and more training available in developed countries. In rural areas of some developing countries, the focus is likely to be on employment in the primary sector and the jobs available may be low skilled. Promotion prospects may be greater in countries which have well developed secondary and tertiary sectors. Access to good training may attract some people to work in developed countries as such training will increase their job prospects and earning potential at home and abroad.

(c) The article supports the view that economic development is largely the result of private actions rather than government policy. It mentions that some of the money sent home by individuals is used to finance the education of children. With more years of education and better quality education, children should have an increased chance of gaining a well-paid job, better health care and more choices. The extract also states that investment financed by Chinese living abroad and remittances from Ghana and Nigeria have been significant factors in the development of the countries. In addition, it shows that money sent home by workers to Bangladesh was more significant than aid in 2005. Data, however, is not available for a range of countries over time.

The article also states that in Mexico local governments match any private sector money and shows that money given as aid was greater than workers' remittances in the case of Kenya in 2005.

To draw a firm conclusion as to whether private actions rather than government policy play a more important role in economic development, it would be necessary to examine data covering more countries and for more than a year.

(d) The most traditional indicators about production and consumption are real GDP and real GDP per head. The latter is better indicator of welfare as it takes into account population size. Figures on real GDP per head are available to compare over time and between countries. The figures are not always accurate, however, and do not provide a full picture of the quality of people's lives.

Care has to be taken in measuring real GDP. In using the output measure it is important to avoid double counting, in the income method transfer payments should not be included and in the expenditure method exports must be added and imports deducted. Real GDP per head figures may understate production and consumption to a significant extent if there is a large informal economy with people not declaring economic activity either because it is illegal or because they want to evade paying tax. Real GDP per head figures do not include all the aspects which affect welfare. As a result, economists use a number of other measures including the Human Development Index, which as well as GDP per head, also includes education and life expectancy. The HDI, however, also fails to include a number of key factors that influence welfare including hours of leisure time. Both real GDP per head and HDI would also be increased by a number of changes which might not result in higher welfare. For instance, a rise in the number of policeman and policewomen due to higher crime would increase both indicators but would be unlikely to raise welfare – most people would rather have less crime and so less people in the police force.

The traditional indicators about production and consumption are of some use as measures of welfare. They are starting points but they can be added to although some factors that influence welfare, including the quality of relationships are difficult to measure.

3. (a) A budget deficit arises when government spending exceeds tax revenue. A cyclical budget deficit is an excess of government spending over tax revenue caused by a downturn in economic activity. When an economy enters a recession, direct and indirect tax revenue falls as income and expenditure declines. Government spending on unemployment related benefits increases as more people lose their jobs.

(b) Economic growth is usually associated with a rise in the value of products a country imports. As an economy's output increases, its firms may buy more imported raw materials to produce products to sell at home. As incomes rise, households produce more finished products including some from abroad. Higher the marginal propensity to import, the greater will be the rise in the value of imports purchased. It is, however, possible that for a period of time a government may seek to promote economic growth by means of an import-substitution policy. In this case, a government may subsidise domestic producers and impose trade restrictions.

(c) An open economy is one which engages in international trade. The more open an economy is, the greater its exports and imports will be as a proportion of its gross domestic product (GDP).

China and Russia appear to be more open economies than Brazil and India as they export a higher proportion of their GDP. No specific information, however, is given on the proportion the countries import although as Brazil and India have only a small current account deficits, the value of their imports must be relatively close to the value of their exports.

The extract mentions that all of the BRICs impose trade restrictions which limits international trade. It does, however, state that the BRICs are exporting and importing more as economic growth occurs but no data is given of how exports as a percentage of GDP has changed over time.

All that can be concluded from the extract is that at one point in time, China and Russia sold a higher proportion of their output to people in other countries, that all four BRICs restrict international trade but all are trading more.

(d) There is only limited evidence in the extract to support the view that the BRICs have a similar macroeconomic performance. It mentions that all the BRICs have high growth rates and this comment is supported by the economic growth rate figures given in Table 1 for three of the economies but not for Russia. It is also predicted that all four will become important economies by 2050. In addition, Table 1 shows that all four had relatively high unemployment rates and the extract mentions that all four economies impose trade restrictions, had avoided going into a recession but had a cyclical budget deficit.

Table 1 does, however show a variation in inflation rates and a noticeable difference in income per head with, for instance, people in Russia enjoying an income level nearly nine times higher than that experienced in India. The data is also restricted to one year and it is not stated what measures of inflation and unemployment are being used. For instance, one country might be measuring the inflation rate using a consumer price index whilst another might be using the GDP deflator and one country might be measuring unemployment using the labour force survey whilst another may be using the claimant count.

In addition, the extract mentions that Brazil and India have recently experienced small current account deficits whilst China and Russia have experienced large current account surpluses. It suggests China and Russia are more open economies than Brazil and India.

To decide the extent to which the BRICs have a similar macroeconomic performance, more information would be needed on the sources of data provided, figures for more than one year would be necessary and data on, for instance, exchange rate movements would be useful.

Answers to essay questions

1. Whether the aims of government policy will conflict will be affected by the time period under consideration, the current state of the economy and the type of economic policies pursued. If there is a conflict, which aim should be given priority will again be influenced by the current state of the economy, future predictions, the costs involved and which aim the government thinks it will be most effective in achieving.

   In the short run, if an economy is operating close to full employment, there may be a conflict between reducing the inflation rate or at least ensuring the inflation rate remains stable. If a government seeks to reduce unemployment by expansionary fiscal policy, the higher aggregate demand may result in inflationary pressure. In the figure as aggregate demand (AD) increases, output is increased rises closer to the vertical part of the aggregate supply curve (AS) and the price level rises from P to P1.

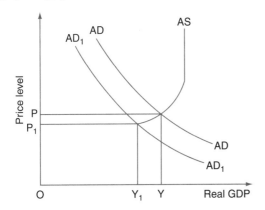

   If, however, the economy has considerable spare capacity, with a noticeable output gap, it may be possible to reduce unemployment by a noticeable amount without causing inflationary pressure.

   In the short run, the objectives of economic growth and a stable inflation rate may also conflict. As the previous diagram shows, higher AD causes real GDP to rise but it also pushes up the price level. As an economy's AD increases, more resources are used and they become in shorter supply and the rising competition pushes up their prices. Firms know they can change more for their products and so they are willing to increase their prices.

   Economic growth may also conflict with a balance on the current account of the balance of payments. If economic growth occurs, incomes will rise. This may result in demand for goods and services. As a result imports of finished products and raw materials may increase. Some products originally intended for the export market may also be diverted to the home market. Of course, it is possible that economic growth may be export led and in such a circumstance may be accompanied by a declining current account deficit.

   In the long run, if a government follows effective supply-side policy measures it may be able to achieve all of its objectives. This is because such policies, such as improved education and training, will shift the aggregate supply (AS) curve to the right

enabling AD to increase, lowering unemployment, achieving economic growth without causing inflation. Effective supply side policy measures can also improve the current account position by improving the quality and price competitiveness of domestic products. Of course, a government may also have to ensure that both AD and AS increase in line with each other. In the figure below, both AD and AS rise. These shifts keep the price level at P and raises real GDP from Y to Y1.

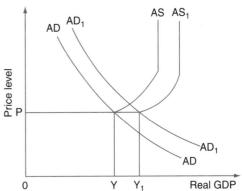

   In practice, a government is likely to need to use a range of policy measures to achieve its objectives. In the short run, whilst unemployment and economic growth are likely to be helped by an increase in AD, a reduction in AD may be needed to reduce demand-pull inflation and a current account deficit. In the long run, however, all the objectives should benefit from an increase in AS which matches an increase in AD. This is why some economists claim that sustained economic growth should be given priority. If both aggregate demand and aggregate supply shift to the right, inflation and unemployment should be kept low and the current account close to balance. In the short run, if there is significant and rising unemployment whilst inflation is low and stable, there is still positive economic growth and the current account position is close to balance, a government is likely to prioritise reducing unemployment. If, however, unemployment is of a short-term duration and on a downturn trend whilst inflation is unanticipated and of a cost-push nature, a government may decide to concentrate on reducing inflation.

2. (a) An increase in taxes would reduce national income. Higher income tax would reduce disposable income. Lower disposable income would be likely to reduce consumer expenditure. Higher indirect taxes would also probably result in lower consumer expenditure. A decline in consumer expenditure, ceteris paribus, will reduce aggregate demand. Lower aggregate demand can reduce real GDP as shown in the figure here.

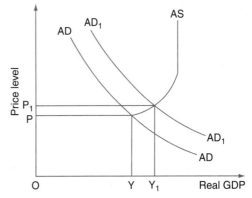

The extent to which an increase in taxes will reduce national income will be influenced by the size of the rise in taxes, which taxes are raised and how firms and households react. A small increase in taxes will not have much effect. A rise in the top rates of progressive taxes may also not cause national income to fall significantly. If the rich have to pay higher taxes they may reduce their savings rather than their spending. Higher direct taxes may increase tax evasion and, as a result, not increase tax revenue. A rise in indirect taxes on products with inelastic demand or on products with substitutes that are not taxed may also mean that consumer spending does not fall by much.

It is even possible that during periods of high inflation or large government debt, a rise in taxes may increase consumer and business confidence and so raise aggregate demand.

(b) A fall in the level of national income may be accompanied by a decline in the standard of a living in a country but this will not necessarily occur. A decline in real national income with the same population size will result in a fall in income per head. Lower income will reduce people's material standard of living as people, on average, will be able to buy fewer goods and services.

It is, however, possible that national income may fall but real national income may rise. If the price level has fallen, it is possible that more products have been produced and so people may be able to experience a rise in material living standards. It is also possible that national income may have declined but the population might have fallen by a greater extent. In this case, national income per head would have risen.

In addition, there are a number of reasons why a decline in national income per head might be accompanied by a rise in the standard of living for most people. One is that income may have become very evenly distributed so that whilst some people may have experienced a decline in their material living standards, the majority may have been able to enjoy more goods and services. Another is that the size of the informal economy may have increased so despite the official figures, more goods and services may have been produced.

The quality of people's lives may also have increased if the composition of output has changed. For instance, a rise in the output of leisure goods and a fall in the output of weapons may make people happier. There is a range of factors not measured by national income that affect the quality of people's lives including working conditions, working hours and pollution.

Whilst GDP per head is a common measure of the standard of living over time and across countries, economists use a number of other measures. One is the Human Development Index which, in addition to GDP per head, includes life expectancy and education as measured by adult literacy and primary, secondary and tertiary school enrolment ratios. Another measure is the Index of Sustainable Economic Welfare which seeks to measure welfare by deducting items that reduce the quality of life including the costs of crime, loss of wildlife habitats and pollution, whilst adding items that improve the quality of life such as unpaid housework and takes into account income inequality.

Changes in national income can have a significant influence on the standard of living but they have to be interpreted carefully and it has to be recognised that the quality of people's lives is influenced by far more than income.

3. (a) There is a range of causes of unemployment, some more serious than others. Even in a booming economy there is likely to be some frictional unemployment. This type of unemployment occurs when workers are in between jobs. Some people leave or lose one job before taking up another job. This type of unemployment may not be much of a problem and can be reduced by improving information about job vacancies.

Seasonal unemployment is a form of frictional unemployment. It arises when people are out of work because demand for their services occurs only at certain times of the year. For instance, tour guides and some hotel workers may lose their jobs when holiday periods end and fruit pickers may be out of work during certain seasons of the year. This type of unemployment might be reduced by finding workers jobs during off peak periods or by increasing the length of time when their services are in demand. For instance, hotels may develop conference facilities in order to create demand out of season.

Another form of frictional unemployment in search unemployment. Some of the unemployed may not accept the first jobs offered to them but may spend time looking for a better job. The existence of relatively generous unemployment benefits may encourage them to do this. They may also encourage voluntary unemployment with some people preferring to live on benefits rather than working. To tackle this type of unemployment, a government may cut unemployment benefits.

Structural unemployment tends to be more serious than frictional unemployment as it can involve more people and can be long lasting. Structural unemployment is unemployment which results from changes in the structure of the economy. These may be from the demand or the supply side. Workers may lose their jobs because some products may become less popular or because changes in methods of production, often caused by advances in technology, mean that fewer workers are required. At any one time, the number of jobs in some industries is likely to be declining. These job losses may be matched by new jobs in other industries but if workers are either geographically or occupationally immobile, unemployment is likely to persist. To reduce this type of unemployment, a government may seek to promote training.

The most serious type of unemployment is cyclical unemployment. This arises due to a lack of aggregate demand and can result in large numbers of people losing their jobs across all industries and staying unemployed for some time. Cyclical unemployment will be associated with a negative output gap. During a recession, cyclical unemployment may reach a high level. To reduce cyclical unemployment, a government is likely to try to increase aggregate demand by means of expansionary fiscal or monetary policy.

(b) A rise in the rate of interest may reduce demand-pull inflation. Indeed, interest rate changes are the main policy measure used by central banks to influence changes in the price level. It may, however, not always be successful in reducing demand-pull inflation and may add to cost-push inflation.

Demand-pull inflation occurs when aggregate demand increases by more than aggregate supply. As an economy approaches full employment, it becomes increasingly difficult to meet higher aggregate demand with increased output as resources become scarcer. A rise in the rate of interest is likely to reduce aggregate demand by reducing or lowering the growth of three of the components of aggregate demand. Consumer expenditure is likely to be reduced because saving will become more rewarding, borrowing will become more expensive and anyone who has borrowed before on a variable interest rate will have less money to spend. Investment may be discouraged as again borrowing will be more expensive, firms may decide to save their money rather than spend it on capital goods and may expect consumer expenditure to be lower. Net exports may also fall. This is because a rise in the rate of interest may stimulate an inflow of hot money in search of a high return. The resulting increase in demand for the currency would raise the value of a floating exchange rate. This, in turn, would increase the price of exports and lower the price of imports. Given elastic demand for exports and imports, export revenue would fall and import expenditure would rise.

A decline in aggregate demand, or at least a decline in the growth of aggregate demand would reduce demand-pull inflation. The following figure shows that restricting the growth of aggregate demand to AD1 would result in a smaller rise in the price level than would be caused by a larger rise in aggregate demand to AD2.

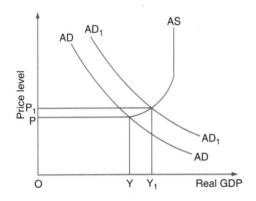

The decision to raise the rate of interest charged by the central bank does, however, take time to work through the economy to affect aggregate demand. Commercial banks have to change their interest rates in response (and there is no guarantee they will do so). Firms and households then have to respond to any change. By the time aggregate demand is reduced, demand-pull inflation may have disappeared.

Higher interest rates may also not discourage firms from investing and households from spending if they are very optimistic about the future.

Higher interest rates may occur at a time when another influence on aggregate demand is changing. For instance, if incomes abroad are rising, export revenue may increase despite a higher exchange rate.

If the government has made a mistake in analysing the cause of inflation and it is actually cost-push, raising interest rates may increase inflation. Higher interest rates will increase the costs of any firms that have borrowed on variable interest rates and they may raise their prices to cover their higher costs.

Interest policy has the potential to prevent a rise in demand-pull inflation by restricting the growth of consumer expenditure, investment and net exports. It will, however, work with a time lag and may not be very effective during a consumer boom when confidence is high.

Index